Lecture Notes in Computer Science 4086

Guang Gong Tor Helleseth
Hong-Yeop Song Kyeongcheol Yang (Eds.)

Sequences and Their Applications – SETA 2006

4th International Conference
Beijing, China, September 24-28, 2006
Proceedings

Springer

Volume Editors

Guang Gong
University of Waterloo
Department of Electrical and Computer Engineering
200 University Avenue West, Waterloo, ON, N2L 3G1, Canada
E-mail: ggong@calliope.uwaterloo.ca

Tor Helleseth
University of Bergen
Department of Informatics
Thormohlensgate 55, 5020 Bergen, Norway
E-mail: tor.helleseth@ii.uib.no

Hong-Yeop Song
Center for Information Technology of Yonsei University
School of Electrical and Electronics Engineering
Seoul,120-749, Korea
E-mail: hy.song@coding.yonsei.ac.kr

Kyeongcheol Yang
Pohang University of Science and Technology (POSTECH)
Dept. of Electronic and Electrical Engineering
Pohang, Gyungbuk 790-784, Korea
E-mail: kcyang@postech.ac.kr

Library of Congress Control Number: 2006932045

CR Subject Classification (1998): E.4, F.2, I.1, E.3, F.1, G.1

LNCS Sublibrary: SL 1 – Theoretical Computer Science and General Issues

ISSN 0302-9743
ISBN-10 3-540-44523-4 Springer Berlin Heidelberg New York
ISBN-13 978-3-540-44523-4 Springer Berlin Heidelberg New York

Springer is a part of Springer Science+Business Media

springer.com

Printed in Germany

Typesetting: Camera-ready by author, data conversion by Scientific Publishing Services, Chennai, India
Printed on acid-free paper SPIN: 11863854 06/3142 5 4 3 2 1 0

Preface

This volume contains the refereed proceedings of the Fourth International Conference on Sequences and Their Applications (SETA 2006), held in Beijing, China during September 24–28, 2006. The previous three conferences SETA '98, SETA 2001, and SETA 2004 were held in Singapore, Bergen, and Seoul, respectively. The SETA conferences are motivated by the numerous applications of sequences in modern communication systems. These applications include pseudorandom sequences in spread spectrum, code-division-multiple-access, stream ciphers in cryptography, and several connections to coding theory and boolean functions.

The Technical Program Committee of SETA 2006 refereed 70 submitted papers. This represented more submissions than to any of the previous SETA conferences. The committee therefore had the challenging task of selecting 32 papers to be presented at the conference in addition to 4 invited papers.

The Co-chairs of the Technical Program Committee for SETA 2006, were Guang Gong (University of Waterloo) and Tor Helleseth (University of Bergen), with Hong-Yeop Song (Yonsei University, Korea) and Kyeongcheol Yang (Pohang University of Science and Technology, Korea) as the co-editors for these proceedings.

The editors wish to thank the other members of the Technical Program Committee: Anne Canteaut (INRIA, France), Claude Carlet (INRIA and University of Paris 8, France), Habong Chung, (Hongik University, Korea), Zongduo Dai (University of Science and Technology of China, Beijing, China), Cunsheng Ding (Hong Kong University of Science and Technology, Hong Kong), Pingzhi Fan (Southwest Jiaotong University, China), Dengguo Feng (Chinese Academy of Sciences, China), Solomon W. Golomb (University of Southern California, USA), Kyoki Imamura (Kyushu Institute of Technology, Japan), Jonathan Jedwab (Simon Fraser University, Canada), Thomas Johansson (University of Lund, Sweden), Andrew Klapper (University of Kentucky, USA), P. Vijay Kumar (University of Southern California, USA), Wai Ho Mow (Hong Kong University of Science and Technology, Hong Kong), Harald Niederreiter (National University of Singapore, Singapore), Jong-Seon No (Seoul National University, Korea), Matthew G. Parker (University of Bergen, Norway), Kenneth G. Paterson (Royal Holloway, University of London, UK), Alexander Pott (Otto-von-Guericke-University Magdeburg, Germany), Hans Schotten (Qualcomm Germany, Nuremberg, Germany), Parampalli Udaya (University of Melbourne, Australia), and Amr Youssef (Concordia University, Canada) for providing clear, insightful, and prompt reviews of the submitted papers.

The editors are also grateful to Serdar Boztas, Jin-Ho Chung, Deepak Kumar Dalai, Frédéric Didier, Gary Greenfield, Yun-Kyoung Han, Tom Høholdt, Alexander Kholosha, Margreta Kuijper, Gohar Kyureghyan, Cedric Lauradoux, Subhamoy Maitra, Joe Rushanan, Frank Ruskey, Igor Semaev, Jean-Pierre

Tillich, and Nam Yul Yu for their help and assistance in the reviewing of papers for SETA 2006. A special thanks goes to Sondre Rønjom for handling all the submissions and the web-review software during the review process.

In addition to the contributed papers, there are four invited papers. These papers provide a historical overview as well as new developments in important areas of the design and analysis of sequences. The invited contribution by Solomon Golomb presents a retro-perspective of some selected results on sequences. The invited paper by Harald Niederreiter includes an updated overview and some recent important results on the complexity of multisequences. Vijay Kumar provides an overview and new results on optical orthogonal codes. This topic is motivated by applying code division multiple access (CDMA) techniques in optical networks. Zongduo Dai presents an overview of multi-continued fraction algorithms and their applications to sequences.

We wish to thank Pingzhi Fan and Dengguo Feng for their support as General Co-chairs of SETA 2006, and Chuan-Kun Wu for local arrangements and updating the web site of SETA '06. We also thank Yi Qin for her support as secretary of SETA 2006, and Shi Zhang for her support as treasurer of SETA 2006. Last but not least, we thank all the authors of the papers for their help and collaboration in preparing this volume. Finally, we would like to thank the National Science Foundation of China (NSFC) and the Chinese Academy of Sciences (CAS) for their financial support.

September 2006

Guang Gong
Tor Helleseth
Hong-Yeop Song
Kyeongcheol Yang

Organization

SETA 2006

September 24-28, 2006, Beijing, China

General Co-chairs

Pingzhi Fan, Southwest Jiaotong University, China
Dengguo Feng, Chinese Academy of Sciences, China

Program Co-chairs

Guang Gong, University of Waterloo, Canada
Tor Helleseth, University of Bergen, Norway

Local Arrangements

Chuan-Kun Wu, Chinese Academy of Sciences, China

Secretary and Registration

Yi Qin, Chinese Academy of Sciences, China

Treasurer

Shi Zhang, Chinese Academy of Sciences, China

Proceedings Co-editors

Guang Gong, University of Waterloo, Canada
Tor Helleseth, University of Bergen, Norway
Hong-Yeop Song, Yonsei University, Korea
Kyeongcheol Yang, Pohang Univ. of Science and Technology, Korea

Technical Program Committee for SETA 2006

Program Co-chairs

Guang Gong .... University of Waterloo, Canada
Tor Helleseth .... University of Bergen, Norway

Program Committee

Anne Canteaut .... INRIA, France
Claude Carlet .... INRIA and University of Paris 8, France
Habong Chung .... Hongik University, Korea
Zongduo Dai .... University of Science and Technology of China, China
Cunsheng Ding .... Hong Kong University of Science and Technology, China
Pingzhi Fan .... Southwest Jiaotong University, China
Dengguo Feng .... Chinese Academy of Sciences, China
Solomon W. Golomb .... University of Southern California, USA
Kyoki Imamura .... Kyushu Institute of Technology, Japan
Jonathan Jedwab .... Simon Fraser University, Canada
Thomas Johansson .... University of Lund, Sweden
Andrew Klapper .... University of Kentucky, USA
P. Vijay Kumar .... University of Southern California, USA
Wai Ho Mow .... Hong Kong University of Science and Technology, China
Harald Niederreiter .... National University of Singapore, Singapore
Jong-Seon No .... Seoul National University, Korea
Matthew G. Parker .... University of Bergen, Norway
Kenneth G. Paterson .... Royal Holloway, University of London, UK
Alexander Pott .... Otto-von-Guericke University Magdeburg, Germany
Hans Schotten .... Qualcomm Germany, Germany
Hong-Yeop Song .... Yonsei University, Korea
Parampalli Udaya .... University of Melbourne, Australia
Kyeongcheol Yang .... Pohang University of Science and Technology, Korea
Amr Youssef .... Concordia University, Canada

Table of Contents

Invited Papers

Linear Complexity of Sequences

Correlation of Sequences

Stream Ciphers and Transforms

Topics in Complexities of Sequences

Linear/Nonlinear Feedback Shift Register Sequences

Multi-sequence Synthesis

Filtering Sequences and Pseudorandom Sequence Generators

Sequences and Combinatorics

FCSR Sequences

Aperiodic Correlation and Applications

Boolean Functions

Shift Register Sequences – A Retrospective Account

Solomon W. Golomb

University of Southern California
Viterbi School of Engineering
Los Angeles, CA 90089-2565
milly@usc.edu

Abstract. Binary feedback shift registers, with applications to reliable communications, stream cipher cryptography, radar signal design, pseudorandom number generation, digital wireless telephony, and many other areas, have been studied for more than half a century. The maximum-length binary linear feedback shift registers, called *m-sequences* or *PN sequences*, are the best-known and most thoroughly understood special case.

The m-sequences have several important *randomness properties*, and are known as *pseudo-random sequences*. They are characterized by the *cycle-and-add property*, whereby the term-by-term sum of two cyclic shifts is a third cyclic shift. Along with other families of binary sequences that correspond to *cyclic Hadamard difference sets*, they have the *two-level autocorrelation property*. The m-sequences share the *span-n property* (all subsequences of length n, except n zeroes, occur in each period of length $2^n - 1$) with a far larger class of nonlinear shift register sequences. No counterexample has been found to the conjecture that only the m-sequences have both the two-level autocorrelation and the span-n properties.

The class of m-sequences is too small, and has too many regularities, to provide useful cryptographic security as key sequences for stream ciphers. For this purpose, nonlinear shift register sequences which have large linear span and a sufficiently high degree of *correlation immunity* may be employed.

1 Linear Shift Register Sequences

Let $S_0 = \{a_1, a_2, \ldots, a_p\} = \{a_i\}$ be a binary sequence of period p, and $S_j = \{a_{1+j}, a_{2+j} \ldots, a_j\}$for all $0 \leq j \leq p-1$. Then S_0 is an m-sequence if and only if $S_i + S_j = S_k$ for all $0 \leq i < j \leq p-1$, where addition of sequences is term-by-term and modulo 2. Equivalently, if a p-component binary vector, together with all its cyclic shifts and the p-component zero vector, form a subspace of $GF(2^p)$, then $p = 2^n - 1$ for some n, and the binary sequence is an m-sequence; and conversely, every m-sequence has this property.

There are $\frac{\phi(p)}{n} = \frac{\phi(2^n-1)}{n}$ cyclically distinct m-sequences of degree n and period p, where ϕ is Euler's phi-function. Of the many additional properties

G. Gong et al. (Eds.): SETA 2006, LNCS 4086, pp. 1–4, 2006.

possessed by these sequences, two of the most important are the *span-n property* and the *two-level autocorrelation property*. The *span-n property* here refers to a binary sequence of period $p = 2^n - 1$ in which every possible n-bit subsequence except for n zeroes occurs exactly once in each period. These sequences are in direct one-to-one correspondence with the *de Bruijn sequences* of span n, which have period 2^n, and in which every possible n-bit subsequence appears exactly once in each period; and the exact number of cyclically distinct de Bruijn sequences is well known [1] to be $2^{2^{n-1}-n}$ for span n, for every positive integer n. The *two-level autocorrelation property* for m-sequences asserts that between S_i and S_j, for all $0 \le i < j \le p-1$, there are $\frac{p-1}{2} = 2^{n-1} - 1$ term-by-term agreements and $\frac{p+1}{2} = 2^{n-1}$ term-by-term disagreements. The cyclically distinct binary two-level autocorrelation sequences of period p are in one-to-one correspondence with *cyclic Hadamard difference sets* modulo p, which are the *perfect* (v, k, λ) *difference sets* having $(v, k, \lambda) = (p, \frac{p-1}{2}, \frac{p-3}{4}) = (2^n - 1, 2^{n-1} - 1, 2^{n-2} - 1)$.

All cyclic Hadamard difference sets with these parameters have been found by exhaustive computer searches for each degree n, $2 \le n \le 10$. All the examples thus found belong to *families* of cyclic Hadamard difference sets, for multiple values of p, which were all known before the complete search at $p = 2^{10} - 1 = 1023$ was undertaken. There is some optimism that all the constructions which yield cyclic Hadamard difference sets are now known (see [2]), but this has not been proved.

Every m-sequence of period $p = 2^n - 1$ is simultaneously a span-n sequence and a $(p, \frac{p-1}{2}, \frac{p-3}{4})$ cyclic Hadamard sequence. It has been conjectured that the converse is also true: that is, that if a binary sequence has both span-n and two-level autocorrelation, then it must be an m-sequence. While the truth of this conjecture has been verified for all $n \le 10$, and for certain two-level autocorrelation sequences with $n > 10$, the general case of this conjecture remains open.

The m-sequences of period $p = 2^n - 1$ are in one-to-one correspondence with the irreducible polynomials of degree n over $GF(2)$ whose roots are primitive p^{th}-roots of unity. It is conjectured that there are infinitely many such polynomials with only three terms, $x^n + x^a + 1$, called *primitive trinomials* over $GF(2)$. It has even been conjectured that $x^n + x + 1$ is primitive for infinitely many values of n. By a theorem of Richard Swan, there are no primitive (or even irreducible) trinomials $x^n + x^a + 1$ where the degree n is a multiple of 8. By a theorem of Øystein Ore, if $f(x) = \sum_{i=0}^{n} a_i x^i$ is a primitive irreducible polynomial over $GF(2)$, then $F(x) = \sum_{i=0}^{n} a_i x^{2^i - 1}$ is irreducible (though not necessarily primitive). While primitive trinomials fail to exist for infinitely many degrees n, it is conjectured that primitive pentanomials (five-term polynomials) exist for every degree $n \ge 5$. However, it has not even been proved that there are infinitely many degrees n having a primitive polynomial with no more than t terms, for *any* specific positive integer t.

2 Nonlinear Shift Register Sequences

In contrast to *linear* binary feedback shift registers, which are well-understood mathematically, the much larger family of *nonlinear* feedback shift registers has far fewer regularities. The most general feedback function for an n-stage binary shift register is an arbitrary one of the 2^{2^n} boolean functions $f(x_1, x_2, \ldots, x_n)$ of n binary variables, where the variables are taken from the n stages of the shift register. The 2^n possible states of an n-stage shift register become the vertices of a directed graph ("digraph") whose directed edges go from each state to its successor state. For any particular shift register, this digraph is a subgraph of the *de Bruijn graph*, whose edges indicate all the possible shift register transitions from one state to the next. In the de Bruijn graph for the general n-stage shift register there are 2^n vertices and 2^{n+1} directed edges, showing the two possible predecessors and the two possible successors of each state of the shift register.

For any specific nonlinear shift register, its digraph will decompose entirely into one or more disjoint cycles (i.e. "cycles without branches") if and only if one can write the feedback function $f(x_1, x_2, \ldots, x_n)$ in the form $g(x_1, x_2, \ldots, x_{n-1}) + x_n$, where x_n comes from the "oldest" stage of the shift register, and is added modulo 2 to an arbitrary boolean function $g(x_1, x_2, \ldots, x_{n-1})$ of the other $n-1$ stages. In this case of "pure cycles without branches", the number of cycles has the same parity (even or odd) as the number of *ones* in the truth table of $g(x_1, x_2, \ldots, x_{n-1})$, for all $n > 2$. In particular, for $n > 2$, in order to get all 2^n possible states of the shift register to lie on a single cycle, the truth table for $g(x_1, x_2, \ldots, x_{n-1})$ must have an odd number of ones, which requires that all $n-1$ variables occur, and occur nonlinearly, in the computation of $g(x_1, x_2, \ldots, x_{n-1})$.

3 Applications

When a nonlinear shift register is used to generate a key stream for use in a stream cipher, cryptanalytic attacks which attempt to reconstruct, in whole or in part, the structure of the shift register being used, are usually based on multi-dimensional correlations of the key sequence. These correlation values correspond directly to the *invariants*, described in [1], of the boolean function $g(x_1, x_2, \ldots, x_{n-1})$, which can be obtained from the Walsh function expansion coefficients of the truth table of the function g.

Shift registers are also used for both the encoding and the decoding of both block codes and convolutional codes. They are used to generate the pseudo-random binary chip sequences needed for spectral spreading in "direct-sequence spread spectrum" secure communications and for mutual non-interference between callers in code division multiple access (CDMA) wireless telephony. The two-level correlation property makes m-sequences very well suited for use as modulation patterns in radar and sonar applications.

For a more detailed account of the properties of both linear and nonlinear shift register sequences, see [1]. For a description of the various constructions now known for $(2^n - 1, 2^{n-1} - 1, 2^{n-2} - 1)$ cyclic (Hadamard) difference sets, as well as a concluding chapter briefly mentioning various applications, see [2].

References

1. Shift Register Sequences, S.W. Golomb, Holden-Day, Inc. San Francisco, CA, 1967. Revised second edition, Aegean Park Press, Laguna Hills, CA, 1982.
2. Signal Design for Good Correlation, for Wireless Communication, Cryptography, and Radar, S.W. Golomb and G. Gong, Cambridge University Press, Cambridge (U.K.), New York (U.S.A.), etc., 2005.

The Probabilistic Theory of the Joint Linear Complexity of Multisequences

Harald Niederreiter

Department of Mathematics, National University of Singapore,
2 Science Drive 2, Singapore 117543, Republic of Singapore
nied@math.nus.edu.sg

Abstract. The joint linear complexity and the joint linear complexity profile are standard complexity measures for multisequences in the context of word-based stream ciphers. The last few years have seen major advances in the theory of these complexity measures, especially with regard to probabilistic results on the behavior of random (periodic and nonperiodic) multisequences. This paper presents a survey of these developments as well as the necessary background for the results.

Keywords: Word-based stream ciphers, Multisequences, Linear complexity, Joint linear complexity, Joint linear complexity profile.

1 Introduction

A central issue in the security analysis of stream ciphers is the quality assessment of keystreams. In other words, we need to know how close the keystream is to "true randomness". Keystreams guaranteeing an adequate security level must meet various requirements such as possessing good statistical randomness properties and a high complexity in a suitable sense. In the system-theoretic approach to the quality assessment of keystreams, the basic complexity measure is the linear complexity (see [33]). The yardstick in the assessment of keystreams by means of linear complexity is the behavior of random sequences (in an appropriate stochastic model) with respect to the linear complexity.

Recent developments in stream ciphers point towards an interest in word-based or vectorized stream ciphers (see e.g. [3], [7], [17], and the proposals DRAGON, NLS, and SSS to the ECRYPT stream cipher project [10]). The theory of word-based stream ciphers requires the study of multisequences, i.e., of parallel streams of finitely many sequences. In the framework of linear complexity theory, the appropriate complexity measure for multisequences is the joint linear complexity.

Let $\mathbb{F}_q$ be the finite field of order q, where q is an arbitrary prime power. By a (finite or infinite) sequence *over* $\mathbb{F}_q$ we mean a sequence of elements of $\mathbb{F}_q$. More generally, for an integer $m \geq 1$, an *m-fold multisequence over* $\mathbb{F}_q$ is an m-tuple of sequences over $\mathbb{F}_q$. In other words, an m-fold multisequence over $\mathbb{F}_q$ is given by $\mathbf{S} = (S_1, \ldots, S_m)$, where each S_j, $j = 1, \ldots, m$, is a sequence over

G. Gong et al. (Eds.): SETA 2006, LNCS 4086, pp. 5–16, 2006.

$\mathbb{F}_q$. For the purposes of this paper, when considering an m-fold multisequence $\mathbf{S} = (S_1, \ldots, S_m)$ over $\mathbb{F}_q$, we agree that all sequences S_j, $j = 1, \ldots, m$, have the same (finite or infinite) length and we call it the *length* of the multisequence. The following definitions are fundamental for this paper.

Definition 1. *Let n be a positive integer and let $\mathbf{S} = (S_1, \ldots, S_m)$ be an m-fold multisequence over $\mathbb{F}_q$ of length at least n. Then the* nth joint linear complexity *$L_n^{(m)}(\mathbf{S})$ of $\mathbf{S}$ is the least order of a linear recurrence relation over $\mathbb{F}_q$ that simultaneously generates the first n terms of each sequence S_j, $j = 1, \ldots, m$. If $\mathbf{S}$ has infinite length, then the sequence $L_1^{(m)}(\mathbf{S}), L_2^{(m)}(\mathbf{S}), \ldots$ of nonnegative integers is called the* joint linear complexity profile *of $\mathbf{S}$.*

Definition 2. *Let $\mathbf{S} = (S_1, \ldots, S_m)$ be an ultimately periodic m-fold multisequence over $\mathbb{F}_q$, that is, each sequence S_j, $j = 1, \ldots, m$, is ultimately periodic. Then the* joint linear complexity *$L^{(m)}(\mathbf{S})$ of $\mathbf{S}$ is defined by*

$$L^{(m)}(\mathbf{S}) = \sup_{n \geq 1} L_n^{(m)}(\mathbf{S}).$$

We always have $0 \leq L_n^{(m)}(\mathbf{S}) \leq n$ and $L_n^{(m)}(\mathbf{S}) \leq L_{n+1}^{(m)}(\mathbf{S})$. If $\mathbf{S}$ is ultimately periodic, then $L^{(m)}(\mathbf{S}) < \infty$.

This paper surveys the theory of the joint linear complexity and the joint linear complexity profile of multisequences, with a special emphasis on recent developments. The last few years have indeed seen dramatic progress in this theory which was undoubtedly motivated by the increased activity in the design of word-based stream ciphers. To create a suitable backdrop for the landscape of joint linear complexity, we first present a concise review of the case of single sequences in Section 2. The core of the paper is Section 3 which covers recent work on the joint linear complexity profile of multisequences and, in particular, the resolution of a conjecture that provides a description of the asymptotic behavior of the joint linear complexity profile of random multisequences. Section 4 is devoted to recent results on the joint linear complexity of periodic multisequences.

For general background on linear complexity and joint linear complexity, we refer to the monographs [2] and [9] and to the survey articles [27], [28], and [33].

2 The Linear Complexity Profile of Single Sequences

In the case $m = 1$ we get a single sequence S over $\mathbb{F}_q$. To simplify the notation, we write $L_n(S)$ for $L_n^{(1)}(S)$ and $L(S)$ for $L^{(1)}(S)$. Note that in the case of a single sequence, we speak of the linear complexity and the linear complexity profile instead of the joint linear complexity and the joint linear complexity profile, respectively.

For a positive integer n, let $\mathbb{F}_q^n$ be the set of n-tuples of elements of $\mathbb{F}_q$, or equivalently the set of sequences over $\mathbb{F}_q$ of length n. If L is an integer with $0 \leq L \leq n$, then $N_n(L)$ denotes the number of sequences $S \in \mathbb{F}_q^n$ with $L_n(S) = L$.

It is trivial that $N_n(0) = 1$. According to the classical formula of Gustavson [16], we have

$$N_n(L) = (q-1)q^{\min(2L-1,2n-2L)} \qquad \text{for } 1 \le L \le n. \tag{1}$$

A proof of this formula can be found also in the book of Niederreiter and Xing [31, Theorem 7.1.6].

We use a canonical stochastic model whereby finite sequences over $\mathbb{F}_q$ of the same length are equiprobable. In detail, let μ_q be the uniform probability measure on $\mathbb{F}_q$ which assigns the measure q^{-1} to each element of $\mathbb{F}_q$. This measure induces the complete product measure μ_q^∞ on the set $\mathbb{F}_q^\infty$ of infinite sequences over $\mathbb{F}_q$. For any integer $n \ge 1$, the expected value E_n of the nth linear complexity $L_n(S)$ of random infinite sequences S over $\mathbb{F}_q$ is then given by

$$E_n = \frac{1}{q^n} \sum_{L=1}^{n} L N_n(L) = \frac{n}{2} + \theta_q - \frac{n}{(q+1)q^n} - \frac{1}{(q+1)^2 q^{n-1}}, \tag{2}$$

where $\theta_q = \frac{q}{(q+1)^2}$ if n is even and $\theta_q = \frac{q^2+1}{2(q+1)^2}$ if n is odd. This formula was proved by Rueppel [32, Chapter 4] for $q = 2$ and by Smeets [36] for arbitrary q. The same sources contain formulas for the variance V_n of $L_n(S)$ for $q = 2$ and arbitrary q, respectively, and the formula in [36] shows that $V_n = O(q^{-1})$ with an absolute implied constant.

The linear complexity profile of a single infinite sequence S over $\mathbb{F}_q$ can be described completely in terms of the continued fraction expansion of the generating function of S (see [24] and [31, Theorem 7.1.4]). In particular, the jumps in the linear complexity profile are given exactly by the degrees of the partial quotients — which are polynomials over $\mathbb{F}_q$ — in this continued fraction expansion. Equivalently, the linear complexity profile can be described in terms of the Berlekamp-Massey algorithm.

The connection between linear complexity profiles and continued fraction expansions is the basis for a powerful method of analyzing the asymptotic behavior of the linear complexity profile of random single infinite sequences over $\mathbb{F}_q$ (see [25]). According to this approach, the continued fraction expansion gives rise to a dynamical system on the set of generating functions over $\mathbb{F}_q$. This dynamical system is isomorphic to a Bernoulli shift on the set P_q^∞ of all infinite sequences of elements of $P_q := \{f \in \mathbb{F}_q[x] : \deg(f) \ge 1\}$. By virtue of this isomorphism, the continued fractions dynamical system inherits all the dynamical properties of the Bernoulli shift, such as being ergodic, and these properties lead, in turn, to probabilistic results on the linear complexity profile. For instance, the individual ergodic theorem yields the following theorem from [25].

Theorem 1. *We have μ_q^∞-almost everywhere*

$$\lim_{n\to\infty} \frac{L_n(S)}{n} = \frac{1}{2}.$$

Recall that "almost everywhere" means the same as "with probability 1" under the ambient probability measure, in this case μ_q^∞ on the set $\mathbb{F}_q^\infty$. A more detailed analysis leads to the following stronger result from [25]. Here and in the sequel, we write $\log_q$ for the logarithm to the base q.

Theorem 2. *We have* μ_q^∞*-almost everywhere*

$$\liminf_{n\to\infty} \frac{L_n(S) - \frac{n}{2}}{\log_q n} = -\frac{1}{2}$$

and

$$\limsup_{n\to\infty} \frac{L_n(S) - \frac{n}{2}}{\log_q n} = \frac{1}{2}.$$

Theorem 1 can be proved also by a relatively elementary argument which uses only the formula (1) as well as the Borel-Cantelli lemma from probability theory. This method of proof was delineated in [26]. However, this simpler method does not seem capable of yielding a proof of the stronger Theorem 2. Note that Theorem 2 says, in particular, that for a random single infinite sequence S over $\mathbb{F}_q$ we have

$$L_n(S) = \frac{n}{2} + O(\log n) \qquad \text{for all } n \geq 2,$$

and furthermore deviations from $\frac{n}{2}$ of the order of magnitude $\log n$ must appear for infinitely many n in both the positive and the negative direction.

3 The Joint Linear Complexity Profile of Multisequences

As outlined in Section 1, it is crucial for the quality assessment of multisequences to determine the behavior of the joint linear complexity profile of random multisequences of infinite length. This turns out to be a challenging task. The strategy should of course be to generalize the tools in the analysis of single sequences to the case of multisequences. The only known powerful method in the case of single sequences is based on the dynamical system that results from the continued fraction or Berlekamp-Massey algorithm (see Section 2). Although various multidimensional versions of the Berlekamp-Massey algorithm have been developed (see e.g. [1], [4], [8], [9, Appendix A], [11], [12], [18], [34], [38]), it is not clear whether any of these can be linked with a dynamical system possessing favorable properties such as ergodicity. The second path to follow is to try to generalize the formula (1) to multisequences and then to extend the method in [26]. This was the approach chosen by Niederreiter and Wang and is described in the sequel.

The principal aim is to find an analog of Theorem 1 for the case of multisequences. There is a folklore conjecture mentioned e.g. in [39] which says that for m-fold multisequences $\mathbf{S}$ over $\mathbb{F}_q$ of infinite length we should have

$$\lim_{n\to\infty} \frac{L_n^{(m)}(\mathbf{S})}{n} = \frac{m}{m+1} \tag{3}$$

with probability 1. This conjecture has now been settled (see Theorem 3 below).

The first step in the proof of (3) is the generalization of the formula (1) to multisequences. For integers $m \geq 1$, $n \geq 1$, and $0 \leq L \leq n$, let $N_n^{(m)}(L)$ denote the number of m-fold multisequences $\mathbf{S}$ over $\mathbb{F}_q$ of length n with $L_n^{(m)}(\mathbf{S}) = L$. It is trivial that $N_n^{(m)}(0) = 1$. For $m \geq 2$, Niederreiter [28] found the first instance of an explicit formula for the counting function $N_n^{(m)}(L)$ by treating the special case of the "lower half" of the range for L, thus obtaining

$$N_n^{(m)}(L) = (q^m - 1)q^{(m+1)L-m} \qquad \text{for } 1 \leq L \leq \frac{n}{2}. \tag{4}$$

This case is considered easier because in this range for L, there is a unique minimal polynomial for any m-fold multisequence $\mathbf{S}$ over $\mathbb{F}_q$ of length n with $L_n^{(m)}(\mathbf{S}) = L$, and this allows a reduction of the enumeration problem for multisequences to an enumeration problem for polynomials over $\mathbb{F}_q$ which can be handled by the theory of arithmetic functions on polynomial rings over finite fields (see [28] for the details). In view of (1), the formula (4) is of course also valid for $m = 1$.

In the range $n/2 < L \leq n$, it seems much more difficult to get a formula for $N_n^{(m)}(L)$. Wang and Niederreiter [37] developed an approach to this enumeration problem which is founded on the sophisticated multisequence shift-register synthesis algorithm of Wang, Zhu, and Pei [38]. The latter algorithm is, in turn, based on a lattice basis reduction algorithm in function fields due to Schmidt [35].

For any integers $m \geq 1$ and $L \geq 1$, let $P(m; L)$ be the set of m-tuples $\mathbf{I} = (i_1, \ldots, i_m) \in \mathbb{Z}^m$ with $i_1 \geq i_2 \geq \cdots \geq i_m \geq 0$ and $i_1 + \cdots + i_m = L$. For any $\mathbf{I} = (i_1, \ldots, i_m) \in P(m; L)$, let $\lambda(\mathbf{I})$ be the number of positive entries in $\mathbf{I}$. Then $\mathbf{I}$ can be written in the form

$$\mathbf{I} = (\underbrace{i_1, \ldots, i_{s_{\mathbf{I},1}}}_{s_{\mathbf{I},1}}, \underbrace{i_{s_{\mathbf{I},1}+1}, \ldots, i_{s_{\mathbf{I},1}+s_{\mathbf{I},2}}}_{s_{\mathbf{I},2}}, \ldots, \underbrace{i_{s_{\mathbf{I},1}+\cdots+s_{\mathbf{I},t-1}+1}, \ldots, i_{s_{\mathbf{I},1}+\cdots+s_{\mathbf{I},t}}}_{s_{\mathbf{I},t}},$$
$$\ldots, \underbrace{i_{s_{\mathbf{I},1}+\cdots+s_{\mathbf{I},\mu(\mathbf{I})-1}+1}, \ldots, i_{s_{\mathbf{I},1}+\cdots+s_{\mathbf{I},\mu(\mathbf{I})}}}_{s_{\mathbf{I},\mu(\mathbf{I})}}, \underbrace{i_{\lambda(\mathbf{I})+1}, \ldots, i_m}_{s_{\mathbf{I},\mu(\mathbf{I})+1}}),$$

where $i_{s_{\mathbf{I},1}+\cdots+s_{\mathbf{I},t-1}+1} = \cdots = i_{s_{\mathbf{I},1}+\cdots+s_{\mathbf{I},t}} > i_{s_{\mathbf{I},1}+\cdots+s_{\mathbf{I},t}+1}$ for $1 \leq t \leq \mu(\mathbf{I})$, $i_{\lambda(\mathbf{I})+1} = \cdots = i_m = 0$, and $\lambda(\mathbf{I}) = s_{\mathbf{I},1} + \cdots + s_{\mathbf{I},\mu(\mathbf{I})}$. If $\lambda(\mathbf{I}) = m$, then $s_{\mathbf{I},\mu(\mathbf{I})+1} = 0$. Furthermore, we define

$$c(\mathbf{I}) = \prod_{i=1}^{\lambda(\mathbf{I})} \frac{(q^{m+1-i} - 1)(q^i - 1)}{q - 1},$$

$$d(\mathbf{I}) = \prod_{j=1}^{\mu(\mathbf{I})} \prod_{i=1}^{s_{\mathbf{I},j}} \frac{q^i - 1}{q - 1}.$$

Put

$$e_{\lambda(\mathbf{I})} = 2 \times (0, 1, 2, \ldots, \lambda(\mathbf{I}), 0, \ldots, 0) \in \mathbb{Z}^{m+1}.$$

For $\mathbf{I} \in P(m; L)$, let $[\mathbf{I}, n-L]$ denote the vector obtained by arranging the $m+1$ numbers between the square brackets in nonincreasing order. Let $\cdot$ denote the standard inner product in $\mathbb{R}^{m+1}$. We define $b(\mathbf{I}, n-L)$ as follows. If $0 \leq n-L < i_{\lambda(\mathbf{I})}$, then we put

$$b(\mathbf{I}, n-L) = e_{\lambda(\mathbf{I})} \cdot [\mathbf{I}, n-L] - \frac{\lambda(\mathbf{I})(\lambda(\mathbf{I})-1)}{2}.$$

If $i_{s_{\mathbf{I},1}+\cdots+s_{\mathbf{I},w+1}} \leq n-L < i_{s_{\mathbf{I},1}+\cdots+s_{\mathbf{I},w}}$ for some integer w with $1 \leq w \leq \mu(\mathbf{I})-1$, then

$$b(\mathbf{I}, n-L) = e_{\lambda(\mathbf{I})} \cdot [\mathbf{I}, n-L] - \left(\frac{\lambda(\mathbf{I})(\lambda(\mathbf{I})+1)}{2} - (s_{\mathbf{I},1} + \cdots + s_{\mathbf{I},w}) \right).$$

Finally, if $n - L \geq i_1$, then

$$b(\mathbf{I}, n-L) = e_{\lambda(\mathbf{I})} \cdot [\mathbf{I}, n-L] - \frac{\lambda(\mathbf{I})(\lambda(\mathbf{I})+1)}{2}.$$

The formula for $N_n^{(m)}(L)$ in [37] is now given as follows.

Proposition 1. *For any integers $m \geq 1$, $n \geq 1$, and $1 \leq L \leq n$, we have*

$$N_n^{(m)}(L) = \sum_{\mathbf{I} \in P(m;L)} \frac{c(\mathbf{I})}{d(\mathbf{I})} q^{b(\mathbf{I}, n-L)}.$$

Convenient closed-form expressions for $N_n^{(m)}(L)$ were shown in the special cases $m = 2$ (see [37]) and $m = 3$ (see [30]). The derivation of such expressions for $N_n^{(m)}(L)$ is exceedingly more complicated the larger the value of m.

We extend the stochastic model in Section 2 from single sequences to multisequences. The assumptions are the following: (i) finite sequences over $\mathbb{F}_q$ of the same length are equiprobable; (ii) corresponding terms in the m streams making up an m-fold multisequence over $\mathbb{F}_q$ are statistically independent. Let $\mathbb{F}_q^m$ be the set of m-tuples of elements of $\mathbb{F}_q$ and let $(\mathbb{F}_q^m)^\infty$ be the set of infinite sequences with terms from $\mathbb{F}_q^m$. It is obvious that $(\mathbb{F}_q^m)^\infty$ can be identified with the set of m-fold multisequences over $\mathbb{F}_q$ of infinite length, and henceforth we will use this identification. Let $\mu_{q,m}$ be the uniform probability measure on $\mathbb{F}_q^m$ which assigns the measure q^{-m} to each element of $\mathbb{F}_q^m$. Furthermore, let $\mu_{q,m}^\infty$ be the complete product measure on $(\mathbb{F}_q^m)^\infty$ induced by $\mu_{q,m}$. The following theorem, which proves the conjecture (3), was shown by Niederreiter and Wang [29] on the basis of Proposition 1.

Theorem 3. *For any integer $m \geq 1$ we have $\mu_{q,m}^\infty$-almost everywhere*

$$\lim_{n \to \infty} \frac{L_n^{(m)}(\mathbf{S})}{n} = \frac{m}{m+1}.$$

This result was refined in [30] to obtain the following weak analog of Theorem 2.

Theorem 4. *For any integer $m \geq 1$ we have $\mu_{q,m}^{\infty}$-almost everywhere*

$$-\frac{1}{m+1} \leq \liminf_{n\to\infty} \frac{L_n^{(m)}(\mathbf{S}) - \frac{mn}{m+1}}{\log_q n} \leq \limsup_{n\to\infty} \frac{L_n^{(m)}(\mathbf{S}) - \frac{mn}{m+1}}{\log_q n} \leq 1.$$

Theorem 4 shows, in particular, that $\mu_{q,m}^{\infty}$-almost everywhere we have

$$L_n^{(m)}(\mathbf{S}) = \frac{mn}{m+1} + O(\log n) \qquad \text{as } n \to \infty.$$

We note also that Dai, Imamura, and Yang [5] have given a sufficient condition for an m-fold multisequence $\mathbf{S}$ over $\mathbb{F}_q$ of infinite length to satisfy

$$\lim_{n\to\infty} \frac{L_n^{(m)}(\mathbf{S})}{n} = \frac{m}{m+1}.$$

For any integers $m \geq 1$ and $n \geq 1$, let $E_n^{(m)}$ be the expected value of the nth joint linear complexity of random m-fold multisequences over $\mathbb{F}_q$ of infinite length. We can write $E_n^{(m)}$ in the form

$$E_n^{(m)} = \frac{1}{q^{mn}} \sum_{L=1}^{n} L N_n^{(m)}(L).$$

For $m = 1$ we have the explicit formula (2) for this expected value. For $m = 2$ and $q = 2$ it was proved by Feng and Dai [13] and for $m = 2$ and arbitrary q it was shown by Wang and Niederreiter [37] that

$$E_n^{(2)} = \frac{2n}{3} + O(1) \qquad \text{as } n \to \infty.$$

For $m = 3$ we have

$$E_n^{(3)} = \frac{3n}{4} + O(1) \qquad \text{as } n \to \infty$$

according to a result of Niederreiter and Wang [30]. The following theorem holds for arbitrary m and can be easily derived from Theorem 3 by using the dominated convergence theorem (see [29]).

Theorem 5. *For any integer $m \geq 1$ we have*

$$E_n^{(m)} = \frac{mn}{m+1} + o(n) \qquad \textit{as } n \to \infty.$$

Niederreiter and Wang [30] conjectured that for any integer $m \geq 1$ we have

$$E_n^{(m)} = \frac{mn}{m+1} + O(1) \qquad \text{as } n \to \infty.$$

According to the results mentioned above, this conjecture is settled for $m \leq 3$. Another result supporting this conjecture is the inequality

$$E_n^{(m)} \geq \frac{mn}{m+1} + c \qquad \text{for all } m \geq 1 \text{ and } n \geq 1$$

with an absolute constant c established in [30].

In analogy with a definition for single sequences (see [24]), an m-fold multisequence $\mathbf{S}$ over $\mathbb{F}_q$ of infinite length is said to have an *almost perfect joint linear complexity profile* if there exists a constant $C(\mathbf{S})$ such that

$$L_n^{(m)}(\mathbf{S}) \geq \frac{mn}{m+1} + C(\mathbf{S}) \qquad \text{for all } n \geq 1.$$

Such multisequences were investigated and constructed by Xing [39] and Xing, Lam, and Wei [40] and more recently by Feng, Wang, and Dai [14]. Meidl and Winterhof [23] proved bounds for the joint linear complexity profile of a family of multisequences generated by an inversive method.

4 Periodic Multisequences

The case of periodic sequences over $\mathbb{F}_q$ has received a lot of attention in the theory of stream ciphers since all keystreams used in practice are periodic. We recall that for a positive integer N, a sequence S with terms $s_0, s_1, \dots$ in $\mathbb{F}_q$ is called *N-periodic* if $s_{i+N} = s_i$ for all $i \geq 0$. Note that N need not necessarily be the least period of the sequence. According to a classical formula (see e.g. [9, Section 5.1.2]), the linear complexity $L(S)$ of an N-periodic sequence S over $\mathbb{F}_q$ is given by

$$L(S) = N - \deg(\gcd(x^N - 1, S^N(x))), \tag{5}$$

where $S^N(x) := s_0 + s_1 x + \cdots + s_{N-1}x^{N-1} \in \mathbb{F}_q[x]$. Another description of $L(S)$ for an N-periodic sequence S over $\mathbb{F}_q$ is possible by means of the generalized discrete Fourier transform and the Günther-Blahut theorem (see [19], [21]).

An important question for stream ciphers is the following: if we pick an N-periodic sequence over $\mathbb{F}_q$ at random, what is the expected linear complexity? For fixed q and $N \geq 1$, the number of N-periodic sequences over $\mathbb{F}_q$ is equal to q^N. In the canonical stochastic model where all these q^N sequences are considered equally likely, the expected value G_N of the linear complexity is given by

$$G_N = \frac{1}{q^N} \sum_S L(S),$$

where S runs through all N-periodic sequences over $\mathbb{F}_q$. The first general formula for G_N was stated by Dai and Yang [6] without proof. Later, a detailed proof of an equivalent formula was given by Meidl and Niederreiter [21]. These formulas correspond to the special case $m = 1$ of the formulas in (7) and Theorem 6, respectively, given below.

These considerations can be generalized to periodic multisequences. If $\mathbf{S} = (S_1, \dots, S_m)$ is an m-fold multisequence over $\mathbb{F}_q$ with each sequence S_j, $j = 1, \dots, m$, being periodic, then we can assume w.l.o.g. that the sequences S_j have the common period N. We say in this case that $\mathbf{S}$ is *N-periodic*. The analog of (5) is the formula

$$L^{(m)}(\mathbf{S}) = N - \deg(\gcd(x^N - 1, S_1^N(x), \dots, S_m^N(x))) \tag{6}$$

for the joint linear complexity $L^{(m)}(\mathbf{S})$ of $\mathbf{S}$, where $S_j^N(x)$ is the polynomial corresponding to the sequence S_j for $j = 1, \ldots, m$ (see [22]). It is a trivial consequence of the definition and follows also from (6) that for any m-fold N-periodic multisequence $\mathbf{S}$ over $\mathbb{F}_q$ we have $L^{(m)}(\mathbf{S}) \leq N$.

Let $G_N^{(m)}$ denote the expected value of the joint linear complexity of m-fold N-periodic multisequences over $\mathbb{F}_q$, that is,

$$G_N^{(m)} = \frac{1}{q^{mN}} \sum_{\mathbf{S}} L^{(m)}(\mathbf{S})$$

with the sum being extended over all q^{mN} m-fold N-periodic multisequences $\mathbf{S}$ over $\mathbb{F}_q$. Meidl and Niederreiter [22] established a formula for $G_N^{(m)}$ by using the generalized discrete Fourier transform for multisequences and cyclotomy. We recall that if $w \geq 1$ and l are integers with $\gcd(w, q) = 1$, then the *cyclotomic coset* C_l mod w (relative to powers of q) is defined by

$$C_l = \{0 \leq k \leq w - 1 : k \equiv lq^r \pmod{w} \text{ for some integer } r \geq 0\}.$$

The different cyclotomic cosets mod w form a partition of the set $\{0, 1, \ldots, w-1\}$.

Theorem 6. *Let $N = p^v w$, where the prime p is the characteristic of $\mathbb{F}_q$ and the integers v and w satisfy $v \geq 0$, $w \geq 1$, and* $\gcd(p, w) = 1$. *Let $B_1, \ldots, B_h$ be the different cyclotomic cosets mod w and put $b_i = |B_i|$ for $1 \leq i \leq h$. Then for any $m \geq 1$,*

$$G_N^{(m)} = N - \sum_{i=1}^{h} \frac{b_i(1 - q^{-mb_i p^v})}{q^{mb_i} - 1}.$$

An equivalent formula for $G_N^{(m)}$ was later noted by Fu, Niederreiter, and Su [15, Remark 2]. Let ϕ be Euler's totient function and for any positive integer d with $\gcd(d, q) = 1$ let $\kappa(d)$ be the order of q mod d, i.e., the least positive integer e such that $q^e \equiv 1 \pmod{d}$. Then we have

$$G_N^{(m)} = N - \sum_{d|w} \frac{\phi(d)(1 - q^{-m\kappa(d)p^v})}{q^{m\kappa(d)} - 1}, \tag{7}$$

where the sum is over all positive divisors d of w and where p, v, and w are as in Theorem 6. For $m = 2$ the formula (7) leads to the lower bound

$$G_N^{(2)} \geq N - O(\log\log(w + 2))$$

with an absolute implied constant. For $m \geq 3$ we obtain

$$G_N^{(m)} \geq N - O(1)$$

with an absolute implied constant. We refer to [15, Remark 3] for proofs of these lower bounds. Further investigations on expected values of the joint linear complexity of periodic multisequences can be found in Meidl [20].

By more involved arguments, Fu, Niederreiter, and Su [15] were also able to obtain an expression for the variance $W_N^{(m)}$ of the joint linear complexity of m-fold N-periodic multisequences over $\mathbb{F}_q$.

Theorem 7. *With the notation in Theorem 6, we have for any $m \geq 1$,*

$$W_N^{(m)} = \sum_{i=1}^{h} b_i^2 \frac{(2p^v + 1)(a_i^{p^v+2} - a_i^{p^v+1}) - a_i^{2p^v+2} + a_i}{(1 - a_i)^2},$$

where $a_i = q^{-mb_i}$ for $1 \leq i \leq h$.

In [15, Remark 4] an alternative formula for $W_N^{(m)}$ is given which bears the same relationship to Theorem 7 as (7) bears to Theorem 6. For $m = 1$ this alternative formula was stated without proof by Dai and Yang [6]. The following bounds are shown in [15]: for $m = 2$ we have

$$W_N^{(2)} = O((\log(w + 1)) \log\log(w + 2))$$

with an absolute implied constant and for $m \geq 3$ we have $W_N^{(m)} = O(1)$ with an absolute implied constant.

The results in this section show that for random m-fold N-periodic multisequences over $\mathbb{F}_q$, the joint linear complexity is close to N — which is the trivial upper bound on the joint linear complexity — with a small variance.

Acknowledgments

This research is partially supported by the DSTA grant R-394-000-025-422 with Temasek Laboratories in Singapore and by a collaborative research grant from the National Institute of Information and Communications Technology in Japan.

References

1. M.A. Armand, Multisequence shift register synthesis over commutative rings with identity with applications to decoding cyclic codes over integer residue rings, *IEEE Trans. Inform. Theory* **50**, 220–229 (2004).
2. T.W. Cusick, C. Ding, and A. Renvall, *Stream Ciphers and Number Theory*, Elsevier, Amsterdam, 1998.
3. J. Daemen and C. Clapp, Fast hashing and stream encryption with PANAMA, *Fast Software Encryption* (S. Vaudenay, ed.), Lecture Notes in Computer Science, Vol. **1372**, pp. 60–74, Springer, Berlin, 1998.
4. Z.D. Dai, X.T. Feng, and J.H. Yang, Multi-continued fraction algorithm and generalized B-M algorithm over F_2, *Sequences and Their Applications – SETA 2004* (T. Helleseth et al., eds.), Lecture Notes in Computer Science, Vol. **3486**, pp. 339–354, Springer, Berlin, 2005.
5. Z.D. Dai, K. Imamura, and J.H. Yang, Asymptotic behavior of normalized linear complexity of multi-sequences, *Sequences and Their Applications – SETA 2004* (T. Helleseth et al., eds.), Lecture Notes in Computer Science, Vol. **3486**, pp. 129–142, Springer, Berlin, 2005.

6. Z.D. Dai and J.H. Yang, Linear complexity of periodically repeated random sequences, *Advances in Cryptology – EUROCRYPT '91* (D.W. Davies, ed.), Lecture Notes in Computer Science, Vol. **547**, pp. 168–175, Springer, Berlin, 1991.
7. E. Dawson and L. Simpson, Analysis and design issues for synchronous stream ciphers, *Coding Theory and Cryptology* (H. Niederreiter, ed.), pp. 49–90, World Scientific, Singapore, 2002.
8. C. Ding, Proof of Massey's conjectured algorithm, *Advances in Cryptology – EUROCRYPT '88* (C.G. Günther, ed.), Lecture Notes in Computer Science, Vol. **330**, pp. 345–349, Springer, Berlin, 1988.
9. C. Ding, G. Xiao, and W. Shan, *The Stability Theory of Stream Ciphers*, Lecture Notes in Computer Science, Vol. **561**, Springer, Berlin, 1991.
10. ECRYPT stream cipher project; available at `http://www.ecrypt.eu.org/stream`.
11. G.-L. Feng and K.K. Tzeng, A generalized Euclidean algorithm for multisequence shift-register synthesis, *IEEE Trans. Inform. Theory* **35**, 584–594 (1989).
12. G.-L. Feng and K.K. Tzeng, A generalization of the Berlekamp-Massey algorithm for multisequence shift-register synthesis with applications to decoding cyclic codes, *IEEE Trans. Inform. Theory* **37**, 1274–1287 (1991).
13. X.T. Feng and Z.D. Dai, Expected value of the linear complexity of two-dimensional binary sequences, *Sequences and Their Applications – SETA 2004* (T. Helleseth et al., eds.), Lecture Notes in Computer Science, Vol. **3486**, pp. 113–128, Springer, Berlin, 2005.
14. X.T. Feng, Q.L. Wang, and Z.D. Dai, Multi-sequences with d-perfect property, *J. Complexity* **21**, 230–242 (2005).
15. F.-W. Fu, H. Niederreiter, and M. Su, The expectation and variance of the joint linear complexity of random periodic multisequences, *J. Complexity* **21**, 804–822 (2005).
16. F.G. Gustavson, Analysis of the Berlekamp-Massey linear feedback shift-register synthesis algorithm, *IBM J. Res. Develop.* **20**, 204–212 (1976).
17. P. Hawkes and G.G. Rose, Exploiting multiples of the connection polynomial in word-oriented stream ciphers, *Advances in Cryptology – ASIACRYPT 2000* (T. Okamoto, ed.), Lecture Notes in Computer Science, Vol. **1976**, pp. 303–316, Springer, Berlin, 2000.
18. N. Kamiya, On multisequence shift register synthesis and generalized-minimum-distance decoding of Reed-Solomon codes, *Finite Fields Appl.* **1**, 440–457 (1995).
19. J.L. Massey and S. Serconek, Linear complexity of periodic sequences: a general theory, *Advances in Cryptology – CRYPTO '96* (N. Koblitz, ed.), Lecture Notes in Computer Science, Vol. **1109**, pp. 358–371, Springer, Berlin, 1996.
20. W. Meidl, Discrete Fourier transform, joint linear complexity and generalized joint linear complexity of multisequences, *Sequences and Their Applications – SETA 2004* (T. Helleseth et al., eds.), Lecture Notes in Computer Science, Vol. **3486**, pp. 101–112, Springer, Berlin, 2005.
21. W. Meidl and H. Niederreiter, On the expected value of the linear complexity and the k-error linear complexity of periodic sequences, *IEEE Trans. Inform. Theory* **48**, 2817–2825 (2002).
22. W. Meidl and H. Niederreiter, The expected value of the joint linear complexity of periodic multisequences, *J. Complexity* **19**, 61–72 (2003).
23. W. Meidl and A. Winterhof, On the joint linear complexity profile of explicit inversive multisequences, *J. Complexity* **21**, 324–336 (2005).
24. H. Niederreiter, Sequences with almost perfect linear complexity profile, *Advances in Cryptology – EUROCRYPT '87* (D. Chaum and W.L. Price, eds.), Lecture Notes in Computer Science, Vol. **304**, pp. 37–51, Springer, Berlin, 1988.

25. H. Niederreiter, The probabilistic theory of linear complexity, *Advances in Cryptology – EUROCRYPT '88* (C.G. Günther, ed.), Lecture Notes in Computer Science, Vol. **330**, pp. 191–209, Springer, Berlin, 1988.
26. H. Niederreiter, A combinatorial approach to probabilistic results on the linear-complexity profile of random sequences, *J. Cryptology* **2**, 105–112 (1990).
27. H. Niederreiter, Some computable complexity measures for binary sequences, *Sequences and Their Applications* (C. Ding, T. Helleseth, and H. Niederreiter, eds.), pp. 67–78, Springer, London, 1999.
28. H. Niederreiter, Linear complexity and related complexity measures for sequences, *Progress in Cryptology – INDOCRYPT 2003* (T. Johansson and S. Maitra, eds.), Lecture Notes in Computer Science, Vol. **2904**, pp. 1–17, Springer, Berlin, 2003.
29. H. Niederreiter and L.-P. Wang, Proof of a conjecture on the joint linear complexity profile of multisequences, *Progress in Cryptology – INDOCRYPT 2005* (S. Maitra, C.E. Veni Madhavan, and R. Venkatesan, eds.), Lecture Notes in Computer Science, Vol. **3797**, pp. 13–22, Springer, Berlin, 2005.
30. H. Niederreiter and L.-P. Wang, The asymptotic behavior of the joint linear complexity profile of multisequences, *Monatsh. Math.*, to appear.
31. H. Niederreiter and C.P. Xing, *Rational Points on Curves over Finite Fields: Theory and Applications*, London Math. Soc. Lecture Note Series, Vol. **285**, Cambridge University Press, Cambridge, 2001.
32. R.A. Rueppel, *Analysis and Design of Stream Ciphers*, Springer, Berlin, 1986.
33. R.A. Rueppel, Stream ciphers, *Contemporary Cryptology: The Science of Information Integrity* (G.J. Simmons, ed.), pp. 65–134, IEEE Press, New York, 1992.
34. S. Sakata, Extension of the Berlekamp-Massey algorithm to N dimensions, *Inform. and Comput.* **84**, 207–239 (1990).
35. W.M. Schmidt, Construction and estimation of bases in function fields, *J. Number Theory* **39**, 181–224 (1991).
36. B. Smeets, The linear complexity profile and experimental results on a randomness test of sequences over the finite field $\mathbf{F}_q$, Technical Report, Department of Information Theory, University of Lund, 1988.
37. L.-P. Wang and H. Niederreiter, Enumeration results on the joint linear complexity of multisequences, *Finite Fields Appl.*, to appear; available online as document `doi:10.1016/j.ffa.2005.03.005`.
38. L.-P. Wang, Y.-F. Zhu, and D.-Y. Pei, On the lattice basis reduction multisequence synthesis algorithm, *IEEE Trans. Inform. Theory* **50**, 2905–2910 (2004).
39. C.P. Xing, Multi-sequences with almost perfect linear complexity profile and function fields over finite fields, *J. Complexity* **16**, 661–675 (2000).
40. C.P. Xing, K.Y. Lam, and Z.H. Wei, A class of explicit perfect multi-sequences, *Advances in Cryptology – ASIACRYPT '99* (K.Y. Lam, E. Okamoto, and C.P. Xing, eds.), Lecture Notes in Computer Science, Vol. **1716**, pp. 299–305, Springer, Berlin, 1999.

Multi-Continued Fraction Algorithms and Their Applications to Sequences*

Zongduo Dai

State Key Laboratory of Information Security, Institute of Software,
Chinese Academy of Sciences, Beijing, 100080, China
daizongduo@is.ac.cn

1 Introduction

Pseudorandom sequences have a wide applications. In stream ciphers, the key stream usually is a pseudorandom sequence over a finite field F_q:

$$\alpha = (a_1, a_2, \cdots, a_i, \cdots),\ a_i \in F_q$$

In the recent years, multi-sequences:

$$\underline{r} = \begin{pmatrix} r_1 \\ r_2 \\ \vdots \\ r_m \end{pmatrix} = \begin{pmatrix} r_{1,1} & r_{1,2} & \cdots & r_{1,n} & \cdots \\ r_{2,1} & r_{2,2} & \cdots & r_{2,n} & \cdots \\ \vdots & \vdots & & \vdots & \\ r_{m,1} & r_{m,2} & \cdots & r_{m,n} & \cdots \end{pmatrix},$$

where $r_{j,i} \in F_q$, $1 \leq j \leq m$ and $i \geq 1$, are applied to the design of stream ciphers. The appearance of such kind of stream ciphers stimulates the study of multi-sequences. One of the interested problem is that on their linear complexity.

A sequence α can be identified to a formal power series:

$$\alpha = a_1 z^{-1} + a_2 z^{-2} + \cdots + a_i z^{-i} + \cdots$$

The problem on the linear complexity of sequences is essentially the optimal rational approximation problem of formal power series.

Continued fraction [1,3,4,5] is a useful tool in dealing with optimal rational approximation problems. It is well-known that the simple continued fraction expansion of a single real number gives the optimal rational approximations. Many people have contrived to construct multidimensional continued fraction in dealing with the rational approximation problem for multi-reals. One construction is the Jacobi-Perron algorithm (JPA) see [6]. This algorithm and its modification are extensively studied [7,8,9,10]. These algorithms are adapted to study the same problem for multi-formal Laurent series [11,12]. But none of these algorithm guarantee optimal rational approximation in general.

Recently we proposed an algorithm [13,14], which realizes the optimal rational approximation (ORA) of multi-formal power series (multi-series, in short).

* This work is partly supported by NSFC (Grant No. 60473025 and 90604011).

G. Gong et al. (Eds.): SETA 2006, LNCS 4086, pp. 17–33, 2006.

We call it the multi-universal continued fraction algorithm (m-UCFA, in short). It can produce all possible continued fraction expansions (CFE) for each multi-series $\underline{r}$. Here by a CFE C of $\underline{r}$ we mean C provides the ORA of $\underline{r}$. We made two specialization of m-UCFA. One is called the multi-continued fraction algorithm (m-CFA, in short), which is the simplest one in the operation, but it seems that there is no way to characterize the set of all possible outputs of m-CFA, and another is called the multi-strict continued fraction algorithm (m-SCFA, in short), which is the most significant in many applications since it has the advantage that its output set can be characterized. In this talk, for easy understanding, we first introduce m-CFA, then m-UCFA and m-SCFA, and finally their applications to sequences.

2 Linear Complexity and Optimal Rational Approximation

The formal power series are special formal Laurent series, and the latter make a field called formal Laurent series:

$$F((z^{-1})) = \left\{ \sum_{i \geq t} a_i z^{-i} \;\middle|\; t \in Z \right\}$$

where F is an arbitrary field. In order to introduce the m-CFA, here we start with the arithmetic operations and concepts in $F((z^{-1}))$.

For any given $\alpha \in F((z^{-1}))$:

$$\alpha = a_t z^{-t} + a_{t+1} z^{-t-1} + \cdots + a_{-1} z + a_0 + a_1 z^{-1} + a_2 z^{-2} + \cdots + a_i z^{-i} + \cdots$$

define

$$\lfloor \alpha \rfloor = a_t z^{-t} + a_{t+1} z^{-t-1} + \cdots + a_{-1} z + a_0$$
$$\{ \alpha \} = a_1 z^{-1} + a_2 z^{-2} + \cdots + a_i z^{-i} + \cdots$$

which are called the polynomial part and the remaining part of α respectively, and

$$v(\alpha) = \begin{cases} \min \{ i \mid a_i \neq 0 \} & \text{if } \alpha \neq 0 \\ \infty & \text{if } \alpha = 0 \end{cases}$$

which is called the valuation of α.

Fact 2.1. *That a sequence $\alpha = \{a_k\}_{k \geq 1}$ satisfies a linear recurrence $\underline{c} = (c_0, c_1, \cdots, c_{l-1})$ can be stated by means of formal series as follow:*

$$\alpha \cdot f(z) = p(z) \in F[z] \subset F((z^{-1}))$$

Hence

$$\alpha = \frac{p(z)}{f(z)} \in F(z) \subset F((z^{-1}))$$

where l is a positive integer, $F[z]$ and $F(z)$ are the polynomial ring and the fraction field over F respectively, and

$$\alpha = \sum_{i\geq 1} a_i z^{-i}$$

$$f(z) = z^l + \sum_{i=0}^{l-1} c_i z^i$$

$$p(z) = \lfloor \alpha \cdot f(z) \rfloor$$

For any given positive integer n, denote by $\alpha^{(n)}$ the prefix of α of length n: $(a_1, a_2, \cdots, a_n)$. The l-tuple $\underline{c} = (c_0, c_1, \cdots, c_{l-1})$ is called an l-level linear relation of $\alpha^{(n)}$ if

$$a_{l+k} = c_{l-1}a_{l+k-1} + c_{l-2}a_{l+k-2} + \cdots + c_0 a_k, \ 1 \leq k \leq n-l$$

The smallest level among the linear relations of $\alpha^{(n)}$ is called the linear complexity (LC) of α^n, and the corresponding linear relation is called a minimal relation, the corresponding polynomial

$$f(z) = z^l + \sum_{i=0}^{l-1} c_i z^i$$

is called a minimal polynomial of $\alpha^{(n)}$. Denote by l_n and $f_n(z)$ the linear complexity and a minimal polynomial of $\alpha^{(n)}$ respectively. Then $\{ l_n \}_{n\geq 1}$ is called the linear complexity profile of α and $\{ f_n \}_{n\geq 1}$ a minimal polynomial profile.

Definition 2.2. *We say $\frac{g(z)}{f(z)}$ is an optimal rational approximant of the series α, and $f(z)$ is an optimal denominator of α, if for an arbitrary rational fraction $\frac{p(z)}{q(z)}$, it satisfies*

$$\begin{cases} v(\alpha - \frac{p(z)}{q(z)}) < v(\alpha - \frac{g(z)}{f(z)}) \text{ if } \deg(q(z)) < \deg(f(z)) \\ v(\alpha - \frac{p(z)}{q(z)}) \leq v(\alpha - \frac{g(z)}{f(z)}) \text{ if } \deg(q(z)) = \deg(f(z)) \end{cases}$$

Fact 2.3. *If $\frac{g(z)}{f(z)}$ is an optimal rational approximant of α, then $g(z) = \lfloor \alpha f(z) \rfloor$.*

Proposition 2.4 (Relation between ORA and LC). *If $f(z)$ is an optimal denominator of α and*

$$v(\alpha - \frac{\lfloor \alpha f(z) \rfloor}{f(z)}) = n+1$$

then

$$\begin{cases} \deg(f(z)) = l_n \\ l_n < l_{n+1} \\ f(z) \text{ is a minimal polynomial of } \alpha^{(n)} \end{cases}$$

And vice versa.

As for multi-sequences, we first introduce the concept of the order on $Z_m \times Z$, where $Z_m = \{1, 2, \cdots, m\}$.

Definition 2.5 (Linear order on $Z_m \times Z$). *For any (j, n) and (j', n') in $Z_m \times Z$, define $(j, n) < (j', n')$ if and only if $n < n'$ or $n = n'$ but $j < j'$.*

For any given multi-sequence $\underline{r}$ and integers j and n, where $1 \le j \le m$ and $n \ge 1$, we denote by $\underline{r}^{(j,n)}$ by the prefix of $\underline{r}$ of length (j, n):

$$\underline{r}^{(j,n)} = \begin{pmatrix} r_1^{(n)} \\ \cdots \\ r_j^{(n)} \\ r_{j+1}^{(n-1)} \\ \cdots \\ r_m^{(n-1)} \end{pmatrix} = \begin{pmatrix} r_{1,1} & r_{1,2} & \cdots & r_{1,n-1} & r_{1,n} \\ \cdots & \cdots & & \cdots & \cdots \\ r_{j,1} & r_{j,2} & \cdots & r_{j,n-1} & r_{j,n} \\ r_{j+1,1} & r_{j+1,2} & \cdots & r_{j+1,n-1} & \\ \cdots & \cdots & \cdots & \cdots & \\ r_{m,1} & r_{m,2} & \cdots & r_{m,n-1} & \end{pmatrix}$$

We say an l-tuple $\underline{c} = (c_0, c_1, \cdots, c_{l-1})$ is a linear relation of $\underline{r}^{(j,n)}$ if it is a linear relation of both $r_h^{(n)}$ for $1 \le h \le j$ and $r_h^{(n-1)}$ for $j < h \le m$.

The smallest level among the linear relations of $\underline{r}^{(j,n)}$ is called the linear complexity (LC) of $\underline{r}^{(j,n)}$, and the corresponding linear relation is called a minimal relation, the corresponding polynomial

$$f(z) = z^l + \sum_{i=0}^{l-1} c_i z^i$$

is called a minimal polynomial of $\underline{r}^{(j,n)}$. Denote by $l_{j,n}$ and $f_{j,n}(z)$ the linear complexity and a minimal polynomial of $\underline{r}^{(j,n)}$ respectively. Then $\{l_{j,n}\}_{(j,n)\ge(1,1)}$ is called the linear complexity profile of α and $\{f_{j,n}\}_{(j,n)\ge(1,1)}$ a minimal polynomial profile.

A multi-sequence $\underline{r}$ can be identified to a multi-series:

$$\underline{r} = \begin{pmatrix} \sum_{i\ge1} r_{1,i} z^{-i} \\ \sum_{i\ge1} r_{2,i} z^{-i} \\ \vdots \\ \sum_{i\ge1} r_{m,i} z^{-i} \end{pmatrix} = \sum_{1\le j\le m,\ i\ge1} r_{j,i} z^{-i} \underline{e}_j$$

where $\underline{e}_j = (0, \cdots, 0, \overset{j}{1}, 0, \cdots, 0)^\tau$ is the j-th standard basis, τ means transport. $\underline{r}$ can be viewed as a column vector in the vector space $F((z))^m$ of dimension m.

Definition 2.6 (Support set and maximal support point). *For any $\underline{r} = \sum_{1\le j\le m,\ i\ge t} a_{j,i} z^{-i} \underline{e}_j \in F((z^{-1}))^m$, define*

$$Supp(\underline{r}) = \{(j, i) \mid r_{j,i} \ne 0, 1 \le j \le m, i \ge t\}$$

and call it the support set of $\underline{r}$. *If* $Supp(\underline{r})$ *is a finite set, we call*

$$Supp^{+}(\underline{r}) = \max Supp(\underline{r})$$

the maximal support point of $\underline{r}$.

Definition 2.7 (Indexed valuation). *For any* $\underline{r} \in F((z^{-1}))^m$, *let*

$$(h, v) = \min Supp(\underline{r})$$

We call (h, v) *the indexed valuation of* $\underline{r}$, *denoted by* $Iv(\underline{r})$, h *the index of* $\underline{r}$, *denoted by* $I(\underline{r})$, *and* v *the valuation of* $\underline{r}$, *denoted by* $v(\underline{r})$. *By convention,* $Iv(\underline{0}) = (1, \infty)$ *and* $v(\underline{0}) = \infty$, *where* $\underline{0} \in F((z^{-1}))^m$.

For any $\underline{r} \in F((z))^m$, denote

$$\lfloor \underline{r} \rfloor = (\lfloor (\rfloor r_1), \lfloor r_2 \rfloor, \cdots, \lfloor r_m \rfloor)^{\tau}$$
$$\{ \underline{r} \} = (\{ r_1 \}, \{ r_2 \}, \cdots, \{ r_m \})^{\tau}$$

and call them the polynomial part and the remaining part of $\underline{r}$ respectively.

Definition 2.8. *Given* $\underline{g}(z) = (g_1(z), g_2(z), \cdots, g_m(z))^{\tau} \in F[z]^m$ *and* $f(z) \in F[z]$. *We say* $\frac{\underline{g}(z)}{f(z)}$ *is an optimal rational approximant of* $\underline{r}$ *and* $deg(f(z))$ *is an optimal denominator degree if for any rational fraction* $\frac{\underline{p}(z)}{q(z)}$, *it satisfies*

$$\begin{cases} Iv(\underline{r} - \frac{\underline{p}(z)}{q(z)}) < Iv(\underline{r} - \frac{\underline{g}(z)}{f(z)}) \text{ if } \deg(q(z)) < \deg(f(z)) \\ Iv(\underline{r} - \frac{\underline{p}(z)}{q(z)}) \leq Iv(\underline{r} - \frac{\underline{g}(z)}{f(z)}) \text{ if } \mathrm{dog}(q(z)) = \deg(f(z)) \end{cases}$$

Proposition 2.9 (Relation between ORA and LC). *If* $f(z)$ *is an optimal denominator of* $\underline{r}$ *and*

$$Iv(\underline{r} - \frac{\lfloor \underline{r} f(z) \rfloor}{f(z)}) = (j, n)^{+}$$

then

$$\begin{cases} \deg(f(z)) = l_{j,n} \\ l_{j,n} < l_{(j,n)^{+}} \\ f(z) \text{ is a minimal polynomial of } \underline{r}^{(j,n)} \end{cases}$$

where $(j, n)^{+}$ *means the successive element of* (j, n) *in* $Z_m \times Z$, *i.e.,*

$$(j, n)^{+} = \begin{cases} (j+1, n) \text{ if } 1 \leq j < m \\ (1, n+1) \text{ if } j = m \end{cases}$$

And vice versa.

Definition 2.10. *We say a rational fraction sequence* $\{\frac{\underline{p}_k(z)}{q_k(z)}\}_{0\le k\le\mu}$, $\mu=\infty$ *or* $\mu<\infty$, *is an optimal rational approximation profile of* $\underline{r}$ *if it satisfies the following conditions:*

1. *Each* $\frac{\underline{p}_k(z)}{q_k(z)}$ *is an optimal rational approximant of* $\underline{r}$, *and* $deg(q_k(z)) < deg(q_{k+1}(z))$ *for all k's.*
2. *If d is an optimal denominator degree, then* $d = deg(q_k(z))$ *for some k.*

Correspondingly, we call the sequence $\{deg(q_k(z))\}_{0\le k\le\mu}$ *the optimal denominator degree profile, and the sequence* $\{Iv(\underline{r}-\frac{\underline{p}_k(z)}{q_k(z)})\}_{0\le k\le\mu}$ *the optimal precision profile of* $\underline{r}$. *It is clear that* $\underline{p}_0(z)=\underline{0}$ *and* $q_0(z)=1$ *for every multi-series.*

3 Multi-Continued Fraction Algorithm

Input: $\underline{r}\in F((z^{-1}))^m$, $v(\underline{r})>0$.
Initially, set

$$\underline{\alpha}_0=\underline{r}$$
$$D_0=I_m=Diag(\cdots,z^{-c_{0,j}},\cdots),\ c_{0,j}=0, 1\le j\le m$$

Suppose

$$\begin{cases} D_{k-1}=Diag(\cdots,z^{-c_{k-1,j}},\cdots)\\ \underline{\alpha}_{k-1}=(\cdots,\underline{\alpha}_{k-1,j},\cdots)^\tau \end{cases}$$

have been obtained for $k(\ge 1)$. Then the computation for the k-th round are defined by the following two steps:

1. Take inverse: $\underline{\rho}_k=(\cdots,\rho_{k,j},\cdots)^\tau$, where

$$\rho_{k,j}=\begin{cases}\frac{\alpha_{k-1,j}}{\alpha_{k-1,h_k}} & \text{if } j\ne h_k\\ \frac{1}{\alpha_{k-1,h_k}} & \text{if } j=h_k\end{cases}$$

and

$$h_k=I(D_{k-1}\underline{\alpha}_{k-1})$$

2. Take polynomial part:

$$\begin{cases}\underline{a}_k=\lfloor\underline{\rho}_k\rfloor\\ \underline{\alpha}_k=\underline{\rho}_k-\underline{a}_k\end{cases}$$

Let $\mu=k$ if $\underline{\alpha}_k=\underline{0}$, and the algorithm terminates; otherwise, take $D_k=Diag(\cdots,z^{-c_{k,j}},\cdots)$, where

$$c_{k,j}=\begin{cases}c_{k-1,j} & \text{if } j\ne h_k\\ v(D_{k-1}\underline{\alpha}_{k-1}) & \text{if } j=h_k\end{cases}$$

and go to the next round if $\underline{\alpha}_k \neq \underline{0}$. Let $\mu = \infty$ if the above procedure never terminates.

The output of the m-CFA on input $\underline{r}$, denoted by m-CFA($\underline{r}$), is a sequence pair $C = (\underline{h}, \underline{a})$, where

$$\begin{cases} \underline{h} = \{\, h_k \,\}_{1 \leq k \leq \mu}, & 1 \leq h_k \leq m \\ \underline{a} = \{\, \underline{a}_k \,\}_{1 \leq k \leq \mu}, & \underline{a}_k \in F[z]^m \end{cases}$$

which provides optimal rational approximation to $\underline{r}$ by the following procedures.

In the sequel, we will call any $(\underline{h}, \underline{a})$ an HA-pair of length μ if

$$\begin{cases} \underline{h} = \{\, h_k \,\}_{1 \leq k \leq \mu}, & 1 \leq h_k \leq m \\ \underline{a} = \{\, \underline{a}_k \,\}_{1 \leq k \leq \mu}, & \underline{a}_k \in F[z]^m \end{cases}$$

Definition 3.1. *Let $C = (\underline{h}, \underline{a})$ be an HA-pair of length μ. Define*

$$\underline{d}(C) = \{\, d_k \,\}_{0 \leq k \leq \mu}$$
$$\underline{n}(C) = \{\, n_k \,\}_{0 \leq k \leq \mu}$$
$$Q(C) = \{\, \frac{\underline{p}_k}{q_k} \,\}_{0 \leq k \leq \mu}$$

where

$$d_k = \sum_{1 \leq i \leq k} t_i, \ d_0 = 0$$
$$t_i = deg(a_{i,h_i})$$
$$n_k = d_{k-1} + v_k$$
$$v_k = v_{k,h_k}$$
$$v_{k,j} = \sum_{\substack{1 \leq i \leq k \\ h_i = j}} t_i, 1 \leq j \leq m$$
$$\begin{pmatrix} \underline{p}_k \\ q_k \end{pmatrix} = B_k \begin{pmatrix} \underline{0} \\ 1 \end{pmatrix}, \ \underline{p}_k \in F[z]^m, q_k \in F[z]$$
$$B_k = B_{k-1} E_{h_k} A(\underline{a}_k), \ B_0 = I_{m+1}$$
$$E_{h_k} = \begin{pmatrix} I & & & \\ & 0 & 1 & \\ & & I & \\ & 1 & & 0 \end{pmatrix} \begin{matrix} \\ \rightarrow h_k \\ \\ \\ \end{matrix}$$
$$\qquad\qquad\quad \underset{h_k}{\downarrow}$$
$$A(\underline{a}_k) = \begin{pmatrix} I_m & \underline{a}_k \\ \mathbf{0} & 1 \end{pmatrix}$$

Theorem 3.2. *For any given multi-series $\underline{r}$, let $C = (\underline{h}, \underline{a}) =$ m-CFA$(\underline{r})$. Then*

1. *The sequence $Q(C) = \{\, \frac{\underline{p}_k}{q_k} \,\}_{0 \leq k \leq \mu}$ is an optimal rational approximation profile of $\underline{r}$. As a consequence, when $\mu = \infty$, $Q(C)$ is convergent.* Denote*

$$\lim Q(C) = \begin{cases} \dfrac{\underline{p}_\mu}{q_\mu} & \mathit{if}\ \mu < \infty \\ \lim\limits_{k\to\infty} \dfrac{\underline{p}_k}{q_k} & \mathit{if}\ \mu = \infty \end{cases}$$

Then $\underline{r} = \lim Q(C)$.

2. *The sequence* $\underline{d}(C)$ *is the optimal denominator degree profile of* $\underline{r}$.
3. *The sequence* $(\underline{h}, \underline{n}(C))$ *is the optimal precision profile of* $\underline{r}$.

Remark*: Here we say that a sequence $\{\underline{x}_k\}_{k\geq 0}$ in $F((z^{-1}))^m$ is convergent if there exists an element $\underline{x} \in F((z^{-1}))^m$ (called the limit of the sequence) such that for any $(h, v) \in Z_m \times Z$, there is a positive integer k_0 such that $Iv(\underline{x}_k - \underline{x}) \geq (h, v)$ whenever $k \geq k_0$.

Corollary 3.3. *For any given multi-series* $\underline{r}$*, Denote* $C = (\underline{h}, \underline{a}) =$ *m-CFA*$(\underline{r})$. *Let*

$$l_{j,n} = d_k,\ (h_k, n_k) \leq (j, n) < (h_{k+1}, n_{k+1})$$
$$f_{j,n} = q_k,\ (h_k, n_k) \leq (j, n) < (h_{k+1}, n_{k+1})$$

Then $\{\, l_{j,n} \,\}_{(j,n)\geq(1,1)}$ *is the linear complexity profile of* $\underline{r}$ *and* $\{\, f_{j,n} \,\}_{(j,n)\geq(1,1)}$ *is a minimal polynomial profile of* $\underline{r}$.

From Theorem 3.2 we see that for any multi-sequence $\underline{r}$, the output of the m-CFA on input $\underline{r}$ shows up its linear structure completely.

Remark: When $m = 1$, the m-CFA is exactly the classical continued fraction algorithm [2,6] for formal power series. In fact, when $m = 1$, we have $h_k = 1$ for all k's. Hence the computation for both h_k and D_k are unnecessary. Now the 1-CFA is as follows (we write $\underline{r} = r$ and $\underline{a}_k = a_k$): Initially, set $a_0 = 0$, $\alpha_0 = r$. Suppose $\alpha_{k-1} \neq 0$ has been obtained. The computation for the k-th round are defined by the following steps:

1. Take the inverse: $\rho_k = \alpha_{k-1}^{-1}$;
2. Take $a_k = \lfloor \rho_k \rfloor$ and $\alpha_k = \rho_k - a_k$.

If $\alpha_k = 0$, set $\mu = k$, and the algorithm terminates. Define $\mu = \infty$ if the above procedure never terminates.

Comparisons among JPA, MJPA and m-CFA: The basic construction of m-CFA is the same as JPA or MJPA. It is an iterative algorithm, and consists of two steps for each round. One step is computing inverse, and another step is taking the polynomial and the remaining part. The main point that m-CFA is different from JPA or MJPA is the way of computing inverse at each round. In JPA, the way of computing the inverse at every round is never changed, as shown below:

$$\rho_{k,j} = \begin{cases} \dfrac{\alpha_{k-1,j+1}}{\alpha_{k-1,1}} & \text{if } j \neq m \\ \dfrac{1}{\alpha_{k-1,1}} & \text{if } j = m \end{cases}$$

In MJPA, the way of computing the inverse at every round is not unchanged. In fact, it depends on the parameter h_k, just like the m-CFA, but the definition of index h_k in MJPA is different to that in m-CFA, as shown below:

$$\rho_{k,j} = \begin{cases} \frac{\alpha_{k-1,j}}{\alpha_{k-1,h_k}} & \text{if } j \neq h_k \\ \frac{1}{\alpha_{k-1,h_k}} & \text{if } j = h_k \end{cases}$$

where

$$h_k = \begin{cases} I(\underline{\alpha}_{k-1}) & \text{in MJPA} \\ I(D_{k-1}\underline{\alpha}_{k-1}) & \text{in m-CFA} \end{cases}$$

It has been proved that in m-CFA

$$I(D_{k-1}\alpha_{k-1}) = I(L(h_1, -\underline{\alpha}_0) \cdots L(h_{k-1}, -\underline{\alpha}_{k-2})\underline{\alpha}_{k-1})$$

where $L(h, -\underline{\alpha})$ is a matrix of order m of the following form:

$$L(h, -\underline{\alpha}) = \begin{pmatrix} I_{h-1} & \boxed{\begin{matrix} \\ -\underline{\alpha} \\ \\ \end{matrix}} & \\ & & I_{m-h} \end{pmatrix}$$
$$\qquad\qquad\qquad\qquad\downarrow$$
$$\qquad\qquad\qquad\qquad h$$

In MJPA, h_k depends only on the k-th remainder $\underline{\alpha}_{k-1}$; to compare with MJPA, in m-CFA, the index h_k depends on all indices and all remainders appeared at the first $k-1$ rounds.

4 Multi-Universal Continued Fraction Algorithm

We call a sequence pair $(\underline{h}, \underline{a})$ a continued fraction expansion of a multi-series $\underline{r}$, or just call it a multi-continued fraction, if it has the same properties as m-CFA($\underline{r}$), i.e., the three properties expressed in Theorem 3.2. We consider the following questions:

1. (Uniqueness?) Whether the CFE of a multi-series $\underline{r}$ is unique?

If NOT, then we have the next two questions:

2. (Generating) Given multi-series $\underline{r}$, how to generate all its CFEs?
3. (Characterization) How to characterize the set of all multi-continued fractions?

The following example gives a negative answer to the first problem.

Example 4.1. *Let $m = 2$ and*

$$\underline{r} = \begin{pmatrix} r_1 \\ r_2 \end{pmatrix} = \begin{pmatrix} \frac{z^3}{z^5+1} \\ \frac{1}{z^7+z^2} \end{pmatrix} \in F_2((z^{-1}))^2$$

Then one can check both

$$C = \begin{bmatrix} 1 & 1 & 2 \\ \binom{z^2}{0} & \binom{z^3}{0} & \binom{0}{z^2} \end{bmatrix}$$

and

$$C' = \begin{bmatrix} 1 & 1 & 2 \\ \binom{z^2}{0} & \binom{z^3+z^2}{0} & \binom{z^4}{z^2} \end{bmatrix}$$

are the CFEs of $\underline{r}$.

For the question on generating, we proposed the following algorithm, the multi-universal continued fraction algorithm (m-UCFA, in short). It is a non-deterministic algorithm.

m-UCFA:
Input: $\underline{r} \in F((z^{-1}))^m$, $v(\underline{r}) > 0$.
Initially, set

$$\begin{aligned} &\underline{\alpha}_0 = \underline{r} \\ &D_0 = I_m = Diag(\cdots, z^{-c_{0,j}}, \cdots),\ c_{0,j} = 0, 1 \le j \le m \end{aligned}$$

Suppose

$$\begin{cases} D_{k-1} = Diag(\cdots, z^{-c_{k-1,j}}, \cdots) \\ \underline{\alpha}_{k-1} = (\cdots, \alpha_{k-1,j}, \cdots)^\tau \end{cases}$$

have been obtained for $k (\ge 1)$. Then the computation for the k-th round are defined by the following two steps:

1. Take inverse: $\underline{\rho}_k = (\cdots, \rho_{k,j}, \cdots)^\tau$, where

$$\rho_{k,j} = \begin{cases} \dfrac{\alpha_{k-1,j}}{\alpha_{k-1,h_k}} & \text{if } j \ne h_k \\ \dfrac{1}{\alpha_{k,h_k}} & \text{if } j = h_k \end{cases}$$

   and

$$h_k = I(D_{k-1}\alpha_{k-1})$$

2. Take polynomial part:

$$\begin{cases} \underline{a}_k = \lfloor \underline{\rho}_k \rfloor - \underline{\epsilon}_k \\ \underline{\alpha}_k = \underline{\rho}_k - \underline{a}_k \end{cases}$$

   where $\underline{\epsilon}_k \in F[z]^m$ such that

$$Iv(D_k\{\underline{\rho}_k\}) < Iv(D_k\underline{\epsilon}_k)$$

   where

$$\begin{cases} D_k = Diag(\cdots, z^{-c_{k,j}}, \cdots) \\ c_{k,j} = \begin{cases} c_{k-1,j} & \text{if } j \ne h_k \\ v(D_{k-1}\underline{\alpha}_{k-1}) & \text{if } j = h_k \end{cases} \end{cases}$$

Let $\mu = k$ if $\underline{\alpha} = \underline{0}$, and the algorithm terminates; otherwise, go to the next round if $\underline{\alpha} \neq \underline{0}$. Let $\mu = \infty$ if the above procedure never terminates.

The output of m-UCFA on input $\underline{r}$ is non-deterministic. Denote by m-UCFA($\underline{r}$) the set of all possible outputs.

Theorem 4.2 (Generating). *Let $C = (\underline{h}, \underline{a})$ be an HA-pair and $\underline{r}$ be a multi-series. Then C is a CFE of $\underline{r}$ if and only if C is a possible output of m-UCFA on the input $\underline{r}$.*

Theorem 4.3 (Characterization). *Let $C = (\underline{h}, \underline{a})$ be an HA-pair of length μ. Then C is a multi-continued fraction if and only if it satisfies the following three properties:*

1. $t_k \geq 1$ *for* $1 \leq k \leq \mu$;
2. $(h_k, v_k - t_k) < (h_{k+1}, v_{k+1})$ *for* $1 \leq k < \mu$;
3. $Iv(\Delta_k \underline{a}_k) = (h_k, v_k - t_k)$ *for* $1 \leq k \leq \mu$, *where* $\Delta_k = Diag(z^{-v_{k,1}}, z^{-v_{k,2}}, \cdots, z^{-v_{k,m}})$.

5 Multi-Strict Continued Fraction Algorithm

Furthermore, we consider a question: whether one may choose one and only one CFE for each multi-series $\underline{r}$ such that it can be generated by a deterministic algorithm and the set of all these chosen multi-continued fractions can be characterized.

The set of all classical continued fractions does contain one and only one CFE for each multi-series, but it seems there is no way to characterize the set made of them, though they can be generated by m-CFA. For solving this question, we propose an algorithm, the multi-strict continued fraction algorithm (m-SCFA, in short). It is a deterministic algorithm:

m-SCFA:

Input: $\underline{r} \in F((z^{-1}))^m$, $v(\underline{r}) > 0$.

Initially, set

$$\underline{\alpha}_0 = \underline{r}$$
$$D_0 = I_m = Diag(\cdots, z^{-c_{0,j}}, \cdots),\ c_{0,j} = 0, 1 \leq j \leq m$$

Suppose

$$\begin{cases} D_{k-1} = Diag(\cdots, z^{-c_{k-1,j}}, \cdots) \\ \underline{\alpha}_{k-1} = (\cdots, \alpha_{k-1,j}, \cdots)^\tau \end{cases}$$

have been obtained for $k(\geq 1)$. Then the computation for the k-th round are defined by the following two steps:

1. Take inverse: $\underline{\rho}_k = (\cdots, \rho_{k,j}, \cdots)^\tau$, where

$$\rho_{k,j} = \begin{cases} \dfrac{\alpha_{k-1,j}}{\alpha_{k-1,h_k}} & \text{if } j \neq h_k \\ \dfrac{1}{\alpha_{k,h_k}} & \text{if } j = h_k \end{cases}$$

and

$$h_k = I(D_{k-1}\alpha_{k-1})$$

2. Take polynomial part:

$$\begin{cases} \underline{a}_k = \lfloor \underline{\rho}_k \rfloor - \underline{\epsilon}_k \\ \underline{\alpha}_k = \underline{\rho}_k - \underline{a}_k \end{cases}$$

where if $\lfloor \underline{\rho}_k \rfloor = \sum A_{k,j,i} z^i \underline{e}_j$, $A_{k,j,i} \in F$, Then

$$\underline{\epsilon}_k = \sum_{(j,\, c_{k,j}-i) > Iv(D_k\{\,\underline{\rho}_k\,\}), 1 \le j \le m} A_{k,j,i} z^i \underline{e}_j$$

and

$$\begin{cases} D_k = Diag(\cdots, z^{-c_{k,j}}, \cdots) \\ c_{k,j} = \begin{cases} c_{k-1,j} & \text{if } j \neq h_k \\ v(D_{k-1}\underline{\alpha}_{k-1}) & \text{if } j = h_k \end{cases} \end{cases}$$

Let $\mu = k$ if $\underline{\alpha} = \underline{0}$, and the algorithm terminates; otherwise, go to the next round if $\underline{\alpha} \neq \underline{0}$. Let $\mu = \infty$ if the above procedure never terminates.

Denote by $\mathbb{S}^m$ the set of all multi-series of dimension m. For any $\underline{r} \in \mathbb{S}^m$, denote by m-SCFA($\underline{r}$) the output of m-SCFA on input $\underline{r}$. Let m-SCFA($\mathbb{S}^m$) be the set of all possible outputs of m-SCFA with inputs in $\mathbb{S}^m$. Then we have

Theorem 5.1. *1. For any $\underline{r} \in \mathbb{S}^m$, m-SCFA($\underline{r}$) is a CFE of $\underline{r}$, which we call a multi-strict continued fraction. As a consequence, there is a 1-1 correspondence between $\mathbb{S}^m$ and m-SCFA($\mathbb{S}^m$).*

2. Let $C = (\underline{h}, \underline{a})$ be an HA-pair of length μ. Then $C \in$ m-SCFA ($\mathbb{S}^m$) if and only if it satisfies the following four properties:
   (a) $t_k \geq 1$ for $1 \leq k \leq \mu$;
   (b) $(h_k, v_k - t_k) < (h_{k+1}, v_{k+1})$ for $1 \leq k < \mu$;
   (c) $Iv(\Delta_k \underline{a}_k) = (h_k, v_k - t_k)$ for $1 \leq k \leq \mu$;
   (d) $Supp^+(\Delta_k \underline{a}_k) < (h_{k+1}, v_{k+1})$ for $1 \leq k \leq \mu$.

Theorem 5.1 establishes a 1-1 correspondence between multi-series and the multi-strict continued fraction by means of the m-SCFA, and characterize the set of all multi-strict continued fractions.

Denote by $\mathbb{S}^{(m,n)}$ the set of all m-tuple of sequences of length n over the field F, and by $\mathbb{C}^{(m,n)}$ the set of all m-dimensional multi-strict continued fraction fractions of finite length satisfying the following property:

$$Supp^+(z^{-d_\mu} \Delta_\mu \underline{a}_\mu) \leq (m, n) \tag{5.1}$$

The sequence set $\mathbb{S}^{(m,n)}$ is identified to a multi-series set in the natural way. The following theorem shows that there is a 1-1 correspondence between the sequence set $\mathbb{S}^{(m,n)}$ and the multi-strict continued fraction set $\mathbb{C}^{(m,n)}$. To show the 1-1 correspondence, we need the concept of the (m,n)-prefix of an m-CF.

For any m-SCF $C = (\underline{h}, \underline{a})$ of length μ, the (h,n)-prefix of C, denoted by $C^{(h,n)}$, is defined as:

$$C^{(h,n)} = \begin{bmatrix} h_1 \; h_2 \cdots h_{w-1} & h_w \\ \underline{a}_1 \; \underline{a}_2 \cdots \underline{a}_{w-1} & \underline{a}_w^{(h,n)} \end{bmatrix} \tag{5.2}$$

where w is determined by the following condition:

$$(h_w, n_w) \leq (h, n) < (h_{w+1}, n_{w+1})$$

where $(h_{w+1}, n_{w+1}) = (1, \infty)$ by convention if $w = \mu < \infty$, and $\underline{a}_k^{(h,n)}$ is defined as below:

$$\underline{a}_k^{(h,n)} = \sum_{\substack{1 \leq j \leq m \\ (j, d_k + v_{k,j} - i) \leq (h,n) \\ a_{k,j,i} \neq 0}} a_{k,j,i} z^i \underline{e}_j \tag{5.3}$$

Theorem 5.2. *The mapping f:*

$$f: \quad \underline{r} \mapsto C^{(m,n)}, \forall\, \underline{r} \in \mathbb{S}^{(m,n)}$$

where C = m-SCFA$(\underline{r})$, is injective from $\mathbb{S}^{(m,n)}$ onto $\mathbb{C}^{m,n}$, and its inverse is the mapping:

$$f^{-1}: \quad C \mapsto \underline{r}^{(m,n)}, \forall\, C \in \mathbb{C}^{(m,n)}$$

where $\underline{r} = \lim Q(C)$.

As a consequence, for any given integer d, where $1 \leq d \leq n$, let

$$\mathbb{S}^{(m,n)}(d) = \left\{ \underline{r} \in \mathbb{S}^{(m,n)} \;\middle|\; L(\underline{r}) = d \right\}$$

$$\mathbb{C}^{(m,n)}(d) = \left\{ C \in \mathbb{C}^{(m,n)} \;\middle|\; d(C) = d \right\}$$

where $d(C)$ denotes d_μ, μ is the length of C, and $L(\underline{r})$ denote the linear complexity of $\underline{r}$. Then restricted on the set $\mathbb{S}^{(m,n)}(d)$, the mapping f establishes a 1-1 correspondence between $\mathbb{S}^{(m,n)}(d)$ and $\mathbb{C}^{(m,n)}(d)$.

6 Applications

6.1 Proof of the Fact That JPA or MJPA Can't Guarantee Optimal Rational Approximation

For any given multi-series $\underline{r}$, assume we have the parameters associated to m-CFA and MJPA as below:

$$\text{m-CFA}: C = \text{m-CFA}(\underline{r}),\ Q(C) = \{\tfrac{\underline{p}_k}{q_k}\}_{k \geq 0},\ \underline{h} = \{h_k\},\ \underline{d} = \{d_k\},\ \underline{n} = \{n_k\};$$

$$\text{MJPA}: \hat{C} = MJPA(\underline{r}),\ Q(\hat{C}) = \{\tfrac{\hat{\underline{p}}_k}{\hat{q}_k}\}_{k \geq 0},\ \hat{\underline{h}} = \{\hat{h}_k\},\ \hat{\underline{d}} = \{\hat{d}_k\},\ \hat{\underline{n}} = \{\hat{n}_k\}.$$

Based on multi-continued fraction theories, it is easy to see that

1. if there is some integer $\hat{d}_i \notin \{d_k\}_{k\geq 0}$, or
2. $\hat{d}_i = d_k$ but $(\hat{h}_{i+1}, \hat{n}_{i+1}) \neq (h_{k+1}, n_{k+1})$

then $\frac{\hat{\underline{p}}_i}{\hat{q}_i}$ is not the optimal rational approximant of $\underline{r}$.

By a counterexample it is shown [24] that for $\underline{r} = (\gamma, \gamma^2)^\tau$ both JPA and MJPA can't guarantee the optimal rational approximation to $\underline{r}$, where γ is the root of certain algebraic function of degree three over $F[z]$ such that $\gamma \in F((z^{-1}))$.

6.2 From m-SCFA to GBMA

The well-known GBMA can be derived from the m-SCFA (see [17,18]). In particular, the minimal polynomial profile, denoted by $\{ f_{j,n} \}_{(j,n)\geq(1,1)}$, and the discrepancy profile, denoted by $\{ \delta_{j,n} \}_{(j,n)\geq(1,1)}$, which are obtained by acting GBMA on a multi-sequence $\underline{r}$, are expressed explicitly by data associated to the multi-strict continued fraction expansion of $\underline{r}$ respectively. In fact, let $C = (\underline{h}, \underline{a})$ be the multi-strict continued fraction expansion of $\underline{r}$. Denote $\underline{a}_k = \sum a_{k,j,i} z^i \underline{e}_j$. Then

$$\sum_{(j,n)\geq(1,1)} \delta_{j,n} z^{-n} \underline{e}_j = -\pi_{k,h_k} z^{-n_k} \underline{e}_{h_k} + \sum_{k=1}^{\mu} z^{-d_k} \Lambda_k \Delta_k (\underline{a}_k - a_{k,h_k,t_k} z^{t_k} \underline{e}_{h_k}) \quad (6.1)$$

where

$$\begin{cases} \Lambda_k = Diag(\cdots, \pi_{k,j}, \cdots) \\ \pi_{k,j} = \prod_{\substack{1\leq i\leq k,\ h_i=j \\ a_{k,j,i}\neq 0}} (-a_{k,j,i})^{-1},\ 1 \leq j \leq m \end{cases}$$

and

$$f_{j,n} = (\underline{0}\ 1) B_{k-1} E_{h_k} \begin{pmatrix} \underline{a}_k^{(j,n)} \\ 1 \end{pmatrix} \quad (6.2)$$

In particular, when $F = F_2$, we have

$$\sum_{(j,n)\geq(1,1)} \delta_{j,n} z^{-n} \underline{e}_j = \sum_{k=1}^{\mu} z^{-d_k} \Delta_k \underline{a}_k \quad (6.3)$$

6.3 *d*-Perfect Multi-sequences

In [19] Xing introduce the concept of d-perfect multi-sequences and bring forward the following conjecture.

Conjecture 6.1 ([19]). *Let $\underline{r}$ be a multi-sequence and d be an integer. If for any positive integer n, it satisfies*

$$l_n \geq \frac{m(n+1) - d}{m+1} \quad (6.4)$$

Then

$$l_n \leq \frac{mn+d}{m+1},\ \forall\, n \geq 1 \tag{6.5}$$

where l_n *denotes the linear complexity of* $\underline{r}^{(m,n)}$.

Based on multi-continued fraction theories, we reduce the above conjecture to the following combinatorial problem:

Equivalent form of Conjecture 6.1. For any given multi-sequence $\underline{r}$, let C be one of its continued fraction expansions. If there is an integer d such that

$$\frac{mn_{k+1}-d}{m+1} \leq d_k,\ \forall\, k \geq 0 \tag{6.6}$$

then

$$d_k \leq \frac{mn_k+d}{m+1},\ \forall\, k \geq 0 \tag{6.7}$$

We demonstrate an example which disproves the above conjecture [23].

6.4 Expected Value of the Normalized Linear Complexity of Multi-sequences

For any multi-sequence $\underline{r}$ and positive integer n, we call $\frac{l_n}{n}$ the normalized linear complexity of $\underline{r}^{(m,n)}$. Given integers m and n, denote by $e(m,n)$ the expected value of the normalized linear complexity of multi-sequences of dimension m and length n. Then it can be expressed as

$$e(m,n) = \sum_{d=1}^{n} \frac{d}{n} \frac{|\mathbb{S}^{(m,n)}(d)|}{2^{mn}} \tag{6.8}$$

where $|\mathbb{S}^{(m,n)}(d)|$ denotes the size of the set $\mathbb{S}^{(m,n)}(d)$. As for the limit of $e(m,n)$, Niederreiter and Ding proposed the following conjecture:

Conjecture 6.2 ([19,20,21]). *For any positive integers* m *and* n, *let* $e(m,n)$ *be defined as above. Then*

$$\lim_{n\to\infty} e(m,n) = \frac{m}{m+1} \tag{6.9}$$

By Theorem 5.2 we have

$$|\mathbb{S}^{(m,n)}(d)| = |\mathbb{C}^{(m,n)}(d)| \tag{6.10}$$

The conjecture is proved for the case $m=2$ in [25], and for the general case later in [26] by evaluating the size of the multi-continued fraction set $\mathbb{C}^{(m,n)}(d)$.

We should mention that the above conjecture has been proved earlier than [26] for the general case in [22] by Niederreiter and Wang by using the lattice basis reduction theory.

6.5 Asymptotic Behavior of the Normalized Linear Complexity of Multi-sequences

We once discussed the asymptotic behavior of the normalized linear complexity of single sequences in [15] by means of classical continued fractions. For any given multi-sequence $\underline{r}$, based on multi-continued fraction theories, we obtain a formula for $\limsup_{n\to\infty} \frac{l_n}{n}$ together with a lower bound and a formula for $\liminf_{n\to\infty} \frac{l_n}{n}$ together with an upper bound, and provide a sufficient and necessary condition on the existence of $\lim_{n\to\infty} \frac{l_n}{n}$. For details, see [16].

References

1. J.L. Massey, Shift-register synthesis and BCH decoding, IEEE Trans. Information Theory, Vol.15, pp.173-180, 1969
2. W.H. Mills, Continued fractions and linear recurrences, Math. Computation, 29(129), pp173-180, 1975
3. A. Lasjaunias, Diophantine approximation and continued fraction expansions of algebraic power series in positive characteristic, J. Number Theorey, Vol.65, pp.206-225, 1997
4. William B. Jones and W.J. Thron, Continued fractions, analytic theory and applications, Encyclopedia of Mathematics and Its Applications (Giancarlo Rota. Editor), Volume 11
5. Wolfgang M. Schmidt, On continued fraction and diophantine approximation in power series fields, Acta Arith. Vol.95, pp.139-166, 2000
6. L. Bernstein, The Jacobi-Perron algorithm: its theory and application, LNM207, Springer-Verlag, Berlin, 1971
7. E.V. Podsypanin, A generalization of continued fraction alogrithm that is related to ViggoBorun algorithm (Russian), Studies in Number Theory (LOMI), Vol.4, Zap. Naucn. Sem. Leningrad. Otdel. Mat. Inst. Steklov 67, pp.184-194, 1977
8. S. Ito, M. Keane, M. Ohtsuki, Almost everywhere exponential convergence of the modified Jocobi-Perron algorithm,Ergod.Th. & Dynam. Sys. *13*, pp.319–334, 1993
9. S. Ito, J. Fujii, H. Higashino, S-I. Yasutomi, On simultaneous approximation to (α, α^2) with $\alpha^3 + k\alpha - 1 = 0$, J. Number Theory, Vol.99, pp.255-283, 2003
10. R. Meester, A Simple proof of the exponential convergence of the modified Jacobi-Perron algorithm, Ergod.Th. &Dynam. Sys. *19*, pp.1077–1083, 1999
11. K. Feng and F. Wang, The Jacobi-Perron algorithm on function fields, Algebra Colloq. 1:2, pp.149-158, 1994
12. K.Inoue, On the exponential convergence of Jacobi-Perron algorithm over $F(x)^d$, JP Journal of Algebra, Number Theory and Application, (3)**1**, pp.27–41, 2003
13. Z.D. Dai, K.P. Wang and D.F. Ye, m-Continued fraction expansions of multi-Laurent series, ADVANCE IN MATHEMATICS(CHINA), Vol.33, No.2, pp.246-248, 2004
14. Z.D. Dai, K.P. Wang and D.F. Ye, Multi-continued fraction algorithm on multi-formal Laurent Series, ACTA ARITHMETICA, pp.1-21, 2006
15. Z.D. Dai, S.Q. Jiang, K. Imamura and G. Gong, Asymptotic behavior of normalized linear complexity of ultimately non-periodic binary sequences, IEEE Trans. Infor. Theory, Vol.50, pp2911-2915, 2004

16. Z.D. Dai, K. Imamura and J.H. Yang, Asymptotic behavior of normalized linear complexity of multi-sequences, SETA 2004, LNCS 3486, pp126-142, Springer, Berlin, 2005
17. Z.D. Dai, X.T. Feng and J.H. Yang, Multi-continued fraction algorithm and generalized B-M algorithm over F_2, SETA 2004, LNCS 3486, pp339-354, Springer, Berlin, 2005
18. Z.D. Dai and J.H. Yang, Multi-continued fraction algorithm and generalized B-M algorithm over F_q, Finite Fields and Their Application, accepted
19. C. Xing, Multi-sequences with almost perfect linear complexity profile and function fields over finite fields, Journal of Complexity, Vol.16, pp.661-675, 2000
20. H. Niederreiter, Some computable complexity measure for binary sequences, Sequences and Their Applications, pp67-78, Springer, London, 1999
21. L.P. Wang and H. Niederreiter, Enumeration results on the joint linear complexity profile of multisequences, Finite Fields and Their Application, doi:10.1016/j.ffa.2005.03.005
22. H. Niederreiter and L.P. Wang, Proof of a conjecture on the joint linear complexity profile of multisequences, INDOCRYPT2005, LNCS3797, pp.13-22, 2005
23. X.T. Feng, Q.L. Wang and Z. D. Dai, Multi-sequences with d-perfect property, Journal of Complexity, Volume 21, Issue 2, pp.230-242, 2005
24. Q.L. Dai and Z.D. Dai, A proof that JPA and MJPA on multi-formal Layrent series can not garrentee the optimal rational approximation, Journal of the Graduate School of the Chinese Academy of Science (IN Chines)Vol.22, pp51-58, 2005.
25. X.T. Feng and Z.D. Dai, Expected value of the linear complexity of two-dimensional binary sequences, SETA 2004, LNCS 3486, pp113-128, 2005
26. Z.D. Dai and X.T. Feng, Expected value of the normalized linear complexity of multi-sequences over the binary fields, preprint.

Codes for Optical CDMA

Reza Omrani[1] and P. Vijay Kumar[2]

[1] EE-Systems, University of Southern California, Los Angeles, CA 90089-2565
omrani@usc.edu
[2] ECE Department, Indian Institute of Science, Bangalore
on leave of absence from
EE-Systems, University of Southern California, Los Angeles, CA 90089-2565
vijayk@usc.edu

Abstract. There has been a recent upsurge of interest in applying Code Division Multiple Access (CDMA) techniques to optical networks. Conventional spreading codes for OCDMA, known as optical orthogonal codes (OOC) spread the signal in the time domain only, which often results in the requirement of a large chip rate. By spreading in both time and wavelength using two-dimensional OOCs, the chip rate can be reduced considerably. This paper presents an overview of 1-D and 2-D optical orthogonal codes as well as some new results relating to bounds on code size and code construction.

Keywords: Optical orthogonal codes, OOC, constant weight codes, optical CDMA, OCDMA, 2-D OOC, MWOOC, wavelength time codes.

1 Introduction

Recently there has been an upsurge of interest in applying code division multiple access (CDMA) techniques to optical networks [1], at least in part due to the increase in security afforded by optical CDMA (OCDMA) as measured for instance, by the increased effort needed to intercept an OCDMA signal, and in part due to the flexibility and simplicity of network control afforded by OCDMA.

As in conventional CDMA, each of the users in an optical CDMA system is assigned a unique spreading code that enables the user to distinguish his signal from that of the other users. In optical CDMA (OCDMA), the typical modulation scheme used is On-Off Keying (OOK) and as a result, the spreading codes are binary, with symbols in $\{0, 1\}$. These spreading codes are termed optical orthogonal codes (OOC). Traditionally, as in the case of wireless communication, the spreading has been carried out in time and we will refer to this class of OOC as one-dimensional OOCs (1-D OOCs).

One drawback of one-dimensional (1-D) OOCs is the requirement of a large chip rate. For example, consider the situation when it is desired to assign codes to 8 potential users, each transmitting data at 1 Gbit/sec of which at most 5 users are active at any given time. If one attempts to design a 1-D OOC to meet this requirement, one will end up with a chip-rate on the order of 161 Gchips

G. Gong et al. (Eds.): SETA 2006, LNCS 4086, pp. 34–46, 2006.

per second (Gcps) (See Example 1). By employing two-dimensional (2-D) OOCs that spread in both wavelength and time, it turns out that the above requirement can be met using a chip rate of just 6 Gcps.

Section 2 reviews bounds on the size of 1-D codes as well as some of the better known code construction techniques. 2-D OOCs are introduced in Section 3. Bounds on the size of a 2-D OOC are treated in Section 4, while Section 5 presents a new construction technique based on the use of rational functions.

2 One-Dimensional Optical Orthogonal Codes

An (n, ω, κ) Optical Orthogonal Code (OOC) $\mathcal{C}$ where $1 \leq \kappa \leq \omega \leq n$, is a family of $\{0,1\}$-sequences of length n and Hamming weight ω satisfying:

$$\sum_{k=0}^{n-1} x(k)y(k \oplus_n \tau) \leq \kappa \tag{1}$$

for every pair of sequences $\{x, y\}$ in $\mathcal{C}$ whenever either $x \neq y$ or $\tau \neq 0$. We have used $\oplus_n$ to denote addition modulo n. We will refer to κ as the maximum collision parameter(MCP).

For a given set of values of n, ω, κ, let $\Phi(n, \omega, \kappa)$, denote the largest possible cardinality of an (n, ω, κ) OOC code.

2.1 Bounds on the Size of 1-D OOCs

If $\mathcal{C}$ is an (n, ω, κ) 1-D OOC, then by including every cyclic shift of each codeword in $\mathcal{C}$ one can construct a constant weight code with parameters (n, ω, κ) of size $= n \mid \mathcal{C} \mid$. This observation allows us to translate the Johnson bounds A, B, C on constant weight codes [2] [3] as well as the improvement of Johnson bound B due to Agrell et. al. [4] and the improvement of Johnson bound C due to Moreno et. al. [5] into bounds on the cardinality of an OOC. These are reproduced below:

Johnson Bound A:

$$\Phi(n, \omega, \kappa) \leq \left\lfloor \frac{1}{\omega} \left\lfloor \frac{n-1}{\omega-1} \cdots \left\lfloor \frac{n-\kappa}{\omega-\kappa} \right\rfloor \right\rfloor \right\rfloor := J_A(n, \omega, \kappa), \tag{2}$$

first noted by Chung, Salehi, and Wei in [6].

Improved Johnson Bound B: Provided $\omega^2 > n\kappa$

$$\phi(n, \omega, \kappa) \leq \min(1, \left\lfloor \frac{\omega-\kappa}{(\omega^2-n\kappa)} \right\rfloor) := J_B(n, \omega, \kappa). \tag{3}$$

The observation that $\Phi(n, \omega, \kappa) \leq 1$ for $\omega^2 > n\kappa$ first appears in [7], its constant weight code equivalent is proved in [4].

Improved Johnson Bound C:

$$\Phi(n, \omega, \kappa) \leq \left\lfloor \frac{1}{\omega} \left\lfloor \frac{n-1}{\omega-1} \cdots \left\lfloor \frac{n-(\ell-1)}{\omega-(\ell-1)} h \right\rfloor \cdots \right\rfloor \right\rfloor \quad := \quad J_C(n, \omega, \kappa), \tag{4}$$

$$\text{where } h = \min(n-\ell, \left\lfloor \frac{(n-\ell)(\omega-\kappa)}{(\omega-\ell)^2-(n-\ell)(\kappa-\ell)} \right\rfloor),$$

and, where ℓ is any integer, $1 \leq \ell \leq \kappa - 1$, such that $(\omega - \ell)^2 > (n - \ell)(\kappa - \ell)$. The Improved Johnson Bound B and C as applied to OOC appeared for the first time, to the best of our knowledge in [5]. However the observation implicit in the Improved Johnson Bound B that $\Phi(n, \omega, \kappa) \leq 1$ for $\omega^2 > n\kappa$ may be found in [7].

Since $\Phi(n, \omega, \kappa)$ denotes the largest possible size of a 1-D OOC, an OOC $\mathcal{C}$ of size P is said to be *optimal* when $P = \Phi(n, w, \kappa)$ and *asymptotically optimum* if: $\lim_{n\to\infty} \frac{P}{\Phi(n,w,\kappa)} = 1$.

2.2 Constructions

There is a large literature on constructions of optical orthogonal code, see for instance [6,7,8,9,10,11,12,13,14,15,16,17,18,19,20,21,22,23,24,25,26,27,28,29,5]. Algebraic constructions for families of OOCs can be found in [6, 7, 8, 9, 12, 20, 28, 29, 5]. Recursive constructions appear in [10, 13, 19, 22, 27]. Constructions specific to a particular choice of weight parameter ω can be found in [16, 15, 17, 18, 21, 23, 24, 25, 26]. Optimum constructions are known only for $\kappa = 1$ [6, 29, 16, 15, 17, 18, 21, 23, 24, 25] and $\kappa = 2$ [7, 26]. Constructions that are asymptotically optimum can be found in [8, 9, 28].

Most papers on 1-D OOC only make use of Johnson Bound A. There are examples of 1-D OOCs which do not achieve Johnson Bound A with equality, which however, are optimal with respect to Johnson Bound B, see [5].

From the point of view of application to fiber-optic communications, a principal drawback of one-dimensional (1-D) OOCs is the requirement of a large chip rate. We illustrate this with an example.

Example 1. Consider the situation where it is desired to assign codes to 8 potential users of which at most 5 users are active at any given time. We assume that the data rate of each user is set to 1 Gbit/sec. A natural attempt at meeting this requirement might be to set the maximum-collision parameter (MCP) κ equal to 1 and set $\omega = 5 > (5-1)\kappa$ where $(5-1)\kappa$, represents the maximum possible interference presented by the 4 other active users under a uniform power assumption. When $\omega = 5$ and $\kappa = 1$, the Johnson bound A (Equation (2)) yields

$$\Phi(n, 5, 1) \leq \left\lfloor \frac{1}{\omega} \left\lfloor \frac{n-1}{\omega - 1} \right\rfloor \right\rfloor \leq \frac{n-1}{\omega(\omega-1)},$$

from which it follows that

$$n - 1 \geq 8 \times 5 \times 4$$

i.e., $n \geq 161$. Thus in this case, the chip-rate must necessarily equal or exceed 161 G chips per second (Gcps) which is currently infeasible to implement. Even if it were feasible to implement, the chip-rate is still large in relation to the data rate supplied to each user. As we shall see, by spreading in both wavelength and time, this chip-rate requirement can be reduced substantially.

A tabular listing of algebraically constructed OOCs appears in Table 1 of Appendix A. The codes appearing in this table can be used in the construction of 2-D OOCs which are based on 1-D OOCs (see [33]).

3 Two-Dimensional Optical Orthogonal Codes

The advent of Wavelength-Division-Multiplexing (WDM) and dense-WDM (D-WDM) technology has made it possible to spread in both wavelength and time [34]. The corresponding codes, are variously called, wavelength-time hopping codes, and multiple-wavelength codes. Here we will refer to these codes as two-dimensional OOCs (2-D OOCs).

A 2-D $(\Lambda \times T, \omega, \kappa)$ OOC $\mathcal{C}$ is a family of $\{0,1\}$ $(\Lambda \times T)$ arrays of constant weight ω. Every pair $\{A, B\}$ of arrays in $\mathcal{C}$ is required to satisfy:

$$\sum_{\lambda=1}^{\Lambda} \sum_{t=0}^{T-1} A(\lambda, t) B(\lambda, (t \oplus_T \tau)) \leq \kappa \tag{5}$$

where either $A \neq B$ or $\tau \neq 0$. We will refer to κ as the maximum collision parameter(MCP). Note that asynchronism is present only along the time axis.

It can be shown that it is possible to construct a 2-D $(\Lambda \times T = 6 \times 6, \omega = 5, \kappa = 1)$ OOC of size 8. Figure 1 shows such a 2-D OOC. Thus in comparison with the earlier 1-D OOC of Example 1 which required a chip-rate in excess of 161 Gcps, with this 2-D code, one can accommodate the same number of users with a chip rate of 6 Gcps.

Practical considerations often place restrictions on the placement of pulses within an array. With this in mind, we introduce the following terminology:

- arrays with one-pulse per wavelength (OPPW): each row of every $(\Lambda \times T)$ code array in $\mathcal{C}$ is required to have Hamming weight $= 1$
- arrays with at most one-pulse per wavelength (AM-OPPW): here each row of any $(\Lambda \times T)$ code in $\mathcal{C}$ is required to have Hamming weight ≤ 1
- arrays with one-pulse per time slot (OPPTS): here each column of every $(\Lambda \times T)$ code array in $\mathcal{C}$ is required to have Hamming weight $= 1$
- arrays with at most one-pulse per time slot (AM-OPPTS): here each column of any $(\Lambda \times T)$ array in $\mathcal{C}$ is required to have Hamming weight ≤ 1.

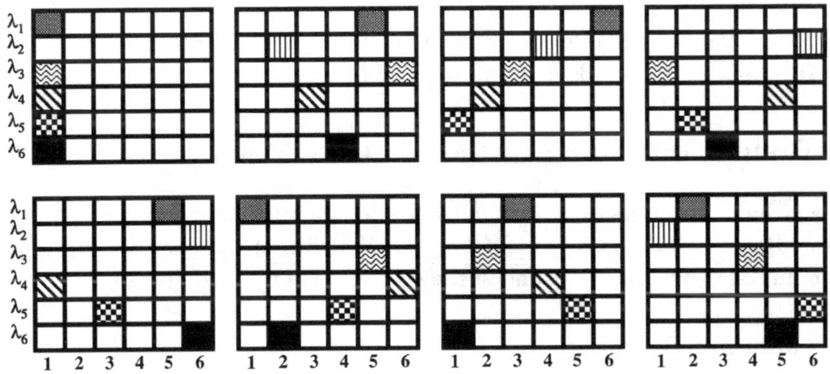

Fig. 1. A (6 × 6,5,1) 2-D OOC. In this figure each row shows a different wavelength, and each column is a different chip time.

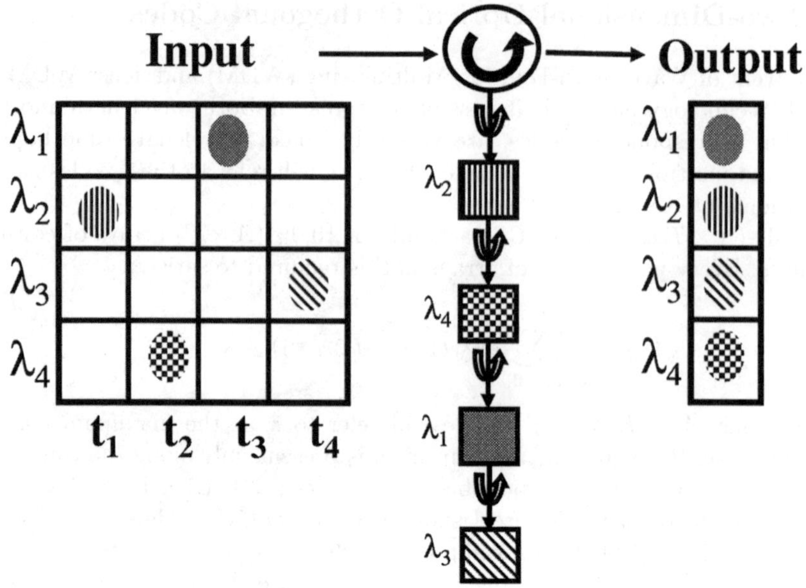

Fig. 2. All optical correlator

A simple means of implementing an optical correlator appears in Fig. 2. Each rectangular box represents an optical filter implemented using for example either a fiber-Bragg grating or an optical micro-resonator. This filter reflects light of the wavelength shown alongside the box and allows light of all other wavelengths to pass through. Filters placed further along the reflection path will suffer an increased delay and in this manner, the placement of the filters can be adjusted to bring the pulses of all the different wavelengths in the desired code matrix into time alignment at the output of the correlator. Implicit in this implementation, is the assumption that the desired code matrix satisfies both the AM-OPPW and AM-OPPTS restrictions.

Constructions for frequency-hopping spreading codes [35,36,37,38,39,40] can often be used to provide 2-D OOCs that satisfy the OPPTS or AM-OPPTS restriction. Papers in the literature dealing with the design of 2-D OOCs include [33,41,42,43,44,45,34,46,47,48,49,50,51,52].

4 Bounds on the Size of a 2-D OCDMA Code

For a given set of values of $\Lambda, T, \omega, \kappa$, let $\Phi(\Lambda \times T, \omega, \kappa)$, denote the largest possible cardinality of a $(\Lambda \times T, \omega, \kappa)$ 2-D OOC code. We define optimal and asymptotically optimum 2-D OOCs as was done in the case of 1-D OOCs in Section 2.

4.1 Johnson Bound

If $\mathcal{C}$ is a $(\Lambda \times T, \omega, \kappa)$ 2-D OOC, then by including every column-cyclic shift of each codeword in $\mathcal{C}$ one can construct a constant weight code using any mapping that reorders the elements of a $\Lambda \times T$ array to form a 1-D string of length ΛT. The resultant constant weight code has parameters $(\Lambda T, \omega, \kappa)$ and size $= T \mid \mathcal{C} \mid$. This observation allows us to translate bounds on constant weight codes to bounds on 2-D OOCs as it is done in Section 2:

We shall refer to these bounds as Johnson bounds for unrestricted 2-D OOCs. They are given by:

Johnson Bound A:

$$\Phi(\Lambda \times T, \omega, \kappa) \leq= \left\lfloor \frac{\Lambda}{\omega} \left\lfloor \frac{\Lambda T - 1}{\omega - 1} \cdots \left\lfloor \frac{\Lambda T - \kappa}{\omega - \kappa} \right\rfloor \right\rfloor \right\rfloor := J_A(\Lambda \times T, \omega, \kappa) \qquad (6)$$

This bound was first pointed out by Yang, and Kwong in [34].

Improved Johnson Bound B: Provided $\omega^2 > \Lambda T \kappa$

$$\Phi(\Lambda \times T, \omega, \kappa) \leq= \min(\Lambda, \left\lfloor \frac{\Lambda(\omega - \kappa)}{(\omega^2 - n\kappa)} \right\rfloor) := J_B(\Lambda \times T, \omega, \kappa). \qquad (7)$$

Improved Johnson Bound C:

$$\Phi(\Lambda \times T, \omega, \kappa) \leq= \left\lfloor \tfrac{\Lambda}{\omega} \left\lfloor \tfrac{\Lambda T-1}{\omega-1} \cdots \left\lfloor \tfrac{\Lambda T-(\ell-1)}{\omega-(\ell-1)} h \right\rfloor \cdots \right\rfloor \right\rfloor := J_C(\Lambda \times T, \omega, \kappa), \qquad (8)$$

$$\text{where } h = \min(\Lambda T - \ell, \left\lfloor \tfrac{(\Lambda T-\ell)(\omega-\kappa)}{(\omega-\ell)^2-(\Lambda T-\ell)(\kappa-\ell)} \right\rfloor)$$

and where ℓ is any integer, $1 \leq \ell \leq \kappa - 1$, such that $(\omega - \ell)^2 > (\Lambda T - \ell)(\kappa - \ell)$.

Theorem 1. *When $n = \Lambda T$ the bound J_i, $i \in \{A, B, C\}$ on the size of a $(\Lambda \times T, \omega, \kappa)$ 2-D OOC satisfies the following inequality compared to the bound on the size of $(\Lambda T, \omega, \kappa)$ 1-D OOC:*

$$\Lambda J_i(\Lambda T, \omega, \kappa) \leq J_i(\Lambda \times T, \omega, \kappa) \leq \Lambda J_i(\Lambda T, \omega, \kappa) + (\Lambda - 1)$$

where $J_i(\Lambda T, \omega, \kappa)$ denotes the upper bound for 1-D OOC stated in equations (2),(3),(4), *and $J_i(\Lambda \times T, \omega, \kappa)$ denoted the upper bound for 2-D OOC stated in equations* (6),(7),(8).

Roughly speaking the theorem suggests that by going to 2-D case, we gain an increase in family size by a factor of Λ.

4.2 Bounds on 2-D AM-OPPW OOCs

Theorem 2. *For any maximally one pulse per wavelength OOC $\mathcal{C}$:*

$$\Phi_{AM-OPPW}(\Lambda \times T, \omega, \kappa) \leq \left\lfloor \frac{\Lambda}{\omega} \left\lfloor \frac{T(\Lambda - 1)}{\omega - 1} \cdots \left\lfloor \frac{T(\Lambda - \kappa)}{\omega - \kappa} \right\rfloor \right\rfloor \right\rfloor$$

For the special case, $\Lambda = \omega$, i.e., for the case when there is exactly one pulse per wavelength, the bound in the Theorem above reduces to

$$\Phi(\Lambda \times T, \omega, \kappa) \leq T^{\kappa}$$

which is the Singleton bound.

5 2-D OOC Code Construction Using Rational Functions

A 2-D OOC can be regarded as the graph of a function $\lambda = f(t)$, $0 \le t \le T-1$, $0 \le \lambda \le \Lambda - 1$ mapping time into wavelength or vice-versa, $t = f(\lambda)$. We define an operator $\mathcal{S}_\tau$ such that $\mathcal{S}_\tau(\cdot, f(\cdot))$ is a τ unit cyclic shifted version of the graph, $(\cdot, f(\cdot))$, along the time axis, t. By this definition the construction is a valid OOC with MCP equal to κ iff the equation $\mathcal{S}_\tau(x, f(x)) = (x, g(x))$ has maximally κ solutions for any f and g when either $f \neq g$ or $\tau \neq 0$. Both polynomial and rational functions can be used as the functions $f(\cdot)$. Constructions employing polynomials were reported in [33]. Here we report on rational function constructions.

5.1 Preliminaries

A function of the form $\frac{f(x)}{g(x)}$ in which both $f(x)$ and $g(x)$ are polynomial functions over $GF(q)$ is called a rational function over $GF(q)$. We assume that f and g are relatively prime and that both numerator and denominator are not simultaneously equal to 0 for any value of x.

Since both $f(x)$ and $g(x)$ can take any value in $GF(q)$ (apart from $f(x) = g(x) = 0$) the value of the rational function will be of the form $\frac{a}{b}$ with $a, b \in GF(q)$, either $a \neq 0$ or $b \neq 0$. The fraction $\frac{a}{0}$ is permissible and we define this fraction to equal the symbol ∞. It is shown in [9] that:

Lemma 1. *The number of rational functions $\frac{f(x)}{g(x)}$ satisfying:*

- *f and g are both nonzero and of degree $\le t$,*
- *f and g are relatively prime,*
- *$\frac{f(x)}{g(x)} \neq a$, for any a and*
- *f is monic.*

is given by:

$$c(t) = \begin{cases} q^{2t+1} - q, & t = 1, 2, 3, 4, 5, 6 \\ \ge q^{2t+1} - q^{2t-6}/7, & t \ge 7 \end{cases}$$

It will be found convenient to identify the pair of symbols $(f(x), g(x))$ for given $x \in GF(q)$ with the element $[f(x), g(x)]^t$ in two-dimensional projective geometry $\mathbb{P}^1(GF(q))$ over $GF(q)$. We have that

$$\mathbb{P}^1(GF(q)) = \left\{ \begin{bmatrix} a \\ b \end{bmatrix} | a, b \in GF(q) \quad \text{and} \quad \begin{bmatrix} a \\ b \end{bmatrix} \neq \begin{bmatrix} 0 \\ 0 \end{bmatrix} \right\}$$

Two elements $\left[\begin{smallmatrix} a \\ b \end{smallmatrix}\right], \left[\begin{smallmatrix} c \\ d \end{smallmatrix}\right] \in \mathbb{P}^1(GF(q))$ are equal provided that there exists an element $\eta \in GF(q)$ such that $\left[\begin{smallmatrix} a \\ b \end{smallmatrix}\right] = \eta \left[\begin{smallmatrix} c \\ d \end{smallmatrix}\right]$.

Theorem 3. *If $f(x) = x^2 + a_1 x + a_0$ with $a_1, a_0 \in GF(q)$ is a primitive polynomial over $GF(q)$, then*

$$\Gamma = \begin{bmatrix} 0 & -a_0 \\ 1 & -a_1 \end{bmatrix}$$

is a matrix having the property that the smallest exponent i for which $\Gamma^i[a,b]^t = \eta[a,b]^t$ for some $\eta \in GF(q)$, is $i = (q+1)$.

It follows that the elements of $\mathbb{P}^1(GF(q))$ can be arranged so as to form an orbit of size $(q+1)$:

$$\mathbb{P}^1(GF(q)) = \left\{ \Gamma^i \begin{bmatrix} a \\ b \end{bmatrix} | i = 0, 1, \cdots, q; \begin{bmatrix} a \\ b \end{bmatrix} \in \mathbb{P}^1(GF(q)) \right\}$$

5.2 Constructions

All of the constructions below employ rational functions $\frac{f(x)}{g(x)}$ satisfying the conditions of Lemma 1 with both $f(x)$ and $g(x)$ of degree $\leq \kappa'$. In these constructions, we have $\kappa = 2\kappa'$.

Mapping Wavelength to Time, $T = q + 1$, q a Power of a Prime: Let $1 \leq \Lambda \leq q$, and $\lambda \in$ some subset of $GF(q)$ of size Λ. Here we consider rational functions $\frac{f(\lambda)}{g(\lambda}$ mapping wavelength into time. Associate to each time slot t, the tth element of a cyclic representation of $\mathbb{P}^1(GF(q))$. Let us define:

$$\Gamma\left(\frac{f(x)}{g(x)}\right) = \mathcal{N}\left(\frac{-a_0 g(x)}{f(x) - a_1 g(x)}\right)$$

where, given a rational function, the operator $\mathcal{N}$ divides out the common factors between numerator and denominator and in addition, scales the two so as to make the numerator monic [9].

Considering f and g are relatively prime, and $a_0 \neq 0$, it is obvious that the operator $\mathcal{N}$ results in some rational function which satisfies the conditions of Lemma 1.

We need to discard all rational functions which are of the form:

$$\frac{f(x)}{g(x)} = \Gamma^k\left(\frac{f(x)}{g(x)}\right) \quad 0 < k \leq q$$

We note that such a rational function doesn't exist since it means that for a certain x_0, $[f(x_0), g(x_0)]^t = \Gamma^k[f(x_0), g(x_0)]^t$ for some $0 < k \leq q$ which is impossible by Theorem 3. Amongst the functions satisfying the conditions of Lemma 1, we have discarded all constant functions, but as this step is unnecessary here, we can add them back.

For any $0 \leq k \leq q$, we declare two rational functions $\frac{f(x)}{g(x)}, \Gamma^k \frac{f(x)}{g(x)}$ to be equivalent. Then the different code matrices correspond to choosing precisely one polynomial from each equivalence class. For each polynomial $f(\cdot)$ the $(\Lambda \times T)$ code array C is given by $C(\lambda, t) = 1$ iff $\frac{f(\lambda)}{g(\lambda)} = t$. This results in a $(\Lambda \times (q+1), \Lambda, 2\kappa')$ 2-D OOC with size $\frac{c(\kappa')}{q+1} + 1$, and $\kappa = 2\kappa' \leq \Lambda \leq q$.

Mapping Time to Wavelength, $\Lambda = q + 1$, q a Power of a Prime: Let $T \mid (q-1)$, and $\beta \in GF(q)$ have multiplicative order T. Take the wavelengths as $\mathbb{P}^1(GF(q))$(the order of elements doesn't matter). Here we consider rational functions mapping time into wavelength. Let us associate to time slot t, the element β^t. We define two rational functions $\frac{f_1(x)}{g_1(x)}, \frac{f_2(x)}{g_2(x)}$ in $\mathcal{F}_\kappa$ to be equivalent if $\frac{f_1(\beta^i x)}{g_1(\beta^i x)} = \frac{f_2(x)}{g_2(x)}$ for some $i \in \mathbb{Z}_T$. First discard all rational functions $\frac{f(x)}{g(x)}$, which satisfy $\frac{f(\beta^i x)}{g(\beta^i x)} = \frac{f(x)}{g(x)}$ for $i \neq 0$. The number of remaining rational functions is computed in [9] as: $\sum_{i|(q-1)} \mu(i) c\left(\lfloor \frac{t}{i} \rfloor\right)$.

Choosing one function $\frac{f(\cdot)}{g(\cdot)}$ from each of the remaining equivalence classes and associating to it, the $(\Lambda \times T)$ code array C by letting $C(\lambda, t) = 1$ iff $\frac{f(\beta^t)}{g(\beta^t)} = \lambda$ where $t \in \mathbb{Z}_T$ and $\lambda \in \mathbb{P}^1(GF(q))$ results in a $((q+1) \times T, T, 2\kappa')$ 2-D OOC of size $\frac{1}{T} \sum_{i|(q-1)} \mu(i) c\left(\left\lfloor \frac{\kappa'}{i} \right\rfloor\right)$.

Note 1. In the function plot constructions if one maps wavelength into time, the resulting 2-D OOC will be of maximally OPPW-type.

Theorem 4. *All of the above constructions are asymptotically optimal with respect to the Johnson bound.*

References

1. J. A. Salehi, "Code division multiple-access techniques in optical fiber networks-part I: Fundamental principles," *IEEE Trans. Commununication*, vol. 37, pp. 824–833, Aug. 1989.
2. S. M. Johnson, "A new upper bound for error-correcting codes," *IRE Trans. Information Theory*, pp. 203–207, Apr. 1962.
3. F. J. MacWilliams and N. J. A. Sloane, *The Theory of Error-Correcting Codes*. New York: North-Holland, 1977.
4. E. Agrell, A. Vardy, and K. Zeger, "Upper bounds for constant-weight codes," *IEEE Trans. Information Theory*, vol. 46, pp. 2373–2395, Nov. 2000.
5. O. Moreno, R. Omrani, and P. V. Kumar, "New bounds on the size of optical orthogonal codes, and constructions," *Preprint to be Submitete to The IEEE Transactions on Information Theory.*
6. F. R. K. Chung, J. A. Salehi, and V. K. Wei, "Optical orthogonal codes: Design, analysis, and applications," *IEEE Trans. Information Theory*, vol. 35, pp. 595–604, May 1989.
7. H. Chung and P. V. Kumar, "Optical orthogonal codes - new bounds and an optimal construction," *IEEE Trans. Information Theory*, vol. 36, pp. 866–873, July 1990.
8. Q. A. Nguyen, L. Györfi, and J. L. Massey, "Constructions of binary constant-weight cyclic codes and cyclically permutable codes," *IEEE Trans. Information Theory*, vol. 38, pp. 940–949, May 1992.
9. O. Moreno, Z. Zhang, P. V. Kumar, and V. A. Zinoviev, "New constructions of optimal cyclically permutable constant weight codes," *IEEE Trans. Information Theory*, vol. 41, pp. 448–455, Mar. 1995.

10. S. Bitan and T. Etzion, "Constructions for optimal constant weight cyclically permutable codes and difference families," *IEEE Trans. Information Theory*, vol. 41, pp. 77–87, Jan. 1995.
11. G. Yang and T. E. Fuja, "Optical orthogonal codes with unequal auto- and cross-correlation constraints," *IEEE Trans. Information Theory*, vol. 41, pp. 96–106, Jan. 1995.
12. M. Buratti, "A powerful method for constructing difference families and optimal optical orthogonal codes," *Designs, Codes and cryptography*, vol. 5, pp. 13–25, 1995.
13. J. Yin, "Some combinatorial constructions for optical orthogonal codes," *Discrete Mathematics*, vol. 185, pp. 201–219, 1998.
14. R. Fuji-Hara and Y. Miao, "Optical orthogonal codes: Their bounds and new optimal constructions," *IEEE Trans. Information Theory*, vol. 46, pp. 2396–2406, Nov. 2000.
15. G. Ge and J. Yin, "Constructions for optimal $(v, 4, 1)$ optical orthogonal codes," *IEEE Trans. Information Theory*, vol. 47, pp. 2998–3004, Nov. 2001.
16. R. Fuji-Hara, Y. Miao, and J. Yin, "Optimal $(9v, 4, 1)$ optical orthogonal codes," *SIAM Journal on Discrete Mathematics*, vol. 14, pp. 256–266, 2001.
17. Y. Tang and J. Yin, "The combinatorial construction for a class of optimal optical orthogonal codes," *Science in China (Series A)*, vol. 45, pp. 1268–1275, Oct. 2002.
18. M. Buratti, "Cyclic designs with block size 4 and related optimal optical orthogonal codes," *Designs, Codes and cryptography*, vol. 26, pp. 111–125, 2002.
19. W. Chu and S. W. Golomb, "A new recursive construction for optical orthogonal codes," *IEEE Trans. Information Theory*, vol. 49, pp. 3072–3076, Nov. 2003.
20. C. Ding and C. Xing, "Several classes of $(2^m - 1, w, 2)$ optical orthogonal codes," *Discrete Applied Mathematics*, vol. 128, pp. 103–120, 2003.
21. Y. Chang, R. Fuji-Hara, and Y. Miao, "Combinatorial constructions of optimal optical orthogonal codes with weight 4," *IEEE Trans. Information Theory*, vol. 49, pp. 1283–1292, May 2003.
22. Y. Chang and Y. Miao, "Constructions for optical orthogonal codes," *Discrete Mathematics*, vol. 261, pp. 127–139, 2003.
23. Y. Chang and L. Ji, "Optimal $(4up, 5, 1)$ optical orthogonal codes," *Journal of Combinatorial Designs*, vol. 12, pp. 346–361, 2004.
24. R. J. R. Abel and M. Buratti, "Some progress on $(v, 4, 1)$ difference families and optical orthogonal codes," *Journal of Combinatorial Theory., Series A*, vol. 106, pp. 59–75, 2004.
25. Y. Chang and J. Yin, "Further results on optimal optical orthogonal codes with weight 4," *Discrete Mathematics*, vol. 279, pp. 135–151, 2004.
26. W. Chu and C. J. Colbourn, "Optimal $(n, 4, 2)$-OOC of small orders," *Discrete Mathematics*, vol. 279, pp. 163–172, 2004.
27. W. Chu and C. J. Colbourn, "Recursive constructions for optimal $(n, 4, 2)$-oocs," *Journal of Combinatorial Designs*, vol. 12, pp. 333–345, 2004.
28. N. Miyamoto, H. Mizuno, and S. Shinohara, "Optical orthogonal codes obtained from conics on finite projective planes," *Finite Fields and Their Applications*, vol. 10, pp. 405–411, 2004.
29. O. Moreno, P. V. Kumar, H. Lu, and R. Omrani, "New construction for optical orthogonal codes, distinct difference sets and synchronous optical orthogonal codes," in *Proc. Int. Symposium on Information Theory*, p. 60, 2003.
30. J. Singer, "A theorem in finite projective geometry and some applications to number theory," *Trans. Amer. Math. Soc.*, vol. 43, pp. 377–385, May 1938.

31. R. C. Bose and S. Chowla, "On the construction of affine difference sets," *Bull. Calcutta Math. Soc.*, vol. 37, pp. 107–112, 1945.
32. R. C. Bose, "An affine analogue of Singer's theorem," *J Indian. Math. Soc.*, vol. 6, pp. 1–15, 1942.
33. R. Omrani, P. Elia and P. V. Kumar, "New constructions and bounds for 2-D optical orthogonal codes," *Proc. of SETA 2004, Springer-Verlag's Lecture Notes in Computer Science Series*, vol. 3486, pp. 389-395, 2005.
34. G. C. Yang and W. C. Kwong, "Performance comparison of multiwavelength CDMA and WDMA+CDMA for fiber-optic networks," *IEEE Trans. Commununication*, vol. 45, pp. 1426–1434, Nov. 1997.
35. A. Lempel and H. Greenberger, "Families of sequences with optimal hamming correlation properties," *IEEE Trans. Information Theory*, vol. 20, pp. 90–94, Jan. 1974.
36. M. K. Simon, J. K. Omura, R. A. Scholtz, and B. K. Levitt, *Spread Sprctrum Communications Handbook.* McGraw-Hill, Inc., 2002.
37. G. Einarsson, "Address assignment for a time-frequency-coded, spread-spectrum system," *Bell System Tech. Journal*, vol. 59, pp. 1241–1255, Sept. 1980.
38. P. V. Kumar, "Frequency-hopping code sequence design having large linear span," *IEEE Trans. Information Theory*, vol. 34, pp. 146–151, Jan. 1988.
39. D. V. Sarwate, "Reed-Solomon codes and the design of sequences for spread-spectrum multiple-access communications," in *Reed-Solomon Codes and Their Applications* (S. B. Wicker and V. K. Bhargava, eds.), Piscataway, NJ: IEEE Press, 1994.
40. O. Moreno and S. V. Maric, "A new family of frequency-hop codes," *IEEE Trans. Commununication*, vol. 48, pp. 1241–1244, Aug. 2000.
41. E. Park, A. J. Mendez, and E. M. Garmire, "Temporal/spatial optical CDMA networks-design, demonstration, and comparison with temporal networks," *IEEE Photon. Technol. Lett.*, vol. 4, pp. 1160–1162, Oct. 1992.
42. K. Kitayama, "Novel spatial spread spectrum based fiber optic CDMA networks for image transmission," *IEEE Journal on Selected Areas in Communications*, vol. 12, pp. 762–772, May 1994.
43. L. Tančevski and I. Andonovic, "Hybrid wavelength hopping/time spreading schemes for use in massive optical networks with increased security," *IEEE Journal of Lightwave Tech.*, vol. 14, pp. 2636–2647, Dec. 1996.
44. L. Tančevski, I. Andonovic, M. Tur, and J. Budin, "Massive optical LAN's using wavelength hopping/time spreading with increased security," *IEEE Photon. Technol. Lett.*, vol. 8, pp. 935–937, July 1996.
45. G. C. Yang and W. C. Kwong, "Two-dimensional spatial signature patterns," *IEEE Trans. Commununication*, vol. 44, pp. 184–191, Feb. 1996.
46. E. S. Shivaleela, K. N. Sivarajan, and A. Selvarajan, "Design of a new family of two-dimensional codes for fiber-optic CDMA networks," *IEEE Journal of Lightwave Tech.*, vol. 16, pp. 501–508, Apr. 1998.
47. H. Fathallah, L. A. Rusch, and S. LaRochelle, "Passive optical fast frequency-hop CDMA communications system," *IEEE Journal of Lightwave Tech.*, vol. 17, pp. 397–405, Mar. 1999.
48. A. J. Mendez, R. M. Gagliardi, H. X. C. Feng, J. P. Heritage, and J. M. Morookian, "Strategies for realizing optical CDMA for dense, high-speed, long span, optical network applications," *IEEE Journal of Lightwave Tech.*, vol. 168, pp. 1685–1695, Dec. 2000.

49. R. M. H. Yim, L. R. Chen, and J. Bajcsy, "Design and performance of 2-D codes for wavelength-time optical CDMA," *IEEE Photon. Technol. Lett.*, vol. 14, pp. 714–716, May 2002.
50. A. J. Mendez, R. M. Gagliardi, V. J. hernandez, C. V. Bennett, and W. j. Lennon, "Design and performance analysis of wavelength/time (W/T) matrix codes for optical CDMA," *IEEE Journal of Lightwave Tech.*, vol. 21, pp. 2524–2533, Nov. 2003.
51. W. C. Kwong and G. C. Yang, "Extended carrier-hopping prime codes for wavelength-time optical code-division multiple access," *IEEE Trans. Commununication*, vol. 52, pp. 1084–1091, 2004.
52. W. C. Kwong, G. C. Yang, V. Baby, C. S. Bres, and P. R. Prucnal, "Multiple-wavelength optical orthogonal codes under prime-sequence permutations for optical CDMA," in *IEEE Trans. Commununication*, vol. 53, pp. 117–123, Jan. 2005.
53. R. Lidl and H. Niederreiter, *Encyclopedia of Mathematics and Its Applications 20: Finite Fields.* Cambridge University Press, 1996.
54. E. R. Berlekamp and O. Moreno, "Extended double-error-correcting binary goppa codes are cyclic," *IEEE Trans. Information Theory*, vol. 19, pp. 817–818, Nov. 1973.
55. G. C. Yang, W. C. Kwong, and C. Y. Chang, "Multiple-wavelength optical orthogonal codes under prime-sequence permutations," in *Proc. Int. Symposium on Information Theory*, p. 367, 2004.

Appendix A: Constructions of 1-D OOCs

Table 1. Different Known Constructions of 1-D Optical Orthogonal Codes, here p denotes a prime and q denotes a power of a prime

Construction Name	Parameters	Code Size
Singer* [6] [30]	$(q^2+q+1, q+1, 1)$	$\mid C \mid = 1$
Projective Geometry* [6]	$(\frac{q^{d+1}-1}{q-1}$,q+1,1)	$\mid C \mid = \begin{cases} \frac{q^d-1}{q^2-1}, & d\ even, \\ \frac{q^d-q}{q^2-1}, & d\ odd. \end{cases}$
Combinatorial Method* [6]	$(n, 3, 1)$, $n \neq 2 \pmod 6$	$\mid C \mid = \lfloor \frac{n-1}{6} \rfloor$
Chung-Kumar* [7]	$(p^{2m}-1, p^m+1, 2)$	$\mid C \mid = p^m - 2$
Chung-Kumar* [7] (via Wilson difference sets)	$(p, \omega, 1)$, $p = \omega(\omega-1)r + 1$ $\omega = 2m+1$, or $\omega = 2m$	$\mid C \mid = r$
MZKZ Family $\mathcal{A}$ [9]	(pm, m, t) $m \mid (p-1)$,$1 \leq t \leq m$	$\mid C \mid = \frac{1}{mp} \sum_{d \mid (p-1)} p^{\lceil (t+1)/d \rceil} \mu(d)$
MZKZ Family $\mathcal{B}$ [9]	$((q-1)p, (p-t), t)$ $1 \leq t \leq (p-t)$	$\mid C \mid = \frac{q}{p} \left(\frac{q^t-1}{q-1}\right)$
MZKZ Family $\mathcal{C}$ [9]	$(m(q+1), m, 2t)$ $m \mid (q-1)$, $(m, q+1) = 1$ $1 \leq t \leq m/2$	$\mid C \mid = \frac{1}{(q+1)m} \sum_{d \mid (q-1)} \mu(d) c([t/d])$ $c(t)$ as defined in Lemma 1
Bose-Chowla* Construction [29] [31] [32]	$(q^2-1, q, 1)$	$\mid C \mid = 1$
Generalized* [29] Bose-Chowla (lines)	$(q^a-1, q, 1)$	$\mid C \mid = q^{a-2} + q^{a-3} + \cdots + 1$
Generalized* [5] Bose-Chowla (hyperplanes)	$(q^a-1, q^{a-1}, q^{a-2})$	$\mid C \mid = 1$
Conics on [28] Finite Projective Plane	$(q^3+q^2+q+1, q+1, 2)$	$\mid C \mid = q^3 - q^2 + q$

An * in the table indicates that the corresponding construction is optimal with respect to some bound.

On the Linear Complexity of Sidel'nikov Sequences over $\mathbb{F}_d$*

Nina Brandstätter[1] and Wilfried Meidl[2]

[1] Johann Radon Institute for Computational and Applied Mathematics,
Austrian Academy of Sciences, Altenbergerstrasse 69,
4040 Linz, Austria
nina.brandstaetter@oeaw.ac.at

[2] Sabanci University, Orhanli, Tuzla,
34956 Istanbul, Turkey
wmeidl@sabanciuniv.edu

Abstract. We study the linear complexity of sequences over the prime field $\mathbb{F}_d$ introduced by Sidel'nikov. For several classes of period length we can show that these sequences have a large linear complexity. For the ternary case we present exact results on the linear complexity using well known results on cyclotomic numbers. Moreover, we prove a general lower bound on the linear complexity profile for all of these sequences. The obtained results extend known results on the binary case. Finally we present an upper bound on the aperiodic autocorrelation.

Keywords: Sidel'nikov sequence; Linear complexity; Linear complexity profile; Aperiodic autocorrelation.

1 Introduction

For an odd prime power q let $\mathbb{F}_q$ be the finite field of order q and let d be a prime divisor of $q-1$. The *cyclotomic classes of order* d give a partition of $\mathbb{F}_q^* := \mathbb{F}_q \setminus \{0\}$ defined by

$$D_0 := \{\alpha^{dn} \ : \ 0 \leq n \leq (q-1)/d - 1\} \quad \text{and} \quad D_j := \alpha^j D_0, \ 1 \leq j \leq d-1,$$

for a generating element α of $\mathbb{F}_q^*$.

In [14] Sidel'nikov introduced the $q-1$-periodic sequence $S = s_0, s_1, \ldots$ with terms in $\mathbb{F}_d$ defined by

$$\begin{aligned} s_n = j \quad &\Leftrightarrow \quad \alpha^n + 1 \in D_j, \quad n = 0, \ldots, q-2, n \neq (q-1)/2, \\ s_{(q-1)/2} = 0, \quad &\text{and} \\ s_{n+q-1} = s_n, \quad &n \geq 0. \end{aligned} \tag{1}$$

* The first author has been supported by the Austrian Science Fund (FWF) grant S83 and by the Austrian Academy of Sciences.

G. Gong et al. (Eds.): SETA 2006, LNCS 4086, pp. 47–60, 2006.

Independently in [9] Lempel, Cohn and Eastman studied the sequence (1) for $d = 2$.

The *linear complexity profile* of a sequence $S = s_0, s_1, \ldots$ over the field $\mathbb{F}_d$ is the function $L(S, N)$ defined for every positive integer N, as the least order L of a linear recurrence relation over $\mathbb{F}_d$

$$s_n = c_1 s_{n-1} + \ldots + c_L s_{n-L}, \tag{2}$$

for all $L \leq n \leq N-1$, which S satisfies. We use the convention that $L(S, N) = 0$ if the first N elements of S are all zero and $L(S, N) = N$ if the first $N - 1$ elements of S are zero and $s_{N-1} \neq 0$. The value

$$L(S) = \sup_{N \geq 1} L(S, N)$$

is called the *linear complexity* of the sequence S. For the linear complexity of any periodic sequence of period t one easily verifies that $L(S) = L(S, 2t) \leq t$. Alternatively, the linear complexity of a periodic sequence with terms in $\mathbb{F}_d$ is the length of the shortest linear recurrence relation (2) the sequence satisfies for all $n \geq L$.

In Section 2 we recall some concepts and facts from the theory of linear recurring sequences over finite fields (see [10, Chapter 6] and [3]), and present a technique for determining the linear complexity of sequences of the form (1). Roughly speaking, we can determine the exact linear complexity whenever we know the value of certain cyclotomic numbers and the factorization of $X^{q-1} - 1$ over $\mathbb{F}_d$. Unconditionally we prove two results which yield good lower bounds on the linear complexity of sequences of the form (1) for several classes of period length. In Section 3 we use the results of Section 2 to obtain exact results on the linear complexity of the ternary Sidel'nikov sequence. In Section 4 we prove a general lower bound on the linear complexity profile. The results on the linear complexity and the linear complexity profile complement and extend results in previous works on the binary case by Helleseth and Yang [6], Kyureghyan and Pott [8], and Meidl and Winterhof [12]. Finally, in Section 5 we prove an upper bound on the aperiodic autocorrelation of the Sidel'nikov sequence which complements the results of [7] on the autocorrelation distribution.

2 Preliminaries

Let $S = s_0, s_1, \ldots$ be an N-periodic sequence over $\mathbb{F}_d$, then we can identify S with the polynomial $S(X) := s_0 + s_1 X + \ldots + s_{N-1} X^{N-1} \in \mathbb{F}_d[X]$ of degree at most $N - 1$. The following well known lemma [3, Lemma 8.2.1] describes the computation of the linear complexity of a periodic sequence.

Lemma 1. *Let S be a sequence of period N over $\mathbb{F}_d$ and*

$$S(X) := s_0 + s_1 X + \ldots + s_{N-1} X^{N-1}.$$

Then the linear complexity of S is given by

$$N - \deg(\gcd(X^N - 1, S(X))).$$

If $N = d^s r$ with $\gcd(d,r) = 1$, then we have $X^N - 1 = (X^r - 1)^{d^s}$. Consequently, in order to calculate the linear complexity of S we are interested in the multiplicities of the rth roots of unity as roots of the polynomial $S(X)$. For the determination of the multiplicity of roots of the polynomial $S(X)$ we can employ the kth Hasse derivative (cf. [5]) $S(X)^{(k)}$ of $S(X)$, which is defined to be

$$S(X)^{(k)} = \sum_{n=k}^{N-1} \binom{n}{k} s_n X^{n-k}.$$

The multiplicity of ξ as root of $S(X)$ is v if $S(\xi) = S(\xi)^{(1)} = \ldots = S(\xi)^{(v-1)} = 0$ and $S(\xi)^{(v)} \neq 0$ (cf. [10, Lemma 6.51]).

In order to obtain results on the linear complexity of the sequence (1) we are interested in the Hasse derivatives of the polynomial $S(X)$ which corresponds to the sequence (1).

The binomial coefficients modulo d appearing in $S(X)^{(k)}$ can be evaluated with *Lucas' congruence* (cf. [4,11])

$$\binom{n}{k} \equiv \binom{n_0}{k_0} \cdots \binom{n_l}{k_l} \bmod d,$$

if $n_0, ..., n_l$ and $k_0, ..., k_l$ are the digits in the d-ary representation of n and k, respectively. We immediately see that

$$\binom{n}{k} \equiv \binom{i}{k} \bmod d \tag{3}$$

for $k < d^l$ and $n \equiv i \bmod d^l$.

As before we denote the cyclotomic classes of order δ by D_j, $j = 0, \ldots \delta - 1$, for a divisor δ of $q-1$. The *cyclotomic numbers* $(i,j)_\delta$ of order δ are defined by

$$(i,j)_\delta = |(D_i + 1) \cap D_j|, \quad 0 \leq i,j \leq \delta - 1.$$

(For monographs on cyclotomic numbers see [2,15].)

Put $l = 1$ if $k = 0$ and $l = \lfloor \log_d(k) \rfloor + 1$ if $k \geq 1$. For the sequence S defined by (1) we can express $S(1)^{(k)}$, $k = 0, 1, \ldots, d^l - 1$, in terms of cyclotomic numbers of order d^l using (3), namely

$$\begin{aligned} S(1)^{(k)} &= \sum_{n=k}^{q-2} \binom{n}{k} s_n = \sum_{i=k}^{d^l-1} \binom{i}{k} \sum_{n \equiv i \bmod d^l} s_n = \sum_{i=k}^{d^l-1} \binom{i}{k} \sum_{n \equiv i \bmod d^l} \sum_{m=1}^{d-1} \sum_{s_n = m} m \\ &= \sum_{i=k}^{d^l-1} \binom{i}{k} \sum_{j=0}^{d^{l-1}-1} \sum_{m=1}^{d-1} (i, dj+m)_{d^l} m. \end{aligned} \tag{4}$$

More general, if r is a divisor of $q-1$ with $\gcd(r,d) = 1$, and ξ is a primitive rth root of unity over $\mathbb{F}_d$ then for the sequence S defined by (1) we can express $S(\xi)^{(k)}$ in terms of cyclotomic numbers of order $d^l r$, namely

$$\begin{aligned}S(\xi)^{(k)} &= \sum_{n=k}^{q-2}\binom{n}{k}s_n\xi^{n-k} = \sum_{h=0}^{r-1}\sum_{\substack{n=k\\ n\equiv h+k \bmod r}}^{q-2}\binom{n}{k}s_n\xi^h\\ &= \sum_{h=0}^{r-1}\sum_{i=k}^{d^l-1}\binom{i}{k}\sum_{\substack{n\equiv i \bmod d^l\\ n\equiv h+k \bmod r}} s_n\xi^h\\ &= \sum_{h=0}^{r-1}\sum_{i=k}^{d^l-1}\binom{i}{k}\sum_{j=0}^{d^{l-1}r-1}\sum_{m=1}^{d-1}(u(h,i),dj+m)_{d^l r}m\xi^h, \qquad (5)\end{aligned}$$

where $u(h,i)$ is (by the Chinese-Remainder-Theorem) the unique integer u with $0 \le u \le d^l r - 1$, $u \equiv h+k \bmod r$, and $u \equiv i \bmod d^l$.

Since in general the determination of cyclotomic numbers of order δ is difficult if δ is not small, we can utilize the above relations solely for small r. The following propositions on large prime factors r of $q-1$ enables us to obtain good lower bounds on the linear complexity for several classes of period length $q-1$. For certain classes of period length the propositions reduce the problem of determining the exact linear complexity to the problem of finding the multiplicity of ± 1 as a root of $S(X)$.

Proposition 1. *Let $r \neq d$ be a prime divisor of $q-1$. If d is a primitive root mod r and $r \ge q^{1/2}+1$ then for each r-th root of unity $\beta \neq 1$ we have $S(\beta) \neq 0$.*

Proof. Since $\beta^r = 1$ we get

$$S(\beta) = \sum_{n=0}^{q-2} s_n\beta^n = \sum_{h=0}^{r-1}\sum_{j=0}^{(q-1)/r-1} s_{h+jr}\beta^h.$$

Note that the least residue of $(q-1)/2$ modulo r is 0. Since d is a primitive root mod r the polynomial $\Phi_r(X) = 1 + X + \ldots + X^{r-1}$ is irreducible and thus the minimal polynomial of β over $\mathbb{F}_d$. Consequently $S(\beta) = 0$ implies

$$\sum_{j=0}^{(q-1)/r-1} s_{h+jr} = \sum_{j=0}^{(q-1)/r-1} s_{jr}, \quad h = 1, \ldots, r-1.$$

Note that for $n \neq (q-1)/2$ we have that

$$\varepsilon_d^{s_n} = \chi_d(\alpha^n + 1), \qquad (6)$$

where χ_d denotes the nontrivial multiplicative character with $\chi_d(\alpha^k) = \mathrm{e}^{2\pi\sqrt{-1}k/d}$ and $\varepsilon_d = \mathrm{e}^{2\pi\sqrt{-1}/d}$.

Furthermore, note that

$$\prod_{j=0}^{(q-1)/r-1}(\alpha^{jr}X + 1) = 1 - X^{(q-1)/r}.$$

Hence,

$$\varepsilon_d^{\sum_{j=0}^{(q-1)/r-1} s_{h+jr}} = \prod_{j=0}^{(q-1)/r-1} \chi_d(\alpha^{h+jr}+1) = \chi_d(1-\alpha^{h(q-1)/r})$$

has the same value for all $h = 1, \ldots, r-1$. Now

$$r - 1 = \left|\sum_{h=0}^{r-1} \chi_d(1-\alpha^{h(q-1)/r})\right| = \frac{r}{q-1}\left|\sum_{h=0}^{q-2} \chi_d(1-\alpha^{h(q-1)/r})\right|$$
$$\leq \frac{r}{q-1}\left(\left(\frac{q-1}{r}-1\right)q^{1/2}+1\right) < q^{1/2}$$

by Weil's bound for character sums (see e.g. [10, Theorem 5.41]) contradicting our assumption on r.

Proposition 2. *Let $r \neq d$ be a prime divisor of $q-1$ and $q \equiv 3 \bmod 4$. If d is a primitive element mod r and*

$$r \geq q^{1/2}\frac{1}{\min_{0\leq a\leq d-1}|\cos 2\pi a/d|}+1 \tag{7}$$

then for each $2r$-th root of unity $\beta \neq \pm 1$ we have $S(\beta) \neq 0$.

Proof. For $\beta^r = 1$ the statement follows from Proposition 1.

If $\beta^r = -1$ we get

$$S(\beta) = \sum_{n=0}^{q-2} s_n \beta^n = \sum_{h=0}^{r-1}\sum_{j=0}^{(q-1)/r-1} (-1)^j s_{h+jr}\beta^h.$$

Again from the irreducibility of $\Phi_r(X) = 1 - X + \ldots - X^{r-2} + X^{r-1}$ we conclude that $\Phi_r(X)$ is the minimal polynomial of β over $\mathbb{F}_d$, and that $S(\beta) = 0$ implies

$$\sum_{j=0}^{(q-1)/r-1} (-1)^j s_{h+jr} = (-1)^h \sum_{j=0}^{(q-1)/r-1} (-1)^j s_{jr}, \quad h = 1, \ldots, r-1.$$

Denote the sum on the left side by $T(h)$. Then it is obvious that $T(h+r) = -T(h)$ and that $T(0) = T(2) = \ldots = T(2r-2) = -T(1) = -T(3) = \ldots = -T(2r-1)$.

Hence,

$$2(r-1)\min_{0\leq a\leq d-1}|\cos 2\pi a/d| \leq \left|(r-1)\left(\varepsilon_d^{T(0)}+\varepsilon_d^{-T(0)}\right)\right|$$
$$= \left|\sum_{\substack{h=1\\h\neq r}}^{2r-1} \varepsilon_d^{\sum_{j=0}^{(q-1)/r-1}(-1)^j s_{h+jr}}\right|. \tag{8}$$

Note that, provided that $q \equiv 3 \bmod 4$, we have

$$\prod_{j=0}^{(q-1)/r-1} (\alpha^{jr}X+1)^{(-1)^j} = \left(1+X^{(q-1)/2r}\right)\left(1-X^{(q-1)/2r}\right)^{-1},$$

where we denote the function on the right side by $f(X)$. Hence, for $1 \le h \le 2r-1$ except for $h=r$, it follows together with (6) that

$$\varepsilon_d^{\sum_{j=0}^{(q-1)/r-1}(-1)^j s_{h+jr}} = \prod_{j=0}^{(q-1)/r-1} \chi_d(\alpha^{h+jr}+1)^{(-1)^j} = \chi_d(f(\alpha^h)).$$

Now, together with (8) this yields

$$\begin{aligned} 2(r-1) \min_{0\le a\le d-1} |\cos 2\pi a/d| &\le \left|\sum_{h=0}^{2r-1} \chi_d(f(\alpha^h))\right| = \frac{2r}{q-1}\left|\sum_{h=0}^{q-2} \chi_d(f(\alpha^h))\right| \\ &\le \frac{2r}{q-2}\left(\left(\frac{q-2}{r}-1\right)q^{1/2}+1\right) < 2q^{1/2} \end{aligned}$$

by Weil's bound for character sums contradicting our assumption on r.

Propositions 1 and 2 immediately yield the lower bound $L(S) \ge 2(r-1)d^s$ for the sequence (1) over $\mathbb{F}_d$ with period length of the form $q-1 = 2ud^sr$, $u \ne d$ odd, d is a primitive root modulo the prime r and r satisfies (7). For instance, for $d=5$ condition (7) equals $r \ge q^{1/2}\frac{1}{\cos 2\pi/d}+1 \approx 3.236q^{1/2}+1$.

3 The Ternary Case $d = 3$

From Propositions 1 and 2 we know that a $2r$th root of unity $\beta \ne \pm 1$ is not a root of the polynomial $S(X)$ if r is a prime such that 3 is a primitive element modulo r and $r \ge 2q^{1/2}+1$, $q \equiv 3 \bmod 4$. If $q = 3^s2r+1$ is a prime power such that r is a prime and 3 is a primitive element modulo r, then we can obtain exact values for the linear complexity of the sequence (1) for the ternary case if we know the multiplicity of 1 and -1 as a root of $S(X)$. In the following we establish general results on the multiplicity of 1 and -1 as a root of $S(X)$. First we focus on the multiplicity of 1 and remark that $X-1$ will always be a divisor of $\gcd(X^{q-1}-1, S(X))$.

For the proof of our first result we will need cyclotomic numbers of order 3. For $q = 3t+1$ let L^2 and M^2 be the uniquely determined integers such that

$$4q = L^2 + 27M^2, \quad L \equiv 1 \bmod 3. \tag{9}$$

We remark that the sign of M is ambiguously determined, depending on the choice of the primitive element α. Then we have [3, p.92]

$$\begin{aligned} (1,1)_3 &= (2q-4-L-9M)/18, \\ (2,1)_3 &= (1,2)_3 = (q+1+L)/9 \quad \text{and} \\ (2,2)_3 &= (2q-4-L+9M)/18. \end{aligned} \tag{10}$$

Proposition 3. *(i)* $(X-1)^2$ *divides* $\gcd(X^{q-1}-1, S(X))$ *if and only if* $q \equiv 1 \bmod 9$.
(ii) $(X-1)^3$ *divides* $\gcd(X^{q-1}-1, S(X))$ *if and only if* $q \equiv 1 \bmod 9$ *and* $M \equiv 0 \bmod 3$, *where* M *is determined (up to sign) from the representation (9) of* q.

Proof. First we note that $(X-1)^2$ and $(X-1)^3$ divide $X^{q-1}-1$. To estimate the multiplicity of 1 as a root of $S(X)$ we employ the Hasse derivatives. With (4) we obtain

$$S(1)^{(1)} = (1,1)_3 + 2(1,2)_3 + 2(2,1)_3 + (2,2)_3, \text{ and}$$
$$S(1)^{(2)} = (2,1)_3 + 2(2,2)_3.$$

With (10) this yields

$$S(1)^{(1)} = (1,1)_3 + (1,2)_3 + (2,2)_3 = \frac{2q-4-L-9M}{18} + \frac{q+1+L}{9} + \frac{2q-4-L+9M}{18} = \frac{q-1}{3} \equiv 0 \bmod 3$$

if and only if $q \equiv 1 \bmod 9$. For $S(1)^{(2)}$ we obtain

$$S(1)^{(2)} = \frac{q+1+L}{9} + 2\frac{2q-4-L+9M}{18} = \frac{q-1}{3} + M \equiv 0 \bmod 3.$$

Since we have to assume that $q \equiv 1 \bmod 9$ this yields $S(1)^{(2)} \equiv 0 \bmod 3$ if and only if $M \equiv 0 \bmod 3$.

The subsequent proposition presents results on the multiplicity of 2 as a root of $\gcd(X^{q-1}-1, S(X))$. Note that 6 divides $q-1$ and that 2 is a root of $X^{q-1}-1$ with multiplicity at least 3. The proof of the proposition uses the same technique as the proof of Proposition 3. For the sake of completeness the proof is added in the Appendix. Instead of cyclotomic numbers of order 3 we have to employ cyclotomic numbers of order 6 which depend upon the decomposition

$$q = 6f + 1 = A^2 + 3B^2 \tag{11}$$

of q with $A \equiv 1 \bmod 3$ and additionally $\gcd(A, q) = 1$ if $q = p^m$ and $p \equiv 1 \bmod 6$. The sign of B is ambiguously determined, depending on the choice of the primitive element α.

Proposition 4. *(i)* $X+1$ *and* $(X+1)^2$ *divide* $\gcd(X^{q-1}-1, S(X))$ *if and only if* $B \equiv 0 \bmod 3$,
(ii) $(X+1)^3$ *divides* $\gcd(X^{q-1}-1, S(X))$ *if and only if* $B \equiv 0 \bmod 9$, *where* B *is determined from the representation (11) of* q.

Remark 1. The condition $B \equiv 0 \bmod 3$ is satisfied if and only if 2 is a cube in $\mathbb{F}_q$ (cf. [2, Corollary 2.6.4]).

With the Propositions 1 – 4 we immediately obtain the following exact values for the linear complexity of the ternary Sidel'nikov sequence.

Theorem 1. *Let S be the ternary Sidel'nikov sequence (1) with period $q-1$ for a prime power q of the form $q = 3^s 2r + 1$, where r is a prime such that 3 is a primitive root modulo r, and suppose that $r \geq 2q^{1/2} + 1$. If*

- $q \not\equiv 1 \bmod 9$, $B \not\equiv 0 \bmod 3$ *then* $L(S) = q - 2$,
- $q \equiv 1 \bmod 9$, $M \not\equiv 0 \bmod 3$, $B \not\equiv 0 \bmod 3$ *then* $L(S) = q - 3$,
- $q \not\equiv 1 \bmod 9$, $B \equiv 0 \bmod 3$, $B \not\equiv 0 \bmod 9$ *then* $L(S) = q - 4$,
- $q \equiv 1 \bmod 9$, $M \not\equiv 0 \bmod 3$, $B \equiv 0 \bmod 3$, $B \not\equiv 0 \bmod 9$ *then* $L(S) = q - 5$.

A Remark to Higher Derivatives

In [1] Baumert and Fredricksen presented formulas for the cyclotomic numbers of order 9 and 18 for the case of a prime field $\mathbb{F}_p$. More precisely, if $p = 3^s 2r + 1$ with $s \geq 2$ and (γ being a 9th root of unity)

$$p = \left(\sum_{i=0}^{5} c_i \gamma^i\right)\left(\sum_{i=0}^{5} c_i \gamma^{-i}\right)$$

is a factorization of p in the field of 9th roots of unity, then each cyclotomic number of order 9 respectively of order 18 is expressed as a constant plus a linear combination of $p, L, M, c_0, \ldots, c_5$. We will indicate how we can use this results to obtain more information on the linear complexity of the Sidel'nikov Sequence.

With the knowledge of the cyclotomic numbers of order 9 and 18 we are able to determine $S^{(k)}(1)$ and $S^{(k)}(2)$ for $k = 3, \ldots, 8$ from (4) and (5).

Here, we restrict ourselves to the 4th derivatives for the special casc that $\operatorname{ind} 2 \equiv 0 \bmod 9$ and $\operatorname{ind} 3 \equiv 1 \bmod 3$. Applying the results of [1] with straightforward but longsome calculations we get

$$S^{(3)}(1) = c_2 \quad \text{and} \quad S^{(3)}(2) = \frac{c_2 - c_5}{2}.$$

Hence we obtain the following proposition for the considered special case.

Proposition 5. *(i)* $(X-1)^4$ *divides* $\gcd(X^{p-1} - 1, S(X))$ *if and only if* $p \equiv 1 \bmod 9$, $M \equiv 0 \bmod 3$ *and* $c_2 \equiv 0 \bmod 3$,
(ii) $(X+1)^4$ *divides* $\gcd(X^{p-1} - 1, S(X))$ *if and only if* $B \equiv 0 \bmod 9$ *and* $c_2 - c_5 \equiv 0 \bmod 6$.

Consequently for this special case we can extend Theorem 1 as follows.

Theorem 2. *Let S and p satisfy the conditions of* Theorem 1. *Let* $\operatorname{ind} 2 \equiv 0 \bmod 9$ *and* $\operatorname{ind} 3 \equiv 1 \bmod 3$. *If*

- $p \equiv 1 \bmod 9$, $M \equiv 0 \bmod 3$, $c_2 \not\equiv 0 \bmod 3$, $B \equiv 0 \bmod 3$, $B \not\equiv 0 \bmod 9$ *then* $L(S) = p - 6$,
- $p \equiv 1 \bmod 9$, $M \not\equiv 0 \bmod 3$, $B \equiv 0 \bmod 9$, $c_2 - c_5 \not\equiv 0 \bmod 6$ *then* $L(S) = p - 6$,
- $p \equiv 1 \bmod 9$, $M \equiv 0 \bmod 3$, $c_2 \not\equiv 0 \bmod 3$, $B \equiv 0 \bmod 9$, $c_2 - c_5 \not\equiv 0 \bmod 6$ *then* $L(S) = p - 7$.

4 A Lower Bound on the Linear Complexity Profile

Theorem 3. *The linear complexity profile* $L(S,N)$ *of the Sidel'nikov sequence* (1) *satisfies*

$$L(S,N) \geq \min\left(\frac{N+1}{q^{1/2}\log q+3}, \frac{q-1}{q^{1/2}\log q+2}\right) - 1.$$

Proof. Suppose that S satisfies the recurrence relation (2) for $L \leq n \leq N-1$. If we put $c_0 = -1$ then we have

$$\sum_{l=0}^{L} c_l s_{n-l} = 0 \in \mathbb{F}_d \quad \text{for } L \leq n \leq \min(N, q-1+L) - 1.$$

Recall that for $m \neq (q-1)/2$ we have

$$\chi_d(\alpha^m + 1) = \varepsilon_d^{s_m}, \tag{12}$$

where χ_d denotes the nontrivial multiplicative character of order d with $\chi_d(\alpha^m)$ $= \mathrm{e}^{2\pi\sqrt{-1}m/d}$ and $\varepsilon_d = \mathrm{e}^{2\pi\sqrt{-1}/d}$.

Thus, for all n satisfying $L \leq n \leq \min(N, q-1+L) - 1$ and $\frac{q-1}{2} \notin \{n, n-1, \ldots, n-L\}$, we get

$$\chi_d\left(\prod_{l=0}^{L}(\alpha^{n-l}+1)^{c_l}\right) = \prod_{l=0}^{L}\chi_d(\alpha^{n-l}+1)^{c_l}$$
$$= \prod_{l=0}^{L}\varepsilon_d^{c_l s_{n-l}} = \varepsilon_d^{\sum_{l=0}^{L} c_l s_{n-l}} = 1.$$

Consequently,

$$\min(N-L, q-1) - 2(L+1) \leq \sum_{n=L}^{\min(N,q-1+L)-1} \chi_d\left(\prod_{l=0}^{L}(\alpha^{n-l}+1)^{c_l}\right)$$
$$\leq (L+1)q^{1/2}\log q,$$

where the last step follows from [13, Lemma 3.3]. The bound immediately follows from the above inequality.

5 An Upper Bound on the Aperiodic Autocorrelation

Let $S = s_0, s_1, \ldots$ be an N-periodic sequence over the finite field $\mathbb{F}_d$. The *autocorrelation* of S is the complex-valued function defined by

$$\mathrm{A}_d(S,t) := \sum_{n=0}^{N-1} \varepsilon_d^{s_{n+t}-s_n}, \quad 1 \leq t \leq N-1,$$

where $\varepsilon_d = \mathrm{e}^{2\pi\sqrt{-1}/d}$.

In [7] Kim et al. presented results on the distribution of the autocorrelation of the Sidel'nikov sequence when t takes different values. In particular the autocorrelation of the Sidel'nikov sequence (1) was determined to be

$$\mathrm{A}_d(S,t) = \chi_d^{-1}(1-\alpha^t) + \chi_d(1-\alpha^{-t}) - \chi_d(\alpha^{-t}) - 1,$$

for $1 \le t \le N-1$.

While the autocorrelation reflects global randomness the *aperiodic autocorrelation*, which is defined by

$$\mathrm{AAC}_d(S,u,v,t) = \sum_{n=u}^{v} \varepsilon_d^{s_n - s_{n+t}}, \quad 0 \le u < v < N,\ 1 \le t < N,$$

reflects local randomness.

If S is a random sequence over $\mathbb{F}_d$ then $|\mathrm{A}_d(S,t)|$ and $|\mathrm{AAC}_d(S,u,v,t)|$ can be expected to be quite small. The security of many cryptographic systems depends upon the generation of pseudorandom, i. e., unpredictable quantities and a low (aperiodic) autocorrelation is a desirable feature for pseudorandom sequences.

Theorem 4. *The aperiodic autocorrelation* $\mathrm{AAC}_d(S,u,v,t)$ *of the Sidel'nikov sequence* (1) *over* $\mathbb{F}_d$ *can be estimated by*

$$|\mathrm{AAC}_d(S,u,v,t)| \le 2q^{1/2}\log q + 2,$$

for $0 \le u < v < q-1$ *and* $1 \le t < q-1$.

Proof. By definition and by (12) we have

$$\begin{aligned} |\mathrm{AAC}_d(S,u,v,t)| &= \left| \sum_{n=u}^{v} \varepsilon_d^{s_n - s_{n+t}} \right| \le \left| \sum_{n=u}^{v} \chi_d(\alpha^n + 1)\chi_d^{d-1}(\alpha^{n+t}+1) \right| + 2 \\ &= \left| \sum_{n=u}^{v} \chi_d\left((\alpha^n+1)(\alpha^{n+t}+1)^{d-1}\right) \right| + 2 \le 2q^{1/2}\log q + 2, \end{aligned}$$

where the last inequality follows from [13, Lemma 3.3].

Remark 2. We remark that the estimate in Theorem 4 accords with

$$\max_{t=1,\dots,q-2} |\mathrm{AAC}_d(S,0,N-1,t)| = \Omega(q^{1/2}),$$

where $N = (1/5 - \varepsilon)q, \varepsilon > 0$.

Acknowledgement

Part of the research was done during a visit of the first author to the Sabanci University. She wishes to thank the university for hospitality.

We would like to thank Arne Winterhof for pointing out Remark 2.

References

1. L.D. Baumert and H. Fredricksen, The cyclotomic numbers of order eighteen with applications to difference sets, Math. Comp. 21 (1967), 204–219.
2. B. C. Berndt, R. J. Evans, and K. S. Williams, Gauss and Jacobi sums, Canadian Mathematical Society Series of Monographs and Advanced Texts. A Wiley-Interscience Publication. John Wiley & Sons, Inc., New York, 1998.
3. T. W. Cusick, C. Ding, and A. Renvall, Stream Ciphers and Number Theory, North-Holland Publishing Co., Amsterdam, 1998.
4. A. Granville, Arithmetic properties of binomial coefficients. I. Binomial coefficients modulo prime powers, in: Organic mathematics, Burnaby, BC, 1995, CMS Conf. Proc. 20, Amer. Math. Soc., Providence, RI, 1997, 253–276.
5. H. Hasse, Theorie der höheren Differentiale in einem algebraischen Funktionenkörper mit vollkommenem Konstantenkörper bei beliebiger Charakteristik, J. Reine Angew. Math. 175 (1936), 50–54.
6. T. Helleseth and K. Yang, On binary sequences with period $n = p^m - 1$ with optimal autocorrelation, In (T. Helleseth, P. Kumar, and K. Yang, eds.), Proceedings of SETA 01, (2002), 209–217.
7. Y.-S. Kim, J.-S. Chung, J.-S. No, and H. Chung, On the autocorrelation distributions of Sidel'nikov sequences, IEEE Trans. Inf. Th. 51 (2005), 3303–3307.
8. G. M. Kyureghyan and A. Pott, On the linear complexity of the Sidelnikov-Lempel-Cohn-Eastman sequences, Designs, Codes, and Cryptography 29 (2003), 149–164.
9. A. Lempel, M. Cohn, and W. L. Eastman, A class of balanced binary sequences with optimal autocorrelation properties. IEEE Trans. Inf. Th. 23 (1977), 38–42.
10. R. Lidl, H. Niederreiter, Finite Fields, Addison-Wesley, Reading, MA, 1983.
11. M. E. Lucas, Sur les congruences des nombres euleriennes et des coefficients differentiels des fuctions trigonometriques, suivant un-module premier, Bull. Soc. Math. France 6 (1878), 122–127.
12. W. Meidl and A. Winterhof, Some notes on the linear complexity of Sidel'nikov-Lempel-Cohn-Eastman sequences, Designs, Codes, and Cryptography 38 (2006), 159–178.
13. I. Shparlinski, Cryptographic Applications of Analytic Number Theory. Complexity Lower Bounds and Pseudorandomness. Progress in Computer Science and Applied Logic. 22, Birkhäuser, Basel, 2003.
14. V. M. Sidel'nikov, Some k-valued pseudo-random sequences and nearly equidistant codes. Problems of Information Transmission 5 (1969), 12–16.; translated from Problemy Peredači Informacii 5 (1969), 16–22 (Russian).
15. T. Storer, Cyclotomy and Difference Sets, Markham Publishing Co., Chicago, III. (1967).

Appendix

For the proof of Proposition 4 we will utilize the following relation between the cyclotomic numbers of order d (cf. [3, p.84]]. Let $q = df + 1$, then

$$(i,j)_d = (d-i, j-i)_d = \begin{cases} (j,i)_d, & f \text{ even} \\ (j+d/2, i+d/2)_d, & f \text{ odd} \end{cases}. \tag{13}$$

We will then need the following cyclotomic numbers of order 6 given in [3, Appendix B]. Let $q \equiv 1 \bmod 6$ with decomposition (11) and let $2 = \alpha^m$.

Case Ia: $q \equiv 1 \bmod 12$, $m \equiv 0 \bmod 3$

$$\begin{aligned}&(0,1)_6 = (q-5+4A+18B)/36,\ (0,2)_6 = (q-5+4A+6B)/36,\\ &(0,4)_6 = (q-5+4A-6B)/36,\ \ (0,5)_6 = (q-5+4A-18B)/36,\\ &(1,2)_6 = (1,3)_6 = (1,4)_6 = (2,4)_6 = (q+1-2A)/36.\end{aligned}$$

Case Ib: $q \equiv 1 \bmod 12$, $m \equiv 1 \bmod 3$

$$\begin{aligned}&(0,1)_6 = (q-5+4A+12B)/36,\ (0,5)_6 = (q-5+4A-6B)/36,\\ &(1,3)_6 = (q+1-2A-6B)/36,\ \ (1,4)_6 = (q+1-2A+12B)/36.\end{aligned}$$

Case Ic: $q \equiv 1 \bmod 12$, $m \equiv 2 \bmod 3$

$$\begin{aligned}&(0,1)_6 = (q-5+4A+6B)/36,\ \ (0,5)_6 = (q-5+4A-12B)/36,\\ &(1,3)_6 = (q+1-2A-12B)/36,\ (1,4)_6 = (q+1-2A+6B)/36.\end{aligned}$$

Case IIa: $q \equiv 7 \bmod 12$, $m \equiv 0 \bmod 3$

$$\begin{aligned}&(1,0)_6 = (q-5+4A+6B)/36,\ (0,1)_6 = (0,2)_6 = (q+1-2A+12B)/36,\\ &(1,1)_6 = (q-5+4A-6B)/36,\ (1,2)_6 = (2,1)_6 = (q+1-2A)/36,\\ &(0,4)_6 = (0,5)_6 = (q+1-2A-12B)/36.\end{aligned}$$

Case IIb: $q \equiv 7 \bmod 12$, $m \equiv 1 \bmod 3$

$$\begin{aligned}&(0,2)_6 = (q+1-2A+12B)/36,\ (0,4)_6 = (q+1-8A-12B)/36,\\ &(1,0)_6 = (q-5-2A+6B)/36,\ \ (1,1)_6 = (q-5+4A-6B)/36.\end{aligned}$$

Case IIc: $q \equiv 7 \bmod 12$, $m \equiv 2 \bmod 3$

$$\begin{aligned}&(0,2)_6 = (q+1-8A+12B)/36,\ (0,4)_6 = (q+1-2A-12B)/36,\\ &(1,0)_6 = (q-5+4A+6B)/36,\ \ (1,1)_6 = (q-5-2A-6B)/36.\end{aligned}$$

Proof of Proposition 4: With (5) we obtain

$$\begin{aligned}S(2) = {} & (0,1)_6 + (0,4)_6 + (4,1)_6 + (4,4)_6 + (2,1)_6 + (2,4)_6\\ & +2(0,2)_6 + 2(0,5)_6 + 2(4,2)_6 + 2(4,5)_6 + 2(2,2)_6 + 2(2,5)_6\\ & +2(3,1)_6 + 2(3,4)_6 + 2(1,1)_6 + 2(1,4)_6 + 2(5,1)_6 + 2(5,4)_6\\ & +(3,2)_6 + (3,5)_6 + (1,2)_6 + (1,5)_6 + (5,2)_6 + (5,5)_6.\end{aligned}$$

If $q \equiv 1 \bmod 12$ with (13) we obtain $S(2) = 2(0,1)_6 + (0,5)_6 + (1,3)_6 + 2(1,4)_6$.

For the Case Ia, i.e. 2 is a cube which implies $B \equiv 0 \bmod 3$, we then get

$$\begin{aligned}S(2) &= 2\frac{q-5+4A+18B}{36} + \frac{q-5+4A-18B}{36} + \frac{q+1-2A}{36}\\ &\quad +2\frac{q+1-2A}{36}\\ &= -\frac{q-5+4A+18B}{36} + \frac{q-5+4A-18B}{36} = -B = 0.\end{aligned}$$

In the Case Ib, where $B \not\equiv 0 \bmod 3$, we obtain

$$S(2) = 2\frac{q-5+4A+12B}{36} + \frac{q-5+4A-6B}{36} + \frac{q+1-2A-6B}{36}$$
$$+2\frac{q+1-2A+12B}{36} = \frac{-18B}{36} + \frac{-18B}{36} = -B \neq 0.$$

Finally for Case Ic (again $B \not\equiv 0 \bmod 3$) we get

$$S(2) = 2\frac{q-5+4A+6B}{36} + \frac{q-5+4A-12B}{36} + \frac{q+1-2A-12B}{36}$$
$$+2\frac{q+1-2A+6B}{36} = \frac{-18B}{36} + \frac{-18B}{36} = -B \neq 0.$$

If $q \equiv 7 \bmod 12$ (13) yields $S(2) = 2(0,4)_6 + 2(1,1)_6 + (0,2)_6 + (1,0)_6$. Consequently for the Case IIa we obtain

$$S(2) = 2\frac{q+1-2A-12B}{36} + 2\frac{q-5+4A-6B}{36} + \frac{q+1-2A+12B}{36}$$
$$+\frac{q-5+4A+6B}{36} = \frac{24B}{36} + \frac{12B}{36} = B = 0.$$

For the Case IIb respectively for the Case IIc we get

$$S(2) = 2\frac{q+1-8A-12B}{36} + 2\frac{q-5+4A-6B}{36} + \frac{q+1-2A+12B}{36}$$
$$+\frac{q-5-2A+6B}{36} = \frac{6A+24B}{36} + \frac{-6A+12B}{36} = B \neq 0,$$

respectively

$$S(2) = 2\frac{q+1-2A-12B}{36} + 2\frac{q-5-2A-6B}{36} + \frac{q+1-8A+12B}{36}$$
$$+\frac{q-5+4A+6B}{36} = \frac{-6A+24B}{36} + \frac{6A+12B}{36} = B \neq 0.$$

Summarizing $S(2) = 0$ if and only if 2 is a cube or equivalently $B \equiv 0 \bmod 3$. With (5) we obtain

$$S(2)^{(1)} = (1,1)_6 + 2(1,2)_6 + (1,4)_6 + 2(1,5)_6 + 2(5,1)_6 + (5,2)_6$$
$$+2(5,4)_6 + (5,5)_6 + 2(4,1)_6 + (4,2)_6 + 2(4,4)_6 + (4,5)_6$$
$$+(2,1)_6 + 2(2,2)_6 + (2,4)_6 + 2(2,5)_6.$$

If $q \equiv 1 \bmod 12$ with (13) this yields $S(2)^{(1)} = (0,5)_6 + (0,1)_6 + 2(2,4)_6 + 2(0,2)_6 + (1,2)_6 + 2(0,4)_6$, and hence for $m \equiv 0 \bmod 3$, the only case of interest, we get

$$S(2)^{(1)} = \frac{q-5+4A-18B}{36} + \frac{q-5+4A+18B}{36} + 2\frac{q+1-2A}{36}$$
$$+2\frac{q-5+4A+6B}{36} + \frac{q+1-2A}{36} + 2\frac{q-5+4A-6B}{36}$$
$$= \frac{-12B}{36} + \frac{12B}{36} = 0.$$

If $q \equiv 7 \bmod 12$ with (13) we have $S(2)^{(1)} = (0,2)_6 + (0,4)_6 + 2(0,5)_6 + 2(2,1)_6 + (1,2)_6 + 2(0,1)_6$, which again vanishes if $m \equiv 0 \bmod 3$ (Case IIa).

Finally (5) yields $S(2)^{(2)} = (2,1)_6 + (2,4)_6 + 2(2,2)_6 + 2(2,5)_6 + 2(5,1)_6 + 2(5,4)_6 + (5,2)_6 + (5,5)_6$. Using (13) for the Case Ia we obtain

$$\begin{aligned} S(2)^{(2)} &= (2,4)_6 + 2(0,4)_6 + 2(1,2)_6 + (0,1)_6 \\ &= \frac{q+1-2A}{36} + 2\frac{q-5+4A-6B}{36} + 2\frac{q+1-2A}{36} \\ &\quad + \frac{q-5+4A+18B}{36} = \frac{2B}{3}, \end{aligned}$$

and for the Case IIa we obtain

$$\begin{aligned} S(2)^{(2)} &= (2,1)_6 + 2(0,1)_6 + 2(1,2)_6 + (0,4)_6 \\ &= \frac{q+1-2A}{36} + 2\frac{q+1-2A+12B}{36} + 2\frac{q+1-2A}{36} \\ &\quad + \frac{q+1-2A-12B}{36} = -\frac{2B}{3}. \end{aligned}$$

Consequently $S(2)^{(2)} = 0$ if and only if $B \equiv 0 \bmod 9$. □

Linear Complexity over F_p of Ternary Sidel'nikov Sequences*

Young-Sik Kim[1], Jung-Soo Chung[1], Jong-Seon No[1], and Habong Chung[2]

[1] School of Electrical Engineering and Computer Science and INMC, Seoul National University, Seoul 151-744, Korea
{kingsi, integer}@ccl.snu.ac.kr, jsno@snu.ac.kr
[2] School of Electronics and Electrical Engineering, Hongik University, Seoul 121-791, Korea
habchung@hongik.ac.kr

Abstract. In this paper, for positive integers m, M, and a prime p such that $M|p^m - 1$, we derive linear complexity over the prime field F_p of M-ary Sidel'nikov sequences of period $p^m - 1$ using discrete Fourier transform. As a special case, the linear complexity of the ternary Sidel'nikov sequence is presented. It turns out that the linear complexity of a ternary Sidel'nikov sequence with the symbol $k_0 \neq 1$ at the $(p^m - 1)/2$-th position is nearly close to the period of the sequence, while that with $k_0 = 1$ shows much lower value.

1 Introduction

Linear complexity of sequences is one of the important properties of sequences employed in the secure communication and cryptography. Having a large linear complexity implies the difficulty in the analysis of the sequence.

For positive integers m, M, and a prime p, such that $M|p^m - 1$, Sidel'nikov [9] constructed M-ary sequences (called *Sidel'nikov sequences*) of period $p^m - 1$, the out-of-phase autocorrelation magnitude of which is upper bounded by 4 [9]. Later, Lempel, Cohn, and Eastman [8] independently rediscovered the binary Sidel'nikov sequences of period $p^m - 1$. These binary sequences have near-ideal autocorrelation property which, under the condition of balancedness, is optimal.

Helleseth and Yang [5] studied the linear complexity over F_2 of the binary Sidel'nikov sequences. And Kyureghyan and Pott [7] extended their results using cyclotomic numbers. But these results are limited only to some special cases.

There has been another approach to the study of the linear complexity of the binary Sidel'nikov sequences. Since Sidel'nikov sequences are constructed based on the finite field F_{p^m}, Helleseth, Kim, and No [3] introduced the linear complexity over F_p of the binary Sidel'nikov sequences. But they showed only for small primes p such as $p = 3, 5$, and 7. Recently, Helleseth, Maas, Mathiassen,

* This research was supported by the MIC, Korea, under the ITRC support program and by the MOE, the MOCIE, and the MOLAB, Korea, through the fostering project of the Laboratory of Excellency.

G. Gong et al. (Eds.): SETA 2006, LNCS 4086, pp. 61–73, 2006.

and Segers [4] derived the linear complexity over the prime field F_p of the binary Sidel'nikov sequences for a prime p. For the balanced Sidel'nikov sequences, Kim, Chung, No, and Chung present the linear complexity over F_p of M-ary Sidel'nikov sequences using discrete Fourier transform [6].

In this paper, the derivation [6] of the linear complexity over F_p of M-ary Sidel'nikov sequences is extended to the general case including unbalanced sequences. It turns out that the linear complexity of a ternary Sidel'nikov sequence with the symbol $k_0 \neq 1$ at the $(p^m-1)/2$-th position is nearly close to the period of the sequence, while that with $k_0 = 1$ shows much lower value.

2 Preliminaries

For a sequence $s(t)$ of period $n = p^m - 1$, the discrete Fourier transform and its inverse Fourier transform are given by

$$A_i = \frac{1}{n}\sum_{t=0}^{n-1} s(t)\alpha^{-it}$$

$$s(t) = \sum_{i=0}^{n-1} A_i\alpha^{it}$$

where α is a primitive element of the finite field F_{p^m} with p^m elements. An M-ary sequence $s(t)$ of period n, $M|n$, is said to be balanced if each element occurs exactly n/M times in a period.

The M-ary Sidel'nikov sequence is defined as follows.

Definition 1. Let m and M be positive integers, and p a prime such that $M|p^m - 1$. Let α be a primitive element of F_{p^m}. For $k = 0, 1, \cdots, M-1$, define

$$S_k = \left\{\alpha^{Ml+k} - 1 \;\middle|\; 0 \le l \le \frac{p^m-1}{M} - 1\right\}.$$

Then the M-ary Sidel'nikov sequence $s(t)$ is defined as

$$s(t) = \begin{cases} k, & \alpha^t \in S_k \\ k_0, & \alpha^t = -1. \end{cases}$$

□

When $k_0 = 0$, the Sidel'nikov sequence is balanced. The following theorem shows some combinatorial relation between a number and its p-ary expansion.

Theorem 1. [Lucas' Theorem] [1] If p is a prime and $N = \sum_{i=0}^{I} N_i p^i$, $0 \le N_i \le p-1$, $K = \sum_{i=0}^{I} K_i p^i$, $0 \le K_i \le p-1$, then we have

$$\binom{N}{K} \equiv \prod_{i=0}^{I}\binom{N_i}{K_i} \bmod p.$$

□

In this paper, we will call $\binom{N_i}{K_i}$ Lucas factor of $\binom{N}{K}$.

3 Linear Complexity of M-ary Sidel'nikov Sequences

From the Blahut's theorem, the linear complexity of periodic sequences can be determined by computing the Hamming weight of their Fourier transform, that is, the number of nonzero values of their Fourier transform.

We will compute the Fourier transform of M-ary Sidel'nikov sequences for an alphabet size M.

Theorem 2. Let $L = (p^m - 1)/M$, $n = p^m - 1$, and $p > M$. The Fourier transform of an M-ary Sidel'nikov sequence is derived as

$$A_{-i} \equiv \Big(\frac{(M-1)}{2} - k_0\Big)(-1)^i - (-1)^i \sum_{v=1}^{M-1} \frac{B_v(i)(-1)^{-vL}}{1-\alpha^{vL}} \bmod p \tag{1}$$

where $B_v(i) = \binom{i}{vL}$.

Proof. From Definition 1, the Fourier transform of $s(t)$ is written as

$$\begin{aligned} nA_{-i} &= k_0(-1)^i + \sum_{\alpha^t \in S_0 \setminus \{0\}} 0 \cdot \alpha^{it} + \sum_{\alpha^t \in S_1} \alpha^{it} + \cdots + \sum_{\alpha^t \in S_{M-1}} (M-1)\alpha^{it} \\ &= k_0(-1)^i + \sum_{u=1}^{M-1} \sum_{l=0}^{L-1} u(\alpha^{Ml+u} - 1)^i \\ &= k_0(-1)^i + \sum_{u=1}^{M-1} u \sum_{l=0}^{L-1} \sum_{r=0}^{i} \binom{i}{r} (-1)^{i-r} \alpha^{(Ml+u)r} \\ &= k_0(-1)^i + \sum_{u=1}^{M-1} \sum_{r=0}^{i} u \binom{i}{r} (-1)^{i-r} \alpha^{ur} \sum_{l=0}^{L-1} \alpha^{Mlr}. \end{aligned}$$

The innermost sum is equal to L for $r = 0, L, \cdots, (M-1)L$, and is equal to zero, otherwise. Therefore, we have

$$\begin{aligned} nA_{-i} &= k_0(-1)^i + \sum_{v=0}^{M-1} \sum_{u=1}^{M-1} Lu \binom{i}{vL} (-1)^{i-vL} \alpha^{uvL} \\ &= k_0(-1)^i + \sum_{v=0}^{M-1} L \binom{i}{vL} (-1)^{i-vL} \sum_{u=1}^{M-1} u\alpha^{uvL}. \end{aligned}$$

For $v = 0$ in the above summation, we have

$$\sum_{u=1}^{M-1} \frac{un}{M} \binom{i}{0} (-1)^i = \frac{n}{M}(-1)^i \sum_{u=1}^{M-1} u = \frac{n(M-1)}{2}(-1)^i$$

and thus

$$nA_{-i} = k_0(-1)^i + \frac{n(M-1)}{2}(-1)^i + \sum_{v=1}^{M-1} L \binom{i}{vL} (-1)^{i-vL} \sum_{u=1}^{M-1} u\alpha^{uvL}. \tag{2}$$

We can modify the inner sum in the last term of (2) as

$$\sum_{u=1}^{M-1} u\alpha^{uvL} = \frac{1}{1-\alpha^{vL}}\left(\sum_{u=1}^{M-1}\alpha^{uvL} - (M-1)\alpha^{vLM}\right) = \frac{-M}{1-\alpha^{vL}}. \quad (3)$$

Applying (3) to (2) and $n \equiv -1 \bmod p$, we have

$$A_{-i} \equiv \left(\frac{(M-1)}{2} - k_0\right)(-1)^i - (-1)^i \sum_{v=1}^{M-1} \frac{B_v(i)(-1)^{-vL}}{1-\alpha^{vL}} \bmod p$$

where $B_v(i) = \binom{i}{vL}$. □

Let F be number of integers i, $0 \le i < n$, satisfying the relation

$$\frac{(M-1)}{2} - k_0 \equiv \sum_{v=1}^{M-1} \frac{B_v(i)(-1)^{-vL}}{1-\alpha^{vL}} \bmod p, \quad (4)$$

which corresponds to $A_{-i} = 0$ in (1). Then the linear complexity over F_p of the M-ary Sidel'nikov sequences of period n is given as

$$L_M(p) = n - F.$$

In order to compute the linear complexity of M-ary Sidel'nikov sequences, we have to find F in (4). $B_v(i)$ in (4) can be factored into Lucas factors using Lucas' theorem. Since the Lucas factors are integers, they can be represented by the primitive element β of the prime field F_p.

Note that for $0 \le i < vL$, we have $B_v(i) = B_{v+1}(i) = \cdots = B_{M-1}(i) = 0$. Let $b_v = (-1)^{-vL}/(1-\alpha^{vL})$. Note that for $p \equiv 1 \bmod M$, $\alpha^L \in F_p$ because $(\alpha^L)^{p-1} = (\alpha^{\frac{p-1}{M}})^{p^m-1} = 1$. Thus, we also have $b_v \in F_p$.

By dividing the range of i into M subranges, (4) can be separately rewritten as

$$\begin{array}{ll} 0 = \frac{M-1}{2} - k_0, & \text{for } 0 \le i < L \\ b_1B_1(i) = \frac{M-1}{2} - k_0, & \text{for } L \le i < 2L \\ b_1B_1(i) + b_2B_2(i) = \frac{M-1}{2} - k_0, & \text{for } 2L \le i < 3L \\ \vdots & \vdots \\ \sum_{v=1}^{M-1} b_vB_v(i) = \frac{M-1}{2} - k_0, & \text{for } (M-1)L \le i < ML. \end{array} \quad (5)$$

For $1 \le l \le M-1$, let $F_l(c_1, c_2, \cdots, c_l)$ be the number of i, $lL \le i < (l+1)L$, such that $(B_1(i), B_2(i), \cdots, B_l(i)) = (c_1, c_2, \cdots, c_l)$. Then the total number of i satisfying the $(l+1)$-th equation is given as

$$\sum_{b_1c_1+b_2c_2+\cdots+b_lc_l=\frac{M-1}{2}-k_0} F_l(c_1, c_2, \cdots, c_l).$$

Let the number of i satisfying the first equation in (5) be denoted by F_0. If $k_0 \neq (M-1)/2$, the solutions for (4) do not exist in the subrange $0 \leq i < L$. And if $k_0 = (M-1)/2$, all i's, $0 \leq i < L$, satisfy (4). That is, we have

$$F_0 = \begin{cases} L, \text{ if } k_0 = \frac{M-1}{2} \\ 0, \text{ otherwise.} \end{cases}$$

Using the above procedure, we can obtain the number of i satisfying (4) as

$$F = F_0 + \sum_{l=1}^{M-1} \sum_{b_1c_1+b_2c_2+\cdots+b_lc_l=\frac{M-1}{2}-k_0} F_l(c_1, c_2, \cdots, c_l),$$

which corresponds to the number of i, $0 \leq i < n$, such that $A_{-i} = 0$. Thus, we have the following theorem.

Theorem 3. The linear complexity over F_p of the M-ary Sidel'nikov sequences of period $n = p^m - 1$ equals

$$L_M(p) =$$
$$\begin{cases} n - L - \sum\limits_{l=1}^{M-1} \sum\limits_{b_1c_1+b_2c_2+\cdots+b_lc_l=\frac{M-1}{2}-k_0} F_l(c_1, c_2, \cdots, c_l), \text{ for } k_0 = \frac{M-1}{2} \\ n - \sum\limits_{l=1}^{M-1} \sum\limits_{b_1c_1+b_2c_2+\cdots+b_lc_l=\frac{M-1}{2}-k_0} F_l(c_1, c_2, \cdots, c_l), \quad \text{otherwise.} \end{cases}$$

□

In general, it is not easy to find $F_l(c_1, c_2, \cdots, c_l)$ for M-ary Sidel'nikov sequences. In the next section, we will find the linear complexity for $M = 3$ as a special case.

4 Linear Complexity of Ternary Sidel'nikov Sequences

Let β be a primitive element of F_p. For $M = 3$, we have to count the number of nonzero A_{-i}'s, $0 \leq i < n$ in (4). Note that for $M = 3$, we have $(-1)^L = 1$ and $(\alpha^L)^3 = 1$. Thus, we have

$$(\alpha^L + 2)(1 - \alpha^L) = 3. \tag{6}$$

Lemma 1. For $M = 3$, (5) is written as

$$3(1 - k_0) = \big(B_1(i) - B_2(i)\big)\alpha^L + 2B_1(i) + B_2(i). \tag{7}$$

Proof. From Theorem 3, for $M = 3$, we have

$$1 - k_0 = \sum_{v=1}^{2} \frac{B_v(i)}{1 - \alpha^{vL}} = \frac{B_1(i)}{1 - \alpha^L} + \frac{B_2(i)}{1 - \alpha^{2L}} = \frac{(1 - \alpha^{2L})B_1(i) + (1 - \alpha^L)B_2(i)}{1 - \alpha^L - \alpha^{2L} + \alpha^{3L}}.$$

From $1 + \alpha^L + \alpha^{2L} = 0$, (7) is easily derived. □

Now, we are going to derive the linear complexities of the ternary Sidel'nikov sequences of period $n = p^m - 1$ for $p = 3d+1$ and $p = 3d+2$, as in the following two theorems.

Theorem 4. Let $n = p^m - 1$ and $p = 3d + 2$ be a prime, where d is a positive integer. Let $3|n$. Let $\beta^h = 1 - k_0$ for $k_0 \neq 1$. For $0 \leq k \leq p-1$, let $\beta^{f_k} \equiv \binom{k}{d} \bmod p$, $\beta^{g_k} \equiv \binom{k}{2d+1} \bmod p$, and v_k and u_k be the numbers of $\binom{k}{d}$ and $\binom{k}{2d+1}$ among the Lucas factors of $B_v(i)$, respectively. Let $V_1 = \sum_{k=2d+1}^{p-1}(v_k f_k + u_k g_k)$ and $V_2 = \sum_{k=2d+1}^{p-1}(v_k g_k + u_k f_k)$. Then the linear complexity $L_3(p)$ over F_p of ternary Sidel'nikov sequences of period n is given as

$$L_3(p) = \begin{cases} n - \displaystyle\sum_{\substack{V_1 \equiv h \bmod (p-1) \\ V_2 \equiv h \bmod (p-1)}} \Big[(v_{2d+1}, \cdots, v_{p-1})!(u_{2d+1}, \cdots, u_{p-1})!\Big] + \frac{1}{2}(2 - k_0), & \text{for } k_0 \neq 1 \\ (d+1)^m(2^{\frac{m}{2}+1} - 1) - 1, & \text{for } k_0 = 1 \end{cases}$$

where $m = 2\sum_{k=2d+1}^{p-1} v_k$ and $(x_1, x_2, \cdots, x_l)!$ is a multinomial coefficient defined as

$$(x_1 x_2, \cdots, x_l)! = \frac{(x_1 + x_2 + \cdots + x_l)!}{x_1! x_2! \cdots x_l!}.$$

Proof. Since $3|n$, m should be even. From (5), (6), and Lemma 1, we have to consider the following three equations.

$$0 = 3(1 - k_0), \qquad \text{for } 0 \leq i < L \tag{8}$$

$$(\alpha^L + 2)B_1(i) = 3(1 - k_0), \qquad \text{for } L \leq i < 2L \tag{9}$$

$$\big(B_1(i) - B_2(i)\big)\alpha^L + 2B_1(i) + B_2(i) = 3(1 - k_0), \qquad \text{for } 2L \leq i < 3L. \tag{10}$$

We will derive the linear complexity for the following two cases.

Case 1) $k_0 \neq 1$;

In this case, we have to consider the following three subcases.

Case 1-a) For $0 \leq i < L$: Certainly, (8) cannot be satisfied. Thus $A_{-i} \neq 0$, for $0 \leq i < L$.

Case 1-b) For $L \leq i < 2L$: Since $p \equiv 2 \bmod 3$, we have $(\alpha^L)^{p-1} \neq 1$, i.e., $\alpha^L \notin F_p$. Then the right hand side of (9) is an element of F_p while its left hand side is not an element of F_p. It is a contradiction. Therefore, $A_{-i} \neq 0$ for $L \leq i < 2L$.

Case 1-c) $2L \leq i < 3L$: In (10), if $B_1(i) - B_2(i) \neq 0$, $A_{-i} \neq 0$ because $\alpha^L \notin F_p$ and $B_1(i)$ and $B_2(i)$ are elements of F_p. If $B_1(i) - B_2(i) = 0$, we have $B_1(i) = B_2(i) \equiv 1 - k_0 \bmod p$.

In order to apply Lucas' theorem, we need to expand L as

$$L = \frac{p^m - 1}{3} = \frac{p^2 - 1}{3} \sum_{j=0}^{(m-2)/2} p^{2j} = \Big[dp + (2d+1)\Big] \sum_{j=0}^{(m-2)/2} p^{2j}$$
$$= dp^{m-1} + (2d+1)p^{m-2} + dp^{m-3} + (2d+1)p^{m-4} + \cdots + dp + (2d+1).$$

Let $i = \sum_{a=0}^{m-1} i_a p^a$. By Lucas' theorem, we have

$$B_1(i) = \binom{i}{L} \equiv \binom{i_{m-1}}{d}\binom{i_{m-2}}{2d+1}\cdots\binom{i_1}{d}\binom{i_0}{2d+1} \equiv 1 - k_0 \bmod p. \quad (11)$$

Similarly, we can expand $2L$ as

$$2L = (2d+1)p^{m-1} + dp^{m-2} + (2d+1)p^{m-3} + dp^{m-4} + \cdots + (2d+1)p + d$$

and we have

$$B_2(i) = \binom{i}{2L} \equiv \binom{i_{m-1}}{2d+1}\binom{i_{m-2}}{d}\cdots\binom{i_1}{2d+1}\binom{i_0}{d} \equiv 1 - k_0 \bmod p. \quad (12)$$

Since β is a primitive element of F_p, $\beta^h = 1 - k_0$, $\beta^{f_k} \equiv \binom{k}{d} \bmod p$, and $\beta^{g_k} \equiv \binom{k}{2d+1} \bmod p$, (11) and (12) can be rewritten as

$$B_1(i) = \beta^{f_{i_{m-1}} + g_{i_{m-2}} \cdots + f_{i_1} + g_{i_0}} \equiv \beta^h \bmod p \quad (13)$$

$$B_2(i) = \beta^{g_{i_{m-1}} + f_{i_{m-2}} + \cdots + g_{i_1} + f_{i_0}} \equiv \beta^h \bmod p. \quad (14)$$

Since all of the Lucas factors of $B_1(i)$ and $B_2(i)$ are not equal to zero, from (13) and (14), we have

$$V_1 = \sum_{i=2d+1}^{p-1} (v_i f_i + u_i g_i) \equiv h \bmod (p-1) \quad (15)$$

$$V_2 = \sum_{i=2d+1}^{p-1} (v_i g_i + u_i f_i) \equiv h \bmod (p-1). \quad (16)$$

In order to count the number of i satisfying $B_1(i) = B_2(i) \equiv 1 - k_0 \bmod p$, we have to count the number of solutions v_i and u_i, $2d+1 \le i \le p-1$, satisfying (15) and (16). For $k_0 = 0$, we must rule out the case, $i_0 = \cdots = i_{m-1} = p-1$, which corresponds to $i = p^m - 1$. Then we have

$$F_2(1-k_0, 1-k_0) = \sum_{\substack{V_1 \equiv h \bmod (p-1) \\ V_2 \equiv h \bmod (p-1)}} (v_{2d+1}, \cdots, v_{p-1})!(u_{2d+1}, \cdots, u_{p-1})! - \frac{1}{2}(2 - k_0).$$

Since the linear complexity $L_3(p)$ of ternary Sidel'nikov sequences is

$$L_3(p) = n - F_2(1-k_0, 1-k_0),$$

we proved this case.

Case 2) $k_0 = 1$;

From (8), we have $F_0 = L$. And for $L \le i < 2L$, from (9), we know that $A_{-i} = 0$ if and only if $B_1(i) = 0$. For $2L \le i < 3L$, (10) tells us that $A_{-i} = 0$

if and only if $B_1(i) = B_2(i) = 0$. Thus, the linear complexity is equal to the number of i satisfying the following three cases.

i) $B_1(i) \neq 0$ and $B_2(i) \neq 0$
ii) $B_1(i) = 0$ and $B_2(i) \neq 0$
iii) $B_1(i) \neq 0$ and $B_2(i) = 0$.

From (11) and (12), the number of i satisfying i) is the number of i such that all i_a's are greater than or equal to $2d+1$, which is given as $(d+1)^m - 1$. Now, let us count the number of i satisfying ii). From (11) and (12), we have $i_a \geq d$, $0 \leq a < m$, because $B_2(i) \neq 0$. Since $B_1(i) = 0$, at least one Lucas factor $\binom{i_a}{2d+1}$ in $B_1(i)$ is equal to 0, i.e., there is at least one Lucas factor satisfying $d \leq i_a < 2d+1$, which can be counted as

$$\sum_{j=1}^{m/2} \binom{\frac{m}{2}}{j} (d+1)^m = (d+1)^m (2^{\frac{m}{2}} - 1).$$

Clearly, ii) and iii) give us the same values. Thus, for $k_0 = 1$, the linear complexity of ternary Sidel'nikov sequences can be derived as in the theorem. □

Example 1. Let $M = 3$. Let $p = 3d+2 = 5$ and $\beta = 3$, where m is even. For $k_0 = 0$, we have

$$v_3 f_3 + v_4 f_4 + u_3 g_3 + u_4 g_4 \equiv 0 \bmod 4$$
$$v_3 g_3 + v_4 g_4 + u_3 f_3 + u_4 f_4 \equiv 0 \bmod 4.$$

Since $f_3 = 1$, $f_4 = 2$, $g_3 = 0$, and $g_4 = 2$, we have

$$V_1 = v_3 + 2v_4 + 2u_4 \equiv 0 \bmod 4$$
$$V_2 = 2v_4 + u_3 + 2u_4 \equiv 0 \bmod 4.$$

Therefore, v_3 and u_3 are multiples of 4 and $v_4 + u_4$ is a multiple of 2. And $v_3 + v_4 = m/2$ and $u_3 + u_4 = m/2$. Then the linear complexity $L_3(5)$ of ternary Sidel'nikov sequences is written as

$$L_3(5) = p^m - \sum_{\substack{V_1 \equiv 0 \bmod 4 \\ V_2 \equiv 0 \bmod 4}} (v_3, v_4)!(u_3, u_4)! = p^m - \left\{ \sum_{j=0}^{\lfloor \frac{m}{8} \rfloor} \binom{\frac{m}{2}}{4j} \right\}^2.$$

For $k_0 = 1$, from the above theorem, it can be easily derived as

$$L_3(5) = 2^{\frac{3m}{2}+1} - 2^m - 1.$$ □

Now, we will derive the linear complexity of ternary Sidel'nikov sequences for the case of $p \equiv 1 \bmod 3$. The following lemma can be used in the calculation of $B_v(i)$.

Lemma 2. Let M and d be positive integers. For $p = Md + 1$ and $1 \leq j \leq d$, we have

$$\binom{(M-1)d-j}{d-j} \equiv \binom{(M-1)d+j}{d+j} \bmod p.$$

Proof. Since $Md \equiv -1 \bmod p$, we have

$$\frac{d-i}{(M-1)d-i} \equiv \frac{(M-1)d+i+1}{d+i+1} \bmod p.$$

Then the proof is done by noting that

$$\binom{(M-1)d-j}{d-j} = \binom{(M-1)d}{d} \prod_{i=0}^{j-1} \frac{d-i}{(M-1)d-i}$$

and

$$\binom{(M-1)d+j}{d+j} = \binom{(M-1)d}{d} \prod_{i=0}^{j-1} \frac{(M-1)d+i+i}{d+i+1}.$$

□

For $M = 3$, it can be easily modified as

$$\binom{2d-j}{d} \equiv \binom{2d+j}{d} \bmod p. \tag{17}$$

The following lemmas are need to derive the linear complexity for $p = 3d+1$.

Lemma 3. Let $p = 3d + 1$ and $k_0 \neq 1$. Let $\beta^{f'} = (1 - \alpha^L)(1 - k_0)$. For $0 \leq k \leq p - 1$, let $\beta^{f_k} = \binom{k}{d}$ and v_k be the number of $\binom{k}{d}$ among the Lucas factors of $B_1(i)$. Let $\sum_{j=k}^{p-1} v_j = m$ and $V(k) = \sum_{j=k}^{p-1} v_j f_j$. For $L \leq i < 3L$, the number of i satisfying $B_1(i) \equiv (1 - \alpha^L)(1 - k_0) \bmod p$ and $B_2(i) = 0$ in (9) and (10) is given as

$$F_1((1-\alpha^L)(1-k_0)) + F_2((1-\alpha^L)(1-k_0), 0) = E(d) - E(2d)$$

where

$$E(k) = \sum_{V(k) \equiv f' \bmod p-1} (v_k, \cdots, v_{p-1})!.$$

Proof. Clearly, we have

$$L = \frac{p^m - 1}{3} = \left(\frac{p-1}{3}\right) \sum_{i=0}^{m-1} p^i = d \sum_{i-0}^{m-1} p^i. \tag{18}$$

By Lucas' theorem, it can be easily derived that

$$B_1(i) = \prod_{a=0}^{m-1} \binom{i_a}{d} = \beta^{f_{i_0}+f_{i_1}+\cdots+f_{i_{m-1}}} = (1-\alpha^L)(1-k_0) \equiv \beta^{f'} \bmod p.$$

Then we have

$$V(d) = \sum_{j=d}^{p-1} v_j f_j = f'$$

$$E(d) = \sum_{V(d)\equiv f' \bmod (p-1)} (v_d, \cdots, v_{p-1})!.$$

Since $B_2(i) = 0$, we have to rule out the case $B_2(i) \neq 0$ from $E(d)$. $E(2d)$ is the number of i such that all the coefficients i_a, $0 \le a < m$, of its p-ary expansion are in the range $2d \le i_a \le p-1$, which corresponds to $B_2(i) \neq 0$ and $B_1(i) \equiv (1-\alpha^L)(1-k_0) \bmod p$. Thus, we prove it. □

Similarly, we can easily obtain the following lemma.

Lemma 4. Let $p = 3d+1$. Let $\beta^{f'} = B_1(i) \neq 0$ and $\beta^{g'} = B_2(i) \neq 0$. For $0 \le k \le p-1$, let $\beta^{f_k} = \binom{k}{d}$ and $\beta^{g_k} = \binom{k}{2d}$. Let $\sum_{j=2d}^{p-1} v_j = m$, $V_1 = \sum_{j=2d}^{p-1} v_j f_j$, and $V_2 = \sum_{j=2d}^{p-1} v_j g_j$. For $2L \le i < 3L$, the number of i satisfying $(B_1(i), B_2(i)) = (c_1, c_2)$ is given as

$$F_2(c_1, c_2) = \sum_{\substack{V_1\equiv f' \bmod (p-1)\\ V_2\equiv g' \bmod (p-1)}} (v_{2d}, \cdots, v_{p-1})!. \tag{19}$$

□

Since $p = 3d+1$, we have $(\alpha^L)^{p-1} = (\alpha^L)^3 = 1$, $\alpha^L \in F_p$, and $\alpha^{2L} + \alpha^L + 1 = 0$. Let $\gamma = \alpha^L$. Note that $(1-\gamma)(\gamma+2) = 2 - \gamma - \gamma^2 = 3$. Then we can derive the linear complexity of ternary Sidel'nikov sequences for $p \equiv 1 \bmod 3$ as in the following theorem.

Theorem 5. Let $n = p^m - 1$ and $p = 3d+1$ be a prime, where d is a positive integer. Let $3|n$. Then the linear complexity $L_3(p)$ over F_p of ternary Sidel'nikov sequences of period n is given as

$$L_3(p) =$$
$$\begin{cases} n - \sum\limits_{V(d)\equiv f' \bmod (p-1)} (v_d, \cdots, v_{p-1})! + \sum\limits_{V(2d)\equiv f' \bmod (p-1)} (v_{2d}, \cdots, v_{p-1})! \\ \quad - \sum\limits_{\substack{(\gamma+2)c_{21}-(\gamma-1)c_{22}=3(1-k_0)\\ c_{22}\neq 0}} \sum\limits_{\substack{V_1\equiv f' \bmod (p-1)\\ V_2\equiv g' \bmod (p-1)}} (v_{2d}, \cdots, v_{p-1})! - \frac{1}{2}(2-k_0), \\ \hfill \text{if } k_0 \neq 1 \\ (2d+1)^m - 1 - \sum\limits_{\substack{(\gamma+2)c_{21}-(\gamma-1)c_{22}=0\\ c_{21}\neq 0,\ c_{22}\neq 0}} \sum\limits_{\substack{V_1\equiv f' \bmod (p-1)\\ V_2\equiv g' \bmod (p-1)}} (v_{2d}, \cdots, v_{p-1})!, \\ \hfill \text{if } k_0 = 1. \end{cases}$$

Proof. Since γ is an element of F_p with order 3, (7) can be represented as

$$3(1-k_0) = (\gamma+2)B_1(i) - (\gamma-1)B_2(i). \tag{20}$$

Similarly to (8), (9), and (10), we have to consider the following three equations.

$$\begin{aligned} 0 &= 3(1-k_0), && \text{for } 0 \le i < L \quad (21)\\ (\gamma+2)B_1(i) &= 3(1-k_0), && \text{for } L \le i < 2L\\ (\gamma+2)B_1(i) + (1-\gamma)B_2(i) &= 3(1-k_0), && \text{for } 2L \le i < 3L. \quad (22) \end{aligned}$$

Case 1) $k_0 \neq 1$;
Case 1-a) $0 \le i < L$: Clearly, (21) cannot be satisfied. Thus $A_{-i} \neq 0$, for $0 \le i < L$.
Case 1-b) $L \le i < 2L$: We have to count the number of i satisfying $B_1(i) \equiv (1-\gamma)(1-k_0) \bmod p$ and $B_2(i) = 0$, i.e., $F_1((1-\gamma)(1-k_0))$.
Case 1-c) $2L \le i < 3L$: We have to count the number $F_2(c_1, c_2)$ of i satisfying (22) for $(B_1(i), B_2(i)) = (c_1, c_2)$. If $c_2 = 0$, we have $c_1 = (1-\gamma)(1-k_0)$, which corresponds to $F_2((1-\gamma)(1-k_0), 0)$. Lemma 3 gives us the value of $F_1((1-\gamma)(1-k_0)) + F_2((1-\gamma)(1-k_0), 0)$, which includes the cases of **Case 1-b)**. If $c_2 \neq 0$, we also have $c_1 \neq 0$ and Lemma 4 gives us the value of $F_2(c_1, c_2)$.

Here, we need to exclude the case that all i_a's are equal to $p-1$, $0 \le a < m$. When $i_a = p-1$, $0 \le a < m$, from (17), we have $f_{p-1} = 0$, $g_{p-1} = 0$, and $v_{p-1} = m$. Thus, counting of i for $(c_1, c_2) = (1, 1)$ contains the case that all i_a's are equal to $p-1$, $0 \le a < m$, which occurs only when $k_0 = 0$.

Therefore, the linear complexity $L_3(p)$ of ternary Sidel'nikov sequences for $p \equiv 1 \bmod 3$ is given as

$$L_3(p) = n - F_1((1-\gamma)(1-k_0)) - \sum_{(\gamma+2)c_1 - (\gamma-1)c_2 = 3(1-k_0)} F_2(c_1, c_2) + \frac{1}{2}(2-k_0).$$

Using Lemmas 3 and 4, we prove this case.

Case 2) $k_0 = 1$;
Case 2-a) $0 \le i < L$: The number of i satisfying (20) is given as $F_0 = L$.
Case 2-b) $L \le i < 2L$: We need to count the number of i satisfying $B_1(i) = B_2(i) \equiv 0 \bmod p$.
Case 2-c) $2L \le i < 3L$: We need to count the number of i satisfying $(B_1(i), B_2(i)) = (c_1, c_2)$, where $(\gamma+2)c_1 - (\gamma-1)c_2 = 0$. Note that $c_1 = c_2 = 0$ is always a solution of it. From $B_1(i) = 0$ and $B_2(i) = 0$, there is at least one Lucas factor $\binom{i_a}{d}$ and $\binom{i_a}{2d}$ such that $0 \le i_a < d$. It is equivalent to subtract the number of i satisfying $B_1(i) \neq 0$ or $B_2(i) \neq 0$ from $2L$. Thus, we can easily find the value, $F_1(0) + F_2(0,0) = 2L - (2d+1)^m + 1$, where all cases in **Case 2-b)** are included but the case of $i = p^m - 1$ is excluded.

Finally, we have to find $F_2(c_1, c_2)$ for nonzero c_1 and c_2, which is given by Lemma 4. Therefore, for $k_0 = 1$, the linear complexity $L_3(p)$ of ternary Sidel'nikov sequences for $p \equiv 1 \bmod 3$ is given as

$$L_3(p) = n - L - 2L + (2d+1)^m - 1 - \sum_{\substack{(\gamma+2)c_1-(\gamma-1)c_2=0 \\ c_1\neq 0,\ c_2\neq 0}} F_2(c_1, c_2).$$

Using Lemma 4, we prove the theorem. □

For $M = 3$ and $p = 7$, the linear complexity of ternary Sidel'nikov sequences is given in the following example.

Example 2. Let $p = 7$, $\beta = 3$, $\gamma = 2$, and $M = 3$. For $k_0 = 0$, we have

$$f_2 = 0, f_3 = 1, f_4 = 3, f_5 = 1, f_6 = 0,$$
$$g_4 = 0, g_5 = 5, g_6 = 0.$$

Then we have $v_2 + \cdots + v_6 = m$ and

$$E(d) = \sum_{v_3+3v_4+v_5\equiv 3 \bmod 6} (v_2, \cdots, v_6)!.$$

Also we have $v_4 + v_5 + v_6 = m$ and

$$E(2d) = \sum_{3v_4+v_5\equiv 3 \bmod 6} (v_4, v_5, v_6)!.$$

We can calculate the numbers c_1 and c_2 satisfying (20). Finally, we have

$$L_3(7) = p^m - \big(E(d) - E(2d)\big) - \sum_{4c_1-c_2=3} \sum_{\substack{3v_4+v_5\equiv f' \bmod 6 \\ 5v_5\equiv g' \bmod 6}} (v_4, v_5, v_6)!.$$

Table 1. Linear complexity of ternary Sidel'nikov sequences for $p = 7$

m	Period	$k_0 = 0$		$k_0 = 1$		$k_0 = 2$	
	$p^m - 1$	$\gamma = 2$	$\gamma = 4$	$\gamma = 2$	$\gamma = 4$	$\gamma = 2$	$\gamma = 4$
3	342	323	315	118	121	322	314
4	2,400	2,301	2,274	607	612	2,307	2,287
5	16,806	16,300	16,236	3,079	3,083	16,300	16,296
6	117,648	115,088	114,988	15,498	15,498	114,956	115,120
7	823,542	810,620	810,633	77,759	77,746	809,863	810,633
8	5,764,800	5,699,809	5,700,521	389,544	389,503	5,697,118	5,699,176
9	40,353,606	40,027,751	40,030,599	1,949,884	1,949,803	40,020,946	40,023,794

For $k_0 = 1$, from the above theorem, the linear complexity of ternary Sidel'nikov sequence is given as

$$L_3(7) = 5^m - \sum_{4c_1 - c_2 = 0} \sum_{\substack{3v_4 + v_5 \equiv f' \bmod 6 \\ 5v_5 \equiv g' \bmod 6}} (v_4, v_5, v_6)! - 1.$$

Table 1 lists the linear complexities $L_3(7)$ over F_7 of some ternary Sidel'nikov sequences. □

References

1. E. R. Berlekamp, *Algebraic Coding Theory.* Laguna Hills, CA: Aegean Park Press, 1987.
2. R. E. Blahut, "Transform techniques for error control codes," *IBM J. Res. Devel.*, vol. 63, pp. 550–560, 1979.
3. T. Helleseth, S.-H. Kim, and J.-S. No, "Linear complexity over F_p and trace representation of Lempel-Cohn-Eastman sequences," *IEEE Trans. Inf. Theory*, vol. 49, no. 6, pp. 1548–1552, June 2003.
4. T. Helleseth, M. Maas, J. E. Mathiassen, and T. Segers, "Linear complexity over F_p of Sidel'nikov sequences," *IEEE Trans. Inf. Theory*, vol. 50, no. 10, pp. 2468–2472, Oct. 2004.
5. T. Helleseth and K. Yang, "On binary sequences of period $n = p^m - 1$ with optimal autocorrelation," in *Proc. SETA 2001*, 2001, pp. 29–30.
6. Y.-S. Kim, J.-S. Chung, J.-S. No, and H. Chung, "On the linear complexity over F_p of M-ary Sidel'nikov sequences," in *Proc. IEEE ISIT 2005*, Sept. 2005, pp. 2007–2011.
7. G. M. Kyureghyan and A. Pott, "On the linear complexity of the Sidelnikov-Lempel-Cohn-Eastman sequences," *Des., Codes and Cryptogr.*, vol. 29, pp. 149–164, 2003.
8. A. Lempel, M. Cohn, and W. L. Eastman, "A class of balanced binary sequences with optimal autocorrelation properties," *IEEE Trans. Inf. Theory*, vol. IT-23, no. 1, pp. 38–42, Jan. 1977.
9. V. M. Sidel'nikov, "Some k-valued pseudo-random sequences and nearly equidistant codes," *Probl. Inf. Transm.*, vol. 5, no. 1, pp. 12–16, 1969.

Bounds on the Linear Complexity and the 1-Error Linear Complexity over F_p of M-ary Sidel'nikov Sequences*

Jin-Ho Chung and Kyeongcheol Yang

Dept. of Electronics and Electrical Engineering
Pohang University of Science and Technology (POSTECH)
Pohang, Gyungbuk 790-784, Korea
{jinho, kcyang}@postech.ac.kr

Abstract. In this paper we derive a lower bound on the linear complexity and an upper bound on the 1-error linear complexity over F_p of M-ary Sidel'nikov sequences of period $p^m - 1$ when $M \geq 3$ and $p \equiv \pm 1 \mod M$. In particular, we exactly compute the 1-error linear complexity of ternary Sidel'nikov sequences when $p \equiv -1 \mod 3$ and $m \geq 4$. Furthermore, we give a tighter lower bound on the linear complexity of ternary and quaternary Sidel'nikov sequences for $p \equiv -1 \mod M$ by a more detailed analysis. Based on these results, we present the ratios of the linear complexity and the 1-error linear complexity to the period asymptotically.

Keywords: M-ary sequences, Sidel'nikov sequences, linear complexity, 1-error linear complexity.

1 Introduction

Sidel'nikov introduced M-ary sequences of period $p^m - 1$ where p is a prime, m is a positive integer and M is a divisor of $p^m - 1$, and got their out-of-phase autocorrelation properties [11]. Later, Lempel, Cohn, and Eastman proposed binary Sidel'nikov sequences independently and verified that it has optimal autocorrelation properties [10].

Linear complexity and k-error linear complexity are considered as important characteristics of the sequences used in communication systems and cryptography. Helleseth and Yang first studied the linear complexity of binary Sidel'nikov sequences [8]. Helleseth, Kim, and No addressed the linear complexity over F_p of binary Sidel'nikov sequences and derived their linear complexity for $p = 3, 5$, and 7 [6]. Here F_q denotes the finite field of q elements. Later, Helleseth *et al.* derived the closed-form expression of the linear complexity over F_p of binary Sidel'nikov sequences for all prime p [7].

* This work was supported by grant No. R01-2003-000-10330-0 from the Basic Research Program of the Korea Science and Engineering Foundation.

G. Gong et al. (Eds.): SETA 2006, LNCS 4086, pp. 74–87, 2006.

Recently, Kim *et al.* studied the linear complexity over F_p of M-ary Sidel'nikov sequences and derived their Fourier transforms. In particular, they found the closed-form expression of the linear complexity of ternary Sidel'nikov sequences [9].

In this paper we derive a lower bound on the linear complexity over F_p of M-ary Sidel'nikov sequences of period $p^m - 1$ when $p \equiv \pm 1 \bmod M$ and $M \geq 3$. We also derive an upper bound on their 1-error linear complexity over F_p in the same case. Especially for $p \equiv -1 \bmod 3$ and $m \geq 4$, their 1-error linear complexity over F_p is exactly computed. Finally we calculate a tighter lower bound on the linear complexity of ternary and quaternary Sidel'nikov sequences for $p \equiv -1 \bmod M$ by a more detailed analysis.

The outline of this paper is as follows. Section 2 gives some preliminaries for our presentation. In Section 3, a lower bound on the linear complexity and an upper bound on the 1-error linear complexity over F_p of M-ary Sidel'nikov sequences are derived by analyzing their discrete Fourier transforms when $p \equiv \pm 1 \bmod M$ and $M \geq 3$. In Section 4, we exactly compute the 1-error linear complexity of ternary Sidel'nikov sequences using the discrete Fourier transforms of the 1-error allowed sequences when $p \equiv -1 \bmod 3$ and $m \geq 4$. Tighter lower bounds on the linear complexity of ternary and quaternary Sidel'nikov sequences are calculated by a more detailed analysis in Section 5. Finally, we summarize those results and give some concluding remarks in Section 6.

2 Preliminaries

Let $S = \{s(t) | t = 0, 1, \ldots, N-1\}$ be a sequence of period $N = p^m - 1$. The linear complexity $LC(S)$ of the sequence S is defined by

$$LC(S) = N - \deg\left(\gcd(x^N - 1, S(x))\right)$$

where $S(x) = \sum_{t=0}^{N-1} s(t)x^t$. For a positive integer k, the k-sphere complexity $SC_k(S)$ of S is given by

$$SC_k(S) = \min\{LC(S') |\ 0 < d_H(S, S') \leq k\}$$

where S' is a sequence of period N and $d_H(X, Y)$ denotes the Hamming distance between X and Y in one period [4]. Then the k-error linear complexity $LC_k(S)$ of S is equal to the minimum between $LC(S)$ and $SC_k(S)$ [12]. The discrete Fourier transform of S is defined by

$$A_i = \frac{1}{N} \sum_{t=0}^{N-1} s(t)\alpha^{-it},$$

where α is a primitive element of the finite field F_{p^m}. Blahut showed that the linear complexity of a periodic sequence is equal to the Hamming weight of its discrete Fourier transform [1], [2].

The M-ary Sidel'nikov sequences are defined as follows:

Definition 1 ([11]). *Let p be a prime such that $M|p^m - 1$, and α a primitive element of the finite field F_{p^m}. For $r = 0, 1, \cdots, M-1$, let*

$$R_r = \left\{ \alpha^{Ml+r} - 1 \,\middle|\, 0 \le l \le \frac{p^m - 1}{M} - 1 \right\}.$$

An M-ary Sidel'nikov sequence $S_M = \{s_M(t) | t = 0, 1, \ldots, p^m - 2\}$ is defined as

$$s_M(t) = \begin{cases} r, & \alpha^t \in R_r, \\ r_0, & \alpha^t = -1 \end{cases}$$

where $r_0 \in \{0, 1, \ldots, M-1\}$. □

In order that S_M is balanced, that is, each symbol in $\{0, 1, \cdots, M-1\}$ appears equally in one period, r_0 should be zero. In this paper we deal with only the case that $r_0 = 0$ and $M \ge 3$.

3 Linear Complexity and 1-Error Linear Complexity of M-ary Sidel'nikov Sequences

Kim *et al.* derived the discrete Fourier transform of M-ary Sidel'nikov sequences as follows.

Theorem 2 ([9]). *Let $L = \frac{p^m-1}{M}$, $N = p^m - 1$, and $p > M$. For $r_0 = 0$, the Fourier transform of the M-ary Sidel'nikov sequence S_M is given as*

$$NA_{-i} = \frac{N(M-1)}{2}(-1)^i - N(-1)^i \sum_{v=1}^{M-1} \frac{B_v(i)}{1 - \alpha^{vL}} \tag{1}$$

where $B_v(i) = \binom{i}{vL}(-1)^{-vL}$. □

Using this Fourier transform, we can derive lower bounds on the linear complexity and upper bounds on the 1-error linear complexity over F_p of M-ary Sidel'nikov sequences when $M \ge 3$ and $p \equiv \pm 1 \bmod M$.

3.1 The Case of $p \equiv -1 \bmod M$

Let $p = Md - 1$ for some integer $d \ge 2$. In this case, m must be even so that $M|p^m - 1$. When $M = 3$, the Lucas expansions [3] of $\binom{i}{L}$ and $\binom{i}{2L}$ have some special forms [9]. The following lemma is deduced by generalizing these forms.

Lemma 3. *Let p be a prime such that $p = Md - 1$ for some integer $d \ge 2$, and let $L = \frac{p^m-1}{M}$. For $0 \le i \le p^m - 2$ and $1 \le v \le M-1$, we have*

$$\binom{i}{vL} = \binom{i_{m-1}}{vd-1}\binom{i_{m-2}}{(M-v)d-1}\cdots\binom{i_1}{vd-1}\binom{i_0}{(M-v)d-1} \bmod p \tag{2}$$

where $i = \sum_{k=0}^{m-1} i_k p^k$ and $0 \le i_k \le p-1$.

Proof. Because m is even, vL can be expanded as follows:

$$vL = v \cdot \frac{p^2-1}{M}(p^{m-2} + p^{m-4} + \cdots + p^2 + 1).$$

Note that

$$\begin{aligned} v \cdot \frac{p^2-1}{M} &= v \cdot [(d-1)p + (M-1)d - 1] \\ &= (vd-1)p + (M-v)d - 1. \end{aligned}$$

Hence

$$vL = (vd-1)p^{m-1} + \{(M-v)d-1\}p^{m-2} + \cdots + (vd-1)p + \{(M-v)d-1\}.$$

Then the form in (2) is easily verified by Lucas' theorem [3]. □

Note that $\binom{i}{j} = 0 \bmod p$ if and only if at least one of its Lucas factor is zero. Using Theorem 2 and Lemma 3, a lower bound on the linear complexity over F_p of M-ary Sidel'nikov sequences can be derived in the following theorem.

Theorem 4. *When $M \geq 3$, let p be a prime such that $p = Md-1$ for some integer $d \geq 2$. Then the linear complexity $LC(S_M)$ over F_p of the M-ary Sidel'nikov sequence S_M of period $p^m - 1$ satisfies the followings:*
(a) *if M is odd,*

$$LC(S_M) \geq p^m - 1 - \epsilon_1; \tag{3}$$

(b) *if M is even,*

$$LC(S_M) \geq p^m - 1 - \epsilon_1 - \epsilon_2 + \epsilon_3 \tag{4}$$

where

$$\epsilon_1 - d^m \cdot \sum_{j=2}^{M-1} \{(M-j)^{m/2} - (M-j-1)^{m/2}\} \cdot (j-1)^{m/2} - 1,$$

$$\epsilon_2 = d^m \cdot \left\{ \left(\frac{M}{2}\right)^{m/2} - \left(\frac{M}{2}-1\right)^{m/2} \right\}^2,$$

$$\epsilon_3 = 2d^m \cdot \left\{ \left(\frac{M}{2}\right)^{m/2} - \left(\frac{M}{2}-1\right)^{m/2} \right\} \cdot \left\{ \left(\frac{M}{2}-1\right)^{m/2} - \left(\frac{M}{2}-2\right)^{m/2} \right\}.$$

Proof. Firstly, we consider the case that M is odd. Since $L = \frac{p^m-1}{M}$ and $p = Md-1$, we have

$$(\alpha^{vL})^{p-1} = \alpha^{(p^m-1)v \cdot \frac{(Md-2)}{M}}. \tag{5}$$

Thus α^{vL} is an element of F_p if and only if $v \cdot \frac{Md-2}{M}$ is an integer. For odd M, α^{vL} is not an element of F_p for all $1 \leq v \leq M-1$. Let $V_i = \{v | B_v(i) \neq 0,\ 1 \leq v \leq M-1\}$ for $i = 0, 1, \cdots, p^m - 2$. If $V_i = \emptyset$, then

$$A_{-i} = (-1)^i \frac{M-1}{2} \neq 0. \tag{6}$$

If $V_i = \{v_0\}$ for some v_0, $1 \le v_0 \le M-1$, we have

$$A_{-i} = (-1)^i \frac{M-1}{2} - (-1)^i \frac{B_{v_0}(i)}{1-\alpha^{v_0 L}}. \tag{7}$$

In this case A_{-i} is not an element of F_p, and so $A_{-i} \neq 0$. Therefore, if we define ϵ_1 as the number of i's such that $|V_i| \ge 2$, then $LC(S_M)$ is greater than or equal to $p^m - 1 - \epsilon_1$. Define $i_a = \min\{i_{m-1}, i_{m-3}, \dots, i_1\}$ and $i_b = \min\{i_{m-2}, i_{m-4}, \dots, i_0\}$ where $i = \sum_{k=0}^{m-1} i_k p^k$, $0 \le i_k \le p-1$. If we let $\delta_a(j)$ denote the number of choices for $(i_{m-1}, i_{m-3}, \cdots, i_1)$ such that $jd - 1 \le i_a < (j+1)d - 1$, then

$$\begin{aligned}\delta_a(j) &= |\{(i_{m-1}, i_{m-3}, \cdots, i_1) | i_a \ge jd-1\}| \\ &\quad - |\{(i_{m-1}, i_{m-3}, \cdots, i_1) | i_a \ge (j+1)d-1\}| \\ &= \{(M-j)d\}^{m/2} - \{(M-j-1)d\}^{m/2},\end{aligned}$$

for $1 \le j \le M-1$. Let $\delta_b(j)$ for $j \ge 1$ denote the number of choices for $(i_{m-2}, i_{m-4}, \cdots, i_0)$ such that $i_b \ge jd - 1$. It is clear that

$$\delta_b(j) = \{(M-j)d\}^{m/2},$$

for $1 \le j \le M-1$. Except for the case of $i = p^m - 1$, ϵ_1 can be given as

$$\epsilon_1 = \sum_{j=2}^{M-1} \delta_a(j)\delta_b(M-j+1) - 1.$$

Next, we consider the case that M is even. From (5), it is checked that α^{vL} is an element of F_p if and only if $v = \frac{M}{2}$ for $1 \le v \le M-1$. Then we can deduce that $A_{-i} \neq 0$ if $|V_i| \le 1$ and $V_i \neq \{\frac{M}{2}\}$, from (6) and (7). Let ϵ_2 be the number of i's such that $V_i = \{\frac{M}{2}\}$. Then

$$\begin{aligned}\epsilon_2 &= \left|\left\{ i \,\middle|\, \frac{M}{2}d - 1 \le i_a, i_b < \left(\frac{M}{2}+1\right)d - 1 \right\}\right| \\ &= \left[\left\{\left(\frac{M}{2}\right)d\right\}^{m/2} - \left\{\left(\frac{M}{2}-1\right)d\right\}^{m/2}\right]^2.\end{aligned}$$

If $V_i = \{\frac{M}{2} - 1, \frac{M}{2}\}$ or $\{\frac{M}{2}, \frac{M}{2}+1\}$, we have

$$A_{-i} = (-1)^i \frac{M-1}{2} - (-1)^i \left\{ \frac{B_{\frac{M}{2}}(i)}{1-\alpha^{\frac{M}{2}L}} + \frac{B_{v_1}(i)}{1-\alpha^{v_1 L}} \right\},$$

where $v_1 = \frac{M}{2} - 1$ or $\frac{M}{2} + 1$. In this case, $A_{-i} \neq 0$. Let ϵ_3 be the number of i's such that $V_i = \{\frac{M}{2} - 1, \frac{M}{2}\}$ or $\{\frac{M}{2}, \frac{M}{2}+1\}$. Then

$$\begin{aligned}\epsilon_3 = 2 \cdot &\left[\left\{\left(\frac{M}{2}\right)d\right\}^{m/2} - \left\{\left(\frac{M}{2}-1\right)d\right\}^{m/2}\right] \\ &\cdot \left[\left\{\left(\frac{M}{2}-1\right)d\right\}^{m/2} - \left\{\left(\frac{M}{2}-2\right)d\right\}^{m/2}\right].\end{aligned}$$

Consequently, $LC(S_M)$ is greater than or equal to $p^m - 1 - \epsilon_1 - \epsilon_2 + \epsilon_3$ for even M. □

Note that

$$\sum_{j=2}^{M-1} (M-j)^{m/2}(j-1)^{m/2} \le (M-2)\left(\frac{M-1}{2}\right)^m,$$

$$\sum_{j=2}^{M-1} (M-j-1)^{m/2}(j-1)^{m/2} \ge (M-3)\cdot(M-3)^{m/2}.$$

Applying the above inequalities, ϵ_1 is less than or equal to

$$d^m \cdot \left\{(M-2)\left(\frac{M-1}{2}\right)^m - (M-3)^{\frac{m}{2}+1}\right\},$$

which becomes very small compared to $p^m - 1$ when m is large.

Eun *et al.* computed the 1-error linear complexity over F_p of binary Sidel'nikov sequences using the discrete Fourier transforms of the one-error allowed sequences [5]. In a similar way, we can derive an upper bound on the 1-error linear complexity over F_p of M-ary Sidel'nikov sequences.

Theorem 5. *Let $p = Md - 1$ be a prime for some integer $d \ge 2$. Then the 1-error linear complexity $LC_1(S_M)$ over F_p of the M-ary Sidel'nikov sequence S_M of period $p^m - 1$ satisfies*

$$LC_1(S_M) \le d^m \sum_{j=1}^{M-1} \{(M-j)^{m/2} - (M-j-1)^{m/2}\} \cdot j^{m/2} - 1. \tag{8}$$

Proof. For $\lambda \subset F_p$ and $0 \le \tau \le p^m - 2$, we define an error sequence $E(\lambda, \tau)$ with Hamming weight ≤ 1 and period $p^m - 1$ as follows:

$$E(\lambda,\tau) = \{e^{\lambda,\tau}(t) | \, e^{\lambda,\tau}(t) = \lambda I(\alpha^{t-\tau}+1),\ 0 \le t \le p^m - 2\},$$

where $I(0) = 1$ and $I(x) = 0$ for $x \ne 0$. Then a one-error allowed sequence $S_M(\lambda, \tau)$ is defined as

$$S_M(\lambda,\tau) = \{s_M(t) + e^{\lambda,\tau}(t) | 0 \le t \le p^m - 2\}.$$

The discrete Fourier transform $A_{-i}(\lambda, \tau)$ of $S_M(\lambda, \tau)$ is given by

$$A_{-i}(\lambda,\tau) = (-1)^i\left(\frac{M-1}{2} - \lambda\alpha^{\tau i}\right) - (-1)^i \sum_{v=1}^{M-1} \frac{B_v(i)}{1-\alpha^{vL}}. \tag{9}$$

Thus

$$A_{-i}\left(\frac{M-1}{2}, 0\right) = -(-1)^i \sum_{v=1}^{M-1} \frac{B_v(i)}{1-\alpha^{vL}}. \tag{10}$$

A necessary condition for $A_{-i}\left(\frac{M-1}{2},0\right) \neq 0$ is that $B_v(i) \neq 0$ for some v, $1 \le v \le M-1$. If we denote the number of such i's as ϵ, then

$$\epsilon = \sum_{j=1}^{M-1} \delta_a(j)\delta_b(M-j) - 1 = d^m \sum_{j=1}^{M-1} \{(M-j)^{m/2} - (M-j-1)^{m/2}\} \cdot j^{m/2} - 1. \tag{11}$$

Therefore,

$$LC_1(S_M) \le LC\left(S_M\left(\frac{M-1}{2},0\right)\right) \le \epsilon.$$

□

From Theorems 4 and 5, it can be deduced that $LC(S_M)/(p^m-1)$ goes to 1 and $LC_1(S_M)/(p^m-1)$ approaches 0 when m increases.

3.2 The Case of $p \equiv 1 \bmod M$

In this case, m can be any positive integer such that $M|p^m-1$. Note that α^{vL} belongs to F_p for all $1 \le v \le M-1$ because $M|(p-1)$. Let $p = Md+1$ for some positive integer d then

$$vL = vd(p^{m-1} + p^{m-2} + \cdots + 1).$$

Therefore,

$$\binom{i}{vL} = \binom{i_{m-1}}{vd}\binom{i_{m-2}}{vd}\cdots\binom{i_0}{vd} \mod p. \tag{12}$$

Using this expression, the following theorem is easily derived.

Theorem 6. *Let $p = Md+1$ be a prime for some positive integer d. Then the linear complexity $LC(S_M)$ over F_p of the M-ary Sidel'nikov sequence S_M of period p^m-1 satisfies*

$$LC(S_M) \ge p^m - \{(M-1)d+1\}^m.$$

The 1-error linear complexity $LC_1(S_M)$ satisfies

$$LC_1(S_M) \le \{(M-1)d+1\}^m - 1.$$

Proof. For $i = \sum_{k=0}^{m-1} i_k p^k$, let $i_c = \min\{i_{m-1}, i_{m-2}, \ldots, i_0\}$. If $i_c < d$ then $B_v(i) = 0$ for all $1 \le v \le M-1$. In that case, $A_{-i} \neq 0$ from (1) and $A_{-i}\left(\frac{M-1}{2},0\right) = 0$ from (10). Let ϵ' be the number of i's such that $i_c < d$. Then

$$\begin{aligned}\epsilon' &= p^m - 1 - |\{i | i_c \ge d,\ 0 \le i \le p^m-2\}| \\ &= p^m - \{(M-1)d+1\}^m.\end{aligned}$$

It is clear that

$$LC(S_M) \ge \epsilon',$$

and

$$LC_1(S_M) \le LC\left(S_M\left(\frac{M-1}{2},0\right)\right) \le p^m - 1 - \epsilon'.$$

□

When $p \equiv 1 \bmod M$, Theorem 6 tells us that $LC(S_M)/(p^m-1)$ goes to 1 and $LC_1(S_M)/(p^m-1)$ approaches 0 asymptotically like the case of $p \equiv -1 \bmod M$.

4 1-Error Linear Complexity of Ternary Sidel'nikov Sequences

In this section we derive the 1-error linear complexity over F_p of ternary Sidel'nikov sequences when $p \equiv -1 \bmod 3$ and $m \ge 4$. The following result has a form similar to that of the binary case [5].

Theorem 7. *Let $p = 3d-1$ be a prime with some integer $d \ge 2$. Then the 1-error linear complexity $LC_1(S_3)$ over F_p of the ternary Sidel'nikov sequence S_3 of period $p^m - 1$ with $m \ge 4$ is given by*

$$LC_1(S_3) = (2 \cdot 2^{m/2} - 1)d^m - 1.$$

Proof. From (9), the discrete Fourier transform $A_{-i}(\lambda, \tau)$ of $S_3(\lambda, \tau)$ is given by

$$A_{-i}(\lambda,\tau) = (-1)^i(1-\lambda\alpha^{\tau i}) - \frac{(-1)^i}{3}\left\{\left(\binom{i}{L} - \binom{i}{2L}\right)\alpha^L + 2\binom{i}{L} + \binom{i}{2L}\right\}. \tag{13}$$

We consider the next four cases with respect to (λ, τ).

Case i) $(\lambda,\tau) = (1,0)$: Note that $A_{-i}(1,0) = 0$ if and only if $\binom{i}{L} = \binom{i}{2L} = 0$ by (13). Thus $LC(S_3(1,0))$ is equal to the number of i's such that $\binom{i}{L}$ is nonzero or $\binom{i}{2L}$ is nonzero. Considering the Lucas expansion of $\binom{i}{L}$, we have

$$\begin{aligned}\left|\left\{i \,\middle|\, \binom{i}{L} \ne 0,\ 0 \le i \le p^m - 2\right\}\right| &= |\{i | i_a \ge d-1 \text{ and } i_b \ge 2d-1\}| \\ &= (2d)^{m/2} \cdot d^{m/2} - 1.\end{aligned}$$

Similarly,

$$\left|\left\{i \,\middle|\, \binom{i}{2L} \ne 0,\ 0 \le i \le p^m - 2\right\}\right| = 2^{m/2} \cdot d^m - 1$$

and

$$\left|\left\{i \,\middle|\, \binom{i}{L} \ne 0 \text{ and } \binom{i}{2L} \ne 0,\ 0 \le i \le p^m - 2\right\}\right| = d^m - 1.$$

Therefore,

$$LC(S_3(1,0)) = (2^{m/2} \cdot d^m - 1) + (2^{m/2} \cdot d^m - 1) - (d^m - 1) = (2 \cdot 2^{m/2} - 1)d^m - 1.$$

Case ii) $\lambda = 0$: Note that $LC(S_3(0,\tau))$ is equal to $LC(S_3)$, which is greater than $p^m - d^m$ by (3). Since $p^m - d^m \geq (2 \cdot 2^{m/2} - 1)d^m - 1$ for $m \geq 4$, we have

$$LC(S_3(0,\tau)) \geq LC(S_3(1,0))$$

for all $0 \leq \tau \leq p^m - 2,\ m \geq 4$.

Case iii) $\lambda \neq 0,\ \alpha^\tau \in F_p$, and $(\lambda,\tau) \neq (1,0)$: Let C^* be the number of i's such that $A_{-i}(\lambda,\tau) = 0$. In this case, by (13),

$$\begin{aligned} C^* &= \left|\left\{ i \left| \binom{i}{L} = \binom{i}{2L} \text{ and } \binom{i}{L} = 1 - \lambda\alpha^{\tau i} \right.\right\}\right| \\ &\leq \left|\left\{ i \left| \binom{i}{L} = \binom{i}{2L} = 0 \text{ and } \alpha^{\tau i} = \lambda^{-1} \right.\right\}\right| \\ &\qquad + \left|\left\{ i \left| \binom{i}{L} = \binom{i}{2L} \neq 0 \text{ and } \alpha^{\tau i} \neq \lambda^{-1} \right.\right\}\right| \\ &\leq \left|\left\{ i \left| \alpha^{\tau i} = \lambda^{-1} \right.\right\}\right| + \left|\left\{ i \left| \binom{i}{L} \neq 0 \text{ and } \binom{i}{2L} \neq 0 \right.\right\}\right|. \end{aligned}$$

Because $\left|\left\{ i \left| \alpha^{\tau i} = \lambda^{-1} \right.\right\}\right| \leq \frac{p^m - 1}{2}$ and $\left|\left\{ i \left| \binom{i}{L} \neq 0 \text{ and } \binom{i}{2L} \neq 0 \right.\right\}\right| = d^m - 1$, we have

$$LC(S_3(\lambda,\tau)) = p^m - 1 - C^* \geq \frac{p^m - 1}{2} - d^m + 1.$$

Therefore, $LC(S_3(\lambda,\tau))$ is greater than or equal to $LC(S_3(1,0))$ for $m \geq 4$.

Case iv) $\lambda \neq 0,\ \alpha^\tau \notin F_p$: If $\alpha^{\tau i} = \lambda^{-1}$ and $\binom{i}{L} = \binom{i}{2L} = 0$ then $A_{-i}(\lambda,\tau)$ is zero. On the other hand, if $\alpha^{\tau i} \neq \lambda^{-1}$ and $\binom{i}{L} = \binom{i}{2L} = 0$ then $A_{-i}(\lambda,\tau) \neq 0$. Therefore,

$$\begin{aligned} C^* &\leq \left|\left\{ i \left| \alpha^{\tau i} = \lambda^{-1} \text{ and } \binom{i}{L} = \binom{i}{2L} = 0 \right.\right\}\right| \\ &\qquad + \left|\left\{ i \left| \alpha^{\tau i} \neq \lambda^{-1} \text{ and } \left(\binom{i}{L} \neq 0 \text{ or } \binom{i}{2L} \neq 0 \right) \right.\right\}\right| \\ &\leq \left|\left\{ i \left| \alpha^{\tau i} = \lambda^{-1} \right.\right\}\right| + \left|\left\{ i \left| \binom{i}{L} \neq 0 \text{ or } \binom{i}{2L} \neq 0 \right.\right\}\right|. \end{aligned}$$

Note that $|\{i | \alpha^{\tau i} = \lambda^{-1}\}| \leq \frac{p^m - 1}{3}$ since α^τ is not an element of F_p. Furthermore, we know that $|\{i | \binom{i}{L} \neq 0 \text{ or } \binom{i}{2L} \neq 0\}| = (2 \cdot 2^{m/2} - 1)d^m - 1$ from the computation of Case i). Therefore,

$$LC(S_3(\lambda,\tau)) = p^m - 1 - C^* \geq \frac{2}{3}(p^m - 1) - (2 \cdot 2^{m/2} - 1)d^m + 1.$$

For $m \geq 4$, $LC(S_3(\lambda,\tau))$ is equal to or greater than $LC(S_3(1,0))$.

From the results for the above four cases, we have

$$LC(S_3(1,0)) \leq LC(S_3(\lambda,\tau)) \text{ for all } \lambda \in F_p,\ 0 \leq \tau \leq p^m - 2$$

when $m \geq 4$. So the 1-error linear complexity over F_p of S_3 is equal to $LC(S_3(1,0))$ under given conditions. □

5 Linear Complexity of Ternary and Quaternary Sidel'nikov Sequences

In this section a more detailed analysis is applied to computation of lower bounds on the linear complexity of ternary and quaternary Sidel'nikov sequences. The following theorem shows a lower bound on the linear complexity over F_p of ternary Sidel'nikov sequences for $p \equiv -1 \bmod 3$, which is closer to the period than the bound calculated from (3).

Theorem 8. *Let $p = 3d-1$ be a prime with some integer $d \geq 2$. Then the linear complexity $LC(S_3)$ over F_p of the ternary Sidel'nikov sequence S_3 of period p^m-1 satisfies*

$$LC(S_3) \geq p^m - 1 - d^{m-1} \cdot \left\lfloor \frac{d+2}{3} \right\rfloor. \tag{14}$$

Proof. From (1), the equivalent condition for $A_{-i} = 0$ is

$$\binom{i}{L} = \binom{i}{2L} = 1 \bmod p. \tag{15}$$

Let $i = \sum_{k=0}^{m-1} i_k p^k$ where $0 \leq i_k \leq p-1$. From the expression in (2), the number of possible choices for $(i_{m-1}, i_{m-2}, \cdots, i_1)$ satisfying (15) is d^{m-1} because $2d-1 \leq i_k \leq 3d-2$ for all $0 \leq k \leq m-1$. For a fixed $(i_{m-1}, i_{m-2}, \cdots, i_1)$, suppose that (15) is satisfied when $i_0 = x$ for some x, $2d-1 \leq x \leq 3d-2$. Then (15) is not satisfied for $i_0 = x+1$, because $\binom{x+1}{2d-1} \neq \binom{x}{2d-1} \bmod p$. Also, (15) is not satisfied for $i_0 = x+2$ since

$$\binom{x+2}{2d-1} = \frac{(x+2)(x+1)}{(x+3-2d)(x+2-2d)} \binom{x}{2d-1} \neq \binom{x}{2d-1} \bmod p$$

for $2d-1 \leq x \leq 3d-2$. Thus the number of i_0's satisfying (15) is less than or equal to $\lfloor \frac{d-1}{3} \rfloor + 1$. Therefore, the number of i's such that $A_{-i} = 0$ is less than or equal to

$$d^{m-1} \left(\left\lfloor \frac{d-1}{3} \right\rfloor + 1 \right).$$

□

Remark: In Theorem 8, if $\binom{2d-1}{2d-1}, \binom{2d}{2d-1}, \cdots, \binom{3d-2}{2d-1}$ are all distinct over F_p, then $LC(S_3)$ satisfies the following:

Table 1. Lower bounds on the normalized linear complexity and upper bounds on the normalized 1-error linear complexity with respect to the period, when $M = 3, 4$ and $p = 11$

m	$l_3/(p^m-1)$	$h_3/(p^m-1)$	$l_4/(p^m-1)$	$h_4/(p^m-1)$
2	0.9333	0.3917	0.9500	0.4417
4	0.9913	0.1223	0.9816	0.1438
6	0.9988	0.0347	0.9931	0.0420
8	0.9998	0.0095	0.9977	0.0118

$$LC(S_3) \geq p^m - 1 - d^{m-1}.$$

In fact, this bound holds for any prime p such that $p \equiv -1 \bmod 3$ and $p \leq 29$. □

From the similar calculation, a lower bound on the linear complexity of quaternary Sidel'nikov sequences with $p \equiv -1 \bmod 4$ can be derived.

Theorem 9. *Let $p = 4d-1$ be a prime with some integer $d \geq 2$. Then the linear complexity $LC(S_4)$ over F_p of the quaternary Sidel'nikov sequence S_4 of period $p^m - 1$ satisfies*

$$LC(S_4) \geq p^m - 1 - \gamma(d) \cdot d^{m-1}, \tag{16}$$

where

$$\gamma(d) = \left\{ (2^{m-1} - 2^{\frac{m}{2}} - 2^{\frac{m}{2}-1} + 1) \cdot \left\lfloor \frac{2d+2}{3} \right\rfloor + 2^{\frac{m}{2}} \cdot \left\lfloor \frac{d+2}{3} \right\rfloor \right\}.$$

Proof. See Appendix. □

Finally we give an example for the cases $M = 3$ and 4.

Example: Let l_3 and l_4 be the lower bounds on the linear complexity given in (14) and (16), respectively. Also, let h_3 and h_4 be the upper bounds on the 1-error linear complexity obtained by putting $M = 3$ and $M = 4$ to (8), respectively. When $p = 11$, we have

$$\begin{aligned} l_3 &= (11^m - 1) - 2 \cdot 4^{m-1}, \\ l_4 &= (11^m - 1) - 2 \cdot 3^{m-1} \cdot (2^{m-1} - 2^{\frac{m}{2}} + 1), \\ h_3 &= (2^{\frac{m}{2}+1} - 1) \cdot 4^m - 1, \\ h_4 &= (2 \cdot 3^{\frac{m}{2}} - 2^{\frac{m}{2}+1} + 2^m) \cdot 3^m - 1. \end{aligned}$$

Table 1 shows their proportions to the period $11^m - 1$ according to m. □

6 Conclusion

We derived a lower bound on the linear complexity over F_p of M-ary Sidel'nikov sequences of period $p^m - 1$ when $M \geq 3$ and $p \equiv \pm 1 \bmod M$. We also derived upper bound on their 1-error linear complexity in the same cases. For both ternary and quaternary cases, we derived tighter lower bounds on their linear complexity than those calculated by putting $M = 3, 4$ to the general results, respectively. For the case that $p \equiv -1 \bmod 3$ and $m \geq 4$, the exact value of the 1-error linear complexity over F_p of ternary Sidel'nikov sequences was derived. From those results, we can deduce that the linear complexity over F_p of M-ary Sidel'nikov sequences is almost the same as the period and that the ratio of the 1-linear complexity over F_p to the period is almost zero, when the period goes to infinity.

References

1. R. E. Blahut, "Transform techniques for error control codes," *IBM J. Res. Develop.*, vol. 23, pp. 299-315, 1979.
2. R. E. Blahut, *Theory and Practice of Error Control Codes*, Addison-Wesley Publishing Company, 1983.
3. P. J. Cameron, *Combinatorics: topics, techniques, algorithms*, Cambridge University Press, 1994.
4. C. Ding, G. Xiao, and W. Shan, *The Stability Theory of Stream Ciphers*, Lecture Notes in Computer Science, Vol. 561, Springer-Verlag, 1991.
5. Y.-C. Eun, H.-Y. Song, and G. M. Kyureghyan, "1-error linear complexity over F_p of Sidelnikov sequences," *Lecture Notes in Computer Science*, vol 3486, *Sequences and Their Applications (SETA '04)*, Springer, pp.154-165, Mar. 2005.
6. T. Helleseth, S.-H. Kim, and J.-S. No, "Linear complexity over F_p and trace representation of Lempel-Cohn-Eastman sequences," *IEEE Trans. Inform. Theory*, vol. 49, no. 6, pp. 1548-1552, June. 2003.
7. T. Helleseth, M. Maas, J. E. Mathiassen, and T. Segers, "Linear complexity over F_p of Sidel'nikov sequences," *IEEE Trans. Inform. Theory*, vol. 50, no. 10, pp. 2468-2472, Oct. 2004.
8. T. Helleseth and K. Yang, "On binary sequences of period $p^m - 1$ with optimal autocorrelation," *Sequences and Their Applications (SETA '01), Discrete Mathemathics and Theoretical Computer Science*, Springer, pp. 209-217, Aug. 2001.
9. Y.-S. Kim, J.-S. Chung, J.-S. No, and H. Chung, "On the linear complexity over F_p of M-ary Sidel'nikov sequences," in *Proc. 2005 IEEE Inter. Symp. Inform. Theory (ISIT 2005)*, Adelaide, Australia, Sep. 4-9, 2005, pp. 2007-2011.
10. A. Lempel, M. Cohn, and W. L. Eastman, "A class of balnaced binary sequences with optimal autocorrelation properties," *IEEE Trans. Inform. Theory*, vol. 23, no. 1, pp.38-42, Jan. 1977.
11. V. M. Sidelnikov, "Some k-valued pseudo-random sequences and nearly equidistant codes," *Probl. Inf. Transm.*, vol. 5, no. 1, pp. 12-16, 1969.
12. M. Stamp and C. Martin, "An algorithm for the k-error linear complexity of binary sequences with period 2^n," *IEEE Trans. Inform. Theory*, vol. 39, no. 4, pp. 1398-1401, July. 1993.

Appendix

Proof of Theorem 9. The equivalent condition for $A_{-i} = 0$ is that

$$3 = \left(\binom{i}{L} - \binom{i}{3L}\right)\alpha^L + \binom{i}{L} + \binom{i}{2L} + \binom{i}{3L}. \tag{17}$$

Define $\binom{i}{L}^*$, $\binom{i}{2L}^*$ and $\binom{i}{3L}^*$ as follows:

$$\begin{aligned}
\binom{i}{L}^* &= \binom{i_{m-1}}{d-1}\binom{i_{m-2}}{3d-1}\cdots\binom{i_1}{d-1}, \\
\binom{i}{2L}^* &= \binom{i_{m-1}}{2d-1}\binom{i_{m-2}}{2d-1}\cdots\binom{i_1}{2d-1}, \\
\binom{i}{3L}^* &= \binom{i_{m-1}}{3d-1}\binom{i_{m-2}}{d-1}\cdots\binom{i_1}{3d-1} \bmod p,
\end{aligned}$$

where $i = \sum_{k=0}^{m-1} i_k p^k$, $0 \le i_k \le p-1$. The set of i's satisfying (17) can be divided into two classes.

Case i) $\binom{i}{L} = \binom{i}{3L} = 0$ and $\binom{i}{2L}$=3: Because $\binom{i}{2L} \ne 0$, we have $2d-1 \le i_k \le 4d-2$ for all $0 \le k \le m-1$.

Subcase a. $\binom{i}{L}^*$=$\binom{i}{3L}^*$=0: Let θ_a be the number of choices for $(i_{m-1}, i_{m-2}, \cdots, i_1)$ such that $\binom{i}{L}^* = \binom{i}{3L}^* = 0$ and $\binom{i}{2L}^* \ne 0$. Then

$$\begin{aligned}
\theta_a = &\left|\left\{(i_{m-1}, \cdots, i_1) \middle| \binom{i}{2L}^* \ne 0\right\}\right| \\
&- \left|\left\{(i_{m-1}, \cdots, i_1) \middle| \binom{i}{2L}^* \ne 0 \text{ and } \left(\binom{i}{L}^* \ne 0 \text{ or } \binom{i}{3L} \ne 0\right)\right\}\right| \\
= &(2d)^{m-1} - \left\{(2d)^{\frac{m}{2}} \cdot d^{\frac{m}{2}-1} + (2d)^{\frac{m}{2}-1} \cdot d^{\frac{m}{2}} - d^{m-1}\right\} \\
= &(2^{m-1} - 2^{\frac{m}{2}} - 2^{\frac{m}{2}-1} + 1) \cdot d^{m-1}.
\end{aligned}$$

From the proof of Theorem 8, it can be easily verified that the number of i_0's satisfying $\binom{i}{2L} = 3$ for a given $(i_{m-1}, i_{m-2}, \cdots, i_1)$ is less than or equal to $\left(\lfloor \frac{2d-1}{3} \rfloor + 1\right)$.

Subcase b. $\binom{i}{L}^* \ne 0$, and $\binom{i}{3L}^* = 0$: Let θ_b be the number of choices for $(i_{m-1}, i_{m-2}, \cdots, i_1)$ such that $\binom{i}{L}^* \ne 0$, $\binom{i}{2L}^* \ne 0$ and $\binom{i}{3L}^* = 0$. Then

$$\begin{aligned}
\theta_b &= (2d)^{\frac{m}{2}} \cdot d^{\frac{m}{2}-1} - d^{m-1} \\
&= (2^{\frac{m}{2}} - 1) \cdot d^{m-1}.
\end{aligned}$$

Note that $2d-1 \le i_0 \le 3d-2$ since $\binom{i}{L} = 0$ holds. Therefore, the number of i_0's satisfying $\binom{i}{2L} = 3$ for a given $(i_{m-1}, i_{m-2}, \cdots, i_1)$ is less than or equal to $\left(\lfloor \frac{d-1}{3} \rfloor + 1\right)$.

Subcase c. $\binom{i}{3L}^* \neq 0$: In this case, $\binom{i}{3L}$ cannot be zero because $\binom{i_0}{d-1} \neq 0$.

Case ii) $\binom{i}{L} = \binom{i}{3L} \neq 0$ and $\binom{i}{2L} = 3 - 2\binom{i}{L}$: The number of choices for $(i_{m-1}, i_{m-2}, \cdots, i_1)$ such that $\binom{i}{L}^* \neq 0$ and $\binom{i}{3L}^* \neq 0$ is equal to d^{m-1}. Note that $3d - 1 \leq i_0 \leq 4d - 2$ since $\binom{i}{L} \neq 0$ and $\binom{i}{3L} \neq 0$. Hence the number of i_0's satisfying $\binom{i}{L} = \binom{i}{3L} \neq 0$ for a given $(i_{m-1}, i_{m-2}, \cdots, i_1)$ is less than or equal to $\left(\lfloor \frac{d-1}{3} \rfloor + 1\right)$.

From the results in Cases i) and ii), we have

$$LC(S_4) \geq p^m - 1 - \gamma(d) \cdot d^{m-1},$$

where

$$\gamma(d) = \left\{(2^{m-1} - 2^{\frac{m}{2}} - 2^{\frac{m}{2}-1} + 1) \cdot \left(\left\lfloor \frac{2d-1}{3} \right\rfloor + 1\right) + 2^{\frac{m}{2}} \cdot \left(\left\lfloor \frac{d-1}{3} \right\rfloor + 1\right)\right\}.$$

□

The Characterization of 2^n-Periodic Binary Sequences with Fixed 1-Error Linear Complexity

Fang-Wei Fu[1,3], Harald Niederreiter[2], and Ming Su[3]

[1] Temasek Laboratories, National University of Singapore,
5 Sports Drive 2, Singapore 117508, Republic of Singapore
tslfufw@nus.edu.sg

[2] Department of Mathematics, National University of Singapore,
2 Science Drive 2, Singapore 117543, Republic of Singapore
nied@math.nus.edu.sg

[3] Department of Mathematics, Nankai University,
Tianjin 300071, P.R. China
suming@nankai.edu.cn

Abstract. The linear complexity of sequences is one of the important security measures for stream cipher systems. Recently, using fast algorithms for computing the linear complexity and the k-error linear complexity of 2^n-periodic binary sequences, Meidl determined the counting function and expected value for the 1-error linear complexity of 2^n-periodic binary sequences. In this paper, we study the linear complexity and the 1-error linear complexity of 2^n-periodic binary sequences. Some interesting properties of the linear complexity and the 1-error linear complexity of 2^n-periodic binary sequences are obtained. Using these properties, we characterize the 2^n-periodic binary sequences with fixed 1-error linear complexity. Along the way, we obtain a new approach to derive the counting function for the 1-error linear complexity of 2^n-periodic binary sequences. Finally, we give new fast algorithms for computing the 1-error linear complexity and locating the error positions for 2^n-periodic binary sequences.

Keywords: Stream cipher systems, Periodic sequences, Linear complexity, k-Error linear complexity, Counting function, Fast algorithms.

1 Introduction

Let $\mathbf{S} = (s_0, s_1, s_2, \ldots)$ be a binary sequence, that is, a sequence with terms in the binary field $\mathbb{F}_2 = \{0, 1\}$. For a positive integer N, the sequence $\mathbf{S}$ is called *N-periodic* if $s_{i+N} = s_i$ for all $i \geq 0$. The N-periodic binary sequence $\mathbf{S}$ can be completely described by the N-tuple

$$\mathbf{S}^{(N)} = (s_0, s_1, \ldots, s_{N-1}).$$

The polynomial corresponding to the N-periodic sequence $\mathbf{S}$ is defined as

$$\mathbf{S}(x) = s_0 + s_1 x + s_2 x^2 + \cdots + s_{N-1} x^{N-1}.$$

G. Gong et al. (Eds.): SETA 2006, LNCS 4086, pp. 88–103, 2006.

Definition 1. *The linear complexity $L(\mathbf{S})$ of an N-periodic binary sequence $\mathbf{S}$ is the smallest nonnegative integer l for which there exist coefficients $d_1, d_2, \ldots, d_l \in \mathbb{F}_2$ such that*

$$s_j + d_1 s_{j-1} + \cdots + d_l s_{j-l} = 0 \qquad \text{for all } j \geq l.$$

Note that $L(\mathbf{S}) = 0$ if $\mathbf{S}$ is the zero sequence. Obviously, we always have $0 \leq L(\mathbf{S}) \leq N$. Note that if $\mathbf{S}$ is not the zero sequence, then $L(\mathbf{S})$ is the length of the shortest linear feedback shift register that can generate $\mathbf{S}$. For a general introduction to the theory of linear feedback shift register sequences, we refer the reader to [8, Chapter 8] and the references therein.

The following lemma gives a relationship between the linear complexity $L(\mathbf{S})$ of an N-periodic binary sequence $\mathbf{S}$ and its corresponding polynomial $\mathbf{S}(x)$ (see [2, pp. 86–87]).

Lemma 1. *The linear complexity $L(\mathbf{S})$ of the N-periodic binary sequence $\mathbf{S}$ with $\mathbf{S}^{(N)} = (s_0, s_1, \ldots, s_{N-1})$ is given by*

$$L(\mathbf{S}) = N - \deg(\gcd(x^N - 1, \mathbf{S}(x))), \tag{1}$$

where $\mathbf{S}(x) = s_0 + s_1 x + s_2 x^2 + \cdots + s_{N-1} x^{N-1}$ is the corresponding polynomial.

Given two binary vectors $\mathbf{a} = (a_0, a_1, \ldots, a_{m-1})$ and $\mathbf{b} = (b_0, b_1, \ldots, b_{m-1})$ in $\mathbb{F}_2^m$, the *Hamming distance* $d_H(\mathbf{a}, \mathbf{b})$ between $\mathbf{a}$ and $\mathbf{b}$ is the number of coordinates in which they differ. The *Hamming weight* $w_H(\mathbf{b})$ of a vector $\mathbf{b}$ is the number of nonzero coordinates in $\mathbf{b}$. For any two N-periodic binary sequences $\mathbf{S}$ and $\mathbf{T}$, the *Hamming distance* $d_H(\mathbf{S}, \mathbf{T})$ between $\mathbf{S}$ and $\mathbf{T}$ is defined as the Hamming distance between $\mathbf{S}^{(N)}$ and $\mathbf{T}^{(N)}$. The *Hamming weight* $w_H(\mathbf{S})$ of $\mathbf{S}$ is defined as the Hamming weight of $\mathbf{S}^{(N)}$.

The notion of k-error linear complexity $L_k(\mathbf{S})$ of an N-periodic binary sequence $\mathbf{S}$ was introduced in [1], [2], and [23] as follows.

Definition 2. *For an integer $0 \leq k \leq N$, the k-error linear complexity $L_k(\mathbf{S})$ of an N-periodic binary sequence $\mathbf{S}$ is defined as*

$$L_k(\mathbf{S}) = \min_{\mathbf{T}} L(\mathbf{T}), \tag{2}$$

where the minimum is over all N-periodic binary sequences $\mathbf{T}$ with $d_H(\mathbf{T}, \mathbf{S}) \leq k$. In other words,

$$L_k(\mathbf{S}) = \min_{\mathbf{E}} L(\mathbf{S} + \mathbf{E}), \tag{3}$$

where the minimum is over all N-periodic binary sequences $\mathbf{E}$ with $w_H(\mathbf{E}) \leq k$.

The linear complexity and the k-error linear complexity of sequences are important security measures for stream cipher systems. A cryptographically strong sequence should not only have a large linear complexity, but also altering a few terms should not cause a significant decrease of the linear complexity. That is, the k-error linear complexity of the sequence should also be large for certain small values of k.

In this paper, we consider 2^n-periodic binary sequences, i.e., $N = 2^n$. Note that

$$x^{2^n} - 1 = (x-1)^{2^n}. \tag{4}$$

Rueppel [21, Chapter 4] determined the counting function for the linear complexity of 2^n-periodic binary sequences, i.e., the number of 2^n-periodic binary sequences with fixed linear complexity. Using (4) and Lemma 1, it is easy to characterize the 2^n-periodic binary sequences with fixed linear complexity.

Lemma 2. *Let $\mathcal{N}(L)$ denote the number of 2^n-periodic binary sequences with given linear complexity L, $0 \le L \le 2^n$. Then*

$$\mathcal{N}(0) = 1 \text{ and } \mathcal{N}(L) = 2^{L-1} \text{ for } 1 \le L \le 2^n. \tag{5}$$

Let $\mathcal{A}(L)$ denote the set of 2^n-periodic binary sequences with given linear complexity L, $0 \le L \le 2^n$. Then $\mathcal{A}(0) = \{(0,0,0,\ldots)\}$ and $\mathcal{A}(L)$, where $1 \le L \le 2^n$, is equal to the set of 2^n-periodic binary sequences $\mathbf{S}$ with the corresponding polynomials

$$\mathbf{S}(x) = (x-1)^{2^n - L} a(x),$$

where $a(x)$ is a binary polynomial with $\deg(a(x)) \le L-1$ and $a(1) \ne 0$.

The counting function and expected value for the k-error linear complexity of periodic sequences have been studied in [9]–[20]. Recently, using fast algorithms for computing the linear complexity and the k-error linear complexity of 2^n-periodic binary sequences, Meidl [10] determined the counting function and expected value for the 1-error linear complexity of 2^n-periodic binary sequences. In this paper, we study the linear complexity and the 1-error linear complexity of 2^n-periodic binary sequences. Some interesting properties of the linear complexity and the 1-error linear complexity of 2^n-periodic binary sequences are obtained. Using these properties, we characterize the 2^n-periodic binary sequences with fixed 1-error linear complexity. Along the way, we obtain a new approach to derive the counting function for the 1-error linear complexity of 2^n-periodic binary sequences. Finally, we give new fast algorithms for computing the 1-error linear complexity and locating the error positions for 2^n-periodic binary sequences.

2 Some Properties of $\mathcal{A}(L)$

Recall that $\mathcal{A}(L)$ is the set of 2^n-periodic binary sequences with given linear complexity L, $0 \le L \le 2^n$. In this section, we present some interesting properties of $\mathcal{A}(L)$. These properties will be used in the next section to characterize the 2^n-periodic binary sequences with fixed 1-error linear complexity. We start with the following simple lemma.

Lemma 3. *Let $\mathbf{S}$ be a 2^n-periodic binary sequence. Then $L(\mathbf{S}) = 2^n$ if and only if the Hamming weight of $\mathbf{S}$ is odd.*

Proof. This result was mentioned in [10]. For completeness, we give a simple proof here. It follows from (4) and Lemma 1 that

$$L(\mathbf{S}) = 2^n - \deg(\gcd((x-1)^{2^n}, \mathbf{S}(x))).$$

Hence, $L(\mathbf{S}) = 2^n$ if and only if $\mathbf{S}(1) \neq 0$, that is, the Hamming weight of $\mathbf{S}$ is odd. □

Now we consider $\mathcal{A}(L)$ with $1 \leq L < 2^{n-1}$.

Theorem 1. *For any two distinct sequences $\mathbf{S}_1, \mathbf{S}_2 \in \mathcal{A}(L)$, where $1 \leq L < 2^{n-1}$, we have $d_H(\mathbf{S}_1, \mathbf{S}_2) \geq 4$.*

Proof. From Lemma 3 we know that $w_H(\mathbf{S}_1)$ and $w_H(\mathbf{S}_2)$ are even numbers. Hence $d_H(\mathbf{S}_1, \mathbf{S}_2)$ is an even number. If $d_H(\mathbf{S}_1, \mathbf{S}_2) = 2$, we can assume that

$$\mathbf{S}_1(x) = \mathbf{S}_2(x) + x^i + x^j, \quad \text{where } 0 \leq i < j \leq 2^n - 1. \tag{6}$$

Since $L(\mathbf{S}_1) = L(\mathbf{S}_2) = L$, by (4) and Lemma 1 we have

$$\gcd(x^{2^n} - 1, \mathbf{S}_1(x)) = \gcd(x^{2^n} - 1, \mathbf{S}_2(x)) = (x-1)^{2^n - L}. \tag{7}$$

Hence

$$\deg(\gcd(x^{2^n} - 1, \mathbf{S}_1(x) + \mathbf{S}_2(x))) \geq 2^n - L > 2^{n-1}. \tag{8}$$

It follows from (6) that

$$\begin{aligned}\gcd(x^{2^n} - 1, \mathbf{S}_1(x) + \mathbf{S}_2(x)) &= \gcd(x^{2^n} - 1, x^i + x^j) \\ &= \gcd(x^{2^n} - 1, x^{j-i} - 1) = x^{\gcd(2^n, j-i)} - 1.\end{aligned}$$

Therefore

$$\deg(\gcd(x^{2^n} - 1, \mathbf{S}_1(x) + \mathbf{S}_2(x))) = \gcd(2^n, j-i). \tag{9}$$

If $0 < j - i \leq 2^{n-1}$, then $\gcd(2^n, j-i) \leq j - i \leq 2^{n-1}$. If $2^{n-1} < j - i \leq 2^n - 1$, assume that $j - i = 2^{n-1} + l$, $0 < l < 2^{n-1}$, then

$$\gcd(2^n, j-i) = \gcd(2^n, 2^{n-1} + l) = \gcd(2^{n-1}, l) \leq l < 2^{n-1}.$$

Hence by (9) we have

$$\deg(\gcd(x^{2^n} - 1, \mathbf{S}_1(x) + \mathbf{S}_2(x))) \leq 2^{n-1}$$

which contradicts (8). Hence $d_H(\mathbf{S}_1, \mathbf{S}_2) \neq 2$, and so $d_H(\mathbf{S}_1, \mathbf{S}_2) \geq 4$. □

For a 2^n-periodic binary sequence $\mathbf{S}$ and two integers i and j with $0 \leq i, j \leq 2^n - 1$, denote by $\mathbf{S}_{i,j}$ the 2^n-periodic binary sequence with corresponding polynomial

$$\mathbf{S}_{i,j}(x) = \mathbf{S}(x) + x^i + x^j. \tag{10}$$

Next we consider $\mathcal{A}(L)$ with $2^n - 2^{n-r} < L < 2^n - 2^{n-r-1}$, where $1 \leq r \leq n-2$.

Theorem 2. *For any sequence* $\mathbf{S} \in \mathcal{A}(L)$, *where* $2^n - 2^{n-r} < L < 2^n - 2^{n-r-1}$ *for some* $1 \le r \le n-2$, *and for any two integers* $0 \le i \le 2^n - 1$ *and* $1 \le t \le 2^r - 1$, *we have*

$$L(\mathbf{S}_{i,j}) = L(\mathbf{S}) \quad \text{for } j = i \oplus t2^{n-r},$$

where $\oplus$ *is the operation of addition modulo* 2^n. *That is,* $\mathbf{S}_{i,j} \in \mathcal{A}(L)$ *for* $j = i \oplus t2^{n-r}$.

Proof. Since $2^n - 2^{n-r} < L(\mathbf{S}) = L < 2^n - 2^{n-r-1}$, we have $2^{n-r-1} < 2^n - L(\mathbf{S}) < 2^{n-r}$. By Lemma 2, we can write the corresponding polynomial of $\mathbf{S}$ as

$$\mathbf{S}(x) = (x-1)^{2^n - L} a(x), \tag{11}$$

where $a(x)$ is a binary polynomial with $\deg(a(x)) \le L - 1$ and $a(1) \ne 0$. Note that

$$x^i + x^{i+t2^{n-r}} = x^i (x^t - 1)^{2^{n-r}} = (x-1)^{2^{n-r}} (1 + \cdots + x^{t-1})^{2^{n-r}} x^i.$$

Therefore

$$\begin{aligned} \deg(\gcd(x^{2^n} - 1, x^i + x^{i+t2^{n-r}})) &= \deg(\gcd((x-1)^{2^n}, x^i + x^{i+t2^{n-r}})) \\ &\ge 2^{n-r} > 2^n - L. \end{aligned} \tag{12}$$

By (11) and (12), we have

$$\deg(\gcd(x^{2^n} - 1, \mathbf{S}(x) + x^i + x^{i+t2^{n-r}})) = 2^n - L. \tag{13}$$

It is easy to see that

$$\gcd(x^{2^n} - 1, \mathbf{S}(x) + x^i + x^{i \oplus t2^{n-r}}) = \gcd(x^{2^n} - 1, \mathbf{S}(x) + x^i + x^{i+t2^{n-r}}). \tag{14}$$

Hence, the theorem follows from Lemma 1 and (13) and (14). □

Theorem 2 tells us that if $2^n - 2^{n-r} < L < 2^n - 2^{n-r-1}$, where $1 \le r \le n-2$, then for any sequence $\mathbf{S} \in \mathcal{A}(L)$ and any integer $0 \le i \le 2^n - 1$, there exist at least $2^r - 1$ sequences among $\mathbf{S}_{i,j}$, where $0 \le j \le 2^n - 1$ and $j \ne i$, such that $\mathbf{S}_{i,j} \in \mathcal{A}(L)$. The next theorem shows that there are exactly $2^r - 1$ such sequences $\mathbf{S}_{i,j} \in \mathcal{A}(L)$, where $j \ne i$.

Theorem 3. *For any sequence* $\mathbf{S} \in \mathcal{A}(L)$, *where* $2^n - 2^{n-r} < L < 2^n - 2^{n-r-1}$ *for some* $1 \le r \le n-2$, *and for any integer* $0 \le i \le 2^n - 1$, *the number of the sequences* $\mathbf{S}_{i,j} \in \mathcal{A}(L)$, *where* $0 \le j \le 2^n - 1$ *and* $j \ne i$, *is exactly* $2^r - 1$. *That is,*

$$\{j : \mathbf{S}_{i,j} \in \mathcal{A}(L), 0 \le j \le 2^n - 1, j \ne i\} = \{j : j = i \oplus t2^{n-r}, 1 \le t \le 2^r - 1\},$$

where $\mathbf{S}_{i,j}$ *is defined by* (10).

Proof. Assume that there is one more $j = j_1$ except $j = i \oplus t2^{n-r}$, $t = 0, 1, \ldots, 2^r - 1$, such that $\mathbf{S}_{i,j_1} \in \mathcal{A}(L)$. By considering the distribution of the 2^r numbers $j = i \oplus t2^{n-r}$, $t = 0, 1, \ldots, 2^r - 1$, over the interval $[0, 2^n - 1]$, we can see that there exists j_2 among these 2^r numbers such that $0 < |j_2 - j_1| \leq 2^{n-r-1}$. By Theorem 2, $\mathbf{S}_{i,j_2} \in \mathcal{A}(L)$. Without loss of generality, we assume that $j_2 > j_1$. Since $\mathbf{S}_{i,j_1}$, $\mathbf{S}_{i,j_2} \in \mathcal{A}(L)$, we obtain by Lemma 2,

$$\mathbf{S}_{i,j_1}(x) = (x-1)^{2^n - L} a_1(x), \quad \mathbf{S}_{i,j_2}(x) = (x-1)^{2^n - L} a_2(x),$$

where $a_1(x)$ and $a_2(x)$ are two distinct binary polynomials with $\deg(a_1(x)) \leq L-1$, $\deg(a_2(x)) \leq L-1$, and $a_1(1) \neq 0$, $a_2(1) \neq 0$. It follows from (10) that $\mathbf{S}_{i,j_1}(x) + \mathbf{S}_{i,j_2}(x) = x^{j_1} + x^{j_2}$. Then

$$\deg(\gcd((x-1)^{2^n}, x^{j_1} + x^{j_2})) \geq 2^n - L > 2^{n-r-1}. \tag{15}$$

On the other hand,

$$\begin{aligned}\deg(\gcd((x-1)^{2^n}, x^{j_1} + x^{j_2})) &= \deg(\gcd((x-1)^{2^n}, 1 + x^{j_2 - j_1})) \\ &\leq j_2 - j_1 \leq 2^{n-r-1},\end{aligned}$$

which contradicts (15). Hence, there is no more j except $j = i \oplus t2^{n-r}$, $t = 0, 1, \ldots, 2^r - 1$, such that $\mathbf{S}_{i,j} \in \mathcal{A}(L)$. This completes the proof. □

3 The Characterization

In this section, we use Theorems 1, 2, and 3 to characterize the 2^n-periodic binary sequences with fixed 1-error linear complexity.

Denote by $merr(\mathbf{S})$ the minimum value k for which the k-error linear complexity of a nonzero 2^n-periodic binary sequence $\mathbf{S}$ is strictly less than the linear complexity of $\mathbf{S}$, that is,

$$merr(\mathbf{S}) = \min\{k : L_k(\mathbf{S}) < L(\mathbf{S})\}. \tag{16}$$

Kurosawa *et al.* [6] determined the exact value of $merr(\mathbf{S})$ as follows.

Lemma 4. *Let $\mathbf{S}$ be a nonzero 2^n-periodic binary sequence. Then*

$$merr(\mathbf{S}) = 2^{w_H(2^n - L(\mathbf{S}))}, \tag{17}$$

where $w_H(j)$, $0 \leq j \leq 2^n - 1$, is the Hamming weight of the binary representation of j.

Next we need the following lemma to establish the theorems in this section.

Lemma 5. *Let $\mathbf{S}$ be a 2^n-periodic binary sequence. If $w_H(\mathbf{S})$ is even, then $L_1(\mathbf{S}) = L(\mathbf{S})$. If $w_H(\mathbf{S})$ is odd, then $L_1(\mathbf{S}) < L(\mathbf{S}) = 2^n$.*

Proof. If $w_H(\mathbf{S})$ is even, then $L(\mathbf{S}) < 2^n$ by Lemma 3. Thus, $w_H(2^n - L(\mathbf{S})) \geq 1$. Hence by Lemma 4 we get $merr(\mathbf{S}) \geq 2$. Therefore $L_1(\mathbf{S}) = L(\mathbf{S})$.

If $w_H(\mathbf{S})$ is odd, then $L(\mathbf{S}) = 2^n$ by Lemma 3. Thus, $w_H(2^n - L(\mathbf{S})) = 0$. Hence by Lemma 4 we get $merr(\mathbf{S}) = 1$. Therefore $L_1(\mathbf{S}) < L(\mathbf{S}) = 2^n$. □

Theorem 4. *Let* $\mathbf{S}$ *be a* 2^n*-periodic binary sequence. If* $L(\mathbf{S}) \neq 2^n$, $2^n - 2^{n-r}$, $r = 1, 2, \ldots, n$, *then* $L_1(\mathbf{S}+\mathbf{E}) = L_1(\mathbf{S}) = L(\mathbf{S})$, *where* $\mathbf{E}$ *is any* 2^n*-periodic binary sequence with* $w_H(\mathbf{E}) = 1$.

Proof. It is easy to see that the result is true if $\mathbf{S}$ is the zero sequence. Next we consider the case where $\mathbf{S}$ is a nonzero sequence. Since $2^n - L(\mathbf{S}) \neq 0, 2^j$, $j = 0, 1, \ldots, n-1$, we have $w_H(2^n - L(\mathbf{S})) \neq 0, 1$. Hence by Lemma 4 we get $merr(\mathbf{S}) \geq 4$. Therefore

$$L_2(\mathbf{S}) = L_1(\mathbf{S}) = L(\mathbf{S}). \tag{18}$$

From the definitions of $L_1(\mathbf{S}+\mathbf{E})$ and $L_2(\mathbf{S})$, we obtain

$$L_1(\mathbf{S}+\mathbf{E}) \geq L_2(\mathbf{S}) = L_1(\mathbf{S}). \tag{19}$$

On the other hand, by the definition of $L_1(\mathbf{S}+\mathbf{E})$ and noting that $\mathbf{S} = \mathbf{S}+\mathbf{E}+\mathbf{E}$, we have

$$L_1(\mathbf{S}+\mathbf{E}) \leq L(\mathbf{S}) = L_1(\mathbf{S}). \tag{20}$$

Combining (19) and (20), we obtain the theorem. □

Denote by $\mathcal{A}_1(L)$ the set of 2^n-periodic binary sequences with given 1-error linear complexity L, $0 \leq L \leq 2^n$. Let $\mathcal{N}_1(L)$ be the number of 2^n-periodic binary sequences with given 1-error linear complexity L, $0 \leq L \leq 2^n$. Denote by $\mathbf{0}$ the zero sequence. For $0 \leq i \leq 2^n - 1$, denote by $\mathbf{E}_i$ the 2^n-periodic binary sequence of weight one with 1 at the position with subscript i in the first period. Note that $\mathbf{E}_i^{(2^n)}$, $i = 0, 1, \ldots, 2^n - 1$, form a standard basis for the binary vector space $\mathbb{F}_2^{2^n}$. It is easy to see that

$$\mathcal{A}_1(0) = \{\mathbf{0}, \mathbf{E}_0, \mathbf{E}_1, \ldots, \mathbf{E}_{2^n-1}\}, \quad \mathcal{N}_1(0) = 2^n + 1. \tag{21}$$

We infer from Lemmas 3 and 5 that

$$\mathcal{A}_1(2^n) = \emptyset, \quad \mathcal{N}_1(2^n) = 0. \tag{22}$$

Theorem 5. *If* $L \neq 0$, 2^n, $2^n - 2^{n-r}$, $r = 1, 2, \ldots, n$, *then*

$$\mathcal{A}_1(L) = \mathcal{A}(L) \cup (\mathbf{E}_0 + \mathcal{A}(L)) \cup (\mathbf{E}_1 + \mathcal{A}(L)) \cup \cdots \cup (\mathbf{E}_{2^n-1} + \mathcal{A}(L)). \tag{23}$$

Furthermore: (i) *if* $1 \leq L < 2^{n-1}$, *then the sets* $\mathcal{A}(L)$, $\mathbf{E}_0 + \mathcal{A}(L)$, $\mathbf{E}_1 + \mathcal{A}(L)$, $\ldots$, $\mathbf{E}_{2^n-1} + \mathcal{A}(L)$ *are disjoint and*

$$\mathcal{N}_1(L) = (1 + 2^n)2^{L-1}; \tag{24}$$

(ii) *if* $2^n - 2^{n-r} < L < 2^n - 2^{n-r-1}$ *for some* $1 \leq r \leq n-2$, *then for any* $0 \leq i \leq 2^n - 1$ *we have*

$$\mathbf{E}_i + \mathcal{A}(L) = \mathbf{E}_{i \oplus t2^{n-r}} + \mathcal{A}(L) \ \textit{for all } t = 0, 1, \ldots, 2^r - 1; \tag{25}$$

moreover, the $2^{n-r}+1$ *sets* $\mathcal{A}(L)$, $\mathbf{E}_0+\mathcal{A}(L)$, $\mathbf{E}_1+\mathcal{A}(L)$, …, $\mathbf{E}_{2^{n-r}-1}+\mathcal{A}(L)$ *are disjoint,*

$$\mathcal{A}_1(L)=\mathcal{A}(L)\cup(\mathbf{E}_0+\mathcal{A}(L))\cup(\mathbf{E}_1+\mathcal{A}(L))\cup\cdots\cup(\mathbf{E}_{2^{n-r}-1}+\mathcal{A}(L)), \quad (26)$$

and

$$\mathcal{N}_1(L)=(1+2^{n-r})2^{L-1}. \quad (27)$$

Proof. For any $\mathbf{S}\in\mathcal{A}(L)$, since $L(\mathbf{S})=L\neq 2^n$, it follows from Lemma 3 that $w_H(\mathbf{S})$ is even. Hence, for any $\mathbf{T}\in\mathbf{E}_i+\mathcal{A}(L)$, $w_H(\mathbf{T})$ is odd. Therefore

$$\mathcal{A}(L)\cap(\mathbf{E}_i+\mathcal{A}(L))=\emptyset,\quad i=0,1,\ldots,2^n-1. \quad (28)$$

Moreover, by Lemma 5 and Theorem 4, we have $\mathbf{S},\mathbf{T}\in\mathcal{A}_1(L)$. Therefore

$$\mathcal{A}_1(L)\supseteq\mathcal{A}(L)\cup(\mathbf{E}_0+\mathcal{A}(L))\cup(\mathbf{E}_1+\mathcal{A}(L))\cup\cdots\cup(\mathbf{E}_{2^n-1}+\mathcal{A}(L)). \quad (29)$$

On the other hand, by the definition of the 1-error linear complexity (see Definition 2), we have

$$\mathcal{A}_1(L)\subseteq\mathcal{A}(L)\cup(\mathbf{E}_0+\mathcal{A}(L))\cup(\mathbf{E}_1+\mathcal{A}(L))\cup\cdots\cup(\mathbf{E}_{2^n-1}+\mathcal{A}(L)). \quad (30)$$

Hence, (23) follows from (29) and (30).
(i) If $1\le L<2^{n-1}$, then by Theorem 1 the sets $\mathcal{A}(L)$, $\mathbf{E}_0+\mathcal{A}(L)$, …, $\mathbf{E}_{2^n-1}+\mathcal{A}(L)$ are disjoint. Hence by (23) and Lemma 2,

$$\mathcal{N}_1(L)=|\mathcal{A}_1(L)|=(1+2^n)|\mathcal{A}(L)|=(1+2^n)\mathcal{N}(L)=(1+2^n)2^{L-1}.$$

(ii) If $2^n-2^{n-r}<L<2^n-2^{n-r-1}$ for some $1\le r\le n-2$, then by Theorem 3, for any sequence $\mathbf{S}\in\mathcal{A}(L)$, the sequence $\mathbf{E}_i+\mathbf{S}$ appears in exactly 2^r sets $\mathbf{E}_{i\oplus t2^{n-r}}+\mathcal{A}(L)$, $t=0,1,\ldots,2^r-1$. This implies that

$$(\mathbf{E}_i+\mathcal{A}(L))\cap(\mathbf{E}_j+\mathcal{A}(L))=\emptyset\quad\text{for any } 0\le i<j\le 2^{n-r}-1 \quad (31)$$

and

$$\mathbf{E}_i+\mathcal{A}(L)=\mathbf{E}_{i\oplus t2^{n-r}}+\mathcal{A}(L),\quad t=0,1,\ldots,2^r-1, \quad (32)$$

since $|\mathbf{E}_i+\mathcal{A}(L)|=|\mathbf{E}_{i\oplus t2^{n-r}}+\mathcal{A}(L)|=|\mathcal{A}(L)|=\mathcal{N}(L)$. Hence, by (23), (28), (31), and (32), the $2^{n-r}+1$ sets $\mathcal{A}(L)$, $\mathbf{E}_0+\mathcal{A}(L)$, $\mathbf{E}_1+\mathcal{A}(L)$, …, $\mathbf{E}_{2^{n-r}-1}+\mathcal{A}(L)$ are disjoint and

$$\mathcal{A}_1(L)=\mathcal{A}(L)\cup(\mathbf{E}_0+\mathcal{A}(L))\cup(\mathbf{E}_1+\mathcal{A}(L))\cup\cdots\cup(\mathbf{E}_{2^{n-r}-1}+\mathcal{A}(L)).$$

This implies that

$$\mathcal{N}_1(L)=\mathcal{N}(L)+2^{n-r}\mathcal{N}(L)=(1+2^{n-r})2^{L-1}.$$

This completes the proof. □

Remark 1: From Lemma 2 and Theorem 5, we deduce that:
(i) if $1 \leq L < 2^{n-1}$, then $\mathcal{A}_1(L)$ is the set of 2^n-periodic binary sequences $\mathbf{S}$ with the corresponding polynomials

$$\mathbf{S}(x) = (x-1)^{2^n - L} a(x) \quad \text{or} \quad x^i + (x-1)^{2^n - L} a(x), \quad 0 \leq i \leq 2^n - 1,$$

where $a(x)$ is a binary polynomial with $\deg(a(x)) \leq L-1$ and $a(1) \neq 0$;
(ii) if $2^n - 2^{n-r} < L < 2^n - 2^{n-r-1}$ for some $1 \leq r \leq n-2$, then $\mathcal{A}_1(L)$ is the set of 2^n-periodic binary sequences $\mathbf{S}$ with the corresponding polynomials

$$\mathbf{S}(x) = (x-1)^{2^n - L} a(x) \quad \text{or} \quad x^i + (x-1)^{2^n - L} a(x), \quad 0 \leq i \leq 2^{n-r} - 1,$$

where $a(x)$ is a binary polynomial with $\deg(a(x)) \leq L-1$ and $a(1) \neq 0$.

The remaining values of L are covered by the following theorem.

Theorem 6. *If* $L = 2^n - 2^{n-r}$, $r = 1, 2, \ldots, n$, *then*

$$\mathcal{A}_1(L) = \mathcal{A}(L), \qquad \mathcal{N}_1(L) = 2^{L-1}. \tag{33}$$

Proof. For any sequence $\mathbf{S} \in \mathcal{A}(L)$, we have $L(\mathbf{S}) = L \neq 2^n$. Hence by Lemma 3, $w_H(\mathbf{S})$ is even. Therefore by Lemma 5, $L_1(\mathbf{S}) = L(\mathbf{S}) = L$. That is, $\mathbf{S} \in \mathcal{A}_1(L)$. Hence,

$$\mathcal{A}_1(L) \supseteq \mathcal{A}(L), \qquad \mathcal{N}_1(L) \geq \mathcal{N}(L) = 2^{L-1}. \tag{34}$$

Assume that

$$\mathcal{N}_1(L) = \mathcal{N}(L) + \beta(L) = 2^{L-1} + \beta(L) \quad \text{for } L = 2^n - 2^{n-r}, r = 1, \ldots, n. \tag{35}$$

By (34), $\beta(L) \geq 0$ for all those L. Now we prove that $\beta(L) = 0$ for all those L. This implies that (33) holds. Note that

$$\begin{aligned} 2^{2^n} &= \sum_{L=0}^{2^n} \mathcal{N}_1(L) \\ &= \sum_{L=0}^{2^{n-1}-1} \mathcal{N}_1(L) + \sum_{r=1}^{n-2} \sum_{L=2^n-2^{n-r}+1}^{2^n-2^{n-r-1}-1} \mathcal{N}_1(L) \\ &\quad + \sum_{r=1}^{n} \mathcal{N}_1(2^n - 2^{n-r}) + \mathcal{N}_1(2^n). \end{aligned} \tag{36}$$

By (21), (22), and (24),

$$\begin{aligned} \sum_{L=0}^{2^{n-1}-1} \mathcal{N}_1(L) + \mathcal{N}_1(2^n) &= (2^n+1)\left(1 + \sum_{L=1}^{2^{n-1}-1} 2^{L-1}\right) \\ &= 2^{2^n}(2^{-2^{n-1}+n-1} + 2^{-2^{n-1}-1}). \end{aligned} \tag{37}$$

It follows from (27) that

$$\sum_{r=1}^{n-2}\sum_{L=2^n-2^{n-r}+1}^{2^n-2^{n-r-1}-1}\mathcal{N}_1(L)=\sum_{r=1}^{n-2}\sum_{L=2^n-2^{n-r}+1}^{2^n-2^{n-r-1}-1}(2^{n-r}+1)2^{L-1}$$
$$=\sum_{r=1}^{n-2}(2^{n-r}+1)2^{2^n-2^{n-r}}(2^{2^{n-r-1}-1}-1)$$
$$=2^{2^n}\sum_{i=2}^{n-1}(2^i+1)(2^{-2^{i-1}-1}-2^{-2^i})$$
$$=2^{2^n}\left[\frac{3}{4}-2^{-2^{n-1}+n-1}-2^{-2^{n-1}-1}-\frac{1}{2}\sum_{i=1}^{n-1}2^{-2^i}\right]. \tag{38}$$

By (35) we have

$$\sum_{r=1}^{n}\mathcal{N}_1(2^n-2^{n-r})=\sum_{r=1}^{n}2^{2^n-2^{n-r}-1}+\sum_{r=1}^{n}\beta(2^n-2^{n-r})$$
$$=2^{2^n}\left[\frac{1}{4}+\frac{1}{2}\sum_{i=1}^{n-1}2^{-2^i}\right]+\sum_{r=1}^{n}\beta(2^n-2^{n-r}). \tag{39}$$

From (36)–(39) we obtain

$$2^{2^n}=2^{2^n}+\sum_{r=1}^{n}\beta(2^n-2^{n-r}),$$

which implies that $\beta(L)=0$ for $L=2^n-2^{n-r}$, $r=1,2,\ldots,n$. □

Remark 2: From Lemma 2 and Theorem 6, we infer that if $L=2^n-2^{n-r}$, $r=1,2,\ldots,n$, then $\mathcal{A}_1(L)$ is the set of 2^n-periodic binary sequences $\mathbf{S}$ with the corresponding polynomials

$$\mathbf{S}(x)=(x-1)^{2^n-L}a(x),$$

where $a(x)$ is a binary polynomial with $\deg(a(x))\le L-1$ and $a(1)\ne 0$.

Remark 3: Using fast algorithms for computing the linear complexity and the k-error linear complexity of 2^n-periodic binary sequences, Meidl [10, Theorem 1] determined the number of 2^n-periodic binary sequences with linear complexity 2^n and given 1-error linear complexity and implicitly obtained $\mathcal{N}_1(L)$ in the proof of [10, Corollary 1]. In this section, we used algebraic and combinatorial methods to characterize the sets $\mathcal{A}_1(L)$ of the 2^n-periodic binary sequences with fixed 1-error linear complexity L. Using the characterizations of $\mathcal{A}_1(L)$, we explicitly determined the counting function $\mathcal{N}_1(L)$ for the 1-error linear complexity of 2^n-periodic binary sequences.

4 New Algorithms

In this section, we present new fast algorithms for computing the 1-error linear complexity and locating the error positions for 2^n-periodic binary sequences.

Games and Chan [3] designed a fast algorithm for computing the linear complexity of a 2^n-periodic binary sequence. The Games-Chan algorithm can be described as a recursive computation method as follows.

Games-Chan Algorithm: Let $\mathbf{S}$ be a 2^n-periodic binary sequence with

$$\mathbf{S}^{(2^n)} = (s_0, s_1, \ldots, s_{2^n-1}).$$

Decompose $\mathbf{S}^{(2^n)}$ into its left and right half by

$$\mathbf{S}_L^{(2^{n-1})} = (s_0, s_1, \ldots, s_{2^{n-1}-1}), \quad \mathbf{S}_R^{(2^{n-1})} = (s_{2^{n-1}}, s_{2^{n-1}+1}, \ldots, s_{2^n-1}).$$

Denote by $\mathbf{S}_L$ and $\mathbf{S}_R$ the two 2^{n-1}-periodic binary sequences with the corresponding vectors $\mathbf{S}_L^{(2^{n-1})}$ and $\mathbf{S}_R^{(2^{n-1})}$, respectively.

(i) If $\mathbf{S}_L^{(2^{n-1})} = \mathbf{S}_R^{(2^{n-1})}$, then $L(\mathbf{S}) = L(\mathbf{S}_L)$.
(ii) If $\mathbf{S}_L^{(2^{n-1})} \neq \mathbf{S}_R^{(2^{n-1})}$, then $L(\mathbf{S}) = 2^{n-1} + L(\mathbf{S}_L + \mathbf{S}_R)$.
(iii) Apply the above procedure recursively to the 2^{n-1}-periodic binary sequence $\mathbf{S}_L$ in (i), or the 2^{n-1}-periodic binary sequence $\mathbf{S}_L + \mathbf{S}_R$ in (ii).

There are several algorithms for computing the k-error linear complexity (for all k) of a 2^n-periodic binary sequence, such as the Stamp-Martin algorithm [23], the Kaida-Uehara-Imamura algorithm [4], the Lauder-Paterson algorithm [7], and the Sălăgean algorithm [22]. In these algorithms, some cost functions are introduced and need to be computed. Below we give a fast algorithm for computing the 1-error linear complexity of a 2^n-periodic binary sequence. In our algorithm, we compute the 1-error linear complexity recursively or reduce the problem to that of computing the linear complexity of certain sequences using the Games-Chan algorithm. Moreover, we do not need to compute a cost function.

Algorithm 1: Let $\mathbf{S}$ be a 2^n-periodic binary sequence. Let $\mathbf{S}^{(2^n)}$, $\mathbf{S}_L^{(2^{n-1})}$, $\mathbf{S}_R^{(2^{n-1})}$, $\mathbf{S}_L$, and $\mathbf{S}_R$ be the same as in the Games-Chan algorithm.

(i) If $w_H(\mathbf{S})$ is even, then $L_1(\mathbf{S}) = L(\mathbf{S})$, and we can use the Games-Chan algorithm.
(ii) If $w_H(\mathbf{S})$ is odd and $d_H(\mathbf{S}_L^{(2^{n-1})}, \mathbf{S}_R^{(2^{n-1})}) = 1$, then $L_1(\mathbf{S}) = L(\mathbf{S}_L)$ if $w_H(\mathbf{S}_L^{(2^{n-1})})$ is even and $L_1(\mathbf{S}) = L(\mathbf{S}_R)$ if $w_H(\mathbf{S}_R^{(2^{n-1})})$ is even, and we can use the Games-Chan algorithm.
(iii) If $w_H(\mathbf{S})$ is odd and $d_H(\mathbf{S}_L^{(2^{n-1})}, \mathbf{S}_R^{(2^{n-1})}) > 1$, then $L_1(\mathbf{S}) = 2^{n-1} + L_1(\mathbf{S}_L + \mathbf{S}_R)$.
(iv) Apply the above procedure recursively to the 2^{n-1}-periodic binary sequence $\mathbf{S}_L + \mathbf{S}_R$ in (iii).

Proof. If $w_H(\mathbf{S})$ is even, then by Lemma 5, $L_1(\mathbf{S}) = L(\mathbf{S})$.

If $w_H(\mathbf{S})$ is odd, then by Lemma 5, $L_1(\mathbf{S}) < L(\mathbf{S}) = 2^n$. From the definition of $L_1(\mathbf{S})$ (see Definition 2), we get

$$L_1(\mathbf{S}) = \min_{0 \le i \le 2^n - 1} L(\mathbf{S} + \mathbf{E}_i). \tag{40}$$

If $d_H(\mathbf{S}_L^{(2^{n-1})}, \mathbf{S}_R^{(2^{n-1})}) = 1$, then $\mathbf{S}_L^{(2^{n-1})} = \mathbf{S}_R^{(2^{n-1})} + \mathbf{E}_t^{(2^{n-1})}$, where $0 \le t \le 2^{n-1} - 1$. From this and the Games-Chan algorithm, we obtain

$$L(\mathbf{S} + \mathbf{E}_t) = L(\mathbf{S}_R), \quad L(\mathbf{S} + \mathbf{E}_{t+2^{n-1}}) = L(\mathbf{S}_L), \tag{41}$$

$$L(\mathbf{S} + \mathbf{E}_i) > 2^{n-1}, \quad i \ne t, t + 2^{n-1}. \tag{42}$$

If $w_H(\mathbf{S}_L^{(2^{n-1})})$ is even, then $w_H(\mathbf{S}_R^{(2^{n-1})})$ is odd. Hence by Lemma 3,

$$L(\mathbf{S}_L) < L(\mathbf{S}_R) = 2^{n-1}. \tag{43}$$

Therefore, by (40)–(43), we have $L_1(\mathbf{S}) = L(\mathbf{S}_L)$. In the same way, if $w_H(\mathbf{S}_R^{(2^{n-1})})$ is even, then $L_1(\mathbf{S}) = L(\mathbf{S}_R)$.

If $w_H(\mathbf{S})$ is odd and $d_H(\mathbf{S}_L^{(2^{n-1})}, \mathbf{S}_R^{(2^{n-1})}) > 1$, then by the Games-Chan algorithm we have

$$L(\mathbf{S} + \mathbf{E}_i) = 2^{n-1} + L(\mathbf{S}_L + \mathbf{S}_R + \mathbf{E}'_i) \text{ for } 0 \le i \le 2^{n-1} - 1, \tag{44}$$

$$L(\mathbf{S} + \mathbf{E}_i) = 2^{n-1} + L(\mathbf{S}_L + \mathbf{S}_R + \mathbf{E}'_{i-2^{n-1}}) \text{ for } 2^{n-1} \le i \le 2^n - 1, \tag{45}$$

where $\mathbf{E}'_j$ is the 2^{n-1}-periodic binary sequence of weight one with 1 at the position with subscript j in the first period. Since $w_H(\mathbf{S})$ is odd, then $w_H(\mathbf{S}_L + \mathbf{S}_R)$ is also odd. Hence by Lemma 5, $L_1(\mathbf{S}_L + \mathbf{S}_R) < L(\mathbf{S}_L + \mathbf{S}_R) = 2^{n-1}$. From the definition of 1-error linear complexity, we get

$$L_1(\mathbf{S}_L + \mathbf{S}_R) = \min_{0 \le j \le 2^{n-1} - 1} L(\mathbf{S}_L + \mathbf{S}_R + \mathbf{E}'_j). \tag{46}$$

Therefore, by (40) and (44)–(46), we have $L_1(\mathbf{S}) = 2^{n-1} + L_1(\mathbf{S}_L + \mathbf{S}_R)$. This completes the proof. □

Recall that the 1-error linear complexity of a 2^n-periodic binary sequence $\mathbf{S}$ is given by $L_1(\mathbf{S}) = \min_{\mathbf{E}} L(\mathbf{S} + \mathbf{E})$, where the minimum is taken over all 2^n-periodic binary sequences $\mathbf{E}$ with $w_H(\mathbf{E}) \le 1$. After computing $L_1(\mathbf{S})$ using Algorithm 1, we want to determine all possible 2^n-periodic binary sequences $\mathbf{E}$ with $w_H(\mathbf{E}) \le 1$ such that

$$L_1(\mathbf{S}) = L(\mathbf{S} + \mathbf{E}). \tag{47}$$

If $w_H(\mathbf{S})$ is even, then by Lemmas 3 and 5,

$$L_1(\mathbf{S}) = L(\mathbf{S}) < L(\mathbf{S} + \mathbf{E}_i) \text{ for all } 0 \le i \le 2^n - 1.$$

Hence, only the zero sequence $\mathbf{E} = \mathbf{0}$ satisfies (47).

Note that if $L_1(\mathbf{S}) = 2^n - 2^r$, where $1 \le r \le n$, then by Theorem 6,

$$L(\mathbf{S}) = L_1(\mathbf{S}) = 2^n - 2^r \neq 2^n.$$

Hence by Lemma 3, $w_H(\mathbf{S})$ is even.

If $w_H(\mathbf{S})$ is odd, then by Lemma 5, $L_1(\mathbf{S}) < L(\mathbf{S}) = 2^n$. Hence, only some $\mathbf{E}_i$ satisfy (47). Such sequences $\mathbf{E}_i$ are called *error sequences* for $\mathbf{S}$. The subscript i of an error sequence $\mathbf{E}_i$ for $\mathbf{S}$ is called an *error position* for $\mathbf{S}$. Note that the error positions for $\mathbf{S}$ range from 0 to $2^n - 1$. Now we give an algorithm for determining all error positions for $\mathbf{S}$.

Algorithm 2: Let $\mathbf{S}$ be a 2^n-periodic binary sequence with odd Hamming weight.

(i) Compute the 1-error linear complexity $L_1(\mathbf{S})$ using Algorithm 1.

(ii) If $1 \le L_1(\mathbf{S}) < 2^{n-1}$, there is only one error position for $\mathbf{S}$. At Step 1 of Algorithm 1, $d_H(\mathbf{S}_L^{(2^{n-1})}, \mathbf{S}_R^{(2^{n-1})}) = 1$. Assume that $\mathbf{S}_L^{(2^{n-1})} = \mathbf{S}_R^{(2^{n-1})} + \mathbf{E}_m^{(2^{n-1})}$, where $0 \le m \le 2^{n-1} - 1$. If $w_H(\mathbf{S}_R^{(2^{n-1})})$ is even, the error position for $\mathbf{S}$ is m. If $w_H(\mathbf{S}_L^{(2^{n-1})})$ is even, the error position for $\mathbf{S}$ is $m + 2^{n-1}$.

(iii) If $2^n - 2^{n-r} < L_1(\mathbf{S}) < 2^n - 2^{n-r-1}$, where $1 \le r \le n-2$, there are exactly 2^r error positions for $\mathbf{S}$. Note that $2^n - 2^{n-r} = 2^{n-1} + 2^{n-2} + \cdots + 2^{n-r}$. Algorithm 1 proceeds recursively in $r+1$ steps. At Step i, $1 \le i \le r+1$, we work with a 2^{n-i+1}-periodic binary sequence $\mathcal{S}_i$ with odd Hamming weight. We decompose the binary 2^{n-i+1}-tuple corresponding to $\mathcal{S}_i$ into its left half $\mathcal{L}_i$ and right half $\mathcal{R}_i$. By Algorithm 1, $\mathcal{L}_i$ and $\mathcal{R}_i$ are binary 2^{n-i}-tuples with $d_H(\mathcal{L}_i, \mathcal{R}_i) > 1$ for $i = 1, 2, \ldots, r$ and $d_H(\mathcal{L}_{r+1}, \mathcal{R}_{r+1}) = 1$. Assume that $\mathcal{L}_{r+1} = \mathcal{R}_{r+1} + \mathbf{E}_m^{(2^{n-r-1})}$, where $0 \le m \le 2^{n-r-1} - 1$. If $w_H(\mathcal{R}_{r+1})$ is even, the error positions for $\mathbf{S}$ are $m + t2^{n-r}$, $t = 0, 1, \ldots, 2^r - 1$. If $w_H(\mathcal{L}_{r+1})$ is even, the error positions for $\mathbf{S}$ are $m + 2^{n-r-1} + t2^{n-r}$, $t = 0, 1, \ldots, 2^r - 1$.

Proof. If $1 \le L_1(\mathbf{S}) < 2^{n-1}$, then by Theorem 5(i) and noting that $w_H(\mathbf{S})$ is odd, there is only one $\mathbf{E}_i$ such that $L_1(\mathbf{S}) = L(\mathbf{S} + \mathbf{E}_i)$. Hence, there is only one error position for $\mathbf{S}$. The remaining conclusions in (ii) follow from Algorithm 1 and its proof.

If $2^n - 2^{n-r} < L_1(\mathbf{S}) < 2^n - 2^{n-r-1}$, where $1 \le r \le n-2$, then by Theorem 5(ii), Theorem 3, and noting that $w_H(\mathbf{S})$ is odd, there exists a unique $0 \le l \le 2^{n-r} - 1$ such that

$$L_1(\mathbf{S}) = L(\mathbf{S} + \mathbf{E}_{l \oplus t2^{n-r}}) \text{ for all } t = 0, 1, \ldots, 2^r - 1,$$
$$L_1(\mathbf{S}) \neq L(\mathbf{S} + \mathbf{E}_i) \text{ for all } i \neq l \oplus t2^{n-r},\ t = 0, 1, \ldots, 2^r - 1.$$

Hence, there are exactly 2^r error positions for $\mathbf{S}$. From Algorithm 1 and its proof, we know that $l = m$ if $w_H(\mathcal{R}_{r+1})$ is even, and $l = m + 2^{n-r-1}$ if $w_H(\mathcal{L}_{r+1})$ is even. This completes the proof. □

Next we give an example to illustrate Algorithms 1 and 2. In this example, we use $\mathcal{S}$ to denote both the binary periodic sequence and its corresponding vector.

Example: $n = 4$ and $\mathcal{S} = 1011010101111101$.
Algorithm 1:
(1) Step 1:

$$\mathcal{S}_1 = \mathcal{S} = 1011010101111101,\ \mathcal{L}_1 = 10110101,\ \mathcal{R}_1 = 01111101,\ d_H(\mathcal{L}_1, \mathcal{R}_1) = 3.$$

So $L_1(\mathcal{S}_1) = 8 + L_1(\mathcal{S}_2)$, where $\mathcal{S}_2 = \mathcal{L}_1 + \mathcal{R}_1 = 11001000$.
(2) Step 2:

$$\mathcal{S}_2 = \mathcal{L}_1 + \mathcal{R}_1 = 11001000,\ \ \mathcal{L}_2 = 1100,\ \ \mathcal{R}_2 = 1000,\ \ d_H(\mathcal{L}_2, \mathcal{R}_2) = 1.$$

Note that the weight of $\mathcal{L}_2$ is even, so $L_1(\mathcal{S}_2) = L(\mathcal{L}_2) = L(1100) = 3$, where $L(1100)$ is computed using the Games-Chan algorithm. Hence,

$$L_1(\mathcal{S}_1) = 8 + L_1(\mathcal{S}_2) = 11.$$

Algorithm 2:
Note that $8 = 2^4 - 2^3 < L_1(\mathcal{S}_1) = 11 < 2^4 - 2^2 = 12$, so there are exactly two error positions for $\mathcal{S}$. Since $\mathcal{L}_2 = \mathcal{R}_2 + 0100$ and the weight of $\mathcal{L}_2$ is even, we have $m = 1$ and the two error positions are $1 + 4 = 5$ and $1 + 4 + 8 = 13$. The two corresponding error vectors are 0000010000000000 and 0000000000000100.

5 Conclusion

In this paper, we studied the linear complexity and 1-error linear complexity of 2^n-periodic binary sequences. First, we gave some interesting properties on the structure of the sets of 2^n-periodic binary sequences with fixed linear complexity. Next, using these properties, we characterized the 2^n-periodic binary sequences with fixed 1-error linear complexity. Along the way, we described a new approach to derive the counting function for the 1-error linear complexity of 2^n-periodic binary sequences. Finally, we presented new fast algorithms for computing the 1-error linear complexity and locating the error positions for 2^n-periodic binary sequences. Recently, using the Kaida-Uehara-Imamura algorithm [5] for computing the k-error linear complexity of a p^n-periodic sequence over the prime field $\mathbb{F}_p$, Meidl and Venkateswarlu [15] determined the counting function for the 1-error linear complexity of p^n-periodic sequences over the prime field $\mathbb{F}_p$. It is worthwhile to note that our methods and results in this paper can be directly generalized to p^n-periodic sequences over the prime field $\mathbb{F}_p$.

Acknowledgments

The research of the first two authors is supported by the DSTA grant R-394-000-025-422 with Temasek Laboratories in Singapore. The second author is supported as well by a collaborative research grant from the National Institute of Information and Communications Technology in Japan. Comments by the anonymous referees are also gratefully acknowledged.

References

1. T. W. Cusick, C. Ding, A. Renvall, Stream Ciphers and Number Theory, North-Holland Mathematical Library, Vol. 55, Elsevier, Amsterdam, 1998.
2. C. Ding, G. Xiao, W. Shan, The Stability Theory of Stream Ciphers, Lecture Notes in Computer Science, Vol. 561, Springer, Berlin, 1991.
3. R. A. Games, A. H. Chan, A fast algorithm for determining the complexity of a binary sequence with period 2^n, IEEE Trans. Inform. Theory 29 (1983) 144–146.
4. T. Kaida, S. Uehara, K. Imamura, Computation of the k-error linear complexity of binary sequences with period 2^n, in: J. Jaffar, R. H. C. Yap (Eds.), Concurrency and Parallelism, Programming, Networking, and Security, Lecture Notes in Computer Science, Vol. 1179, Springer, Berlin, 1996, pp. 182–191.
5. T. Kaida, S. Uehara, K. Imamura, An algorithm for the k-error linear complexity of sequences over GF(p^m) with period p^n, p a prime, Inform. Comput. 151 (1999) 134–147.
6. K. Kurosawa, F. Sato, T. Sakata, W. Kishimoto, A relationship between linear complexity and k-error linear complexity, IEEE Trans. Inform. Theory 46 (2000) 694–698.
7. A. G. B. Lauder, K. G. Paterson, Computing the error linear complexity spectrum of a binary sequence of period 2^n, IEEE Trans. Inform. Theory 49 (2003) 273–280.
8. R. Lidl, H. Niederreiter, Finite Fields, Cambridge University Press, Cambridge, 1997.
9. W. Meidl, How many bits have to be changed to decrease the linear complexity?, Des. Codes Cryptogr. 33 (2004) 109–122.
10. W. Meidl, On the stability of 2^n-periodic binary sequences, IEEE Trans. Inform. Theory 51 (2005) 1151–1155.
11. W. Meidl, H. Niederreiter, Counting functions and expected values for the k-error linear complexity, Finite Fields Appl. 8 (2002) 142–154.
12. W. Meidl, H. Niederreiter, Linear complexity, k-error linear complexity, and the discrete Fourier transform, J. Complexity 18 (2002) 87–103.
13. W. Meidl, H. Niederreiter, On the expected value of the linear complexity and the k-error linear complexity of periodic sequences, IEEE Trans. Inform. Theory 48 (2002) 2817–2825.
14. W. Meidl, H. Niederreiter, Periodic sequences with maximal linear complexity and large k-error linear complexity, Appl. Algebra Engrg. Comm. Comput. 14 (2003) 273–286.
15. W. Meidl, A. Venkateswarlu, Remarks on the k-error linear complexity of p^n-periodic sequences, submitted for publication.
16. H. Niederreiter, Some computable complexity measures for binary sequences, in: C. Ding, T. Helleseth, H. Niederreiter (Eds.), Sequences and Their Applications, Springer, London, 1999, pp. 67–78.
17. H. Niederreiter, Periodic sequences with large k-error linear complexity, IEEE Trans. Inform. Theory 49 (2003) 501–505.
18. H. Niederreiter, Linear complexity and related complexity measures for sequences, in: T. Johansson, S. Maitra (Eds.), Progress in Cryptology — INDOCRYPT 2003, Lecture Notes in Computer Science, Vol. 2904, Springer, Berlin, 2003, pp. 1–17.
19. H. Niederreiter, H. Paschinger, Counting functions and expected values in the stability theory of stream ciphers, in: C. Ding, T. Helleseth, H. Niederreiter (Eds.), Sequences and Their Applications, Springer, London, 1999, pp. 318–329.

20. H. Niederreiter, I. E. Shparlinski, Periodic sequences with maximal linear complexity and almost maximal k-error linear complexity, in: K. G. Paterson (Ed.), Cryptography and Coding, Lecture Notes in Computer Science, Vol. 2898, Springer, Berlin, 2003, pp. 183–189.
21. R. A. Rueppel, Analysis and Design of Stream Ciphers, Springer, Berlin, 1986.
22. A. Sălăgean, On the computation of the linear complexity and the k-error linear complexity of binary sequences with period a power of two, IEEE Trans. Inform. Theory 51 (2005) 1145–1150.
23. M. Stamp, C. F. Martin, An algorithm for the k-error linear complexity of binary sequences with period 2^n, IEEE Trans. Inform. Theory 39 (1993) 1398–1401.

Crosscorrelation Properties of Binary Sequences with Ideal Two-Level Autocorrelation*

Nam Yul Yu and Guang Gong

Department of Electrical and Computer Engineering
University of Waterloo, Waterloo, Ontario, Canada
nyyu@engmail.uwaterloo.ca, ggong@calliope.uwaterloo.ca

Abstract. For odd n, binary sequences of period $2^n - 1$ with ideal two-level autocorrelation are investigated with respect to 3- or 5-valued crosscorrelation property between them. At most 5-valued crosscorrelation of m-sequences is first discussed, which is linked to crosscorrelation of some other binary two-level autocorrelation sequences. Then, several theorems and conjectures are established for describing 3- or 5-valued crosscorrelation of a pair of binary two-level autocorrelation sequences.

1 Introduction

In code-division multiple access (CDMA) communication systems, a binary two-level autocorrelation sequence is needed to acquire accurate timing information of received signals by means of its impulse-like autocorrelation property. In cryptography, the sequence is also required for avoiding correlation attack that exploits pseudorandom sequences having weak autocorrelation property. In the last few years, several new binary two-level autocorrelation sequences have been discovered; Kasami power function (KPF) sequences [3], Welch-Gong (WG) sequences [15], and Maschiettie's hyperoval sequences [12]. Together with traditionally known m-sequences, Gordon-Mills-Welch (GMW) sequences [8], quadratic residue (QR) sequences, and Hall's sextic residue sequences, these are all known binary two-level autocorrelation sequences of period $2^n - 1$.

For theory and practice of sequences, it would be interesting to study crosscorrelation of a pair of binary two-level autocorrelation sequences of period $2^n - 1$. For odd n, the crosscorrelation has been investigated for following pairs of binary sequences.

- An m-sequence and its decimations [6] [11] [13] [9] (Gold, Kasami, Welch, Niho, and some conjectured exponents)
- An m-sequence and a GMW sequence with the same primitive polynomial [5], and a pair of GMW sequences [1] (The crosscorrelations are reduced to crosscorrelations of m-sequences)
- An m-sequence and a decimated KPF sequence with one particular exponent [3]

* This work was supported by NSERC Grant RGPIN 227700-00.

G. Gong et al. (Eds.): SETA 2006, LNCS 4086, pp. 104–118, 2006.

- An m-sequence and a WG sequence without decimation [7]
- An m-sequence and a hyperoval sequence without decimation [4]
- A pair of KPF sequences without decimations [10]

If maximum crosscorrelation of a pair of binary sequences of period $2^n - 1$ is much larger than its optimum value achieving the Welch [17] or the Sidelnikov bound [16], then the pair is not so attractive for communication and cryptographic applications. For odd n, therefore, 3-valued crosscorrelation, i.e., $\{0, \pm 2^{\frac{n+1}{2}}\}$, has been intensively studied by many researchers. In this work, we are also interested in 5-valued crosscorrelation, i.e., $\{0, \pm 2^{\frac{n+1}{2}}, \pm 2^{\frac{n+3}{2}}\}$, which might be suboptimal for some applications.

In this paper, we study the 3- or 5-valued crosscorrelation of a pair of binary two-level autocorrelation sequences of period $2^n - 1$ for odd n, excluding GMW, QR, and Hall's sextic residue sequences. In Section 3, at most 5-valued crosscorrelation of m-sequences is discussed, which is linked to crosscorrelation of some other sequences. In Section 4, the 3- or 5-valued crosscorrelation of following pairs is investigated.

- A 5-term KPF sequence and a decimated WG sequence with one new exponent
- An m-sequence and a decimated WG sequence with one new exponent
- An m-sequence and a decimated hyperoval sequence with several new exponents
- An m-sequence and a decimated 3-term KPF sequence with one new exponent

With the new results as well as already known ones, relations of binary two-level autocorrelation sequences are summarized with respect to 3- or 5-valued crosscorrelation. From our experiments for $n = 13, 15, 17$, and 19, we observed that all 3- or 5-valued crosscorrelations of a pair of binary two-level autocorrelation sequences are completely described by the already known and new results listed above unless both are m-sequences.

2 Preliminaries

In this section, we give preliminary definitions and concepts related to binary two-level autocorrelation sequences. Following notations will be used throughout this paper.

- $\mathbb{F}_q = GF(q)$ is a finite field with q elements and $\mathbb{F}_q^*$ is a multiplicative group of $\mathbb{F}_q$.
- Let n, m be positive integers with $m|n$. A trace function from $\mathbb{F}_{2^n}$ to $\mathbb{F}_{2^m}$ is denoted by $Tr_m^n(x)$, i.e.,

$$Tr_m^n(x) = x + x^{2^m} + \cdots + x^{2^{m(\frac{n}{m}-1)}}, \quad x \in \mathbb{F}_{2^n},$$

  or simply as $Tr(x)$ if $m = 1$ and a context is clear.

2.1 Correspondence Between Binary Periodic Sequences and Functions from $\mathbb{F}_{2^n}$ to $\mathbb{F}_2$

Let $\mathcal{S}$ be a set of all binary sequences of period $2^n - 1$ and $\mathcal{F}$ be a set of all functions from $\mathbb{F}_{2^n}$ to $\mathbb{F}_2$. Any function $f(x)$ in $\mathcal{F}$ can be represented as

$$f(x) = \sum_{i=1}^{r} Tr_1^{n_i}(A_i x^{t_i}), \quad A_i \in \mathbb{F}_{2^{n_i}}$$

where t_i is a coset leader of a cyclotomic coset modulo $2^{n_i} - 1$, and $n_i|n$ is a size of the cyclotomic coset containing t_i. For any sequence $\underline{\mathbf{a}} = \{a_i\} \in \mathcal{S}$, there exists $f(x) \in \mathcal{F}$ such that $a_i = f(\alpha^i)$, $i = 0, 1, \cdots$, where α is a primitive element of $\mathbb{F}_{2^n}$. Then, $f(x)$ is called a *trace representation* of $\underline{\mathbf{a}}$. In particular, $\underline{\mathbf{a}}$ is an m-sequence if $f(x)$ consists of a single trace term. Also, $f(x)$ is called an *orthogonal function* if $\underline{\mathbf{a}}$ is a binary two-level autocorrelation sequence. In this paper, we will always use its trace representation to represent any binary two-level autocorrelation sequence.

2.2 Decimation of Periodic Sequences

Let $\underline{\mathbf{a}}$ be a binary sequence of period $2^n - 1$ and $f(x)$ be a trace representation of $\underline{\mathbf{a}}$. Let $0 < s < 2^n - 1$. Then, a sequence $\underline{\mathbf{b}} = \{b_i\}$ is said to be an *s-decimation* of $\underline{\mathbf{a}}$, denoted by $\underline{\mathbf{a}}^{(s)}$, if elements of $\underline{\mathbf{b}}$ are given by $b_i = a_{si}$, $i = 0, 1, \cdots$, where the multiplication is computed modulo $2^n - 1$. A trace representation of $\underline{\mathbf{a}}^{(s)}$ is $f(x^s)$, denoted by $f^{(s)}$.

2.3 Crosscorrelation

Crosscorrelation of binary sequences $\underline{\mathbf{a}}$ and $\underline{\mathbf{b}}$ of period $2^n - 1$ is defined by

$$C_{\underline{\mathbf{a}},\underline{\mathbf{b}}}(\tau) = \sum_{i=0}^{2^n-2} (-1)^{a_{i+\tau}+b_i} = -1 + \sum_{x \in \mathbb{F}_{2^n}} (-1)^{f(\lambda x)+g(x)} = -1 + C_{f,g}(\lambda)$$

where $\lambda = \alpha^\tau$ with $0 \le \tau \le 2^n - 2$, τ is a phase shift of the sequence $\underline{\mathbf{a}}$, α is a primitive element of $\mathbb{F}_{2^n}$, and $f(x)$ and $g(x)$ are trace representations of $\underline{\mathbf{a}}$ and $\underline{\mathbf{b}}$, respectively. Throughout this paper, we always use $C_{f,g}(\lambda)$ to represent the crosscorrelation of $\underline{\mathbf{a}}$ and $\underline{\mathbf{b}}$ with their trace representations $f(x)$ and $g(x)$.

If $C_{f,g}(\lambda)$ belongs to $\{0, \pm 2^{\frac{n+1}{2}}\}$, then it is called *3-valued.* If it belongs to $\{0, \pm 2^{\frac{n+1}{2}}, \pm 2^{\frac{n+3}{2}}\}$, on the other hand, it is called *5-valued.* (In fact, a term '3- or 5-valued' means the number of kinds of values that $C_{f,g}(\lambda)$ takes, no matter what its actual values are. In this paper, however, we restrict the term '3- or 5-valued' by the above definition.) If $f(x) = Tr(x)$, then $C_{f,g}(\lambda)$ is the *Hadamard transform* of $g(x)$. In particular, $C_{f,g}(\lambda)$ is denoted by $H_d(\lambda)$ if $f(x) = Tr(x)$ and $g(x) = Tr(x^d)$, where the distribution of $H_d(\lambda)$ is determined by d.

2.4 Parseval's Equation

Let $f(x)$, $g(x)$, and $h(x)$ be functions from $\mathbb{F}_{2^n}$ to $\mathbb{F}_2$, respectively, and $h(x)$ be orthogonal. Then,

$$\sum_{x\in\mathbb{F}_{2^n}} (-1)^{g(x)+f(x)} = \frac{1}{2^n}\sum_{x\in\mathbb{F}_{2^n}} \widehat{g_h}(x)\widehat{f_h}(x) \tag{1}$$

where $\widehat{f_h}(x) = \sum_{y\in\mathbb{F}_{2^n}}(-1)^{h(xy)+f(y)}$.

2.5 Recently Constructed Binary Two-Level Autocorrelation Sequences

In this subsection, we briefly introduce three classes of binary two-level autocorrelation sequences of period $2^n - 1$ which have been constructed recently.

Kasami Power Function (KPF) Sequences: Let k be an integer of $1 \leq k < \lfloor \frac{n}{2} \rfloor$ with $\gcd(k,n) = 1$. For $d = 2^{2k} - 2^k + 1$, consider a set

$$B_k = \{(x+1)^d + x^d + 1 |\ x \in \mathbb{F}_{2^n}\}.$$

Then, its characteristic sequence given by

$$a_i = \begin{cases} 0, & \text{if } \alpha^i \in B_k \\ 1, & \text{if } \alpha^i \notin B_k \end{cases}$$

has an ideal two-level autocorrelation, where the sequence is called the *Kasami power function (KPF) sequence* [3]. According to k with $\gcd(k,n) = 1$, there exist $\frac{\phi(n)}{2}$ inequivalent KPF sequences of period $2^n - 1$, where $\phi(\cdot)$ is the Euler-totient function. If $k = 1$, in particular, the KPF sequence is identical to an m-sequence. Let $b_k(x)$ be a trace representation of the KPF sequence. For odd n, the KPF sequence has a Hadamard equivalence given by

$$\sum_{x\in\mathbb{F}_{2^n}} (-1)^{Tr(\lambda x)+b_k(x^{2^k+1})} = \sum_{x\in\mathbb{F}_{2^n}} (-1)^{Tr(\lambda^{\frac{2^k+1}{3}}x)+Tr(x^3)} = H_3(\lambda^{\frac{2^k+1}{3}}) \tag{2}$$

which is 3-valued [3].

Welch-Gong (WG) Sequences: For $n = 3k \pm 1$ and $d = 2^{2k} - 2^k + 1$, consider a map $\delta_k(x) = (x+1)^d + x^d$ and a set

$$W_k = \begin{cases} \delta_k(x), & \text{if } n \text{ is even} \\ \mathbb{F}_{2^n} \setminus \delta_k(x), & \text{if } n \text{ is odd.} \end{cases}$$

Then, its characteristic sequence given by

$$a_i = \begin{cases} 0, & \text{if } \alpha^i \in W_k \\ 1, & \text{if } \alpha^i \notin W_k \end{cases}$$

has an ideal two-level autocorrelation [14]. This sequence is identical to the *Welch-Gong sequence*, which is obtained from the Welch-Gong transformation of the 5-term sequences [15]. Let $w_k(x)$ be a trace representation of the WG sequence. For odd n, the WG sequence has a Hadamard equivalence [7] given by

$$\sum_{x\in\mathbb{F}_{2^n}} (-1)^{Tr(\lambda x)+w_k(x)} = \sum_{x\in\mathbb{F}_{2^n}} (-1)^{Tr(\lambda^{d^{-1}}x)+Tr(x^{2^k+1})} = H_{2^k+1}(\lambda^{d^{-1}}) \quad (3)$$

which is also 3-valued.

Hyperoval Sequences: For odd n, consider a set

$$M_k = \{x + x^k | x \in \mathbb{F}_{2^n}\}$$

where k is given as follows [12].

i) Singer type: $k = 2$, Segre type: $k = 6$.
ii) Glynn type I: $k = 2^\sigma + 2^\tau$ where $\sigma = \frac{n+1}{2}$ and $4\tau \equiv 1 \pmod{n}$.
iii) Glynn type II: $k = 3 \cdot 2^\sigma + 4$ with $\sigma = \frac{n+1}{2}$.

Then, a characteristic sequence of M_k given by

$$a_i = \begin{cases} 0, & \text{if } \alpha^i \in M_k \\ 1, & \text{if } \alpha^i \notin M_k \end{cases}$$

has an ideal two-level autocorrelation, where the sequence is called the *hyperoval sequence.* In this paper, we are only interested in the Glynn type I and II hyperoval sequences because the Singer and Segre type hyperoval sequences are identical to m-sequences and the KPF sequences for $k = 2$, respectively [3].

Let $h_k(x)$ be a trace representation of the hyperoval sequence. For odd n, Dillon derived a Hadamard equivalence of the hyperoval sequence [4], i.e.,

$$\sum_{x\in\mathbb{F}_{2^n}} (-1)^{Tr(\lambda x)+h_k(x)} = \sum_{x\in\mathbb{F}_{2^n}} (-1)^{Tr(\lambda^{\frac{k-1}{k}}x)+Tr(x^k)} = H_k(\lambda^{\frac{k-1}{k}}). \quad (4)$$

If k is the Glynn type I exponent in ii), then (4) is 3-valued because k is quadratic. If k is the Glynn type II exponent in iii), on the other hand, then (4) is conjectured to be at most 5-valued because $k = 3 \cdot 2^{\frac{n+1}{2}} + 4 \equiv 2^{\frac{n-1}{2}} + 2^{\frac{n-3}{2}} + 1 \pmod{2^n - 1}$ is equivalently the inverse of the exponent of Conjecture 4-6 (1) in [13] where $H_{k^{-1}}(\lambda)$ is conjectured to be at most 5-valued. We will restate this in Conjecture 2 of this paper.

3 Some Observations of Crosscorrelation of Binary m-Sequences

In this section, we recall at most 5-valued crosscorrelation of a binary m-sequence and its d-decimation, i.e., $H_d(\lambda) = \sum_{x\in\mathbb{F}_{2^n}} (-1)^{Tr(\lambda x)+Tr(x^d)}$. In terms of 3-valued $H_d(\lambda)$, many exponents d are known, i.e., Gold [6], Kasami [11], Welch, Niho [13] exponents, and their respective inverses. In terms of 5-valued $H_d(\lambda)$, on the other hand, we need to clarify known results.

Proposition 1. *Let n be odd, t be a positive integer of $1 \leq t \leq \frac{n-1}{2}$, and $e = \gcd(n,t)$ with $n/e \geq 4$. Let $d(k,l) = (1+2^k)/(1+2^l)$ with positive integers k and l $(k \neq l)$. Then, $H_{d(k,l)}(\lambda)$ belongs to $\{0, \pm 2^{(n+e)/2}, \pm 2^{(n+3e)/2}\}$ if a pair (k,l) is one of following three cases*

$$(a)\ (k,l) = (5t,t), \quad (b)\ (k,l) = (5t,3t), \quad (c)\ (k,l) = (2t,t)$$

where the multiplication is computed modulo n. If $e = 1$, in particular, $H_{d(k,l)}(\lambda)$ is at most 5-valued, i.e., $\{0, \pm 2^{(n+1)/2}, \pm 2^{(n+3)/2}\}$.

Proposition 1-(a) has been proven by Niho (Lemma 4-1 in [13]). Although he had never stated Proposition 1-(b) and (c) in [13], we believe those have been implicitly known to many coding and sequence experts. In literatures, however, we could not find an explicit proof for (b) and (c) which is not trivial. So, we present it in this section because the result is linked to crosscorrelation of some other binary two-level autocorrelation sequences in Section 4. In order to prove Proposition 1, we need to use the Kasami's Theorem on weight distribution of subcodes of the second order Reed-Muller codes, which was partly used by Niho to prove Proposition 1-(a). In the following, we consider the odd case of his original theorem in [11].

Fact 1 (Kasami [11]). *For odd n, let t and u be positive integers with $1 \leq t \leq \frac{n-1}{2}$ and $1 \leq u \leq \lfloor \frac{n}{2e} \rfloor + 1$ where $e = \gcd(n,t)$. Let $A_t(u)$ be a binary cyclic code of length $2^n - 1$ whose generator polynomial is given by $g_a(x) = \prod_{i=0}^{u-1} m_{1+2^{ti}}(x)$ where $m_i(x)$ is a minimal polynomial of α^i and α is a primitive element of $\mathbb{F}_{2^n}$. Similarly, let $F_t(u)$ be a binary cyclic code of length $2^n - 1$ whose generator polynomial is given by $g_f(x) = \prod_{i=0}^{u-1} m_{1+2^{t(2i+1)}}(x)$. Dual codes of $A_t(u)$ and $F_t(u)$ are denoted by $A_t(u)^\perp$ and $F_t(u)^\perp$, respectively. Then, $A_t(u)^\perp$ and $F_t(u)^\perp$ have the same weight distribution as those of $A_e(u)^\perp$ whose distinct weights are given by*

$$\{0, 2^{n-1}, 2^{n-1} \pm 2^{(n-e)/2+ie-1}\} \text{ for } 1 \leq i \leq u-1.$$

Using Fact 1, we can prove Proposition 1.

Proof of Proposition 1. In (a) and (b), $H_{d(k,l)}(\lambda)$ is represented by

$$H_{d(k,l)}(\lambda) = \sum_{x \in \mathbb{F}_{2^n}} (-1)^{Tr(\lambda x + x^{\frac{1+2^k}{1+2^l}})} = \begin{cases} \sum_{x \in \mathbb{F}_{2^n}} (-1)^{Tr(\lambda x^{1+2^t} + x^{1+2^{5t}})} & \text{for } (a) \\ \sum_{x \in \mathbb{F}_{2^n}} (-1)^{Tr(\lambda x^{1+2^{3t}} + x^{1+2^{5t}})} & \text{for } (b). \end{cases}$$

Then, we can consider codes $\mathcal{R}_5$ and $\mathcal{R}_{5/3}$ given by

$$\mathcal{R}_5 = \{Tr(\alpha x^{1+2^t} + \beta x^{1+2^{5t}}) | \alpha, \beta \in \mathbb{F}_{2^n}\},$$
$$\mathcal{R}_{5/3} = \{Tr(\gamma x^{1+2^{3t}} + \delta x^{1+2^{5t}}) | \gamma, \delta \in \mathbb{F}_{2^n}\}$$

which are subcodes of the dual of $F_t(u)$ for $u = 3$ where $F_t(u)$ has zeros $1 + 2^{t(2i+1)}, i = 0, 1, 2$. For any t of $1 \leq t \leq \frac{n-1}{2}$, therefore, weight distributions of

$\mathcal{R}_5$ and $\mathcal{R}_{5/3}$ are immediate from Fact 1, and consequently $H_{d(k,l)}(\lambda)$ belongs to $\{0, \pm 2^{\frac{n+e}{2}}, \pm 2^{\frac{n+3e}{2}}\}$ for both (a) and (b).

In (c), on the other hand, $A_t(3)$ generated by $g_a(x)$ has zeros $\{2, 1+2^t, 1+2^{2t}\}$, so a code $\mathcal{R}_2$ given by

$$\mathcal{R}_2 = \{Tr(\zeta x^{1+2^t} + \eta x^{1+2^{2t}}) | \zeta, \eta \in \mathbb{F}_{2^n}\}$$

is also a subcode of the dual of $A_t(3)$. From Fact 1, therefore, it is clear that $H_{d(k,l)}(\lambda) = \sum_{x \in \mathbb{F}_{2^n}} (-1)^{Tr(\lambda x^{1+2^t} + x^{1+2^{2t}})}$ belongs to $\{0, \pm 2^{\frac{n+e}{2}}, \pm 2^{\frac{n+3e}{2}}\}$. □

Lemma 1. *For odd n, let k and l be positive integers of $1 \le k, l \le \frac{n-1}{2}$ $(k \ne l)$, and $d(k,l) = \frac{1+2^k}{1+2^l}$. Then, $H_{d(n-k,l)}(\lambda), H_{d(k,n-l)}(\lambda)$, and $H_{d(n-k,n-l)}(\lambda)$ have the same correlation spectrum as $H_{d(k,l)}(\lambda)$. Furthermore, $H_{d(l,k)}(\lambda)$ also belongs to the same correlation spectrum as $H_{d(k,l)}(\lambda)$.*

Proof. Note that $H_{d \cdot 2^j}(\lambda) = H_d(\lambda)$ for any integer j [9]. Since $2^{n-k} \cdot (1 + 2^k) = 2^{n-k} + 2^n \equiv 2^{n-k} + 1 \pmod{2^n - 1}$, we see that $1 + 2^k$ and $1 + 2^{n-k}$ belong to the same cyclotomic coset. Hence, $d(k,l)$ belongs to the same cyclotomic coset as $d(n-k,l)$. Therefore, $H_{d(k,l)}(\lambda)$ and $H_{d(n-k,l)}(\lambda)$ have the same correlation distribution. By the similar way, cases of $H_{d(k,n-l)}(\lambda)$ and $H_{d(n-k,n-l)}(\lambda)$ are simply proved. From $d(l,k) = d(k,l)^{-1}$, furthermore, it is immediate that $H_{d(k,l)}(\lambda)$ and $H_{d(l,k)}(\lambda)$ belong to the same correlation spectrum. □

Table 1 shows (k,l) pairs and $d(k,l) = \frac{1+2^k}{1+2^l}$ corresponding to 5-valued $H_{d(k,l)}(\lambda)$ in computer experiments. We only list pairs of $1 \le l < k \le \frac{n-1}{2}$ which are enough to cover the other possible pairs from Lemma 1. Each pair of '*' is due to (a), '+' due to (b), and 'o' due to (c) in Proposition 1, respectively. For odd $n = 9 - 17$, Proposition 1 is verified from the experiments.

4 Crosscorrelation of a Pair of Binary Two-Level Autocorrelation Sequences

4.1 A Pair of KPF Sequences

In [10], Hertel investigated crosscorrelation of two distinct KPF sequences for odd n. (She called the sequences as Dillon-Dobbertin (DD) sequences after their discoverers' name.)

Fact 2 (Hertel [10]). *For odd n, let k and l be distinct positive integers with $\gcd(n,k) = \gcd(n,l) = 1$. Let $b_k(x)$ and $b_l(x)$ be trace representations of two distinct KPF sequences, respectively. Then,*

$$C_{b_k,b_l}(\lambda) = \sum_{x \in \mathbb{F}_{2^n}} (-1)^{b_k(\lambda x) + b_l(x)} = H_{d(k,l)}(\lambda^{\frac{1}{1+2^k}}), \quad \lambda \in \mathbb{F}_{2^n}$$

where $d(k,l) = \frac{1+2^k}{1+2^l}$. If $(k,l) = (3t,t)$, in particular, $C_{b_k,b_l}(\lambda)$ is 3-valued, i.e., $\{0, \pm 2^{\frac{n+1}{2}}\}$.

Table 1. (k,l) pairs and $d(k,l)$'s for 5-valued crosscorrelation of $Tr(x)$ and $Tr(x^{d(k,l)})$

n	(k,l)	$d(k,l)$	n	(k,l)	$d(k,l)$	n	(k,l)	$d(k,l)$	n	(k,l)	$d(k,l)$
9	$(2,1)^{*,o}$	43	13	$(4,2)^{+,o}$	1645	15	$(7,4)^{o}$	2895	17	$(4,3)^{*}$	14571
9	$(4,1)^{*,o}$	11	13	$(5,3)^{+,o}$	1367	15	$(6,5)^{+}$	1119	17	$(5,3)^{+}$	21847
9	$(4,2)^{*,o}$	109	13	$(6,3)^{+,o}$	939	15	$(7,5)^{*}$	3229	17	$(6,3)^{o}$	14679
11	$(2,1)^{*,+,o}$	171	13	$(5,4)^{+,o}$	1461	17	$(2,1)^{o}$	10923	17	$(7,3)^{o}$	15019
11	$(5,1)^{*,+,o}$	11	13	$(6,4)^{*}$	497	17	$(4,1)^{+}$	2731	17	$(6,4)^{*}$	11567
11	$(4,2)^{*,+,o}$	423	15	$(2,1)^{o}$	2731	17	$(5,1)^{*}$	11	17	$(8,4)^{o}$	7831
11	$(4,3)^{*,+,o}$	235	15	$(5,1)^{*}$	11	17	$(7,1)^{*}$	43	17	$(6,5)^{o}$	12909
11	$(5,3)^{*,+,o}$	343	15	$(7,1)^{o}$	43	17	$(8,1)^{o}$	171	17	$(7,5)^{o}$	13917
13	$(2,1)^{+,o}$	683	15	$(4,2)^{o}$	6567	17	$(3,2)^{*}$	3277	17	$(8,5)^{*}$	4003
13	$(5,1)^{*}$	11	15	$(5,2)^{*}$	205	17	$(4,2)^{o}$	26221	17	$(7,6)^{+}$	10587
13	$(6,1)^{+,o}$	43	15	$(5,3)^{+}$	5463	17	$(7,2)^{*}$	205	17	$(8,6)^{*}$	2143
13	$(3,2)^{*}$	205	15	$(5,4)^{*}$	1943	17	$(8,2)^{+}$	26317			

Corollary 1. *With the notation of Proposition 1 and Fact 2, if a pair (k,l) is one of the pairs in Proposition 1, then $C_{b_k,b_l}(\lambda)$ is at most 5-valued, i.e.,* $\{0, \pm 2^{\frac{n+1}{2}}, \pm 2^{\frac{n+3}{2}}\}$.

Proof. Corollary 1 is immediate from combining Proposition 1 and Fact 2. □

From Corollary 1, it is obvious that crosscorrelation of $b_k(x)$ and $b_l(x)$ with a (k,l) pair in Table 1 is 5-valued.

4.2 5-Term KPF Sequences and Welch-Gong (WG) Sequences

The WG sequences are obtained from the Welch-Gong transformation of KPF sequences for $k = \frac{n\pm 1}{3}$, where the KPF sequences always have five trace terms [3] [15]. By the Parseval's equation exploited in [10], we derive a theorem on crosscorrelation of the 5-term KPF and the WG sequences.

Theorem 1. *Let n be odd and $n = 3k\pm 1$. Let $b_k(x)$ and $w_k(x)$ be trace representations of the KPF sequences and the WG sequences, respectively. For $s = \frac{1}{2^k+1}$, crosscorrelation of the two sequences given by*

$$C_{b_k,w_k^{(s)}}(\lambda) = \sum_{x\in\mathbb{F}_{2^n}} (-1)^{b_k(\lambda x)+w_k(x^s)} = H_{\frac{2^k+1}{3}}(\lambda)$$

is 3-valued, i.e., $\{0, \pm 2^{\frac{n+1}{2}}\}$.

Proof. Applying the Parseval's equation in (1),

$$\begin{aligned} C_{b_k,w_k^{(s)}}(\lambda) =& \frac{1}{2^n}\sum_{x\in\mathbb{F}_{2^n}}\sum_{y\in\mathbb{F}_{2^n}}(-1)^{b_k(\lambda y)+Tr(xy^{\frac{1}{2^k+1}})}\sum_{z\in\mathbb{F}_{2^n}}(-1)^{w_k(z^s)+Tr(xz^{\frac{1}{2^k+1}})} \\ =& \frac{1}{2^{2n}}\sum_{x,z\in\mathbb{F}_{2^n}}\sum_{y\in\mathbb{F}_{2^n}}(-1)^{b_k(y)+Tr(x\lambda^{-\frac{1}{2^k+1}}y^{\frac{1}{2^k+1}})} \\ &\cdot\sum_{u\in\mathbb{F}_{2^n}}(-1)^{w_k(u^s)+Tr(zu^s)}\sum_{v\in\mathbb{F}_{2^n}}(-1)^{Tr(xv^{\frac{1}{2^k+1}})+Tr(zv^s)}. \end{aligned} \tag{5}$$

If $s=\frac{1}{2^k+1}$, then we have

$$\sum_{v\in\mathbb{F}_{2^n}}(-1)^{Tr(xv^{\frac{1}{2^k+1}})+Tr(zv^s)}=\sum_{v\in\mathbb{F}_{2^n}}(-1)^{Tr((x+z)v^{\frac{1}{2^k+1}})}=\begin{cases}2^n, & \text{if } x=z\\ 0, & \text{if } x\neq z.\end{cases}$$

If the Hadamard equivalences (2) and (3) are applied to (5), then we have

$$\begin{aligned} C_{b_k,w_k^{(s)}}(\lambda) =& \frac{1}{2^n}\sum_{x\in\mathbb{F}_{2^n}}\sum_{y\in\mathbb{F}_{2^n}}(-1)^{Tr(\lambda^{-\frac{1}{3}}x^{\frac{2^k+1}{3}}y)+Tr(y^3)}\sum_{u\in\mathbb{F}_{2^n}}(-1)^{Tr(x^bu)+Tr(u^a)} \\ =& \frac{1}{2^n}\sum_{y\in\mathbb{F}_{2^n}}\sum_{u\in\mathbb{F}_{2^n}}(-1)^{Tr(y^3)+Tr(u^a)}\sum_{x\in\mathbb{F}_{2^n}}(-1)^{Tr(\lambda^{-\frac{1}{3}}x^{\frac{2^k+1}{3}}y)+Tr(ux^b)} \end{aligned}$$

where $a=2^k+1$ and $b=(2^{2k}-2^k+1)^{-1}$. From $3k=n\pm1$, it is clear that $b^{-1}\cdot\frac{2^k+1}{3}=\frac{2^{3k}+1}{3}\equiv1 \pmod{2^n-1}$. Thus, we have $b\equiv\frac{2^k+1}{3} \pmod{2^n-1}$. Consequently,

$$\begin{aligned} C_{b_k,w_k^{(s)}}(\lambda) =& \frac{1}{2^n}\sum_{y\in\mathbb{F}_{2^n}}\sum_{u\in\mathbb{F}_{2^n}}(-1)^{Tr(y^3)+Tr(u^a)}\sum_{x\in\mathbb{F}_{2^n}}(-1)^{Tr((\lambda^{-\frac{1}{3}}y+u)x^{\frac{2^k+1}{3}})} \\ =& \sum_{u\in\mathbb{F}_{2^n}}(-1)^{Tr(\lambda u^3+u^a)}=\sum_{u\in\mathbb{F}_{2^n}}(-1)^{Tr(\lambda u)+Tr(u^{\frac{2^k+1}{3}})}=H_{\frac{2^k+1}{3}}(\lambda) \end{aligned} \tag{6}$$

where $y=\lambda^{\frac{1}{3}}u$. In (6), $\frac{2^k+1}{3}\equiv b=(2^{2k}-2^k+1)^{-1}$. Since it is an inverse of the Kasami exponent with $\gcd(n,k)=1$, we see that $H_{\frac{2^k+1}{3}}(\lambda)$ is 3-valued and so is $C_{b_k,w_k^{(s)}}(\lambda)$. □

4.3 *m*-Sequences and Welch-Gong (WG) Sequences

In an effort to search for new two-level autocorrelation sequences, Gong and Golomb proposed the *decimation-Hadamard transform (DHT)* in [7]. With respect to orthogonal functions $f(x)$ and $h(x)$, they defined a *realizable pair* (v,t)

of $g(x)$ in the DHT by generalizing the Hadamard equivalence developed in [3], i.e.,

$$\sum_{x\in\mathbb{F}_{2^n}} (-1)^{h(\lambda^t x)+f(x^v)} = \sum_{x\in\mathbb{F}_{2^n}} (-1)^{h(\lambda x)+g(x)}. \tag{7}$$

They also showed that there exist at most 6 realizable pairs for the realization. Among them, we will use the fact that if (v,t) is a realizable pair of $g(x)$, then $(t,-(vt)^{-1})$ is also a realizable pair of $g(x^{(vt)^{-1}})$ [7] from which we have

$$\sum_{x\in\mathbb{F}_{2^n}} (-1)^{h(\lambda^{-(vt)^{-1}} x)+f(x^t)} = \sum_{x\in\mathbb{F}_{2^n}} (-1)^{h(\lambda x)+g(x^{(vt)^{-1}})}. \tag{8}$$

Using this, we establish a theorem on crosscorrelation of m-sequences and WG sequences.

Theorem 2. *Let n be odd and $n = 3k \pm 1$, and $d = 2^{2k} - 2^k + 1$. Let $w_k(x)$ be a trace representation of the WG sequences. For $s = \frac{d}{2^k+1}$, crosscorrelation of m-sequences and the WG sequences given by*

$$C_{Tr,w_k^{(s)}}(\lambda) = \sum_{x\in\mathbb{F}_{2^n}} (-1)^{Tr(\lambda x)+w_k(x^s)} = H_{d^{-1}}(\lambda^{-s})$$

is 3-valued, i.e., $\{0, \pm 2^{\frac{n+1}{2}}\}$.

Proof. From the Hadamard equivalence of (3), we have a realizable pair $(v,t) = (2^k+1, d^{-1})$ in (7) where $f(x) = h(x) = Tr(x)$ and $g(x) = w_k(x)$. From (8), therefore, we have

$$\sum_{x\in\mathbb{F}_{2^n}} (-1)^{Tr(\lambda^{-\frac{d}{2^k+1}} x)+Tr(x^{d^{-1}})} = \sum_{x\in\mathbb{F}_{2^n}} (-1)^{Tr(\lambda x)+w_k(x^{\frac{d}{2^k+1}})}.$$

Thus, $C_{Tr,w_k^{(s)}}(\lambda) = H_{d^{-1}}(\lambda^{-s})$ for $s = \frac{d}{2^k+1}$. Since d is the Kasami exponent with $\gcd(n,k) = 1$, $H_{d^{-1}}(\lambda^{-s})$ is 3-valued and so is $C_{Tr,w_k^{(s)}}(\lambda)$. □

4.4 m-Sequences and Hyperoval Sequences

Applying (8) to hyperoval sequences with the Hadamard equivalence of (4), we can derive another Hadamard equivalence, i.e.,

$$\sum_{x\in\mathbb{F}_{2^n}} (-1)^{Tr(\lambda x)+h_k(x^{\frac{1}{k-1}})} = \sum_{x\in\mathbb{F}_{2^n}} (-1)^{Tr(\lambda^{-\frac{1}{k-1}} x)+Tr(x^{\frac{k-1}{k}})} = H_{\frac{k-1}{k}}(\lambda^{-\frac{1}{k-1}}). \tag{9}$$

From (9), we consider a theorem for the Glynn type II hyperoval sequences.

Theorem 3. *Let n be odd and $k = 3\cdot 2^\sigma + 4$ where $\sigma = \frac{n+1}{2}$. Let $h_k(x)$ be a trace representation of the Glynn type II hyperoval sequences. For $s = \frac{1}{k-1}$, crosscorrelation of m-sequences and the Glynn type II hyperoval sequences given by*

$$C_{Tr,h_k^{(s)}}(\lambda) = \sum_{x\in\mathbb{F}_{2^n}} (-1)^{Tr(\lambda x)+h_k(x^s)} = H_{\frac{k-1}{k}}(\lambda^{-s}) \tag{10}$$

is at most 5-valued, i.e., $\{0, \pm 2^{\frac{n+1}{2}}, \pm 2^{\frac{n+3}{2}}\}$.

Proof. From (9), $C_{Tr,h_k^{(s)}}(\lambda)$ is determined by a decimation factor $\frac{k-1}{k}$ of a trace function. Note that the cyclotomic coset that $\frac{k-1}{k}$ belongs to does not change by multiplying its numerator and denominator by $2^{\frac{n-1}{2}}$ and $2^{\frac{n-3}{2}}$, respectively. Then,

$$\begin{aligned}\frac{k-1}{k} &\equiv \frac{2^{\frac{n-1}{2}}}{2^{\frac{n-3}{2}}}\cdot\frac{(k-1)}{k} = \frac{2^{\frac{n-1}{2}}}{2^{\frac{n-3}{2}}}\cdot\frac{3\cdot(1+2^{\frac{n+1}{2}})}{(2^{\frac{n+3}{2}}+2^{\frac{n+1}{2}}+4)} \equiv \frac{3\cdot(1+2^{\frac{n-1}{2}})}{(1+2^{\frac{n-1}{2}})^2}\\ &= \frac{1+2}{1+2^{\frac{n-1}{2}}} \pmod{2^n-1}.\end{aligned}$$

Hence, $\frac{k-1}{k} \equiv \frac{1+2^\mu}{1+2^\nu} = d(\mu,\nu)$ in Proposition 1 where $\mu = 1$ and $\nu = \frac{n-1}{2}$. Since $2\nu = n-\mu$, we have $(n-\mu,\nu) = (2t,t)$ with $t = \frac{n-1}{2}$, a pair of Proposition 1-(c). From $e = \gcd(n,t) = \gcd(n, \frac{n-1}{2}) = 1$, we see that $H_{d(n-\mu,\nu)}(\lambda)$ is at most 5-valued and so is $H_{d(\mu,\nu)}(\lambda)$ from Lemma 1. □

In terms of the Glynn type I hyperoval sequences, on the other hand, $k = 2^\sigma + 2^\tau$ where $\sigma = \frac{n+1}{2}$ and $\tau = \frac{n+1}{4}$ or $\tau = \frac{3n+1}{4}$ such that $4\tau \equiv 1 \pmod{n}$. Using the similar approach to the proof of Theorem 3, we can establish the following equivalence of $\frac{k-1}{k}$.

$$\frac{k-1}{k} \equiv \begin{cases} 2^{\frac{n-1}{2}} - 2^{\frac{n+1}{4}} + 1, & \text{if } \tau = \frac{n+1}{4} \\ 2^{\frac{n+1}{2}} - 2^{\frac{n+3}{4}} + 1, & \text{if } \tau = \frac{3n+1}{4}. \end{cases} \tag{11}$$

In (11), we see that $\frac{k-1}{k}$ is equivalent to the decimation factor r in Conjecture 4-6 (3) and (4) of [13], where $H_r(\lambda)$ is conjectured to be at most 5-valued. Together with our experimental results, we establish the following conjecture.

Conjecture 1. Let n be odd and $k = 2^\sigma + 2^\tau$ where $\sigma = \frac{n+1}{2}$ and $4\tau \equiv 1 \pmod{n}$. Let $h_k(x)$ be a trace representation of the Glynn type I hyperoval sequences. For $s = \frac{1}{k-1}$, crosscorrelation of m-sequences and the Glynn type I hyperoval sequences given by $C_{Tr,h_k^{(s)}}(\lambda)$ in (10) is at most 5-valued, i.e., $\{0, \pm 2^{\frac{n+1}{2}}, \pm 2^{\frac{n+3}{2}}\}$.

With respect to crosscorrelation of m-sequences and the Glynn type II hyperoval sequences, we also observed another exponent corresponding to at most 5-valued crosscorrelation. Together with (4) which is conjectured to be at most 5-valued for the Glynn type II hyperoval sequences, we establish Conjecture 2.

Conjecture 2. For odd n, let $h_k(x)$ be a trace representation of the Glynn type II hyperoval sequences. For $s = 1$ or $\frac{1}{3}$, crosscorrelation of m-sequences and the Glynn type II hyperoval sequences given by $C_{Tr,h_k^{(s)}}(\lambda)$ is at most 5-valued, i.e., $\{0, \pm 2^{\frac{n+1}{2}}, \pm 2^{\frac{n+3}{2}}\}$.

Conjectures 1 and 2 have been verified for odd $n = 9 - 19$ through computer experiments.

4.5 m-Sequences and 3-Term KPF Sequences

In [3], the 3-term KPF sequences are represented by

$$b_k(x) = Tr(x + x^{2^k+1} + x^{2^k-1}), \quad k = \frac{n+1}{2}$$

where n is odd. On the other hand, *T3 sequences*, or 3-term sequences with ideal two-level autocorrelation which had been conjectured in [15] are represented by

$$T_3(x) = Tr(x + x^r + x^{r^2}), \quad r = 2^{\frac{n-1}{2}} + 1.$$

With the equivalence under modulo $2^n - 1$, we see that the T3 sequences are decimation of the 3-term KPF sequences, i.e., $T_3(x) = b_k(x^{2^k+1})$ where $k = \frac{n+1}{2}$. Using this relation, we establish the following theorem.

Theorem 4. *Let n be odd and $k = \frac{n+1}{2}$. Let $b_k(x)$ be a trace representation of the 3-term KPF sequences. For $s = 2^k - 1$, crosscorrelation of m-sequences and the 3-term KPF sequences given by*

$$C_{Tr,b_k^{(s)}}(\lambda) = \sum_{x \in \mathbb{F}_{2^n}} (-1)^{Tr(\lambda x) + b_k(x^s)}$$

is at most 5-valued, i.e., $\{0, \pm 2^{\frac{n+1}{2}}, \pm 2^{\frac{n+3}{2}}\}$.

Proof. In [2], Chang *et al.* showed that a binary cyclic code represented by

$$\mathcal{T} = \{Tr(ax + bx^r + cx^{r^2}) | a, b, c \in \mathbb{F}_{2^n}, r = 2^{\frac{n-1}{2}} + 1\}$$

is a dual of a triple error correcting cyclic code and has five nonzero distinct weights. Then, crosscorrelation of m-sequences and the T3 sequences given by $C_{Tr,T_3}(\lambda) = \sum_{x \in \mathbb{F}_{2^n}} (-1)^{Tr(\lambda x) + T_3(x)}$ is at most 5-valued - in fact, 3-valued - because the exponent in the summation is a codeword of $\mathcal{T}$. In the following, we can consider another at most 5-valued crosscorrelation $C_{Tr^{(r^2)},T_3}(\lambda)$ where the exponent is also a codeword of $\mathcal{T}$. Note that $2^{\frac{n+1}{2}} \cdot r = 2^n + 2^{\frac{n+1}{2}} \equiv 1 + 2^k \pmod{2^n - 1}$, and thus $r \equiv 2^k + 1$. Therefore, $T_3(x) = b_k(x^r)$ where $k = \frac{n+1}{2}$. Then,

$$C_{Tr^{(r^2)},T_3}(\lambda) = \sum_{x \in \mathbb{F}_{2^n}} (-1)^{Tr(\lambda^{r^2} x^{r^2}) + T_3(x)} = \sum_{x \in \mathbb{F}_{2^n}} (-1)^{Tr(\lambda^{r^2} x) + T_3(x^{r^{-2}})}$$

$$= \sum_{x \in \mathbb{F}_{2^n}} (-1)^{Tr(\lambda^{r^2} x) + b_k(x^{r^{-1}})} = C_{Tr,b_k^{(s)}}(\lambda^{r^2})$$

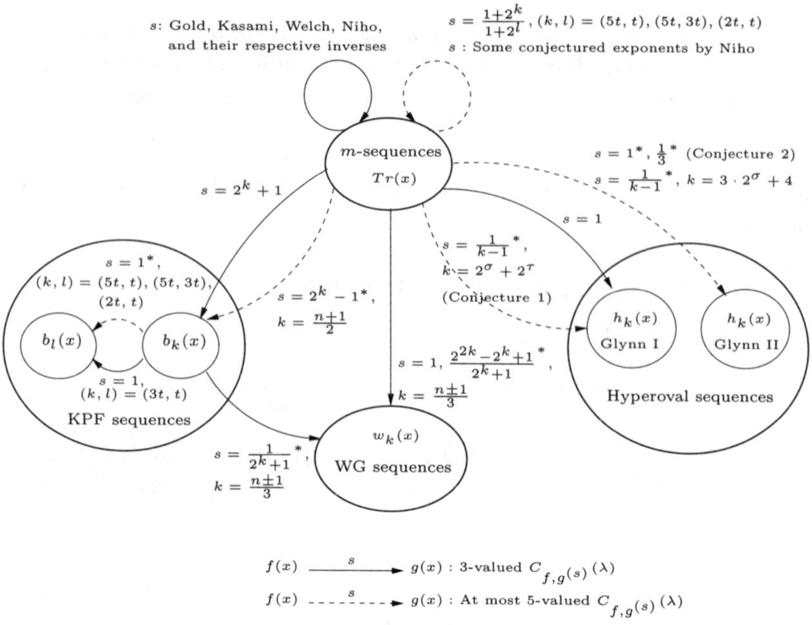

Fig. 1. Relations of binary two-level autocorrelation sequences with respect to 3- or 5-valued crosscorrelation ($\gcd(n,t)=1$, $\sigma = \frac{n+1}{2}$, and $4\tau \equiv 1 \pmod{n}$). The crosscorrelations corresponding to the exponents s with '$*$' are proved or conjectured in this paper.

where $s = r^{-1} = (2^k+1)^{-1} \equiv 2^k - 1 \pmod{2^n - 1}$. Hence, $C_{Tr,b_k^{(s)}}(\lambda)$ is at most 5-valued. □

5 Conclusion and Discussion

In this paper, we have studied 3- or 5-valued crosscorrelation of a pair of binary two-level autocorrelation sequences given by

$$C_{f,g^{(s)}}(\lambda) = \sum_{x \in \mathbb{F}_{2^n}} (-1)^{f(\lambda x) + g(x^s)}$$

where n is odd, and $f(x)$ and $g(x)$ are trace representations of the pair, excluding GMW, QR, and Hall's sextic residue sequences.

If $f(x) = g(x) = Tr(x)$, all known exponents s's of 3- or 5-valued $C_{f,g^{(s)}}(\lambda)$ are (a) Gold, Kasami, Welch, Niho exponents, and their respective inverses; (b) the exponents of Proposition 1; (c) the other exponents conjectured by Niho [13] and their inverses. Otherwise, all known exponents s's of 3- or 5-valued $C_{f,g^{(s)}}(\lambda)$ for the corresponding $f(x)$ and $g(x)$ are (a) $s = 2^k + 1$ from (2), or $s = 1$ from (3) and (4); (b) s's from Fact 2 and Theorems 1 - 4; (c) s's from Conjectures 1 and 2. With the classification, we can summarize relations of binary two-level autocorrelation sequences with respect to 3- or 5-valued crosscorrelation by

Fig. 1, where a solid line is for exactly 3-valued crosscorrelation and a dotted line for at most 5-valued crosscorrelation. (For small n, it may be 3-valued in some cases.) In Fig. 1, the crosscorrelations corresponding to the exponents s with '$*$' are proved or conjectured in this paper.

From the observation of our experiments for $n = 13, 15, 17$, and 19, it is interesting that the exponents and relations in Fig. 1 completely describe all 3- or 5-valued crosscorrelations of binary two-level autocorrelation sequences unless both are m-sequences.

Acknowledgment

The authors would like to thank Professor Tor Helleseth for helpful comments on Proposition 1.

References

1. Antweiler, M.: Cross-correlation of p-ary GMW sequences. IEEE Trans. Inform. Theory, Vol. 40 (1994) 1253-1261
2. Chang, A., Gaal, P., Golomb, S. W., Gong, G., Helleseth, T., Kumar, P. V.: On a conjectured ideal autocorrelation sequence and a related triple-error correcting cyclic code. IEEE Trans. Inform. Theory, Vol. 46-2 (2000) 680-687
3. Dillon, J. F., Dobbertin, H.: New cyclic difference sets with Singer parameters. Finite Fields and Their Applications 10 (2004) 342-389
4. Dillon, J. F.: Multiplicative difference sets via additive characters. Designs, Codes and Cryptography, Vol. 17 (1999) 225-235
5. Games, R. A.: Crosscorrelation of m-sequences and GMW-sequences with the same primitive polynomial. Discrete Applied Mathematics, Vol. 12 (1985) 139-146
6. Gold, R.: Maximal recursive sequences with 3-valued recursive cross-correlation functions. IEEE Trans. Inform. Theory, Vol. 14 (1968) 154-156
7. Gong, G., Golomb, S. W.: The decimation-Hadamard transform of two-level autocorrelation sequences. IEEE Trans. Inform. Theory, Vol. 48-4 (2002) 853-865
8. Gordon, B., Mills, W. H., Welch, L. R.: Some new difference sets. Canadian Journal of Mathematics, Vol. 14-4 (1962) 614-625
9. Helleseth, T., Kumar, P. V.: Sequences with Low Correlation. A chapter in Handbook of Coding Theory. Edited by Pless, V. and Huffmann, C. Elsevier Science Publishers (1998)
10. Hertel, D.: Cross-correlation properties of perfect binary sequences. Lecture Notes in Computer Science, Vol. 3486. Edited by Helleseth, T. et al. Springer-Verlag (2005) 208-219
11. Kasami, T.: Weight enumerators for several classes of subcodes of the 2nd-order Reed-Muller codes. Information and Control, Vol. 18 (1971) 369-394
12. Maschietti, A.: Difference sets and hyperovals. Designs, Codes and Cryptography, Vol. 14 (1998) 89-98
13. Niho, Y.: Multi-valued cross-correlation functions between two maximal linear recursive sequences. Ph.D. Dissertation. University of Southern California (1972)
14. No, J. S., Chung, H. C., Yun, M. S.: Binary pseudorandom sequences of period $2^m - 1$ with ideal autocorrelation generated by the polynomial $z^d + (z+1)^d$. IEEE Trans. Inform. Theory, Vol.44-3 (1998) 1278-1282

15. No, J. S., Golomb, S. W., Gong, G., Lee, H. K., Gaal, P.: Binary pseudorandom sequences of period $2^m - 1$ with ideal autocorrelation. IEEE Trans. Inform. Theory, Vol.44-2 (1998) 814-817
16. Sidelnikov, V. M.: On mutual correlation of sequences. Soviet Math. Dokl, Vol. 12 (1971) 197-201
17. Welch, L. R.: Lower bounds on the maximum cross correlation of signals. IEEE Trans. Inform. Theory, Vol. IT-20 (1974) 397-399

Extended Hadamard Equivalence

Doreen Hertel

Institute for Algebra and Geometry, Faculty of Mathematics
Otto-von-Guericke-University Magdeburg, Postfach 4120, 39016 Magdeburg, Germany
doreen.hertel@mathematik.uni-magdeburg.de

Abstract. Binary sequences with good autocorrelation properties are widely used in cryptography. If the autocorrelation properties are optimum, then the sequences are called perfect. All recently discovered perfect sequences of period $n = 2^k - 1$ are Hadamard equivalent, when k is odd. In this paper we generalise this concept to sequences of period $n = 4m - 1$, where m is not necessarily a power of 2. Using this notion we show, that the Hall and the Legendre sequences are extended Hadamard equivalent.

Keywords: Binary sequence, Perfect sequence, Autocorrelation, Cross-correlation, Hadamard equivalence.

1 Introduction and Definitions

New classes of perfect sequences of period $n = 2^k - 1$ have been found in [2] by Dillon and Dobbertin. For this spectacular result, a new type of "equivalence of functions" has been defined: The powerful tool in [2] is the Hadamard equivalence. The basic is that Hadamard equivalent sequences have the same autocorrelation spectra. This concept has been generalised by Gong and Golomb in [3]. Based on this equivalence, a method is given to construct new perfect sequences of period $n = 2^k - 1$, see [3]. Unfortunately, no new perfect sequences have been found for $k \leq 17$. We will extend the concept of Hadamard equivalence to sequences of period $n = 4m - 1$.

In the first section we give some basic definitions. In the second section we generalise the Hadamard equivalence to sequences of period $n = 4m - 1$: We will call it **extended Hadamard equivalence**. It turns out that extended Hadamard equivalent sequences have the same autocorrelation spectra. In the case $n = 4m - 1$ and m is not a power of 2, three construction for perfect sequences are known: the Legendre ($n \equiv 3 \bmod 4$ prime), the Hall ($n = 4t^2 + 27$ prime) and the twin prime ($n = p(p+2)$, p prime) sequences. In the last section we show, that the Legendre and the Hall sequences of the same period are equivalent under our new notion.

We consider binary sequences $\underline{a} := (a_i)_{i \geq 0}$ of **period** n, i.e. $a_{i+n} = a_i$ for all $i \geq 0$. We define the **autocorrelation (AC)** of a n-periodic sequence $\underline{a}$ by

$$c_t(\underline{a}) = \sum_{i=0}^{n-1} (-1)^{a_i + a_{i+t}}$$

G. Gong et al. (Eds.): SETA 2006, LNCS 4086, pp. 119–128, 2006.

for all $t = 0, ..., n-1$. We consider the indices modulo n. The set of all AC-coefficients are called AC-spectrum. We define the binary complement $\overline{\underline{a}} = (\overline{a}_i)_{i\geq 0}$ of a sequence $\underline{a}$ by $\overline{a}_i := a_i + 1$, the shift $\underline{a}^{[t]} = (a_i^{[t]})_{i\geq 0}$ by $a_i^{[t]} := a_{i+t}$ and the decimation $\underline{a}^{(d)} = (a^{(d)})_{i\geq 0}$ with $\gcd(d,n) = 1$ by $a_i^{(d)} := a_{id}$. Note the autocorrelation is invariant under this three transformations. We call two sequences $\underline{a}$ and $\underline{b}$ **equivalent**, if $\underline{a}$ can be formed into $\underline{b}$ by this transformations.

Binary sequences with low AC-coefficients are important in cryptography. The most known examples for sequences with small AC-coefficient have period $n \equiv 3 \bmod 4$. We call a sequence $\underline{a}$ **perfect**, if the AC-function takes just the two values

$$c_t(\underline{a}) = \begin{cases} -1 & \text{if } t \not\equiv 0 \bmod n \\ n & \text{otherwise.} \end{cases}$$

A sequence is **balanced**, if the number of 1's and 0's in one period differs only by one. We have for perfect sequences

$$\left(\sum_{i=0}^{n-1}(-1)^{a_i}\right)^2 = \sum_{t=0}^{n-1}\sum_{i=0}^{n-1}(-1)^{a_i+a_{i+t}} = (-1)(n-1) + n = 1.$$

Thus, perfect sequences are always balanced. We assume without loss of generality, that perfect sequences have $\frac{n+1}{2}$ entries 1 and $\frac{n-1}{2}$ entries 0 in one period, thus

$$\sum_{i=0}^{n-1}(-1)^{a_i} = -1, \tag{1}$$

otherwise we consider the binary complement.

We define the **crosscorrelation (CC)** between two sequences $\underline{a} := (a_i)_{i\geq 0}$ and $\underline{b} := (b_i)_{i\geq 0}$ by

$$c_t(\underline{a}, \underline{b}) = \sum_{i=0}^{n-1}(-1)^{a_i+b_{i+t}}$$

for all $t = 0, ..., n-1$.

In the next section we will use the crosscorrelation to develop a method to construct sequences with specified AC-properties. This method can also be used to prove that certain sequences are perfect. The basic idea is a generalisation of the Hadamard equivalence introduced in [2]. Hadamard equivalence has been used for sequences of period $n = 2^m - 1$: One can show that certain sequences of period $n = 2^m - 1$ are perfect. The specific feature of sequences with period $n = 2^m - 1$ is that they can be identified with Boolean functions over finite fields of characteristic 2.

We will outline the concept of Hadamard equivalence: We denote the finite field with 2^m elements by $\mathbb{F}_{2^m}$. The reader is referred to [5] for more information on the theory of finite fields. Let α be a primitive element in $\mathbb{F}_{2^m}$ and let $\underline{a}$ be a binary sequence of period $n = 2^m - 1$. Then we can identify $\underline{a}$ with a Boolean function $f : \mathbb{F}_{2^m} \to \mathbb{F}_2$ by

$$f(\alpha^i) := a_i \tag{2}$$

and the value $f(0)$ is irrelevant. If $\underline{a}$ is balanced, then we choose $f(0) \in \{0, 1\}$ such that $\sum_{x \in \mathbb{F}_{2^m}} (-1)^{f(x)} = 0$. We call a Boolean function $f : \mathbb{F}_{2^m} \to \mathbb{F}_2$ perfect, if the corresponding function defined by (2) is perfect. Note, for perfect functions we have

$$\sum_{x \in \mathbb{F}_{2^m}} (-1)^{f(x)+f(yx)} = \begin{cases} 2^m & \text{if } y = 1 \\ 0 & \text{otherwise.} \end{cases}$$

This orthogonality property will be important for the concept of Hadamard equivalence: Let $f, g, h_1, h_2 : \mathbb{F}_{2^m} \to \mathbb{F}_2$ be Boolean functions and d be an integer such that

$$\sum_{x \in \mathbb{F}_{2^m}} (-1)^{f(x)+h_1(y^d x)} = \sum_{x \in \mathbb{F}_{2^m}} (-1)^{g(x)+h_2(yx)} \tag{3}$$

holds for all $y \in \mathbb{F}_{2^m}$. Then

- [1,2]: If $h_1 = h_2$ is the trace function, then the functions f and g are **Hadamard equivalent**. In particularly, if f is perfect, then g is perfect, too. Hadamard equivalence is a powerful tool to prove that functions are perfect. The main idea in the proofs given in [1,2] is, that the functions are Hadamard equivalent.
- [3]: If $h_1 = h_2$ is an arbitrary perfect function, then the functions f and g have the same autocorrelation spectra. In particularly, if f is perfect, then g is perfect, too. Using this slight generalisation of Hadamard equivalence, an algorithm for constructing perfect functions is developed. Unfortunately, no new perfect functions have been found for $m \leq 17$.

In the following we will generalise the concept of Hadamard equivalence to sequences of period $n = 4m-1$, where h_1 and h_2 are arbitrary perfect sequences.

2 Extended Hadamard Equivalence

We generalise the idea of Hadamard equivalence to sequences of period $n = 4m - 1$. We will call this extended Hadamard equivalence. Based on this new equivalence we will give in this section an algorithm to construct perfect sequences of period $n = 4m - 1$.

We define the modified autocorrelation and crosscorrelation for two binary sequences $\underline{a}$ and $\underline{b}$ of period $n = 4m - 1$ by

$$c_t^*(\underline{a}) := c_t(\underline{a}) + 1 \text{ and } c_t^*(\underline{a}, \underline{b}) := c_t(\underline{a}, \underline{b}) + 1$$

for all $t = 0, ..., n-1$. Note, two sequences have the same AC- resp. CC-function if and only if they have the same modified AC- resp. CC-function. For perfect sequences we get the orthogonality property

$$c_t^*(\underline{a}) = \begin{cases} 0 & \text{if } t \not\equiv 0 \bmod n \\ n+1 & \text{if } t \equiv 0 \bmod n. \end{cases}$$

We also define

$$w^*(\underline{a}) := \sum_{i=0}^{n-1} (-1)^{a_i} + 1.$$

If $\underline{a}$ is perfect, using (1) we get $w^*(\underline{a}) = 0$. The next two propositions are well-known for sequences over finite fields (Inverse and Parseval formula).

Proposition 1. *Let $\underline{a} = (a_i)_{i \geq 0}$ and $\underline{d} = (d_i)_{i \geq 0}$ be binary sequences of period $n = 4m - 1$ and $\underline{d}$ be perfect. Then*

$$(-1)^{a_t} = \frac{1}{n+1}\Big(\sum_{k=0}^{n-1} c_k^*(\underline{a}, \underline{d})(-1)^{d_{k+t}} + w^*(\underline{a})\Big). \tag{4}$$

Proposition 1 shows, that we can reconstruct the sequence $\underline{a}$, if we know $\underline{d}$ and the CC-coefficients $c_t^*(\underline{a}, \underline{d}), t = 0, ..., n-1$, since we compute $w^*(\underline{a})$ from the CC-coefficients by

$$\begin{aligned} \sum_{k=0}^{n-1} c_k^*(\underline{a}, \underline{d}) &= \sum_{k=0}^{n-1} \Big(\sum_{i=0}^{n-1} (-1)^{a_i + d_{i+k}} + 1 \Big) \\ &= \sum_{i=0}^{n-1} (-1)^{a_i} \sum_{k=0}^{n-1} (-1)^{d_{i+k}} + n \\ &= -w^*(\underline{a}) + 1 + n. \end{aligned} \tag{5}$$

Proof. We simply transform the right side of equation (4) and we get

$$\begin{aligned} \sum_{k=0}^{n-1} c_k^*(\underline{a}, \underline{d})(-1)^{d_{k+t}} &= \sum_{k=0}^{n-1} (c_k(\underline{a}, \underline{d}) + 1)(-1)^{d_{k+t}} \\ &= \sum_{k=0}^{n-1} \sum_{i=0}^{n-1} (-1)^{a_i + d_{i+k} + d_{k+t}} + \underbrace{\sum_{k=0}^{n-1} (-1)^{d_{k+t}}}_{=-1} \\ &= \sum_{i=0}^{n-1} (-1)^{a_i} \underbrace{\Big(\sum_{k=0}^{n-1} (-1)^{d_{i+k} + d_{k+t}} + 1 \Big)}_{=c_{t-i}^*(\underline{b})} - \sum_{j=0}^{n-1} (-1)^{a_j} - 1 \\ &= (n+1) \cdot (-1)^{a_t} - w^*(\underline{a}), \end{aligned}$$

since $\underline{d}$ is perfect. □

Proposition 2. *Let $\underline{a}, \underline{b}$ and $\underline{d}$ be binary sequences of period $n = 4m - 1$ and $\underline{d}$ be perfect. Then*

$$c_t^*(\underline{a}, \underline{b}) = \frac{1}{n+1}\Big(\sum_{k=0}^{n-1} c_k^*(\underline{a}, \underline{d}) c_{k-t}^*(\underline{b}, \underline{d}) + w^*(\underline{a}) w^*(\underline{b})\Big). \tag{6}$$

Proof. Let $\underline{a} = (a_i)_{i \geq 0}, \underline{b} = (b_i)_{i \geq 0}$ and $\underline{d} = (d_i)_{i \geq 0}$. We expand

$$\begin{aligned}
&\sum_{k=0}^{n-1} c_k^*(\underline{a}, \underline{d}) c_{k-t}^*(\underline{b}, \underline{d}) \\
&= \sum_{k=0}^{n-1} \Big(\sum_{i=0}^{n-1} (-1)^{a_i + d_{i+k}} + 1\Big) \cdot \Big(\sum_{j=0}^{n-1} (-1)^{b_j + d_{j+k-t}} + 1\Big) \\
&= \sum_{i=0}^{n-1} \sum_{j=0}^{n-1} (-1)^{a_i + b_j} \sum_{k=0}^{n-1} (-1)^{d_{i+k} + d_{j+k-t}} \\
&\quad + \sum_{i=0}^{n-1} (-1)^{a_i} \underbrace{\sum_{k=0}^{n-1} (-1)^{d_{i+k}}}_{=-1} + \sum_{j=0}^{n-1} (-1)^{b_j} \underbrace{\sum_{k=0}^{n-1} (-1)^{d_{j+k-t}}}_{=-1} + n \\
&= \sum_{i=0}^{n-1} \sum_{j=0}^{n-1} (-1)^{a_i + b_j} \underbrace{\Big(\sum_{k=0}^{n-1} (-1)^{d_{i+k} + d_{j+k-t}} + 1\Big)}_{=c_{j-t-i}^*(\underline{d})} + n + 1 \\
&\quad - \Big(\sum_{i=0}^{n-1} \sum_{j=0}^{n-1} (-1)^{a_i + b_j} + \sum_{i=0}^{n-1} (-1)^{a_i} + \sum_{j=0}^{n-1} (-1)^{b_j} + 1\Big),
\end{aligned}$$

where we insert $0 = \sum_{i,j=0}^{n-1} (-1)^{a_i + b_j} + 1 - (\sum_{i,j=0}^{n-1} (-1)^{a_i + b_j} + 1)$. Since $\underline{d}$ is perfect we get

$$\begin{aligned}
&\sum_{k=0}^{n-1} c_k^*(\underline{a}, \underline{d}) c_{k-t}^*(\underline{b}, \underline{d}) \\
&= (n+1)\Big(\sum_{i=0}^{n-1} (-1)^{a_i + b_{i+t}} + 1\Big) - \Big(\sum_{i=0}^{n-1} (-1)^{a_i} + 1\Big)\Big(\sum_{j=0}^{n-1} (-1)^{b_j} + 1\Big) \\
&= (n+1) c_t^*(\underline{a}, \underline{b}) - w^*(\underline{a}) w^*(\underline{b}).
\end{aligned}$$

□

We call two binary sequences $\underline{a}$ and $\underline{b}$ of period $n = 4m - 1$ **extended Hadamard equivalent (EH-equivalent)**, if there exist two perfect sequences $\underline{d}$ and $\underline{e}$ and integers s, t with $\gcd(s, n) = 1$ such that

$$c_k(\underline{a}, \underline{d}) = c_{sk+t}(\underline{b}, \underline{e}) \quad (\text{resp. } c_k^*(\underline{a}, \underline{d}) = c_{sk+t}^*(\underline{b}, \underline{e})) \tag{7}$$

holds for all k. Using (5) for EH-equivalent sequences $\underline{a}$ and $\underline{b}$, then $w^*(\underline{a}) = w^*(\underline{b})$.

Proposition 3. *Let $\underline{a}$ and $\underline{b}$ be binary sequences of period $n = 4m - 1$. If $\underline{a}$ and $\underline{b}$ are EH-equivalent, then the AC-spectra of $\underline{a}$ and $\underline{b}$ are equal.*

Proof. If $\underline{a}$ and $\underline{b}$ are EH-equivalent, then there exist two perfect sequences $\underline{d}$ and $\underline{e}$ and integers s, t with $\gcd(s, n) = 1$ such that (7) holds. Since $c_i^*(\underline{a}) = c_i^*(\underline{a}, \underline{a})$ we use formula (6) and get

$$\begin{aligned}(n+1)c_i^*(\underline{a}) &= \sum_{k=0}^{n-1} c_k^*(\underline{a}, \underline{d})c_{k-i}^*(\underline{a}, \underline{d}) + w^*(\underline{a})^2 \\ &= \sum_{k=0}^{n-1} c_{sk+t}^*(\underline{b}, \underline{e})c_{s(k-i)+t}^*(\underline{b}, \underline{e}) + w^*(\underline{b})^2 \\ &= \sum_{k=0}^{n-1} c_k^*(\underline{b}, \underline{e})c_{k-si}^*(\underline{b}, \underline{e}) + w^*(\underline{b})^2 \\ &= (n+1)c_{si}^*(\underline{b}).\end{aligned}$$

□

Let $\underline{a} = (a_i)_{i\geq 0}, \underline{d} = (d_i)_{i\geq 0}$ and $\underline{e} = (e_i)_{i\geq 0}$ be binary sequences of period $n = 4m - 1$ and let $\underline{d}$ and $\underline{e}$ be perfect. Let z_1, z_2, z_3 be integers with $\gcd(z_i, n) = 1$, i=1,2,3, such that

$$\Big(\sum_{k=0}^{n-1} c_k^*(\underline{a}^{(z_1)}, \underline{d}^{(z_2)})(-1)^{e_{k+i}^{(z_3)}} + w^*(\underline{a}^{(z_1)})\Big) \in \{\pm(n+1)\}. \tag{8}$$

Then we call the binary sequence $\underline{b} = (b_i)_{i\geq 0}$ defined by

$$(-1)^{b_i} = \frac{1}{n+1}\Big(\sum_{k=0}^{n-1} c_k^*(\underline{a}^{(z_1)}, \underline{d}^{(z_2)})(-1)^{e_{k+i}^{(z_3)}} + w^*(\underline{a}^{(z_1)})\Big) \tag{9}$$

a **realisation** of $\underline{a}, \underline{d}, \underline{e}$ by the triple (z_1, z_2, z_3).

Note, in the case $\underline{d} = \underline{e}$ there exists always a realisation: the trivial realisation with $z_2 = z_3$. Then it is $\underline{b} = \underline{a}^{(z_1)}$.

Theorem 1. *Let $\underline{a}$, $\underline{d}$ and $\underline{e}$ be binary sequences of period $n = 4m - 1$ and let $\underline{d}$ and $\underline{e}$ be perfect. Let z_1, z_2, z_3 be integers with $\gcd(z_i, n) = 1$, i=1,2,3, such that (8) holds. Then the binary sequence $\underline{b} = (b_i)_{i\geq 0}$ defined by (9) and $\underline{a}$ have the same AC-spectrum.*

Note the sequence $\underline{b}$ is uniquely defined by the perfect sequence $\underline{e}^{(z_3)}$ and its CC-coefficients $c_k^*(\underline{b}, \underline{e}^{(z_3)})$ $(:= c_k^*(\underline{a}^{(z_1)}, \underline{d}^{(z_2)}))$, see Proposition 1.

Proof. We show, the sequences $\underline{a}^{(z_1)}$ and $\underline{b}$ are EH-equivalent. Then $\underline{a}$ and $\underline{b}$ have the same AC-spectrum, since $\underline{a}$ is equivalent to $\underline{a}^{(z_1)}$. We get

$$\begin{aligned}&(n+1)c_i(\underline{b}, \underline{e}^{(z_3)}) \\ &= (n+1)\sum_{j=0}^{n-1}(-1)^{b_j + e_{j+i}^{(z_3)}}\end{aligned}$$

$$= \sum_{j=0}^{n-1} \Big(\sum_{k=0}^{n-1} c_k^*(\underline{a}^{(z_1)}, \underline{d}^{(z_2)})(-1)^{e_{k+j}^{(z_3)}} + w^*(\underline{a}^{(z_1)}) \Big)(-1)^{e_{j+i}^{(z_3)}}$$

$$= \sum_{k=0}^{n-1} c_k^*(\underline{a}^{(z_1)}, \underline{d}^{(z_2)}) \sum_{j=0}^{n-1} (-1)^{e_{k+j}^{(z_3)} + e_{j+i}^{(z_3)}} + w^*(\underline{a}^{(z_1)}) \underbrace{\sum_{j=0}^{n-1} (-1)^{e_{j+i}^{(z_3)}}}_{=-1}$$

$$= \sum_{k=0}^{n-1} c_k^*(\underline{a}^{(z_1)}, \underline{d}^{(z_2)}) \underbrace{\Big(\sum_{j=0}^{n-1} (-1)^{e_{k+j}^{(z_3)} + e_{j+i}^{(z_3)}} + 1 \Big)}_{=c_{i-k}^*(\underline{e}^{(z_3)})} - \sum_{k=0}^{n-1} c_k^*(\underline{a}^{(z_1)}, \underline{d}^{(z_2)}) - w^*(\underline{a}^{(z_1)})$$

$$= (n+1) c_i^*(\underline{a}^{(z_1)}, \underline{d}^{(z_2)}) - \sum_{k=0}^{n-1} c_k^*(\underline{a}^{(z_1)}, \underline{d}^{(z_2)}) - w^*(\underline{a}^{(z_1)}),$$

since $\underline{e}$ is perfect. Since $\underline{d}$ perfect we get from (5) that $(n+1)c_i(\underline{b}, \underline{e}^{(z_3)}) = (n+1)c_i^*(\underline{a}^{(z_1)}, \underline{d}^{(z_2)}) - (n+1) = (n+1)c_i(\underline{a}^{(z_1)}, \underline{d}^{(z_2)})$. □

We have developed a method to construct sequences with specified autocorrelation. Algorithm idea: Take three shift distinct perfect sequences and check for all possible integers $z_i, i = 1, 2, 3$, if there exists a realisation of these sequences. The big handycap by this algorithm is, that we need three perfect sequences, which are pairwise shift distinct.

If $n = 4m-1$ and m is not a power of 2, in this case we only have at least three (known) shift distinct sequences if $n = 4t^2 + 27$ prime: the Hall and Legendre sequences. The algorithm gives by input Hall and Legendre sequences no new perfect sequences for $n = 4t^2 + 27$ with $t \leq 77$. But we get an other interesting result, which we present in the next section.

3 EH-Equivalence of Legendre and Hall Sequences

All known perfect sequences of period $n = 4m - 1$, where m is not a power of 2, are defined by cyclotomic classes.

Let n be a prime, then $\mathbb{Z}_n$ is a finite field with additive group $\mathbb{Z}_n$ and multiplicative group $\mathbb{Z}_n^* = \mathbb{Z}_n \backslash \{0\}$. The multiplicative group is cyclic, thus $\mathbb{Z}_n^* = \langle z \rangle$. In the following we fix z as a primitive element in $\mathbb{Z}_n$.

Let D be a subset of $\mathbb{Z}_n$. We define the **translate** $D+t$ by $D+t := \{ i + t \bmod n \mid i \in D\}$ and the **decimation** sD by $sD := \{ si \bmod n \mid i \in D\}$, where $\gcd(s,n) = 1$. We also define the corresponding sequence $\underline{a}$ of D by $\underline{a} := seq(D)$, where $a_i = 0$ if $i \in D$ and $a_i := 1$ otherwise.

Let $n = ef + 1$ be prime. We define the cyclotomic classes $C_i^{(e)}$ in $\mathbb{Z}_n$ by

$$C_i^{(e)} := \{ z^{es+i} \bmod n \mid s = 0, ..., f-1\}$$

for $i = 0, ..., e-1$. Note, the sets $C_i^{(e)}$ are pairwise disjoint, and their union is $\mathbb{Z}_n^*$. Further $C_{i+ne}^{(e)} = C_i^{(e)}$, thus we consider the indices modulo e. We define the subsets

$$QR := C_0^{(2)} \quad \text{and} \quad H := C_0^{(6)} \cup C_1^{(6)} \cup C_3^{(6)}. \tag{10}$$

The sequences $\underline{s}_{QR} := seq(QR)$ are called **Legendre sequences** and the sequences $\underline{s}_H := seq(H)$ are called **Hall (sextic residue) sequences**. The Legendre sequence is perfect if $n \equiv 3 \bmod 4$, see [6], and the Hall sequence is perfect if $n = 4t^2 + 27$, see [4]. It is easy to see from the definition of QR and H, that the sequences $\underline{s}_{QR}$ and $\underline{s}_H$ are not equivalent.

Theorem 2. *The Hall sequences and the Legendre sequences are EH-equivalent. More precisely, we have*

$$c_{zk}(\underline{s}_H^{(z)}, \underline{s}_H) = c_k(\underline{s}_{QR}, \underline{s}_H) \tag{11}$$

for all $k = 0, ..., n-1$.

In other words, the Legendre sequence is a realisation of the Hall sequences by $(z, z, 1)$.

Proof. Using the well known correspondence between sets and binary sequences (as indicated above), it is easy to see that

$$\begin{aligned} c_{zk}(\underline{s}_H^{(z)}, \underline{s}_H) &= -n + 2 + 4|(H - zk) \cap z^{-1}H| \text{ and} \\ c_k(\underline{s}_{QR}, \underline{s}_H) &= -n + 2 + 4|(H - k) \cap QR| \end{aligned} \tag{12}$$

holds for all $k = 0, ..., n-1$. We simply write C_i for $C_i^{(6)}$. Note, z is the defining primitive element of QR and H, thus

$$z^i QR = C_i \cup C_{i+2} \cup C_{i+4} \quad \text{and} \quad z^j H = C_j \cup C_{j+1} \cup C_{j+3}, \tag{13}$$

where the indices are obtained modulo 6. For $k = 0$ we get $c_0(\underline{s}_H^{(z)}, \underline{s}_H) = -n + 2 + 4|C_0| = c_0(\underline{s}_{QR}, \underline{s}_H)$. Let $k \neq 0$, then $k = -z^{-i}$ for some i, since z is a primitive element in $\mathbb{Z}_n$. We get from (12), that (11) holds if and only if

$$|(H + z^{-i+1}) \cap z^{-1}H| = |(H + z^{-i}) \cap QR| \tag{14}$$

holds for all $i = 0, ..., n-1$. It is $(H + z^{-i+1}) \cap z^{-1}H = z^{-i+1}((z^{i-1}H + 1) \cap z^{i-2}H)$ and $(H + z^{-i}) \cap QR = z^{-i}((z^i H + 1) \cap z^i QR)$. Thus, from (13) it follows that (14) holds if and only if $h_i = q_i$ for all $i = 0, ..., 5$, where

$$h_i := |(z^{i-1}H + 1) \cap z^{i-2}H| \quad \text{and} \quad q_i := |(z^i H + 1) \cap z^i QR|. \tag{15}$$

We will explicitly calculate h_i and q_i. In general we have

$$((C_{i_1} \cup C_{i_2} \cup C_{i_3}) + 1) \cap (C_{j_1} \cup C_{j_2} \cup C_{j_3}) = \dot{\bigcup_{\substack{r = 1, 2, 3 \\ s = 1, 2, 3}}} ((C_{i_r} + 1) \cap C_{j_s})$$

since C_i's are pairwise disjoint. For fixed i and j we define the cyclotomic number (i, j) to be the number of solutions of the equation $z_i + 1 = z_j$ with $z_i \in C_i$ and $z_j \in C_j$, i.e.

$$(i, j) = |(C_i + 1) \cap C_j|.$$

We refer the readers to [8] for more informations about cyclotomic numbers.

We have

$$|((C_{i_1} \cup C_{i_2} \cup C_{i_3}) + 1) \cap (C_{j_1} \cup C_{j_2} \cup C_{j_3})| = \sum_{\substack{r = 1,2,3 \\ s = 1,2,3}} (i_r, j_s)$$

and therefore we get from (13)

$$h_i = \sum_{\substack{r = 0,2,5 \\ s = 1,4,5}} (i+r, i+s) \text{ and } q_i = \sum_{\substack{r = 0,1,3 \\ s = 0,2,4}} (i+r, i+s).$$

We explicitly calculate the cyclotomic numbers for $n = 4t^2 + 27$. If $n = 4t^2 + 27$ is prime, then $\gcd(t, 3) = 1$. We have $n - 1 \equiv 0 \bmod 6$ and $n - 1 \equiv 6 \bmod 12$, since 2 and 3 divides $n-1$ and 4 is not a divider of $n-1$. Thus, $n = 6f+1$ with f odd. In this case the 36 cyclotomic numbers (i, j) are given by

i \ j	0	1	2	3	4	5
0	A	B	C	D	E	F
1	G	H	I	E	C	I
2	H	J	G	F	I	B
3	A	G	H	A	G	H
4	G	F	I	B	H	J
5	H	I	E	C	I	G

where

$$\begin{aligned}
9 \cdot A &:= t^2 - 4 \cdot t' + 4 \\
9 \cdot B &:= t^2 - t' + 16 \\
9 \cdot C &:= t^2 - t' + 16 \\
9 \cdot D &:= t^2 + 8 \cdot t' + 7 \\
9 \cdot E &:= t^2 - t' - 2 \\
9 \cdot F &:= t^2 - t' - 2 \\
9 \cdot G &:= t^2 + 2 \cdot t' + 10 \\
9 \cdot H &:= t^2 + 2 \cdot t' + 1 \\
9 \cdot I &:= t^2 - t' + 7 \\
9 \cdot J &:= t^2 - t' + 7
\end{aligned}$$

and $t' = -t$ if $t \equiv 1 \bmod 3$ and $t' = t$ if $t \equiv 2 \bmod 3$. We get

$$q_0 = A+C+E+G+I+C+A+H+G = t^2 + \tfrac{22}{3} - \tfrac{2t'}{3} = B+E+F+J+I+B+I+I+G = h_0$$
$$q_1 = H+E+I+J+F+B+F+B+J = t^2 + \tfrac{16}{3} - \tfrac{2t'}{3} = A+C+F+G+I+I+A+H+H = h_1$$
$$q_2 = H+G+I+A+H+G+H+E+I = t^2 + \tfrac{13}{3} + \tfrac{t'}{3} = G+H+E+H+J+F+G+F+B = h_2$$
$$q_3 = B+D+F+G+A+H+F+B+J = t^2 + \tfrac{19}{3} + \tfrac{t'}{3} = J+G+I+G+H+G+I+E+I = h_3$$
$$q_4 = G+I+C+G+I+H+H+E+I = t^2 + \tfrac{19}{3} + \tfrac{t'}{3} = C+D+F+H+A+H+I+B+J = h_4$$
$$q_5 = B+D+F+J+F+B+I+C+G = t^2 + \tfrac{25}{3} + \tfrac{t'}{3} = G+E+C+G+B+H+H+C+I = h_5$$

□

Acknowledgement

Part of this research has been carried out when the author was visiting the University of Bergen. This visit was supported by the Deutschen Akademischen Austauschdienst (DAAD).

References

1. J.F. Dillon, "Multiplicative Difference Sets via Additive Characters," *Designs, Codes and Cryptography*, Vol. 17, No. 1-3, pp. 225-235, 1999.
2. J.F. Dillon, H. Dobbertin, "New Cyclic Difference Sets with Singer Parameters", *Finite Fields Appl.*, vol. 10, No. 3, pp. 342-389, 2004.
3. S.W. Golomb, G. Gong, *Signal Design for good Correlation. For Wireless Communication, Cryptography and Radar*, Cambridge: Cambridge University Press, 2005.
4. M. Hall, "A Survey of Difference Sets," *Proc. Am. Math. Soc.*, vol. 7, pp. 975-986, 1957.
5. R. Lidl, H. Niederreiter, *Finite Fields*, 2nd ed., Encyclopedia of Mathematics and its Applications, Vol. 20, Cambridge University Press, 1996,
6. R.E.A.C. Paley, "On Orthogonal Matrices," *J. Math. Phys.*, Mass. Inst. Techn., Vol. 12, pp. 311-320, 1933.
7. A. Pott, "Finite Geometry and Character Theory," *Lecture Notes in Mathematics*, vol. 1601. Berlin: Springer-Verlag, 1995
8. T. Storer, *Cyclotomy and Difference Sets*, Markham Publishing Co, Chicago III, 1967.

Analysis of Designing Interleaved ZCZ Sequence Families*

Jin-Song Wang and Wen-Feng Qi

Department of Applied Mathematics,
Zhengzhou Information Engineering University, P.O.Box 1001-745,
Zhengzhou, 450002, P.R. China
jinsong.wang@126.com, wenfeng.qi@263.net

Abstract. Interleave structure is a well-known period extending method, by which we can extend the period of an original ZCZ sequence family to generate a long period ZCZ sequence family. In this paper, we first present two basic period extending methods: 1. when period extends, the sequence number keeps unchangeable, while the zero correlation zone length extends; 2. when period extends, the zero correlation zone length keeps unchangeable or slightly decreased, while the sequence number extends. Then we propose the concept of D-matrix, by which to determine the shift sequence in the interleaved structure and to calculate the zero correlation zone length of interleaved ZCZ sequence families. In Section 3 and Section 4, two generating algorithms of interleaved ZCZ sequence families and the corresponding optimal D-matrix are proposed.

1 Introduction

In a typical direct sequence (DS) code division multiple access (CDMA) system, all users use the same bandwidth, but each transmitter is assigned a distinct spreading sequence[1]. The well-known binary Walsh sequences or variable length orthogonal sequences have perfect orthogonality at zero time delay, and are ideal for synchronous CDMA (S-CDMA) systems, such as the forward link transmission. Orthogonal spreading sequences can be used if all the users of the same channel are synchronized in time to the accuracy of a small fraction of one chip, because the cross correlation between different shifts of orthogonal sequences is normally not zero. For asynchronous CDMA (A-CDMA) system, no synchronization between transmitted spreading sequences is required, that is, the relative delays between the transmitted spreading sequences are arbitrary. Unfortunately, according to Welch bounds and other theoretical limits, in theory, it is impossible to construct an ideal sequence set with impulsive autocorrelation functions (ACFs) and zero cross correlation functions (CCFs). To overcome these difficulties, zero correlation zone (ZCZ) sequence families are introduced[2,3], which can be employed in quasisynchronous CDMA (QS-CDMA) system to eliminate

* This work was supported by the National Natural Science Foundation of China (Grant 60373092).

G. Gong et al. (Eds.): SETA 2006, LNCS 4086, pp. 129–140, 2006.

the multiple access interference and multipath interference[4]. Generally speaking, the parameters considered in ZCZ sequence families are: period, sequence number and zero correlation zone length. A sequence family composed of M sequences with period L and zero correlation zone length Z_{cz} is denoted by an (L, M, Z_{cz})-ZCZ sequence family. Tang, Fan and Matsufuji[5] derived that the theoretical bound of an (L, M, Z_{cz})-ZCZ sequence family is $Z_{cz} \leq L/M - 1$. The generation of various types of ZCZ sequence families have been reported in [2]-[4] and [6]-[9].

Interleaved structure is a well-known period extending method, which can be used to generate many types of sequence families. Gong[10,11] constructed two classes of sequence families with good correlation property and large linear complexity by interleaving two ideal autocorrelation sequences. Hayashi[6,7] and Torri et al. [9] also use interleaved structure to generate long period ZCZ sequence families.

In this paper, we propose the concept of D-matrix, by which to control the choice of shift sequences and to calculate the zero correlation zone length of the interleaved ZCZ sequence families. Based on the theoretical bound of ZCZ sequence families, we present two optimal period extending methods(the D-matrix corresponding to these period extending methods are called optimal D-matrix):

1. Keeping the sequence number unchangeable, the multiple of zero correlation zone length extending equals to that of period extending;
2. Keeping zero correlation zone length unchangeable or diminishing 1, the multiple of sequence number extending equals to that of period extending.

The rest of this paper is arranged as follows. Section 2 is some preliminaries in the design of interleaved ZCZ sequence families. In Section 3, we propose Algorithm I and the corresponding optimal D-matrix of a class of interleaved ZCZ sequence family. In Section 4, we propose Algorithm II and the corresponding optimal D-matrix of another class of interleaved ZCZ sequence family.

2 Preliminaries

In the following, the preliminaries in the design of interleaved ZCZ sequence families are proposed.

2.1 Left Shift Operation

Let p be prime, $e \geq 1$, $\boldsymbol{a} = (a_0, a_1, \cdots)$ a sequence over Z_{p^e}. (Besides the shift sequence, all the sequences considered in this paper have the elements over Z_{p^e}.) For any $i > 0$, left shift operator L^i acting on $\boldsymbol{a}$ is defined as $L^i(\boldsymbol{a}) = (a_i, a_{i+1}, \cdots)$. In particular, denote $L^0(\boldsymbol{a}) = \boldsymbol{a}$, $L^\infty(\boldsymbol{a}) = \boldsymbol{0}$.

2.2 Correlation

Let $\boldsymbol{a} = (a_0, a_1, \cdots, a_{L-1})$ and $\boldsymbol{b} = (b_0, b_1, \cdots, b_{L-1})$ be two sequences of period L. For any integer $\tau \geq 0$, their (periodic) cross correlation function $C_{a,b}(\tau)$ is defined as

$$C_{a,b}(\tau) = \sum_{i=0}^{L-1} w^{a_i - b_{(i+\tau) \bmod L}}, \tau = 0, 1, \cdots$$

where $w = e^{\frac{2\pi i}{p^e}}$ is a p^eth primitive root. If $\boldsymbol{a} = \boldsymbol{b}$, then $C_{a,a}(\tau)$ is called the autocorrelation function of $\boldsymbol{a}$. Besides, if

$$C_{a,a}(\tau) = \begin{cases} L, & \text{if } \tau \equiv 0 \bmod L \\ 0, & \text{otherwise} \end{cases},$$

then we say $\boldsymbol{a}$ is a perfect sequence. Although perfect sequences are very useful in the design of ZCZ sequence families, the lack of their number prevents them from being widely used.

2.3 Orthogonal Sequence Families

Let $\boldsymbol{A} = \{\boldsymbol{a}^{(0)}, \boldsymbol{a}^{(1)}, ..., \boldsymbol{a}^{(M-1)}\}$ be a sequence family composed of M sequences with period L and $C_{i,j}$ be the cross correlation function of $\boldsymbol{a}^{(i)}$ and $\boldsymbol{a}^{(j)}$. If for all $0 \leq i \neq j \leq M-1$, $C_{i,j}(0) = 0$, then $\boldsymbol{A}$ is defined as an orthogonal sequence family.

2.4 ZCZ Sequence Families

Let $\boldsymbol{A} = \{\boldsymbol{a}^{(0)}, \boldsymbol{a}^{(1)}, \cdots, \boldsymbol{a}^{(M-1)}\}$ be a sequence family composed of M cyclically distinct sequences with period L, where

$$\boldsymbol{a}^{(i)} = (a_0^{(i)}, a_1^{(i)}, ..., a_{L-1}^{(i)}),$$

then the zero correlation zone length Z_{cz} is defined as

$$Z_{cz} = \max\{\ N \mid \text{if } i \neq j, \forall\ |\tau| \leq N, C_{i,j}(\tau) = 0; \\ \text{if } i = j\ ,\ \forall\ 0 < |\tau| \leq N, C_{i,i}(\tau) = 0\}.$$

2.5 Interleaved Sequences

Let $\boldsymbol{u} = (u_0, u_1, \cdots u_{st-1})$ be a sequence of period st, where both s and t are not equal to 1, then arrange it as an $s \times t$ matrix A[10,11], where

$$A = \begin{bmatrix} u_0 & u_1 & \cdots & u_{t-1} \\ u_t & u_{t+1} & \cdots & u_{2t-1} \\ \cdots & \cdots & \cdots & \cdots \\ u_{(s-1)t} & u_{(s-1)t+1} & \cdots & u_{(s-1)t+t-1} \end{bmatrix}.$$

Let A_j be the jth column vector of A. If for $j = 0, 1, \cdots, t-1$, A_j is a phase shift of a sequence, say $\boldsymbol{a}^{(j)}$, that is, $A_j = L^{e_j}(\boldsymbol{a}^{(j)})$, then $\boldsymbol{u} = (L^{e_0}(\boldsymbol{a}^{(0)}), \cdots, L^{e_{t-1}}(\boldsymbol{a}^{(t-1)}))$ is called an (s,t) interleaved sequence. $\boldsymbol{e} = (e_0, e_1, \cdots, e_{t-1})$ is the shift sequence and $\boldsymbol{a}^{(j)}$ is the base sequence of $\boldsymbol{u}$.

3 Construction I of Interleaved ZCZ Sequence Families

In this section, we firstly generalize the construction of Torri, Nakamura and Suehiro in [9] and present Algorithm I of interleaved ZCZ sequence family. Then we propose the concept of D-matrix, by which to calculate the zero correlation zone length and to control the shift sequence in the interleaved structure. Finally we present the optimal D-matrix corresponding to Algorithm I.

3.1 Algorithm I of Interleaved ZCZ Sequence Families

We generalize the two constructions of interleaved ZCZ sequence families in [9] to get Algorithm I, which correspond to the first period extending method.

Algorithm I:

1. Suppose that $\boldsymbol{A} = \{\boldsymbol{a}^{(0)}, \boldsymbol{a}^{(1)}, \cdots, \boldsymbol{a}^{(M-1)}\}$ is a ZCZ sequence family composed of M sequences with period s and zero correlation zone length $Z_{cz}(\boldsymbol{A})$, and $\boldsymbol{B} = \{\boldsymbol{b}^{(0)}, \boldsymbol{b}^{(1)}, \cdots, \boldsymbol{b}^{(t-1)}\}$ is an orthogonal sequence family of period t, where $t|M$, $\boldsymbol{a}^{(r)} = (a_0^{(r)}, a_1^{(r)}, \cdots, a_{s-1}^{(r)})$, $\boldsymbol{b}^{(k)} = (b_0^{(k)}, b_1^{(k)}, \cdots, b_{t-1}^{(k)})$, $0 \le r \le M-1, 0 \le k \le t-1$.
2. Choose a sequence $\boldsymbol{e} = (e_0, e_1, \cdots, e_{t-1})$ of period t over Z_s as the shift sequence.
3. For $h = 0, 1, \cdots, M/t-1$, construct an (s,t) interleaved sequence $\boldsymbol{u}^{(h)} = (u_0^{(h)}, u_1^{(h)}, \cdots, u_{st-1}^{(h)})$, whose jth column is $L^{e_j}(\boldsymbol{a}^{(ht+j)}), j = 0, 1, \cdots, t-1$.
4. For $h = 0, 1, \cdots, M/t-1$, $k = 0, 1, \cdots, t-1$, let $\boldsymbol{s}^{(h,k)} = (s_0^{(h,k)}, s_1^{(h,k)}, \cdots, s_{st-1}^{(h,k)})$ be a sequence of period $s \cdot t$ defined by

$$s_i^{(h,k)} = u_i^{(h)} + b_i^{(k)}, 0 \le i \le st-1,$$

or equivalently, the jth column of $\boldsymbol{s}^{(h,k)}$ is $L^{e_j}(\boldsymbol{a}^{(ht+j)}) + \boldsymbol{b}^{(k,j)}, j = 0, 1, \cdots, t-1$, where $\boldsymbol{b}^{(k,j)} = (b_j^{(k)}, b_j^{(k)}, \cdots, b_j^{(k)})$ is an s-dimension constant vector. Then the interleaved ZCZ sequence family $\boldsymbol{S}$ is defined as $\boldsymbol{S} = \{\boldsymbol{s}^{(h,k)} | h = 0, 1, \cdots M/t - 1, k = 0, 1, \cdots, t-1\}$.

Remark 1. In the construction of Algorithm I, t is the period extending multiple of the sequences in $\boldsymbol{A}$, which must satisfy $t|M$, where M is the sequence number.

3.2 D-Matrix

Set

$$e_{j+t} = e_j + 1, 0 \le j \le t-1,$$

then the sequence $\boldsymbol{e}$ of period t can be extended to a sequence of period $2t$. We also denote $\boldsymbol{e}$ for that sequence.

From Lemma 2 in [11], we can easily get the correlation function of $\boldsymbol{s}^{(h_1,k_1)}$ and $\boldsymbol{s}^{(h_2,k_2)}$ of our proposed ZCZ sequence family, that is,

Theorem 1. *Let $s^{(h_1,k_1)}$ and $s^{(h_2,k_2)}$ be two interleaved ZCZ sequences generated from Algorithm I and $\tau = rt + v$, then the correlation function $C_{(h_1,k_1),(h_2,k_2)}(\tau)$ is*

$$C_{(h_1,k_1),(h_2,k_2)}(\tau) = \sum_{j=0}^{t-1} w^{b_j^{(k_1)} - b_{(j+v) \bmod t}^{(k_2)}} \cdot C_{a^{(h_1 t+j)}, a^{(h_2 t+(j+v) \bmod t)}}(r + e_{j+v} - e_j),$$

where w is a p^eth primitive root.

From Theorem 1, we can calculate the correlation function and the zero correlation zone length of $\boldsymbol{S}$ as follows.

Case 1. If $\tau = 0$, then for $0 \le h_1, h_2 \le M/t - 1, 0 \le k_1, k_2 \le t - 1$,

$$C_{(h_1,k_1),(h_2,k_2)}(0) = \sum_{j=0}^{t-1} w^{b_j^{(k_1)} - b_j^{(k_2)}} \cdot C_{a^{(h_1 t+j)}, a^{(h_2 t+j)}}(0).$$

(1) If $h_1 = h_2, k_1 = k_2$, then $C_{(h_1,k_1),(h_2,k_2)}(0) = st$.

(2) If $h_1 = h_2, k_1 \ne k_2$, as for $j = 0, 1, \cdots, t-1$,

$$C_{a^{(h_1 t+j)}, a^{(h_2 t+j)}}(0) = s,$$

and $\{\boldsymbol{b}^{(0)}, \boldsymbol{b}^{(1)}, \cdots, \boldsymbol{b}^{(t-1)}\}$ is an orthogonal sequence family, then

$$C_{(h_1,k_1),(h_2,k_2)}(0) = s \cdot \sum_{j=0}^{t-1} w^{b_j^{(k_1)} - b_j^{(k_2)}} = 0.$$

(3) If $h_1 \ne h_2$, as for $j = 0, 1, \cdots, t-1$,

$$C_{a^{(h_1 t+j)}, (h_2 t+j)}(0) = 0,$$

then $C_{(h_1,k_1),(h_2,k_2)}(0) = 0$.

Case 2. Since $C_{(h_1,k_1),(h_2,k_2)}(-\tau) = C_{(h_2,k_2),(h_1,k_1)}(\tau)^*$, to calculate the zero correlation zone length of $\boldsymbol{S}$, we only need to consider the case of $\tau > 0$. Let $\tau = rt + v \ne 0, 0 \le r \le s - 1, 0 \le v \le t - 1$. If $|r + e_{j+v} - e_j| \le Z_{cz}(\boldsymbol{A})$ holds for $j = 0, 1, \cdots, t-1$, then

$$C_{a^{(h_1 t+j)}, a^{(h_2 t+(j+v) \bmod t)}}(r + e_{j+v} - e_j) = 0,$$

therefore $C_{(h_1,k_1),(h_2,k_2)}(\tau) = 0$. Obviously, if $v = 0$, $C_{(h_1,k_1),(h_2,k_2)}(\tau) = 0$ holds when $r \le Z_{cz}(\boldsymbol{A})$.

Let $Z_{cz}(\boldsymbol{S})$ be the zero correlation zone length of $\boldsymbol{S}$. Then from above we know

$$Z_{cz}(\boldsymbol{S}) = \max\{N \mid 0 \le \tau \le N, \tau = rt + v \text{ and } (r, v) \text{ satisfying } |r + e_{j+v} - e_j| \le Z_{cz}(\boldsymbol{A})\}$$

For $\tau = rt+v = 1, 2, \cdots, st-1$, $j = 0, 1, \cdots, t-1$, denote $e_{\tau,j} = r + e_{j+v} - e_j$, then we have

$$E = [e_{i,j}]_{(st-1)\times t}$$

$$= \begin{bmatrix} e_1 - e_0 & e_2 - e_1 & \cdots & e_{t-1} - e_{t-2} & e_t - e_{t-1} \\ e_2 - e_0 & e_3 - e_1 & \cdots & e_t - e_{t-2} & e_{t+1} - e_{t-1} \\ \cdots & \cdots & \cdots & \cdots & \cdots \\ e_{t-1} - e_0 & e_t - e_1 & \cdots & e_{2t-3} - e_{t-2} & e_{2t-2} - e_{t-1} \\ 1 & 1 & \cdots & 1 & 1 \\ 1 + e_1 - e_0 & 1 + e_2 - e_1 & \cdots & 1 + e_{t-1} - e_{t-2} & 1 + e_t - e_{t-1} \\ \cdots & \cdots & \cdots & \cdots & \cdots \end{bmatrix}$$

$$= \begin{bmatrix} e_1 - e_0 & e_2 - e_1 & \cdots & e_{t-1} - e_{t-2} & 1 + e_0 - e_{t-1} \\ e_2 - e_0 & e_3 - e_1 & \cdots & 1 + e_0 - e_{t-2} & 1 + e_1 - e_{t-1} \\ \cdots & \cdots & \cdots & \cdots & \cdots \\ e_{t-1} - e_0 & 1 + e_0 - e_1 & \cdots & 1 + e_{t-3} - e_{t-2} & 1 + e_{t-2} - e_{t-1} \\ 1 & 1 & \cdots & 1 & 1 \\ 1 + e_1 - e_0 & 1 + e_2 - e_1 & \cdots & 1 + e_{t-1} - e_{t-2} & 2 + e_0 - e_{t-1} \\ \cdots & \cdots & \cdots & \cdots & \cdots \end{bmatrix} \quad (1)$$

For $\tau = 1, 2, \cdots, st-1$, in order to calculate the zero correlation zone length of $\boldsymbol{S}$, we need to judge whether the elements of the τth row in E belonging to $[-Z_{cz}(\boldsymbol{A}), Z_{cz}(\boldsymbol{A})]$ or not.

Let $d_{0,0} = e_1 - e_0, d_{0,1} = e_2 - e_1, \cdots, d_{0,t-1} = 1 + e_0 - e_{t-1}$, then we have $d_{0,0} + d_{0,1} + \cdots + d_{0,t-1} \equiv 1 (\bmod\ t)$. Obviously given e_0, from $d_{0,0}, d_{0,1}, \cdots, d_{0,t-2}$, we can uniquely determine the shift sequence $\boldsymbol{e}$. Furthermore, we have

Theorem 2. *Let $E = D = [d_{i,j}]_{0 \le i \le st-1, 0 \le j \le t-1}$, then we have the following recursive formulas:*

$$d_{i,j} = d_{i-1,(j+1) \bmod t} + d_{0,j}, \quad (2)$$

and

$$d_{ri,j} = \sum_{k=0}^{r-1} d_{i,(j+ki) \bmod t}. \quad (3)$$

Proof. From (1) every element in E can be divided into two parts: the subtrahend and minuend.

If $0 \le j \le t-2$, then the minuend of the ith row and the jth column element $d_{i,j}$ is equal to the minuend of the $(i-1)$th row $(j+1)$th column element $d_{i-1,j+1}$, while the difference of subtrahend between $d_{i,j}$ and $d_{i-1,j+1}$ is $e_{j+1} - e_j$, which is exactly $d_{0,j}$, so $d_{i,j} = d_{i-1,j+1} + d_{0,j}$.

If $j = t-1$, the difference between the minuend of the ith row and the $(t-1)$th column element $d_{i,t-1}$ and the first row and the $(t-1)$th column element $d_{0,t-1}$ is $e_j - e_0$, which is exactly $d_{i-1,0}$, so $d_{i,t-1} = d_{i-1,0} + d_{0,t-1}$.

Thus we have

$$d_{i,j} = d_{i-1,(j+1) \bmod t} + d_{0,j}.$$

Repeating using (2), then we can represent D as

$$D = \begin{bmatrix} d_{0,0} & d_{0,1} & \cdots & d_{0,t-1} \\ d_{0,0} + d_{0,1} & d_{0,1} + d_{0,2} & \cdots & d_{0,t-1} + d_{0,0} \\ \cdots & \cdots & \cdots & \cdots \\ \sum_{i=0}^{t-2} d_{0,i} & \sum_{i=0}^{t-2} d_{0,(i+1) \bmod t} & \cdots & \sum_{i=0}^{t-2} d_{0,(i+t-1) \bmod t} \\ 1 & 1 & \cdots & 1 \\ 1 + d_{0,0} & 1 + d_{0,1} & \cdots & 1 + d_{0,t-1} \\ \cdots & \cdots & \cdots & \cdots \end{bmatrix}. \tag{4}$$

From (4) we can also have

$$d_{ri,j} = \sum_{k=0}^{r-1} d_{i,(j+ki) \bmod t}.$$

Remark 2. From (2) and $d_{0,0}+d_{0,1}+\cdots+d_{0,t-1} \equiv 1(\bmod\ t)$, we know $d_{it+k,j} = d_{k,j} + i$. Thus to calculate the zero correlation zone length of $\boldsymbol{S}$, we only need to consider the first t rows of D-matrix.

Remark 3. From (3) and Remark 2 we know that the D-matrix can be totally determined by its rth row, where $\gcd(r,t) = 1$.

Remark 4. Denote $L^i(D)$ as the matrix every row of which is cyclically left shift i positions of the corresponding row of D. Then the zero correlation zone length determined by $L^i(D)$ is equal to that determined by D. Next we only consider cyclically inequivalent D-matrix to generate the shift sequence of interleaved ZCZ sequence families.

3.3 The Optimal D-Matrix Corresponding to Algorithm I

Next we present the optimal D-matrix corresponding to the interleaved ZCZ sequence families generated by Algorithm I.

Theorem 3. *Suppose $\boldsymbol{A}$ is an (L, M, Z_{cz})-ZCZ sequence family. For any positive integer t, $t|M$, let D be the D-matrix whose first row is given as $(\underbrace{0, 0, \cdots, 0}_{t-1}, 1)$. Then we can construct an $(L \cdot t, M, Z_{cz} \cdot t)$-ZCZ sequence family generated by Algorithm I.*

Proof. The first row of D is $(\underbrace{0, 0, \cdots, 0}_{t-1}, 1)$, then the first t rows of D can be calculated from recursive formula (2) as

$$\begin{bmatrix} 0 & 0 & \cdots & 0 & 1 \\ 0 & 0 & \cdots & 1 & 1 \\ \cdots & \cdots & \cdots & \cdots & \cdots \\ 1 & 1 & \cdots & 1 & 1 \end{bmatrix}. \tag{5}$$

From (5) and Remark 2 we know that the zero correlation zone length of generated ZCZ sequence family is $Z_{cz} \cdot t$. #

In fact, the interleaved ZCZ sequence family constructed by the D-matrix of Theorem 3 is a generalization of Torri, Nakamura, and Suehiro[9].

Example 1. From a (8, 4, 1)-ZCZ sequence family proposed by Deng and Fan in [3], we can get a (32, 4, 4)-ZCZ sequence family by Algorithm I.

1. Choose a (8, 4, 1)-ZCZ sequence family $\boldsymbol{A} = \{\boldsymbol{a}^{(0)}, \boldsymbol{a}^{(1)}, \boldsymbol{a}^{(2)}, \boldsymbol{a}^{(3)}\} = \{(1,1,0,0,0,1,1,0),(1,1,1,1,0,1,0,1),(0,1,1,0,1,1,0,0),(0,1,0,1,1,1,1,1)\}$ and an orthogonal sequence family $\boldsymbol{B}=\{\boldsymbol{b}^{(0)}, \boldsymbol{b}^{(1)}, \boldsymbol{b}^{(2)}, \boldsymbol{b}^{(3)}\}=\{(0,0,0,0),(0,0,1,1),(0,1,0,1),(0,1,1,0)\}$ of period 4.

2. Let $e_0 = 0, e_1 = 1, e_2 = 1, e_3 = 1$.

3. For $k = 0,1,2,3$, we can construct sequences $\boldsymbol{s}^{(k)}$, whose jth column is given by $L^{e_j}(\boldsymbol{a}^{(j)}) + b_{k,j}$. For example, the matrix form of $\boldsymbol{s}^{(1)}$ is as follows:

$$\begin{pmatrix} 1\,1\,1\,1 \\ 1\,1\,1\,0 \\ 0\,1\,0\,1 \\ 0\,0\,1\,1 \\ 0\,1\,1\,1 \\ 1\,0\,0\,1 \\ 1\,1\,0\,1 \\ 0\,1\,0\,0 \end{pmatrix} + \begin{pmatrix} 0\,0\,1\,1 \\ 0\,0\,1\,1 \\ 0\,0\,1\,1 \\ 0\,0\,1\,1 \\ 0\,0\,1\,1 \\ 0\,0\,1\,1 \\ 0\,0\,1\,1 \\ 0\,0\,1\,1 \end{pmatrix} = \begin{pmatrix} 1\,1\,0\,0 \\ 1\,1\,0\,1 \\ 0\,1\,1\,0 \\ 0\,0\,0\,0 \\ 0\,1\,0\,0 \\ 1\,0\,1\,0 \\ 1\,1\,1\,0 \\ 0\,1\,1\,1 \end{pmatrix}.$$

Thus the obtained sequence family $\boldsymbol{S} = \{\boldsymbol{s}^{(0)}, \boldsymbol{s}^{(1)}, \boldsymbol{s}^{(2)}, \boldsymbol{s}^{(3)}\}$, where

$\boldsymbol{s}^{(0)} = (1,1,1,1,1,1,1,0,0,1,0,1,0,0,1,1,0,1,1,1,1,0,0,1,1,1,0,1,0,1,0,0)$,
$\boldsymbol{s}^{(1)} = (1,1,0,0,1,1,0,1,0,1,1,0,0,0,0,0,0,1,0,0,1,0,1,0,1,1,1,0,0,1,1,1)$,
$\boldsymbol{s}^{(2)} = (1,0,1,0,1,0,1,1,0,0,0,0,0,1,1,0,0,0,1,0,1,1,0,0,1,0,0,0,0,0,0,1)$,
$\boldsymbol{s}^{(3)} = (1,0,0,1,1,0,0,0,0,0,1,1,0,1,0,1,0,0,0,1,1,1,1,1,1,0,1,1,0,0,1,0)$.

Also, we can find out that the zero correlation zone length of $\boldsymbol{S}$ is 4. For example, for $\tau = 0,1,\cdots,31$, the autocorrelation of $\boldsymbol{s}^{(0)}$ is $\{32,0,0,0,0,4,0,-12,0,12,0,-4,0,0,16,0,0,0,16,0,0,-4,0,12,0,-12,0,4,0,0,0,0\}$, the cross correlation of $\boldsymbol{s}^{(0)}$ and $\boldsymbol{s}^{(1)}$ is $\{0,0,0,0,0,4,8,4,0,-4,-8,-4,0,0,0,0,0,0,0,0,0,-4,8,-4,0,4,-8,4,0,0,0,0\}$.

4 Construction II of Interleaved ZCZ Sequence Families

In this section, we present Algorithm II to generate another class of interleaved ZCZ sequence families, which corresponds to the second period extending method. Also the optimal D-matrix is presented according to the cases that period extends prime times and composite times.

4.1 Algorithm II for Interleaved ZCZ Sequence Families

Algorithm II:

1. Let $\boldsymbol{A} = \{\boldsymbol{a}^{(0)}, \boldsymbol{a}^{(1)}, \cdots, \boldsymbol{a}^{(M-1)}\}$ be a ZCZ sequence family composed of M sequences with period s and zero correlation zone length $Z_{cz}(\boldsymbol{A})$, where $\boldsymbol{a}^{(h)} = (a_0^{(h)}, a_1^{(h)}, \cdots, a_{s-1}^{(h)})$, $0 \le h \le M-1$. From $\boldsymbol{A}$ we can generate a sequence

family $\boldsymbol{G} = \{\boldsymbol{G}_0, \boldsymbol{G}_1, \cdots, \boldsymbol{G}_{M-1}\}$ composed of M interleaved sequences, where $\boldsymbol{G}_i = (\boldsymbol{g}^{(i,0)}, \boldsymbol{g}^{(i,1)}, \cdots, \boldsymbol{g}^{(i,t-1)}), \boldsymbol{g}^{(i,k)} \in \boldsymbol{A}, 0 \le i \le M-1, 0 \le k \le t-1$, and different sequences in $\boldsymbol{G}$ are orthogonal to each other, that is,

$$\forall\ 0 \le i \ne j \le M-1, \sum_{k=0}^{t-1} C_{g^{(i,k)},g^{(j,k)}}(0) = 0.$$

2. Let $\boldsymbol{B} = \{\boldsymbol{b}^{(0)}, \boldsymbol{b}^{(1)}, \cdots, \boldsymbol{b}^{(t-1)}\}$ be an orthogonal sequence family of period t, where $\boldsymbol{b}^{(k)} = (b_0^{(k)}, b_1^{(k)}, \cdots, b_{t-1}^{(k)})$, $0 \le k \le t-1$.

3. Choose a shift sequence $\boldsymbol{e} = (e_0, e_1, \cdots, e_{t-1})$ of period t over Z_s.

4. For $h = 0, 1, \cdots, M-1$, $k = 0, 1, \cdots, t-1$, construct an (s,t) interleaved sequence $\boldsymbol{u}^{(h,k)} = (u_0^{(h,k)}, u_1^{(h,k)}, \cdots, u_{st-1}^{(h,k)})$, whose jth column is $L^{e_j}(\boldsymbol{g}^{(h,j)}), j = 0, 1, \cdots, t-1$.

5. For $h = 0, 1, \cdots, M-1$, $k = 0, 1, \cdots, t-1$, let $\boldsymbol{s}^{(h,k)} = (s_0^{(h,k)}, s_1^{(h,k)}, \cdots, s_{st-1}^{(h,k)})$ be a sequence of period $s \cdot t$ defined by

$$s_i^{(h,k)} = u_i^{(h,k)} + b_i^{(k)}, 0 \le i \le st-1,$$

or equivalently, its jth column is $L^{e_j}(\boldsymbol{g}^{(h,j)}) + \boldsymbol{b}^{(k,j)}, j = 0, 1, \cdots, t-1$, where $\boldsymbol{b}^{(k,j)} = (b_j^{(k)}, b_j^{(k)}, \cdots, b_j^{(k)})$ is an s-dimension constant vector. Then the interleaved ZCZ sequence family $\boldsymbol{S}$ is defined as $\boldsymbol{S} = \{\boldsymbol{s}^{(h,k)} | h = 0, 1, \cdots M-1, k = 0, 1, \cdots, t-1\}$.

Remark 5. Similar to the analysis of Algorithm I, for $\tau = 1, 2, \cdots, st-1$, to calculate the zero correlation zone length of the sequences generated by Algorithm II, we need to judge whether the elements of the τth row of D-matrix belonging to $[-Z_{cz}(\boldsymbol{A}), 0) \cup (0, Z_{cz}(\boldsymbol{A})]$ or not.

4.2 The Optimal D-Matrix Corresponding to Period Extending p Times

The optimal D-matrix corresponding to the second construction method is present in this section. First we consider the case when period extending multiple is a prime, that is:

Theorem 4. *There exists a proper D-matrix by which we can generate an $(L \cdot p, M \cdot p, Z'_{cz})$-ZCZ sequence family from an (L, M, Z_{cz})-ZCZ sequence family based on Algorithm II, where p is a prime and*

$$Z'_{cz} = \begin{cases} Z_{cz} & , \text{if } Z_{cz} \not\equiv -1 \bmod p \\ Z_{cz} - 1, & \text{otherwise} \end{cases} .$$

Proof. If $Z_{cz} \equiv r \not\equiv -1 \bmod p$, then we can construct a D-matrix where its $(r+1)$th row is $(\underbrace{-\lfloor \frac{Z_{cz}}{p} \rfloor, -\lfloor \frac{Z_{cz}}{p} \rfloor, \cdots, -\lfloor \frac{Z_{cz}}{p} \rfloor}_{p-1}, (p-1)\lfloor \frac{Z_{cz}}{p} \rfloor + r + 1)$. From (3) we know that the different elements in the $2(r+1)$th row are $-2\lfloor\frac{Z_{cz}}{p}\rfloor$ and $(p-2)\lfloor\frac{Z_{cz}}{p}\rfloor+r+1$, ..., and the different elements in the $(p-1)(r+1)$th row are $-(p-1)\lfloor\frac{Z_{cz}}{p}\rfloor$ and $\lfloor\frac{Z_{cz}}{p}\rfloor+r+1$. Next we calculate the first p rows of D-matrix. That is, if $k(r+1)\equiv i \bmod p$, $(k(r+1)-i)/p=j$, then the elements in the ith row of D-matrix can be determined by those of the $k(r+1)$th row subtracting j. Based on Remark 2 , Remark 3 and Remark 5, we know that $Z'_{cz}=p\lfloor\frac{Z_{cz}}{p}\rfloor+r=Z_{cz}$.

If $Z_{cz}\equiv -1(\bmod\ p)$, then we can construct a D-matrix where its $(p-1)$th row is $(\underbrace{-\lfloor\frac{Z_{cz}}{p}\rfloor,-\lfloor\frac{Z_{cz}}{p}\rfloor,...,-\lfloor\frac{Z_{cz}}{p}\rfloor}_{p-1},(p-1)\lfloor\frac{Z_{cz}}{p}\rfloor+p-1)$ or $(\underbrace{-\lfloor\frac{Z_{cz}}{p}\rfloor,...,-\lfloor\frac{Z_{cz}}{p}\rfloor}_{a},$ $-\lfloor\frac{Z_{cz}}{p}\rfloor-1,\underbrace{-\lfloor\frac{Z_{cz}}{p}\rfloor,...,-\lfloor\frac{Z_{cz}}{p}\rfloor}_{p-2-a},(p-1)\lfloor\frac{Z_{cz}}{p}\rfloor+p)$, where $0\le a\le p-2$. Then from Remark 2, Remark 3 and Remark 5, we know that $Z'_{cz}=Z_{cz}-1$.

4.3 The Optimal D-Matrix Corresponding to Period Extending n Times

For the case of period extending p^e times, from Theorem 4 we know that if $Z_{cz}\not\equiv -1 \bmod p$, then after period extending p times, the zero correlation zone length is $Z'_{cz}=Z_{cz}$. Repeat this operation e times, then we can generate a ZCZ sequence family with zero correlation zone length $Z'_{cz}=Z_{cz}$. If $Z_{cz}\equiv -1 \bmod p$, after period extends p times, the zero correlation zone length is $Z'_{cz}=Z_{cz}-1$. Repeat this operation e times, we can generate a ZCZ sequence family with zero correlation zone length $Z'_{cz}=Z_{cz}-1$. Generally, for $n=p_1^{\alpha_1}...p_k^{\alpha_k}$, we have

Theorem 5. *There exists a proper D-matrix by which we can generate an $(L\cdot n, M\cdot n, Z'_{cz})$-ZCZ sequence family from an (L,M,Z_{cz})-ZCZ sequence family, where $n=p_1^{\alpha_1}...p_k^{\alpha_k}$ and*

$$Z'_{cz}=\begin{cases} Z_{cz} & , \text{ if } \forall\ 1\le i\le k,\ Z_{cz}\not\equiv -1 \bmod p_i \\ Z_{cz}-1, & \text{otherwise} \end{cases}.$$

For binary interleaved ZCZ sequence families, the most ordinary case is period extending 2^e times. There exist two methods of period extending 2^e times, one is period doubled, then repeat this operation e times, the other is period directly extending 2^e times. Next we consider the second method, that is,

Theorem 6. *Suppose $\boldsymbol{A}$ is an (L,M,Z_{cz})-ZCZ sequence family and $Z_{cz}\equiv r \bmod 2^e$.*

(1) If Z_{cz} is even, then there exists a proper D-matrix D_1 by which we can generate a $(2^e\cdot L, 2^e\cdot M, Z_{cz})$-ZCZ sequence family from $\boldsymbol{A}$, where the $(r+1)$th row of D_1 is

$$(\underbrace{-\lfloor\frac{Z_{cz}}{2^e}\rfloor,-\lfloor\frac{Z_{cz}}{2^e}\rfloor,...,-\lfloor\frac{Z_{cz}}{2^e}\rfloor}_{2^e-1},(2^e-1)\lfloor\frac{Z_{cz}}{2^e}\rfloor+r+1).$$

(2) If Z_{cz} is odd, then there exists a proper D-matrix D_2 by which we can generate a $(2^e \cdot L, 2^e \cdot M, Z_{cz} - 1)$-ZCZ sequence family from $\boldsymbol{A}$, where the rth row of D_2 is

$$(\underbrace{-\lfloor \frac{Z_{cz}}{2^e} \rfloor, -\lfloor \frac{Z_{cz}}{2^e} \rfloor, ..., -\lfloor \frac{Z_{cz}}{2^e} \rfloor}_{2^e-1}, (2^e - 1)\lfloor \frac{Z_{cz}}{2^e} \rfloor + r),$$

or

$$(\underbrace{-\lfloor \frac{Z_{cz}}{2^e} \rfloor, ..., -\lfloor \frac{Z_{cz}}{2^e} \rfloor}_{a}, -\lfloor \frac{Z_{cz}}{2^e} \rfloor - 1, \underbrace{-\lfloor \frac{Z_{cz}}{2^e} \rfloor, ..., -\lfloor \frac{Z_{cz}}{2^e} \rfloor}_{2^e-a-2}, (2^e - 1)\lfloor \frac{Z_{cz}}{2^e} \rfloor + r + 1),$$

where $0 \leq a \leq 2^e - 2$.

The proof of Theorem 6 can be got similar to that of Theorem 4.

Example 2. Construct a (64, 8, 4)-ZCZ sequence family from a (32, 4, 4)-ZCZ sequence family by Algorithm II.

1. Let $\boldsymbol{A} = \{\boldsymbol{a}^{(0)}, \boldsymbol{a}^{(1)}, \boldsymbol{a}^{(2)}, \boldsymbol{a}^{(3)}\}$ be a (32, 4, 4)-ZCZ sequence family, $\boldsymbol{B} = \{\boldsymbol{b}^{(0)}, \boldsymbol{b}^{(1)}\} = \{(0,0),(0,1)\}$ be an orthogonal sequence family of period 2, where
$\boldsymbol{a}^{(0)} = (1,1,1,1,1,1,1,0,0,1,0,1,0,0,1,1,0,1,1,1,1,0,0,1,1,1,0,1,0,1,0,0)$,
$\boldsymbol{a}^{(1)} = (1,1,0,0,1,1,0,1,0,1,1,0,0,0,0,0,0,1,0,0,1,0,1,0,1,1,1,0,0,1,1,1)$,
$\boldsymbol{a}^{(2)} = (1,0,1,0,1,0,1,1,0,0,0,0,0,1,1,0,0,0,1,0,1,1,0,0,1,0,0,0,0,0,0,1)$,
$\boldsymbol{a}^{(3)} = (1,0,0,1,1,0,0,0,0,0,1,1,0,1,0,1,0,0,0,1,1,1,1,1,1,0,1,1,0,0,1,0)$.
2. Choose shift sequence $e_0 = 2$, $e_1 = 5$;
3. A ZCZ sequence family $\boldsymbol{S} = \{\boldsymbol{s}^{(i)} | i = 0, 1, \ldots, 7\}$ can be constructed from Algorithm II, where

$\boldsymbol{s}^{(0)} = \{1,1,1,1,1,0,1,0,1,1,0,0,0,1,1,0,0,0,1,1,0,1,0,0,1,1,1,1,0,1,1,1,$
$1,0,1,0,1,1,0,1,0,1,1,0,1,1,1,0,0,1,1,0,0,0,1,1,0,1,0,1,1,1,1,1\}$,
$\boldsymbol{s}^{(1)} = \{1,0,1,0,1,1,1,1,1,0,0,1,0,0,1,1,0,1,1,0,0,0,0,1,1,0,1,0,0,0,1,0,$
$1,1,1,1,1,0,0,0,0,0,1,1,1,0,1,1,0,0,1,1,0,1,1,0,0,0,0,0,1,0,1,0\}$,
$\boldsymbol{s}^{(2)} = \{0,1,0,0,1,1,1,0,0,1,1,1,0,0,1,0,1,0,0,0,0,0,0,0,0,1,0,0,0,0,1,1,$
$0,0,0,1,1,0,0,1,1,1,0,1,1,0,1,0,1,1,0,1,0,1,1,1,1,1,1,0,1,0,1,1\}$,
$\boldsymbol{s}^{(3)} = \{0,0,0,1,1,0,1,1,0,0,1,0,0,1,1,1,1,1,0,1,0,1,0,1,0,0,0,1,0,1,1,0,$
$0,1,0,0,1,1,0,0,1,0,0,0,1,1,1,1,1,0,0,0,0,0,1,0,1,0,1,1,1,1,1,0\}$,
$\boldsymbol{s}^{(4)} = \{1,0,0,1,1,1,0,0,1,0,1,0,0,0,0,0,0,1,0,1,0,0,1,0,1,0,0,1,0,0,0,1,$
$1,1,0,0,1,0,1,1,0,0,0,0,1,0,0,0,0,0,0,0,1,0,1,0,0,1,1,1,0,0,1\}$,
$\boldsymbol{s}^{(5)} = \{1,1,0,0,1,0,0,1,1,1,1,1,0,1,0,1,0,0,0,0,0,1,1,1,1,1,0,0,0,1,0,0,$
$1,0,0,1,1,1,1,0,0,1,0,1,1,1,0,1,0,1,0,1,0,0,0,0,0,1,1,0,1,1,0,0\}$,
$\boldsymbol{s}^{(6)} = \{0,0,1,0,1,0,0,0,0,0,0,1,0,1,0,0,1,1,1,0,0,1,1,0,0,0,1,0,0,1,0,1,$
$0,1,1,1,1,1,1,1,1,0,1,1,1,1,0,0,1,0,1,1,0,0,0,1,1,0,0,0,1,1,0,1\}$,
$\boldsymbol{s}^{(7)} = \{0,1,1,1,1,1,0,1,0,1,0,0,0,0,0,1,1,0,1,1,0,0,1,1,0,1,1,1,0,0,0,0,$
$0,0,1,0,1,0,1,0,1,1,1,0,1,0,0,1,1,1,1,0,0,1,0,0,1,1,0,1,1,0,0,0\}$.

Also we can find out that $\boldsymbol{S}$ is a (64, 8, 4)-ZCZ sequence family. For example, for $\tau = 0, 1, \cdots, 63$, the autocorrelation of $\boldsymbol{s}^{(0)}$ is given by $\{64, 0, 0, 0, 0, 36, 0, 0,$

$0, -12, 8, 0, 0, 12, -24, 4, 0, -4, 24, -12, 0, 0, -8, 28, 0, 0, 0, -4, 32, 0, 0, 16, 0, 16, 0, 0, 32, -4, 0, 0, 0, 28, -8, 0, 0, -12, 24, -4, 0, 4, -24, 12, 0, 0, 8, -12, 0, 0, 0, 36, 0, 0, 0, 0\}$, the cross correlation of $\boldsymbol{s}^{(0)}$ and $\boldsymbol{s}^{(1)}$ is given by $\{0, 0, 0, 0, 0, 28, 0, 0, 0, 12, 0, 0, 0, -12, 0, 4, 0, 4, 0, -12, 0, 0, 0, -4, 0, 0, 0, -4, 0, 0, 0, -16, 0, 16, 0, 0, 0, 4, 0, 0, 0, 4, 0, 0, 0, 12, 0, -4, 0, -4, 0, 12, 0, 0, 0, -12, 0, 0, 0, -28, 0, 0, 0, 0\}$.

5 Conclusions

In this paper, we analyze the construction method of interleaved ZCZ sequence families and propose two generation algorithms. The optimal D-matrix corresponding these algorithms are also presented here. These algorithms can recursively act on an original ZCZ sequence family to generate ZCZ sequence families with both long zero correlation zone length and large sequence number.

References

1. Fan, P.Z., Damell, M.: Sequences Design for Communications Applications. Research Studies Press, London(1996)
2. Fan, P. Z., Suehiro, N., Kuroyanagi, N., Deng,X. M.: Class of Binary Sequences with Zero Correlation Zone. IEE Electron. Lett., vol. 35. (1999) 777–779
3. Deng, X. M., Fan, P. Z.: Spreading Sequence Sets with Zero Correlation Zone. IEE Electron. Lett., vol. 36. (2000) 982–983
4. Suehiro, N.: Approximately Synchronized CDMA System without Cochannel Using Pseudo-periodic Sequences. Proc. Int. Symp. Personal communication'93, Nanjing, China(1994) 179–184
5. Tang, X. H., Fan, P. Z., Matsufuji,S.: Lower bounds on correlation of spreading sequence set with low or zero correlation zone. Electron. Lett., vol. 36, (2000) 551–552
6. Hayashi, T.: Binary sequences with orthogonal subsequences and zero correlation zone. IEICE Trans. Fundamentals, vol. E85 A, 2002 (1420–1425)
7. Hayashi, T. :A generalization of binary zero correlation zone sequence sets constructed from Hadamard matrices. IEICE Trans. Fundamentals, vol. E87-A, 2004 (286–291)
8. Matsufuji, S., Takatsukasa, K.: Ternary ZCZ sequence sets for efficient frequency usage. Prodeedings of SCI2000, 2000 (119–123)
9. Torii, H., Nakamura, M., Suehiro, N.: A new class of zero correlation zone sequences. IEEE Trans. Inform. Theory, vol.50, 2004 (559–565)
10. Gong, G.: Theory and applications of q-ary interleaved sequences. IEEE Trans. Inform. Theory, vol. 41, 1995 (400–411)
11. Gong, G.: New designs for signal sets with low cross correlation, balance property, and large linear span: GF(p) case. IEEE Trans. Inform. Theory, vol. 48, 2002 (2847–2867)

Security of Jump Controlled Sequence Generators for Stream Ciphers

Tor Helleseth[1], Cees J.A. Jansen[2], Shahram Khazaei[3], and Alexander Kholosha[1]

[1] The Selmer Center
Department of Informatics
University of Bergen
P.O. Box 7800, N-5020 Bergen, Norway
{Tor.Helleseth, Alexander.Kholosha}@uib.no
[2] Banksys NV
Haachtsesteenweg 1442, 1130 Brussels, Belgium
cja@iae.nl
[3] Zaeim Electronic Industries Company
P.O. Box 14155-1434, Tehran, Iran
Khazaei@zaeim.com

Abstract. The use of jump control technique provides efficient and secure ways for generating key-stream for stream ciphers. This design approach was recently implemented in some algorithms submitted to eSTREAM, the ECRYPT Stream Cipher Project. However, inappropriately chosen parameters for jumping constructions can completely undermine their security. In this paper we describe a new inherent property of jump registers that allows to construct linear relations in their output. We illustrate our results by building a key-recovery attack on the Pomaranch stream cipher. We also suggest a slight modification to the jump register configuration in Pomaranch that allows to protect against this type of attacks.

Keywords: Cryptanalysis, jump register, key-stream generator, linear relations, Pomaranch, stream cipher.

1 Introduction

Linear feedback shift registers (LFSR's) are known to allow fast implementation and produce sequences with a large period and good statistical properties (if the feedback polynomial is chosen appropriately). But inherent linearity of these sequences results in susceptibility to algebraic attacks. That is the prime reason why LFSR's are not used directly for key-stream generation. A well-known method for increasing the linear complexity preserving at the same time a large period and good statistical properties is to apply clock control, i.e., to irregularly step an LFSR through successive states.

Due to the multiple clocking, key-stream generators that use clock-controlled LFSR's have decreased rate of sequence generation since such generators are

G. Gong et al. (Eds.): SETA 2006, LNCS 4086, pp. 141–152, 2006.

usually stepped a few times to produce just one bit of the key-stream. The efficient way to let an LFSR move to a state that is more than one step further but without having to step though all the intermediate states (so called, jumping) was suggested in [1]. Further, in Section 2.1, we give a brief description of the this technique.

The idea of jump registers was used to design complete stream ciphers MICKEY and Pomaranch that were submitted to the ECRYPT Stream Cipher Project (see [2]). In this paper, we focus on the latter algorithm to illustrate our results in the analysis of key-stream generators that use jump registers. Pomaranch [3] is a stream cipher that follows a classical design of synchronous bit-oriented stream ciphers and consists of a key-stream generator producing a secure sequence of bits that is further XORed with the plain text previously converted into bits. The key-stream generator of Pomaranch is called Cascade Jump Controlled Sequence Generator (CJCSG) and is primarily intended for hardware implementation. The CJCSG uses a one clock pulse cascade construction of so called jump registers [4] being essentially linear finite state machines with a special transition matrix. Moreover, the characteristic polynomial of the transition matrix was made to be primitive and satisfying additional constraints that arise from the need to use the register in a cascade jump control setup.

In this paper, we present our findings in the security analysis of key-stream generators that use jump control technique and illustrate our approach by building a key-recovery attack on Pomaranch that works with the complexity much lover than the exhaustive key search. However, the spotted weakness has a general nature and can potentially be used to attack other stream ciphers that are built on irregularly clocked registers. In particular, this weakness leads to the biases in the distribution of certain linear relations in the output sequence of jump registers. If parameters of the registers are not chosen carefully then this bias can be high enough to allow running a correlation attack with the complexity lower than the exhaustive key search.

In Section 2, we outline some details of jumping technique and of Pomaranch key-stream generator that are important for understanding the analysis that follows. Section 3 contains main theoretical results about finding linear relations in the output sequence of jump registers and calculating corresponding biases. We apply the theory to the concrete configuration of Pomaranch registers in Section 4.1 and build the general framework of the key-recovery attack in Section 4.2. Finally, we suggest slight modification of the Pomaranch jump register configuration that allows to protect against this type of attacks increasing the complexity to $O(2^{132})$ (higher than the exhaustive key search) and this is discussed in Section 5. This modification was actually applied to Pomaranch version 2 (see [5]).

2 Jump Registers and Cascade Construction

We start this section with a brief description of jumping technique and then give an outline of the Pomaranch stream cipher that is built on jump registers combined in a cascade construction.

2.1 Jumping Technique

Consider an autonomous Linear Finite State Machine (LFSM), not necessarily an LFSR, defined by the transition matrix A of size L over GF(2) with a primitive characteristic polynomial $f(x) = \det(xI + A)$, where I is the identity matrix. It is well known that A is similar to the companion matrix of $f(x)$, i.e., there exists a nonsingular matrix M such that $M^{-1}AM = S(f)$. Let z_t $(t = 0, 1, 2, \ldots)$ denote the inner state of the LFSM at time t. Then $z_t = z_0 A^t = z_0 M S(f)^t M^{-1}$ and $z_t M = (z_0 M) S(f)^t$. Thus, LFSMs defined by A and $S(f)$ are equivalent.

Take a matrix representation of the elements of the finite field GF(2^L). Since $f(S(f)) = 0$ and $f(x)$ is primitive, $S(f)$ can play the role of a root of f that is a primitive element in GF(2^L). Then $S(f)+I$ being an element of GF(2^L) is equal to $S(f)^J$ for some power J and, thus, $A^J = MS(f)^J M^{-1} = MS(f)M^{-1} + I = A+I$. Note that identity $S(f)^J = S(f)+I$ is equivalent to $x^J \equiv x+1 \pmod{f(x)}$ and, therefore, such a value of J is called the *jump index* of f. It is important to observe here that changing the transition matrix of the LFSM from A to $A+I$ results in making J steps through the state space of the original LFSM.

Let $f^{\perp}(x)$ denote the characteristic polynomial of the modified transition matrix $A+I$ that is equal to $f^{\perp}(x) = \det(xI+A+I) = f(x+1)$. The polynomial $f^{\perp}(x)$ is called the *dual* of $f(x)$. It is easy to see that $f(x)$ is irreducible if and only if $f^{\perp}(x)$ is irreducible (however, this equivalence does not hold for being primitive). It can also be shown (see [4, Theorem 2]) that if the dual polynomial $f^{\perp}$ is primitive (the jump index of $f^{\perp}$, naturally, exists) then the jump index of f is coprime with $\lambda = 2^L - 1$ and $J^{\perp} \equiv J^{-1} \pmod{\lambda}$.

The transition matrix A that defines the LFSM used in the CJCSG has a very special form, namely,

$$A = \begin{pmatrix} d_L & 0 & 0 & \cdots & 0 & 1 \\ 1 & d_{L-1} & 0 & \cdots & 0 & t_{L-1} \\ 0 & 1 & d_{L-2} & \ddots & \vdots & \vdots \\ 0 & 0 & \ddots & \ddots & 0 & \vdots \\ \vdots & \vdots & \ddots & 1 & d_2 & t_2 \\ 0 & 0 & \cdots & 0 & 1 & d_1 + t_1 \end{pmatrix} \qquad (1)$$

This is the companion matrix of a polynomial of degree L (L is even) with additional $L/2$ ones on the main diagonal.

2.2 Outline of Pomaranch

Pomaranch follows a classical design of a synchronous, additive, bit-oriented stream cipher and consists of a key-stream generator producing a secure sequence of bits that is further bitwise XORed with the plain text previously converted into bits. After the initialization that comprises key setup, IV setup and the runup (see [3] for the details), the key-stream generator of Pomaranch is run in the generation mode showed in Fig. 1.

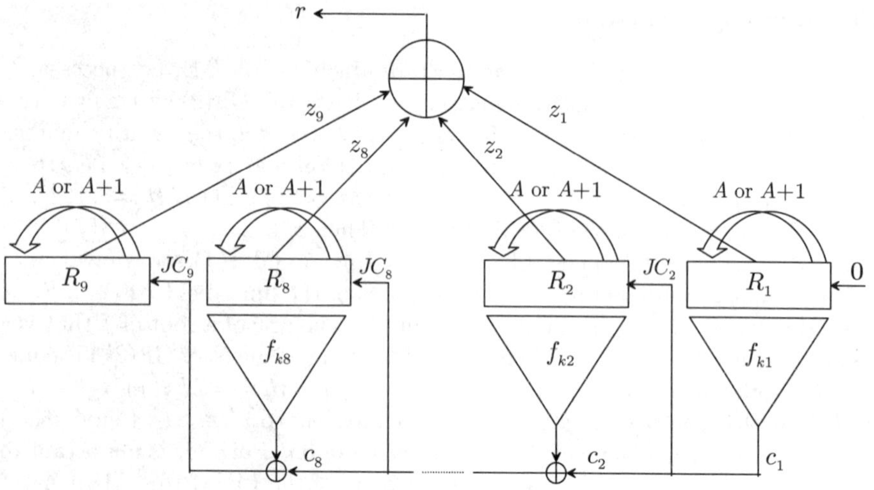

Fig. 1. Key-Stream Generation Mode of Pomaranch

The generator consists of nine irregularly clocked registers R_1 to R_9 (also called Jump Registers (JR)) that are combined in a cascade construction. Each register implements an autonomous LFSM and is built on 14 memory cells, each of them acting either as a simple delay shift cell (S-cell) or feedback cell (F-cell), depending on the value of the Jump Control (JC) bit. At any moment, half of the cells in each register are S-cells, while the others are F-cells which is seen as an important feature against power and side-channel attacks. LFSM implemented by every JR is described by the binary transition matrix A shown in (1), where $t_1, \ldots, t_{L-1}$ are defined by the positions of feedback taps and nonzero $d_1, \ldots, d_L$ correspond to the positions of F-cells in the register. In the particular case of Pomaranch $L = 14$, only $t_6 = 1$ and $d_1 = d_3 = d_7 = d_8 = d_9 = d_{11} = d_{13} = 1$. Transition matrix A is applied if the JC value is zero, otherwise, all cells are switched to the opposite mode which is equivalent to changing the transition matrix to $A+I$ with I being the identity matrix. Let R_i^t denote the state of the register R_i at a time $t \geq 0$. Then

$$R_i^{t+1} = (A + JC_i^t \cdot I)R_i^t \quad (i = 1, \ldots, 9) \ ,$$

where JC_i^t denotes the jump control bit for R_i at time t.

The 128-bit key K is divided into eight 16-bit subkeys k_1 to k_8. The current states of the registers R_1^t to R_8^t are nonlinearly filtered using a function that involves the corresponding subkey k_i $(i = 1, \ldots, 8)$. These functions provide an output of eight bits c_1^t to c_8^t which are used to produce the bits JC_2^t to JC_9^t controlling the registers R_2 to R_9 at time t as follows

$$JC_i^t = c_1^t \oplus \ldots \oplus c_{i-1}^t \quad (i = 2, \ldots, 9) \ .$$

The jump control bit JC_1 of register R_1 is permanently set to zero. The keystream bit generated at time t (denoted r^t) is the XOR of nine bits z_1^t to z_9^t tapped from the second cell of the register states R_1^t to R_9^t so $r^t = z_1^t \oplus \ldots \oplus z_9^t$.

3 Linear Relations in Jump Registers

Configuration of the jump registers in Pomaranch is chosen in such a way that the characteristic polynomial $C(x)$ of the binary transition matrix A in (1) is primitive and is neither self-reciprocal nor self-dual nor dual-reciprocal, i.e., A belongs to a primitive S_6 set, that is a set of six primitive polynomials which are each others reciprocals and duals (for the details see [4]). Obviously, the characteristic polynomial of $A+I$ is the dual $C^{\perp}(x) = C(x+1)$ and is primitive. Clocking of the jump registers is implemented by multiplying the state by the transition matrix A or $A+I$.

Let $Z = \{z^t\}_{t=0}^{\infty}$ denote the output sequence of a jump register being any component in the sequence of register states. Starting from some state R^t, the first output bit z^t is not affected by the jump control bits in $(JC^t, \ldots, JC^{t+L-1})$, the second output bit z^{t+1} is defined by JC^t, the third z^{t+2} is defined by (JC^t, JC^{t+1}) and so on.

Every output bit can be presented as a linear combination of L bits from the initial state R^0 and thus any $L+1$ bits of the output sequence are linearly dependent. The linear relation is defined by the relevant jump control bits and does not depend on the initial state of the register. Take such a relation that holds on $L+1$ consecutive bits of Z at the shift position t. Also assume that this relation holds for every component sequence of the register (i.e., irrespective of position the output sequence is tapped from). This means that for some set of binary coefficients $(\ell_0, \ell_1, \ldots, \ell_L)$ and any initial state we have $\ell_0 z^t + \ell_1 z^{t+1} + \ldots + \ell_L z^{t+L} = 0$ or, equivalently, that the following identity holds

$$\ell_0 I + \sum_{i=1}^{L} \ell_i \prod_{k=0}^{i-1} (A + JC^{t+k} I) = 0 \ .$$

Since $C(x)$, the characteristic polynomial of A, is in particular, irreducible, it coincides with the minimal polynomial of A. Thus, the latter identity holds if and only if

$$\ell_0 + \sum_{i=1}^{L} \ell_i \prod_{k=0}^{i-1} (x + JC^{t+k}) = \sum_{i=0}^{L} \ell_i x^{i-k_i} (x+1)^{k_i} = C(x) \ , \qquad (2)$$

where $0 \leq k_i \leq i$ are defined by the control bits $JC^t, \ldots, JC^{t+L-1}$, namely, $k_0 = 0$ and k_i is equal to the binary weight of vector $(JC^t, \ldots, JC^{t+i-1})$. Thus, if assuming the jump control sequence is purely random, then the values of k_i are binomially distributed. Since the degree of $C(x)$ is L and $C(0) = C(1) = 1$ then the coefficients at the highest-order and the constant term of the polynomial standing on the left hand side of (2) should be nonzero, i.e., $\ell_0 = \ell_L = 1$

for any linear relation in the jump register output. Given an arbitrary jump control sequence (that provides the values of k_i) the solution of (2) for the unknowns ℓ_i can be found applying a simplified version of Gaussian elimination. Such a solution always exists and, in particular, this can be easily seen from the matrix of the system which is triangular and contains ones on the main diagonal.

The complexity of solving the system is linear in L (if counting word operations). Indeed, let the binary coefficients of the binomial expansion of an additive term $x^{i-k_i}(x+1)^{k_i}$ be packed into words. Then, starting with $i = 0$ and $x^{0-k_0}(x+1)^{k_0} = 1$, every next term, depending on the value of JC^{t+i}, is equal to the previous one multiplied by x (shift the coefficient vector by one bit) or multiplied by $x+1$ (shift and add). Thus, expansions of all $L+1$ terms can be computed with $O(L)$ word operations. Further set $\ell_L = 1$ and add the coefficient vector of $x^{L-k_L}(x+1)^{k_L}$ to $C(x)$. If the degree of the obtained polynomial is equal $L-1$ then set $\ell_{L-1} = 1$, otherwise set $\ell_{L-1} = 0$. Proceed further in a similar way till all the unknowns ℓ_i are found. The total complexity remains linear in L.

Take a linear relation defined by the set of binary coefficients $\ell_0, \ldots, \ell_L$ with $\ell_0 = \ell_L = 1$ and take a set of weights $\{k_i \mid i = 0, \ldots, L; \ell_i = 1\}$ with $k_0 = 0$, $k_i \le k_j$ if $i < j$ and $k_j - k_i \le j - i$ such that

$$\sum_{i=0,\ldots,L;\ \ell_i=1} x^{i-k_i}(x+1)^{k_i} = C(x) . \tag{3}$$

Now take two neighboring additive terms from the left hand side of the last identity being $x^{i-k_i}(x+1)^{k_i}$ and $x^{j-k_j}(x+1)^{k_j}$ with $i < j$. Then the number of possible $(j-i)$-long sections of the jump control sequence leading from $x^{i-k_i}(x+1)^{k_i}$ to $x^{j-k_j}(x+1)^{k_j}$ is equal to $\binom{j-i}{k_j-k_i}$ (these are exactly the sequences with the binary weight of $(JC^{t+i}, \ldots, JC^{t+j-1})$ equal to $k_j - k_i$). In a similar manner, starting from the constant term $x^0(x+1)^0$ at $\ell_0 = 1$ and proceeding till the highest-order term at $\ell_L = 1$ is reached we can find the total number of L-long jump control sequences that correspond to the given linear relation and the set of weights. This number is obtained as a product of the relevant binomial coefficients for all $\ell_i \neq 0$ and $i > 0$.

As can be seen from (2), the set of all possible linear relations that correspond to different control sequences and the number of their occurrences only depend on the characteristic polynomial $C(x)$ of the jump register. As the linear relation occurring most often plays an essential role in the key-recovery attack, we will call its occurrence number the *Linear Equivalence Bias* (LEB) of the polynomial. All occurrence numbers together form a *Linear Equivalence Spectrum* (LES) of the polynomial. It can be easily seen by interchanging the roles of x and $x+1$ that $C(x)$ and $C^{\perp}(x)$ have the same LES. The LES value for any linear relation can be calculated as a sum consisting of terms being the product of binomial coefficients. Every set of weights k_i satisfying (3) provides one additive term to the sum.

Again take a linear relation and a set of weights satisfying (3). Applying the following *Doubling Rule*

$$x^a(x+1)^b = \begin{cases} x^{a-1}(x+1)^b + x^{a-1}(x+1)^{b+1}, \\ x^a(x+1)^{b-1} + x^{a+1}(x+1)^{b-1} \end{cases}$$

to different additive terms $x^{i-k_i}(x+1)^{k_i}$ in the left hand side of (3) we can find other relations that have a nonzero LES value. If the original relation has $\ell_{i-1} = 0$ and $\ell_i = 1$ then the new one has $\ell_{i-1} = \ell_i = 1$ that can be seen as doubling of the coefficient ℓ_i. It is not difficult to see that having the LES value of a linear relation that is expressed as a sum of products and applying the doubling rule to any $\ell_i = 1$ (assuming $\ell_{i-1} = 0$) gives us another relation with $\ell_i = \ell_{i-1} = 1$ and a sum with a doubled number of additive terms that is equal to the LES value of a new relation.

The most obvious example is to apply the doubling rule to the highest-order term at the coefficient $\ell_L = 1$ when $\ell_{L-1} = 0$ which leaves ℓ_L unchanged and gives rise to $\ell_{L-1} = 1$. Due to the binomial identity $\binom{n}{k} + \binom{n}{k-1} = \binom{n+1}{k}$ the LES value computed for the new linear relation will be the same as for the old one. This, in particular, implies that all the values in an LES appear even number of times. Applying the doubling rule to other terms results in new relations having higher or lower LES values. This feature will be illustrated in Section 4.1. Applying the doubling rule in the opposite direction results in the merge of two terms.

Note that using the presented technique we can evaluate LES values for some linear relations of length $L+1$ in the output sequence of a jump control register. In some cases this value is equal to the LEB of a polynomial meaning that we have found a relation that belongs to the ones occurring most often. However, we can not currently provide the algorithm for evaluating the LEB with the complexity lower than $O(L\,2^L)$ (checking through all JC sequences of length L and each time implementing a simple version of Gaussian elimination of length L). Finding a less complex algorithm remains an interesting open problem.

4 Key-Recovery Attack Using Linear Relations

In this section we calculate the LEB for the concrete configuration of jump registers in Pomaranch as well as for some minor modifications of the cipher. We also give some intuitive technique for finding the LEB in general. Then we mount a key-recovery attack with the time complexity much less than the exhaustive key search.

4.1 Linear Relations and Biases for Registers in Pomaranch

The characteristic polynomial of the transition matrix (1) can be found directly as follows

$$C(x) = 1 + \sum_{i=0}^{L-1} t_i \prod_{j=i+1}^{L} (d_j + x) \ ,$$

where $t_0 = 1$ is introduced for simplicity of the formula. Now assume L is even, a jump register of length L has two feedback taps (i.e., only $t_0 = t_n = 1$ for some $0 < n < L$), there are k F-cells among the first n cells (i.e., only k values from $d_1, \ldots, d_n$ are nonzero) and the total number of F-cells is $L/2$. Then

$$C(x) = 1 + x^{\frac{L}{2}+k-n}(x+1)^{\frac{L}{2}-k} + x^{\frac{L}{2}}(x+1)^{\frac{L}{2}} \ . \tag{4}$$

Placing this in (2) one immediately spots the evident linear relation $z^t + z^{t+L-n} + z^{t+L} = 0$ that we call *basic*. The corresponding equation coming from (2)

$$1+x^{L-n-k_{L-n}}(x+1)^{k_{L-n}}+x^{L-k_L}(x+1)^{k_L} = 1+x^{\frac{L}{2}+k-n}(x+1)^{\frac{L}{2}-k}+x^{\frac{L}{2}}(x+1)^{\frac{L}{2}}$$

can be shown to be satisfied only by $k_L = L/2$ and $k_{L-n} = L/2 - k$. Thus, this trinomial linear relation has the LES value given by

$$\binom{L-n}{\frac{L}{2}-k}\binom{n}{k} \ . \tag{5}$$

Assuming $n > 1$ and applying the doubling rule to the senior term in (4) we get another relation $z^t + z^{t+L-n} + z^{t+L-1} + z^{t+L} = 0$ having the same LES value.

Restricting our options further to the registers of length $L = 14$ that have a characteristic polynomial belonging to a primitive S_6 set, we are left with the following five alternative (n, k)-pairs of parameters $(6, 2)$, $(7, 2)$, $(7, 3)$, $(8, 3)$ and $(11, 5)$. Note that two polynomials corresponding to parameters (n, k) and $(n, n-k)$ form a dual pair. By (5), the corresponding LES values for the basic trinomial relations are 840, 441, 1225, 840 and 1386 respectively. For all the configurations except $(7, 2)$ these turned out to be the LEB values of the characteristic polynomials (4). For the remaining case $(n, k) = (7, 2)$ the basic relation is $z^t + z^{t+7} + z^{t+14} = 0$. Applying the doubling rule to the middle term here we obtain a new linear relation $z^t + z^{t+6} + z^{t+7} + z^{t+14} = 0$ and a new LES value $\binom{6}{1} \cdot \binom{7}{1} + \binom{6}{2} \cdot \binom{7}{3} = 567$ equal to the LEB in this case. On the other hand, for $(n, k) = (6, 2)$, applying the doubling rule to the middle term in the basic relation $z^t + z^{t+8} + z^{t+14} = 0$ we obtain a new linear relation $z^t + z^{t+7} + z^{t+8} + z^{t+14} = 0$ having the LES value $\binom{7}{5} \cdot \binom{6}{1} + \binom{7}{4} \cdot \binom{6}{3} = 826$, the second largest for this polynomial. We believe that in general, starting from the basic relation and consecutively applying the doubling rule, splitting and merging various terms, one can find all linear relations that hold at least for one control sequence. Tracking the LES values computed after each split or merge one can also find the LEB of the characteristic polynomial.

The concrete parameters initially chosen for Pomaranch are $(6, 2)$ giving the basic trinomial relation $z^t + z^{t+8} + z^{t+14} = 0$. The resulting LEB of $\binom{8}{5} \cdot \binom{6}{2} = 840$ is high enough to mount the key-recovery correlation attack that we are going to present in Section 4.2. Another linear relation $z^t + z^{t+8} + z^{t+13} + z^{t+14} = 0$ with the same LES value 840 is obtained applying the doubling rule to the senior term of the basic relation. The use of both relations makes the attack more efficient. The LES of the corresponding characteristic polynomial contains just 334 linear relations having nonzero occurrence numbers out of $2^{13} = 8192$ possible.

Suppose the LEB value of the characteristic polynomial of a jump register on L memory cells is $F > 0$ that corresponds to the linear relation on the output bits Z defined by $\boldsymbol{\ell} = (\ell_0, \ell_1, \ldots, \ell_L)$. Assume that the jump control sequence $\{JC^t\}_{t=0}^{\infty}$ is a sequence of independent and identically uniformly distributed random variables. In our case it is convenient to define the distribution bias of a binary random variable x as $\epsilon = 1 - 2\Pr\{x = 1\}$. Then the following random binary sequence $e^t = \sum_{i=0}^{L} \ell_i z^{t+i}$ $(t = 0, 1, 2, \ldots)$ has a nonuniform distribution with the bias $\epsilon = F/2^L$. Indeed, let H denote the event that a random subsection $(JC^t, \ldots, JC^{t+L-1})$ is one of those F that correspond to $\boldsymbol{\ell}$. The complementary event of H is denoted by $\overline{H}$. It is clear that $\Pr\{H\} = F/2^L$ and $\Pr\{e^t = 1 \mid H\} = 0$. The probability $\Pr\{e^t = 1 \mid \overline{H}\}$ can be considered equal to 1/2 because in this case e^t has a uniform distribution. Therefore, by the rule of total probability

$$\begin{aligned}\Pr\{e^t = 1\} &= \Pr\{e^t = 1 \mid H\}\Pr\{H\} + \Pr\{e^t = 1 \mid \overline{H}\}\Pr\{\overline{H}\} \\ &= 1/2(1 - F/2^L) \ .\end{aligned}$$

For the characteristic polynomial in Pomaranch the LEB is equal to 840 and $\epsilon = 840/2^{14} \approx 2^{-4.3}$.

4.2 Description of the Attack

Firstly, define the following set of sequences

$$u_i^t = e_i^t \oplus \ldots \oplus e_9^t \quad (t \geq 0,\ 2 \leq i \leq 9) \ ,$$

where $e_i^t = z_i^t \oplus z_i^{t+8} \oplus z_i^{t+14}$ and z_i^t denotes the output from the jump register R_i. Using this notation, the output sequence of the Pomaranch key-stream generator satisfies the following relation

$$r^t \oplus r^{t+8} \oplus r^{t+14} = z_1^t \oplus z_1^{t+8} \oplus z_1^{t+14} \oplus u_2^t \ . \tag{6}$$

Assume that sequences $\{e_i^t\}_{t=0}^{\infty}$ are independent. Then by the Piling-up Lemma, the random sequence $\{u_i^t\}_{t=0}^{\infty}$ has a nonuniform distribution with the bias $\epsilon^{10-i} = (840/2^{14})^{10-i}$ for $2 \leq i \leq 9$.

Equation (6) can be used in a correlation attack to find the correct initial state of R_1. The aim of a correlation attack, first introduced by Siegenthaler in [6], is finding the initial state of a binary LFSR of length L, given the noisy output sequence of a Binary Memoryless Symmetric Channel (BMSC) when the output sequence of the LFSR is applied to the input of this channel. As the first N bits of the output sequence of an LFSR is a codeword of the corresponding truncated cyclic linear code of a given LFSR, the problem is essentially a decoding problem. Siegenthaler solves this problem using Maximum Likelihood (ML) decoding and computes the minimum required output length of the given noisy sequence, denoted by N_0, needed to successfully find the initial state of the LFSR by considering a Hypothesis Testing problem.

However, it is easier to express N_0 using the capacity of the corresponding channel. The ML decoding is performed by searching over all 2^L possible initial states for the LFSR and choosing one that its corresponding output sequence has the smallest (largest) Hamming distance from the given noisy sequence if the channel error probability is less (more) than one half. However, the correlation attack is not limited to LFSR's and can be applied to any generator with M different equally probable initial states provided that the output sequences of different initial states have good statistical properties.

The channel capacity argument shows that the minimum required output length of the given noisy sequence of the generator is determined by

$$N_0 = \log_2(M)/C(p) \tag{7}$$

where p is the error probability and $C(p) = 1 + p\log_2(p) + (1-p)\log_2(1-p)$ is the Channel Capacity of the corresponding BMSC. Since the ML decoding is performed by the exhaustive search over all possible initial states, the required computational complexity is $O(MN_0)$.

The aforementioned discussion on correlation attack shows that the initial state of R_1 (denoted R_1^0) in Pomaranch can successfully be found using $N_0 = 14/C(0.5(1-\epsilon^8)) \approx 2^{73}$ bits of the key-stream with the computational complexity $2^{14}N_0 \approx 2^{87}$. Note that $C(0.5(1-\epsilon)) \approx \epsilon^2/(2\ln 2)$ when $|\epsilon| \ll 1$.

After finding R_1^0, we can eliminate the portion of z_1^t from the output sequence of Pomaranch. Define the sequence r_1^t as the XOR of r^t and z_1^t which is now available. Then, similarly to (6), we have

$$r_1^t \oplus r_1^{t+8} \oplus r_1^{t+14} = z_2^t \oplus z_2^{t+8} \oplus z_2^{t+14} \oplus u_3^t \ . \tag{8}$$

Since z_2^t depends both on the 14-bit initial state of R_2 and 16-bit subkey k_1, it can be considered as an output of a generator with 2^{30} possible output sequences. Similarly, (8) can be used in a correlation attack to find the correct value of initial state of R_2 and subkey k_1. Since the distribution bias of u_3^t is ϵ^7, the required key-stream length and the computational complexity are $N_0 = 30/C(0.5(1-\epsilon^7)) \approx 2^{65}$ and $2^{30}N_0 \approx 2^{95}$ respectively.

Alternatively, the 16-bit subkey k_1 can be found separately in the following way. Test all 2^{16} possible values for k_1 and, knowing the initial state R_1^0 from the first step of the attack, generate the jump control sequence JC_2^t $(t \geq 0)$. Now select only those time instants when the vector $(JC_2^t, \ldots, JC_2^{t+13})$ is one of those 840 that fulfil the LEB relation. Then for these times t automatically holds $z_2^t \oplus z_2^{t+8} \oplus z_2^{t+14} = 0$ and $r_1^t \oplus r_1^{t+8} \oplus r_1^{t+14} = u_3^t$ which we test to fit the distribution bias ϵ^7. Therefore, the needed amount of preselected key-stream bits is equal to $N_0 = 16/C(0.5(1-\epsilon^7)) \approx 2^{65}$. On the average, every 840 bits out of 2^{14} are selected, thus, the total number of key-stream bits is equal to 2^{69}. The computational complexity of finding k_1 is equal to $2^{16}2^{69} = 2^{85}$. Further, knowing the R_1^0 and k_1 (so the JC_2^t is known too) we can proceed with R_2 exactly the same way as before when working with R_1. The initial state of the remaining registers and subkeys can be found similarly with a much lower computational complexity. The total complexity of the attack is dominated by the first step of finding R_1^0.

In this attack we have only used the basic trinomial relation $z^t + z^{t+8} + z^{t+14} = 0$. Using the additional relation $z^t + z^{t+8} + z^{t+13} + z^{t+14} = 0$ with the same LES value allows to halve the required key-stream length. Summarizing the results, we showed that the secret key of Pomaranch can be found using 2^{72} bits of the key-stream with the computational complexity $O(2^{87})$. Note that the required key-stream length can be further reduced if other linear relations present in the LES are used. Assuming that all 334 relations found in the LES have the same probability that is defined by the LEB value of 840, we can obtain the lower bound on the needed amount of key-stream bits that is equal to $2^{73}/334 \approx 2^{64}$. The actual value lies somewhere between 2^{64} and 2^{73}.

5 Modified Jump Registers for Pomaranch

It is clear that the ideal configuration of a jump register should provide a lowest possible LEB value. Note that parameter pair $(7, 2)$ with LEB of 567 would have been a better choice, but even with this configuration our attack recovers the key with the complexity lower than the exhaustive key search. We conclude that all the characteristic polynomials having two feedback taps are not secure enough to counter the attack. Thus, in order to find a characteristic polynomial with a sufficiently low LEB, the Pomaranch jump register has to be changed to have three or more feedback taps.

Consider the registers having exactly three taps. Assume there is one tap, the rightmost, at position n_1 with k_1 feedback cells among cells 1 to n_1. The other tap is at position $n_2 > n_1$, with k_2 feedback cells among cells $n_1 + 1$ to n_2. The modified characteristic polynomial now becomes

$$C(x) = 1 + x^{\frac{L}{2}+k_1+k_2-n_2}(x+1)^{\frac{L}{2}-k_1-k_2} + x^{\frac{L}{2}+k_1-n_1}(x+1)^{\frac{L}{2}-k_1} + x^{\frac{L}{2}}(x+1)^{\frac{L}{2}}$$

for $L = 14$. The LES of this polynomial contains the basic relation $z_t + z_{t+L-n_2} + z_{t+L-n_1} + z_{t+L} = 0$.

Searching through all relevant (n_1, n_2, k_1, k_2) quadruplets results in a set of 16 primitive S_6-set polynomials, amongst which are the five polynomials already obtained for two taps. The polynomial with the least LEB in this set $x^{14} + x^{13} + x^{12} + x^{11} + x^9 + x^7 + x^5 + x^4 + x^2 + x + 1$ is obtained for $n_1 = 4$, $n_2 = 8$, $k_1 = k_2 = 1$ and has an LEB equal to 124 and an LES containing 1088 nonzero values. The linear relation $z_t + z_{t+6} + z_{t+10} + z_{t+14} = 0$ occurs $\binom{6}{1} \cdot \binom{4}{3} \cdot \binom{4}{3} = 96$ times. Performing a doubling operation on the 6th order term yields a relation which occurs 124 times that is equal to the LEB value.

Plugging in the bias of $124/2^{14} \approx 2^{-7.05}$ of the jump register in (7) results in the attack complexity of $O(2^{132})$ with 2^{117} bits of the key-stream required. This complexity exceeds the one of the exhaustive search over the key space containing 2^{128} elements. When using all 1088 linear relation found in the LES the required key-stream length can be reduced but in any case we will need at least $2^{118}/1088 \approx 2^{108}$ bits.

Note that an alternative way to secure Pomaranch against the described key-recovery attack is to make the sections have different characteristic polynomials.

Hereupon each section would have a different most probable linear relation. Thus, adding the outputs from all the sections will compensate for the bias. However, keeping all the sections the same definitively looks like a more "elegant" solution.

6 Conclusion

We considered a jump register arrangement that proved to be a powerful and efficient building block for stream ciphers that use irregular clocking of shift registers. We have identified a new inherent property of such arrangements which should always be observed in the relevant types of cipher design. Jump registers with badly chosen parameters allow building linear relations in the output sequence with significantly biased distribution.

Using the discovered property, we have built an attack on the Pomaranch stream cipher that recovers a 128-bit key with the complexity $O(2^{87})$ requiring less than 2^{72} (but at least 2^{64}) bits of the key-stream. This is the first attack of this type found for this cipher. However, introducing a minor change in the configuration of the jump register section in Pomaranch gives protection against this attack bringing its complexity up to $O(2^{132})$ with less than 2^{117} (but at least 2^{108}) bits of the key-stream required that exceeds the complexity of the exhaustive key search. This was considered in Pomaranch version 2 (see [5]).

Moreover, this new potential weakness can be exploited to attack other stream ciphers that use irregular clocking. The suggested technique has a general character and can be dangerous to other clock-controlled arrangements. This issue will become a focus for our future research.

References

1. Jansen, C.J.A.: Modern stream cipher design: A new view on multiple clocking and irreducible polynomials. In González, S., Martínez, C., eds.: Actas de la VII Reunión Española sobre Criptología y Seguridad de la Información. Volume Tomo I. Servicio de Publicaciones de la Universidad de Oviedo (2002) 11–29
2. eSTREAM: The ECRYPT stream cipher project (2005) `http://www.ecrypt.eu.org/stream/`.
3. Jansen, C.J.A., Helleseth, T., Kholosha, A.: Cascade jump controlled sequence generator (CJCSG). In: Symmetric Key Encryption Workshop, Workshop Record, ECRYPT Network of Excellence in Cryptology (2005) `http://www.ecrypt.eu.org/stream/ciphers/pomaranch/pomaranch.pdf`.
4. Jansen, C.J.A.: Stream cipher design based on jumping finite state machines. Cryptology ePrint Archive, Report 2005/267 (2005) `http://eprint.iacr.org/2005/267/`.
5. Jansen, C.J.A., Helleseth, T., Kholosha, A.: Cascade jump controlled sequence generator and Pomaranch stream cipher (Version 2). eSTREAM, ECRYPT Stream Cipher Project, Report 2006/006 (2006) `http://www.ecrypt.eu.org/stream/papersdir/2006/006.pdf`.
6. Siegenthaler, T.: Decrypting a class of stream ciphers using ciphertext only. IEEE Transactions on Computers **C-34**(1) (1985) 81–85

Improved Rijndael-Like S-Box and Its Transform Domain Analysis*

Seok-Yong Jin, Jong-Min Baek, and Hong-Yeop Song

Coding and Information Theory Lab,
School of Electrical and Electronic Engineering, Yonsei University,
134 Sinchon-dong, Seodaemun-gu, Seoul 120-749, Korea
{sy.jin, jm.back, hy.song}@coding.yonsei.ac.kr

Abstract. In this paper, we propose a simple scheme which produces a new S-box from a given S-box. We use the well-known conversion technique between the polynomial functions over $\mathbb{F}_{2^n}$ and the boolean functions from $\mathbb{F}_2^n$ to $\mathbb{F}_2$. We have applied the scheme to Rijndael S-box and obtained 29 new S-boxes, of which only one is a bijection with better algebraic expression than the original Rijndael S-box and has the same spectral properties as the original Rijndael S-box. All others turned out to be non-bijective, and have different spectral properties, and hence, they all are inequivalent to the original as boolean functions.

Keywords: Rijndael, AES, S-box, Hadamard transform, Avalanche transform.

1 Introduction

It is widely known that the properties of substitution box (S-box) are fundamental to the secrecy of symmetric encryption algorithms after Shannon [10]. Since S-boxes are usually implemented as look up tables, they are attractive for fast software encryption algorithms [3]. Most of popular block ciphers and some of stream ciphers have adopted various S-boxes and a lot of research has been given to designing "better" S-boxes.

There have been proposed [3] several methods to generate cryptographically useful S-boxes, such as the selection of nearly optimal (for differential [2] and linear [9] attacks) boolean functions as components of the S-boxes, random generation, using finite field operations and heuristic algorithms. Among these, finite field power operation based S-boxes achieve [3] several security criteria simultaneously, and have been used in many cipher proposals including Rijndael [14,15], major portfolio of NESSIE [19], ARIA [17] in Korea, and CRYPTREC [18] in Japan, mentioned only a few.

Rijndael was selected as the Advanced Encryption Standard (AES) by the US NIST in October 2000, and published as FIPS-197 [16] in November 2001.

* This work was supported by grant No.(R01-2003-000-10330-0) from the Basic Research Program of the Korea Science & Engineering Foundation.

G. Gong et al. (Eds.): SETA 2006, LNCS 4086, pp. 153–167, 2006.

Rijndael S-box is the finite field inversion together with a bitwise affine transformation. Until Rijndael was selected as AES, it was generally claimed that such S-box would prevent algebraic attacks. There have been some progress in the research of algebraic aspect of Rijndael S-box. It is known [13, 3, 7] that every component function of Rijndael S-box is a single term trace function on finite field GF(256), and has a property of algebraic linear redundancy that is inherent in finite field exponentiation. At the same time, researchers successively have proposed several improved S-boxes. In [7], the research effort has focused on the S-boxes with no simple algebraic expression while Fuller and Millan in [3] concentrates on the S-boxes with no linear redundancy.

This paper is organized as follows. In Section 2, we first introduce some background materials including one-to-one correspondence between the polynomial functions over a finite field and the boolean functions. Some definitions which are frequently used in the cryptanalysis of boolean functions will also be given. Section 3 describes the design scheme which produces a new S-box from a given S-box working on 4-bit inputs and outputs. We apply this scheme in Section 4 to Rijndael S-box and obtain 29 new S-boxes, of which only one is a bijection with better algebraic expression than the original Rijndael S-box and has the same spectral properties as the original Rijndael S-box. All others turned out to be non-bijective, and have different spectral properties, and hence, they all are inequivalent to the original as boolean functions. We give some concluding remarks and open problems in Section 5.

2 Preliminaries

2.1 Sequences, Trace-Represented Polynomial Functions and Boolean Functions

Let $\mathbb{F}_{2^n}$ be a finite field with 2^n elements and $\mathbf{a} = \{a_t\}_{t=0}^{N-1}$ be a sequence over $\mathbb{F}_2$ of period $N = 2^n - 1$. Let α be a primitive element in $\mathbb{F}_{2^n}$. The *discrete Fourier transform* (DFT) of $\mathbf{a}$ is defined as

$$A_k = \sum_{t=0}^{N-1} a_t \alpha^{-tk}, k = 0, 1, \cdots, N-1 .$$

Its inverse formula is given as follows:

$$a_t = \sum_{k=0}^{N-1} A_k \alpha^{kt}, t = 0, 1, \cdots, N-1 .$$

For a given sequence $\mathbf{a}$, there exists a polynomial function $f(x)$ from $\mathbb{F}_{2^n}$ to $\mathbb{F}_2$, associated with $\mathbf{a}$, such that $a_t = f(\alpha^t), t = 0, 1, \cdots, N-1$. We write $\mathbf{a} \leftrightarrow f$, and call $\mathbf{a}$ as an evaluation of the function f at α. By the inverse DFT or Lagrange interpolation, we have [5]:

$$\begin{aligned} a_t &= f(x)\Big|_{x=\alpha^t}, \quad t = 0, 1, ...N-1, \\ &= \sum_{j\in\Gamma(N)} Tr_1^{n_j}\left(A_j x^j\right)\Big|_{x=\alpha^t}, \quad A_j \in \mathbb{F}_{2^n}, \end{aligned} \tag{1}$$

where $\Gamma(N)$ is the set of cyclotomic coset leaders modulo N with respect to 2, C_j is the coset which contains j, $n_j = |C_j|$, $Tr_1^{n_j}(x)$ is the trace [8] function from $\mathbb{F}_{2^{n_j}}$ to $\mathbb{F}_2$, and $A_j \in \mathbb{F}_{2^{n_j}}$ is the DFT coefficient of $\mathbf{a}$. Then the sum of trace functions of (1) is a desired polynomial function and called the *trace representation of sequence* $\mathbf{a}$.

Now, let $g(x_{n-1}, \cdots, x_0)$ be a boolean function in n-variables. By applying the Lagrange interpolation, its polynomial representation $f(x)$ of $g(x_{n-1}, \cdots, x_0)$ can be determined as: (x is just indeterminant)

$$f(x) = \begin{cases} g(0, \cdots, 0) & x = 0, \\ \sum_{j=1}^{2^n-1} d_j x^j & x \in \mathbb{F}_{2^n}^*, \end{cases} \tag{2}$$

with coefficient d_j, $1 \le j \le 2^n - 1$, being

$$d_j = \sum_{\lambda\in\mathbb{F}_{2^n}^*} g(x_{n-1}, \cdots, x_0)\lambda^{-j}, \tag{3}$$

where $\lambda = \sum_{i=0}^{n-1} x_i\alpha_i$, and $\{\alpha_0, \cdots, \alpha_{n-1}\}$ is a basis of $\mathbb{F}_{2^n}$ over $\mathbb{F}_2$, denoted by $\mathbb{F}_{2^n} = \langle\{\alpha_0, \cdots, \alpha_{n-1}\}\rangle$.

A conversion from a polynomial function to a boolean function is given by

$$g(x_{n-1}, \cdots, x_0) = f\left(x_0\alpha_0 + \cdots + x_{n-1}\alpha_{n-1}\right), \text{ where } \mathbb{F}_{2^n} = \langle\{\alpha_0, \cdots, \alpha_{n-1}\}\rangle. \tag{4}$$

In the rest of this paper, by a boolean function f in n variables, we mean two notations $f(\mathbf{x}) = f(x_{n-1}, \cdots, x_0)$, $\mathbf{x} \in \mathbb{F}_2^n$ and $f(x)$, $x \in \mathbb{F}_{2^n}$ interchangeably.

2.2 Transform Domain Analysis Tools

For transform domain analysis of cryptographic functions, see Gong and Golomb [4], for example. The following definitions are mainly from [5, Ch. 6 and 10] with the same notation as above. For $\mathbf{a} \leftrightarrow f(x)$, the *Hadamard transform* (HT) of $\mathbf{a}$ or $f(x)$ is defined by

$$\widehat{f}(\lambda) = \sum_{x\in\mathbb{F}_{2^n}} (-1)^{Tr(\lambda x)+f(x)}, \quad \lambda \in \mathbb{F}_{2^n}.$$

The *Walsh transform* of a boolean function $f(\mathbf{x})$ is defined by

$$\widehat{f}(\mathbf{w}) = \sum_{\mathbf{x}\in\mathbb{F}_2^n} (-1)^{\mathbf{w}\cdot\mathbf{x}+f(\mathbf{x})}, \ \mathbf{w} \in \mathbb{F}_2^n.$$

The Hadamard transform of $f(x)$ and the Walsh transform of $f(\mathbf{x})$ have the relation:

$$\widehat{f}(\mathbf{w}) = \widehat{f}(\lambda), \ \mathbf{w} \in \mathbb{F}_2^n, \ \lambda \in \mathbb{F}_{2^n}, \text{ where } \mathbf{w} \cdot \mathbf{x} = Tr(\lambda x) .$$

Nonlinearity N_f of a boolean function f in n variables is defined as

$$N_f = \min_{\mathbf{w} \in \mathbb{F}_2^n, \ c \in \mathbb{F}_2} d\big(f(\mathbf{x}), \mathbf{w} \cdot \mathbf{x} + c\big) ,$$

where $d(\mathbf{x}, \mathbf{y})$ denotes the Hamming distance between $\mathbf{x}$ and $\mathbf{y}$, and is calculated using Hadamard transform of f:

$$\begin{aligned} N_f &= 2^{n-1} - \frac{1}{2} \max_{\mathbf{w} \in \mathbb{F}_2^n} \big|\widehat{f}(\mathbf{w})\big| \\ &= 2^{n-1} - \frac{1}{2} \max_{\lambda \in \mathbb{F}_{2^n}} \big|\widehat{f}(\lambda)\big| . \end{aligned} \tag{5}$$

The *Avalanche transform* (AT) or *additive correlation (convolution)* of $f(x)$ is defined by

$$(f * f)(w) = F(w) = \sum_{x \in \mathbb{F}_{2^n}} (-1)^{f(x+w)+f(x)}, \quad w \in \mathbb{F}_{2^n} . \tag{6}$$

Avalanche transform analysis of cryptographic functions was first introduced by Webster and Tavares [12]. We say that a boolean function f satisfies *Strict Avalanche Criterion* (SAC) if its Avalanche transform $F(\mathbf{w}) = 0$ for all $\mathbf{w}$ with binary hamming weight $\mathrm{wt}(\mathbf{w}) = 1$.

2.3 Equivalence Classes of Boolean Functions

Let f and g be two boolean functions in n-variables. If there exist a non-singular binary matrix D of order n, two n-tuple binary vectors $\mathbf{a}$ and $\mathbf{b}$, and a binary constant c such that for all $\mathbf{x} \in \mathbb{F}_2^n$

$$g(\mathbf{x}) = f\big(D\mathbf{x}^T \oplus \mathbf{a}^T\big) \oplus \mathbf{b} \cdot \mathbf{x}^T \oplus c ,$$

where $\mathbf{b} \cdot \mathbf{x}^T = b_1 x_1 \oplus b_2 x_2 \oplus \cdots \oplus b_n x_n$ denotes a linear function selected by $\mathbf{b}$, then f and g are said to be *(affine) equivalent* [3].

The absolute values of the Hadamard transform and the correlation transform are both re-arranged by affine transform and thus nonlinearity of a boolean function is unchanged under affine transform [3].

2.4 Description of Rijndael S-Box

An n-bit processing substitution box is a *a vector valued boolean function* $\mathbf{s}(\mathbf{x})$ from $\mathbb{F}_2^n$ to $\mathbb{F}_2^n$. If we let $\mathbf{s}(\mathbf{x}) = \big(s_{n-1}(\mathbf{x}), \cdots, s_1(\mathbf{x}), s_0(\mathbf{x})\big)$, then each $s_i(\mathbf{x})$,

$i = 0, \cdots, n-1$, is an ordinary boolean function in n variables and called a *component function* or *coordinate function* of the given S-box. By (4), $s_i(\mathbf{x})$, $\mathbf{x} \in \mathbb{F}_2^n$ can be identified as $s_i(x)$, $x = \sum_{i=0}^{n-1} x_i b_i \in \mathbb{F}_{2^n}$ where $\{b_0, b_1, \ldots, b_{n-1}\}$ is a basis of $\mathbb{F}_{2^n}$ over $\mathbb{F}_2$.

We take $b_i = \alpha^i$ for $0 \le i < 8$ where α is a root of $z^8 + z^4 + z^3 + z^1 + 1$, which is the defining irreducible (but not primitive) polynomial of $\mathcal{F} = \mathbb{F}_{2^8}$ for the Rijndael cipher. This transforms eight boolean functions into eight polynomial functions from $\mathcal{F}$ to $\mathbb{F}_2$, which are

$$\begin{aligned}
s_0(x) &= Tr(\beta^{166}x^{-1}) + 1 = Tr(\beta^{83}x^{127}) + 1 \\
s_1(x) &= Tr(\beta^{53}x^{-1}) + 1 \ = Tr(\beta^{154}x^{127}) + 1 \\
s_2(x) &= Tr(\beta^{36}x^{-1}) \qquad\ \ = Tr(\beta^{18}x^{127}) \\
s_3(x) &= Tr(\beta^{11}x^{-1}) \qquad\ \ = Tr(\beta^{133}x^{127}) \\
s_4(x) &= Tr(\beta^{72}x^{-1}) \qquad\ \ = Tr(\beta^{36}x^{127}) \\
s_5(x) &= Tr(\beta^{76}x^{-1}) + 1 \ = Tr(\beta^{38}x^{127}) + 1 \\
s_6(x) &= Tr(\beta^{51}x^{-1}) + 1 \ = Tr(\beta^{153}x^{127}) + 1 \\
s_7(x) &= Tr(\beta^{26}x^{-1}) \qquad\ \ = Tr(\beta^{13}x^{127}),
\end{aligned} \tag{7}$$

where $\beta = \alpha + 1$ is a primitive element of $\mathcal{F}$, and $x = \sum_{i=0}^{7} x_i b_i \in \mathcal{F}$. The above algebraic expressions of component functions $s_i(x)$ have been determined by Inverse DFT or Lagrange interpolation (2), dual basis approach [13], or q-polynomial method [7].

3 Proposed Scheme of Designing a New S-Box from a Given S-Box

We will describe a proposed scheme of designing a new S-box from a given one. For convenience, we explain using a smaller size example, e.g., over $\mathbb{F}_{2^4}$.

Consider the following S-box denoted as SB 0 (the left most one in Table 1), defined by $s(x) = x^{-1}$ over the field $\mathcal{F} = \mathbb{F}_{2^4}$ using the irreducible polynomial $g_0(z) = z^4 + z^3 + z^2 + z + 1$. Then, the following algorithm produces SB-1 and SB-2 in the middle and right-most in Table 1, respectively.

Table 1. Three S-boxes (in hexadecimal)

	00	01	10	11
00	0	1	f	a
01	8	6	5	9
10	4	7	3	e
11	d	c	b	2

SB-0

	00	01	10	11
00	0	1	a	f
01	6	8	5	9
10	2	b	d	c
11	3	e	7	4

SB-1

	00	01	10	11
00	0	c	7	0
01	6	7	4	7
10	e	2	e	6
11	8	a	5	a

SB-2

Polynomial functions for each of the 4 coordinate boolean functions of SB-0 over $\mathcal{F}$ can be found using Lagrange interpolation explained in Section 2:

$$\begin{aligned}\mathbf{s}(\mathbf{x}) &= \big(s_3(\mathbf{x}),\ s_2(\mathbf{x}),\ s_1(\mathbf{x}),\ s_0(\mathbf{x})\big), \quad \text{or}\\ s(x) &= \big(s_3(x),\ s_2(x),\ s_1(x),\ s_0(x)\big)\\ &= \big(Tr_1^4(\beta^{14}x^7), Tr_1^4(\beta^7x^7), Tr_1^4(\beta^{10}x^7), Tr_1^4(\beta^8x^7)\big),\end{aligned} \tag{8}$$

where $\mathbf{x} = (x_3, x_2, x_1, x_0)$ is the input to SB-0, $x = \sum_{i=0}^{3} x_i b_i \in \mathcal{F} \cong \langle\{b_i \mid b_i = \alpha^i, 0 \le i < 4\}\rangle$, α is a root of $g_0(z)$ which is the defining polynomial of $\mathcal{F}$, and $\beta = 1 + \alpha$ is a primitive element of $\mathcal{F}$.

Now, we let $\mathcal{K}$ be the field defined by $g_1(z) = z^4 + z^3 + 1$. Then the polynomial functions of SB-0 over $\mathcal{K}$ are determined as

$$\begin{aligned} r(x) &= \big(r_3(x),\ r_2(x),\ r_1(x),\ r_0(x)\big)\\ &= \begin{pmatrix} Tr_1^4(\gamma^{10}x + \gamma^{12}x^3 + \gamma^{14}x^7) + Tr_1^2(\gamma^{10}x^5)\\ Tr_1^4(\gamma^{3}x + \gamma^{4}x^3 + \gamma^{5}x^7) + Tr_1^2(x^5)\\ Tr_1^4(\gamma^{9}x + \gamma^{10}x^3 + \gamma^{13}x^7) + Tr_1^2(\gamma^{5}x^5)\\ Tr_1^4(\gamma^{2}x + \gamma^{13}x^3 + \gamma^{6}x^7) + Tr_1^2(\gamma^{5}x^5) \end{pmatrix}^T \end{aligned} \tag{9}$$

where $x = \sum_{i=0}^{3} x_i c_i \in \mathcal{K} \cong \langle\{c_i \mid c_i = \gamma^i, 0 \le i < 4\}\rangle$ and γ is a root of $g_1(z)$.

To obtain polynomial functions for the new S-box, which we call SB-1, we simply replace the coefficients (some powers of γ in (9)) with the corresponding powers of β. This gives new polynomial functions from (9), which are

$$\begin{aligned} h_3(x) &= Tr_1^4(\beta^{10}x + \beta^{12}x^3 + \beta^{14}x^7) + Tr_1^2(\beta^{10}x^5),\\ h_2(x) &= Tr_1^4(\beta^{3}x + \beta^{4}x^3 + \beta^{5}x^7) + Tr_1^2(x^5),\\ h_1(x) &= Tr_1^4(\beta^{9}x + \beta^{10}x^3 + \beta^{13}x^7) + Tr_1^2(\beta^{5}x^5),\\ h_0(x) &= Tr_1^4(\beta^{2}x + \beta^{13}x^3 + \beta^{6}x^7) + Tr_1^2(\beta^{5}x^5). \end{aligned}$$

Finally, to construct SB-1 shown in the middle of Table 1, we evaluate the above polynomial functions over $\mathcal{F} = \mathbb{F}_{2^4}$ with multiplication mod $g_0(z)$.

There is another irreducible polynomial of degree 4 over $\mathbb{F}_2$, which is $g_2(z) = z^4 + z + 1$. We denote $\mathcal{E}$ by the field defined by $g_2(z)$. Then, similarly, over $\mathcal{E}$, the polynomial functions of SB-0 are determined as

$$\begin{aligned} t(x) &= \big(t_3(x),\ t_2(x),\ t_1(x)\ , t_0(x)\big)\\ &= \begin{pmatrix} Tr_1^4(\delta^{2}x + \gamma^{9}x^3 + \delta^{10}x^7) + Tr_1^2(\delta^{5}x^5)\\ Tr_1^4(\delta^{4}x + \delta^{12}x^3 + \delta^{12}x^7) + Tr_1^2(x^5)\\ Tr_1^4(\delta^{6}x + \delta^{2}x^3 + \delta^{14}x^7) + Tr_1^2(x^5)\\ Tr_1^4(\delta^{11}x + \delta^{11}x^3 + \delta^{2}x^7) + Tr_1^2(\delta^{10}x^5) \end{pmatrix}^T \end{aligned} \tag{10}$$

where $x = \sum_{i=0}^{3} x_i d_i \in \mathcal{E} \cong \langle\{d_i \mid d_i = \delta^i, 0 \le i < 4\}\rangle$ and δ is a root of $g_2(z)$.

By replacing δ in (10) with β, we obtain another set of polynomial functions from (10):

$$
\begin{aligned}
u_3(x) &= Tr_1^4(\beta^2 x + \beta^9 x^3 + \beta^{10} x^7) + Tr_1^2(\beta^5 x^5), \\
u_2(x) &= Tr_1^4(\beta^4 x + \beta^{12} x^3 + \beta^{12} x^7) + Tr_1^2(x^5), \\
u_1(x) &= Tr_1^4(\beta^6 x + \beta^2 x^3 + \beta^{14} x^7) + Tr_1^2(x^5), \\
u_0(x) &= Tr_1^4(\beta^{11} x + \beta^{11} x^3 + \beta^2 x^7) + Tr_1^2(\beta^{10} x^5).
\end{aligned}
$$

This, in turn, gives a third S-box, SB-2, shown in the right-most of Table 1, when we evaluate the above polynomial functions over $\mathcal{F} = \mathbb{F}_{2^4}$ with multiplication mod $g_0(z)$.

Remark 1. Observe that SB-1 is a bijection but SB-2 is not. The reason why they are so different would be a topic of further research.

Remark 2. A simple calculation shows that all three S-boxes in Table 1 have the same spectral properties. That is, they have the same profiles of Hadamard transform and Avalanche transform, where the transform is applied to each of the coordinate boolean functions. It turned out that the spectral properties do not have to be all the same when this scheme is applied to larger S-boxes, which we will discuss in the next section.

4 Application of Proposed Scheme to Rijndael S-Box

4.1 Using $z^8 + z^4 + z^3 + z^2 + 1$

We apply the proposed design scheme explained in Section 3 to the original Rijndael S-box, which we denote by BOX-0. From now on, we use the parallel notations in Section 3, but $g_0(z)$ and $g_1(z)$ are changed to:

$$g_0(z) = z^8 + z^4 + z^3 + z^1 + 1, \quad \text{and} \quad g_1(z) = z^8 + z^4 + z^3 + z^2 + 1,$$

where $g_0(z)$ is the defining polynomial of $\mathbb{F}_{2^8}$ for the Rijndael cipher and $g_1(z)$ is a primitive polynomial of degree 8 over $\mathbb{F}_2$.

Recall that the polynomial functions $s_i(x)$, $0 \le i < 8$, for the coordinate boolean functions of BOX-0 were determined as in (7) over $\mathcal{F} = \mathbb{F}_{2^8}$ defined by $g_0(z)$, where $\beta = 1 + \alpha$ is a primitive element of $\mathcal{F}$, where α is a root of $g_0(z)$, and $x = \sum_{i=0}^{7} x_i b_i \in \mathcal{F} \cong \langle \{b_i | b_i = \alpha^i, 0 \le i < 8\} \rangle$.

Now, over $\mathcal{K} = \mathbb{F}_{2^8}$ defined by $g_1(z)$, the same boolean functions give some other polynomial functions $r_i(x)$, $0 \le i < 8$, where, for example,

$$
\begin{aligned}
r_7(x) = {} & Tr_1^2(\gamma^{85} x^{85}) + Tr_1^4(\gamma^{238} x^{17} + \gamma^{34} x^{51} + \gamma^{136} x^{119}) \\
& + Tr_1^8(\gamma^4 x^1 + \gamma^{43} x^3 + \gamma^{60} x^5 + \gamma^3 x^7 + \gamma^{54} x^9 + \gamma^{155} x^{11}) \\
& + Tr_1^8(\gamma^{86} x^{13} + \gamma^{157} x^{15} + \gamma^{157} x^{19} + \gamma^{48} x^{21} + \gamma^{163} x^{23} + \gamma^{98} x^{25}) \\
& + Tr_1^8(\gamma^{50} x^{27} + \gamma^{92} x^{29} + \gamma^{67} x^{31} + \gamma^{69} x^{37} + \gamma^{181} x^{39} + \gamma^1 x^{43}) \\
& + Tr_1^8(\gamma^2 x^{45} + \gamma^{194} x^{47} + \gamma^{110} x^{53} + \gamma^{145} x^{55} + \gamma^{105} x^{59} \gamma^{246} x^{61}) \\
& + Tr_1^8(\gamma^{192} x^{63} + \gamma^{45} x^{87} + \gamma^{20} x^{91} + \gamma^{160} x^{95} + \gamma^{144} x^{111} + \gamma^{13} x^{127}),
\end{aligned} \quad (11)
$$

Table 2. Polynomial functions r_i's of BOX-0 over $\mathcal{K}$ (h_i's of BOX-1 over $\mathcal{F}$)

k	n_k	r_7	r_6	r_5	r_4	r_3	r_2	r_1	r_0
const.		–	**1**	**1**	–	–	–	**1**	**1**
85	2	85	0	170	0	170	170	0	85
17	4	238	0	102	136	136	68	17	119
51	4	34	102	238	17	85	17	17	85
119	4	136	0	187	85	0	∞	187	51
1	8	4	129	65	213	52	83	14	127
3	8	43	251	43	12	233	23	174	30
5	8	60	163	162	197	79	57	166	24
7	8	3	19	50	233	134	193	246	119
9	8	54	221	120	97	33	139	159	33
11	8	155	31	242	163	92	∞	2	226
13	8	86	80	199	91	17	151	208	153
15	8	157	143	74	56	242	41	86	214
19	8	157	∞	231	16	99	148	65	251
21	8	48	28	69	3	190	33	106	136
23	8	163	48	100	173	16	198	248	120
25	8	98	78	37	9	197	242	225	72
27	8	50	29	25	115	16	157	189	167
29	8	92	74	21	220	162	25	71	174
31	8	67	49	69	157	233	130	107	35
37	8	69	253	52	155	32	6	219	230
39	8	181	145	68	145	114	121	12	91
43	8	1	125	168	228	244	242	217	58
45	8	2	253	127	200	25	64	133	164
47	8	194	246	233	173	43	102	108	119
53	8	110	23	129	77	16	133	245	136
55	8	145	173	74	35	6	143	159	64
59	8	105	65	121	186	228	90	182	108
61	8	246	176	111	176	17	161	213	100
63	8	192	252	141	80	142	81	213	178
87	8	45	7	157	61	230	6	98	78
91	8	20	239	73	76	251	20	123	94
95	8	160	236	186	66	236	222	156	248
111	8	144	41	149	35	167	32	154	210
127	8	13	141	14	91	90	220	166	71
LS		254	247	255	254	254	242	255	255

where γ is a root of $g_1(z)$ in this section, and is a primitive element of $\mathcal{K}$. For the other $r_i(x)$, see Table 2.

The first and second column of Table 2 represents cyclotomic coset leaders and sizes, respectively. The values in the third column are the exponents of the coefficients of x^k in the trace representation of $r_7(x)$, with the convention of $\gamma^\infty = 0$, where γ is a primitive element in $\mathcal{K}$. The bottom row of Table 2 shows the number of nonzero terms in each $r_i(x)$. Note that these values are very large (255 is the maximum) compared to that of the expression in (7).

By replacing the coefficients (which are the powers of γ in (11)) with the corresponding powers of β, as described in Section 3, we obtain a set of 8 new polynomial functions $h_i(x)$, $0 \le i < 8$, one of which is

$$\begin{aligned} h_7(x) = {} & Tr_1^2(\beta^{85}x^{85}) + Tr_1^4(\beta^{238}x^{17} + \beta^{34}x^{51} + \beta^{136}x^{119}) \\ & + Tr_1^8(\beta^4x^1 + \beta^{43}x^3 + \beta^{60}x^5 + \beta^3x^7 + \beta^{54}x^9 + \beta^{155}x^{11}) \\ & + Tr_1^8(\beta^{86}x^{13} + \beta^{157}x^{15} + \beta^{157}x^{19} + \beta^{48}x^{21} + \beta^{163}x^{23} + \beta^{98}x^{25}) \\ & + Tr_1^8(\beta^{50}x^{27} + \beta^{92}x^{29} + \beta^{67}x^{31} + \beta^{69}x^{37} + \beta^{181}x^{39} + \beta^1x^{43}) \\ & + Tr_1^8(\beta^2x^{45} + \beta^{194}x^{47} + \beta^{110}x^{53} + \beta^{145}x^{55} + \beta^{105}x^{59} + \beta^{246}x^{61}) \\ & + Tr_1^8(\beta^{192}x^{63} + \beta^{45}x^{87} + \beta^{20}x^{91} + \beta^{160}x^{95} + \beta^{144}x^{111} + \beta^{13}x^{127}), \end{aligned} \tag{12}$$

where $\beta = 1 + \alpha$ is the primitive element of $\mathcal{F}$, where α is a root of $g_0(z)$. Now, evaluating these polynomials over $\mathcal{F} = \mathbb{F}_{2^8}$ with multiplication mod $g_0(z)$ gives a new S-box, BOX-1, shown in Table 3.

Table 3. BOX-1 (in hexadecimal)

	0	1	2	3	4	5	6	7	8	9	a	b	c	d	e	f
0	63	7c	7b	77	6b	f2	6f	c5	76	ab	fe	d7	67	2b	01	30
1	82	ca	c9	7d	fa	59	f0	47	72	c0	a4	9c	af	a2	ad	d4
2	c3	23	04	c7	05	9a	96	18	eb	27	75	b2	12	07	80	e2
3	93	26	fd	b7	cc	f7	36	3f	d8	71	31	15	34	a5	f1	e5
4	fc	20	b1	5b	53	d1	ed	00	be	39	cb	6a	cf	58	4a	4c
5	1b	6e	a0	5a	83	09	2c	1a	b3	d6	52	3b	2f	84	e3	29
6	33	85	4d	43	fb	aa	d0	ef	f9	45	02	7f	50	3c	a8	9f
7	f5	38	92	9d	40	8f	a3	51	bc	b6	21	da	ff	10	f3	d2
8	16	bb	b0	54	2d	0f	99	41	8c	a1	0d	89	e6	bf	42	68
9	28	df	55	ce	e9	87	9b	1e	f8	e1	98	11	69	d9	94	8e
a	4b	bd	8a	8b	dd	e8	74	1f	2e	25	ba	78	b4	c6	a6	1c
b	c1	86	1d	9e	61	35	b9	57	b5	66	3e	70	0e	f6	48	03
c	ac	62	d3	c2	79	e4	91	95	06	49	24	5c	e0	32	0a	3a
d	ea	f4	6c	56	ae	08	7a	65	8d	d5	a9	4e	c8	e7	37	6d
e	ee	46	b8	14	de	5e	db	0b	90	88	2a	22	dc	4f	60	81
f	c4	a7	3d	7e	5d	64	19	73	17	44	5f	97	13	ec	0c	cd

We now list some cryptographic properties of BOX-1 in parallel with those of BOX-0. We will use $h_i(x)$ in Table 2 for BOX-1 and $s_i(x)$ in (7) for BOX-0.

1. BOX-1 is a bijective map. So is BOX-0.
2. The component boolean functions of BOX-1 are balanced. So is BOX-0.
3. It is not difficult to show that the highest degree in its algebraic normal form (ANF) of a boolean function f is the maximum binary Hamming weight $\mathrm{wt}(k)$ as k runs through all the exponents in the trace representation of f [5]. For $k = 127$, $\mathrm{wt}(k) = 7$ and every coordinate function $h_i(x)$, $i = 0, \cdots, 7$,

has the term θx^{127} in its trace representation for some nonzero $\theta \in \mathbb{F}_{2^n}^*$. The ANF of any boolean function can be found by exhaustive "truth table summation" [11]. In fact, the number of linear and highest degree terms in the ANF of $h_i(x)$ and $s_i(x)$ turns out to be given as follows:

	h_0	h_1	h_2	h_3	h_4	h_5	h_6	h_7	s_0	s_1	s_2	s_3	s_4	s_5	s_6	s_7
Number of linear terms	4	3	4	4	6	3	3	3	6	4	6	4	6	2	4	4
Number of degree 7 terms	4	4	5	1	5	4	3	3	5	4	2	4	2	3	4	4

4. Since the linear span of a function or a sequence is just the number of nonzero terms in its polynomial function [5], we have:

	h_0	h_1	h_2	h_3	h_4	h_5	h_6	h_7	s_0	s_1	s_2	s_3	s_4	s_5	s_6	s_7
Linear span	255	255	242	254	254	255	247	254	9	9	8	8	8	9	9	8

5. Hadamard transform of a boolean function has a connection (5) with nonlinearity and with the first-order correlation immunity [11]. Hadamard transform profile of component functions of BOX-1 and BOX-0 are determined as:

Absolute HT value	0	4	8	12	16	20	24	28	32	Total
h_i for all $0 \le i < 8$	17	48	36	40	34	24	36	16	5	256
s_i for all $0 \le i < 8$	17	48	36	40	34	24	36	16	5	256

6. From the above calculation, it is easy to see that nonlinearity of every coordinate function of BOX-1 is 112, which is the same as that of BOX-0, the original Rijndael S-box.
7. The frequency distribution of Avalanche (additive correlation) transform of each component function of BOX-1 and BOX-0 is determined as:

Absolute AT value	0	8	16	24	32	Total
h_i for all $0 \le i < 8$	32	84	74	52	13	255
s_i for all $0 \le i < 8$	32	84	74	52	13	255

8. It is interesting to observe that for all $i = 0, 1, \cdots, 7$, h_i and s_i have the same Hadamard and Avalanche transform spectrum (as a profile), which is not an accident due to the following theorem.

Theorem 1. *Let $\Gamma = \{s_0, s_1, \cdots, s_7, h_0, h_1, \cdots, h_7\}$ be the set consisting of all the component functions of* BOX-0 *and* BOX-1*. Then any two boolean functions in Γ are pairwise equivalent.*

Proof. Since $s_i(x) = Tr(\theta_i x^{-1}) + e_i$ for some $\theta_i \in \mathbb{F}_{2^8}$, $i = 0, \cdots, 7$, and e_i is either 1 or 0 as shown in (7), it is easily shown [3, Theorem 3] that s_i and s_j are equivalent for any $0 \le i, j \le 7$.

Now it is enough to establish the affine equivalence between s_0 and h_i for all $i = 0, 1, \cdots, 7$. Some calculation shows that $h_0(\mathbf{x}) = s_0(\mathcal{D}_0 \mathbf{x}^T)$, where binary 8×8 square matrix $\mathcal{D}_0$ is given as

$$\mathcal{D}_0 = [\ 11_d\ \ 148_d\ \ 182_d\ \ 82_d\ \ 224_d\ \ 8_d\ \ 105_d\ \ 31_d\ ],$$

where the first column 11_d is the decimal form of $[00001011]^T$. Similarly, for $i = 1, 2, \cdots, 7$, we have $h_i(\mathbf{x}) = s_0(\mathcal{D}_i \mathbf{x}^T) + c_i$, where

$$\begin{array}{lcl} \mathcal{D}_1 & = & [\ 51_d\ 150_d\ 235_d\ 156_d\ 223_d\ 77_d\ 28_d\ 1_d\ ] \\ \mathcal{D}_2 & = & [\ 47_d\ 78_d\ 142_d\ 86_d\ 149_d\ 164_d\ 62_d\ 240_d\ ] \\ \mathcal{D}_3 & = & [\ 35_d\ 112_d\ 68_d\ 4_d\ 213_d\ 186_d\ 121_d\ 129_d\ ] \\ \mathcal{D}_4 & = & [\ 26_d\ 94_d\ 156_d\ 1_d\ 172_d\ 55_d\ 85_d\ 124_d\ ] \\ \mathcal{D}_5 & = & [\ 42_d\ 101_d\ 4_d\ 220_d\ 237_d\ 35_d\ 247_d\ 191_d\ ] \\ \mathcal{D}_6 & = & [\ 47_d\ 90_d\ 18_d\ 241_d\ 151_d\ 137_d\ 143_d\ 122_d\ ] \\ \mathcal{D}_7 & = & [\ 67_d\ 146_d\ 81_d\ 29_d\ 161_d\ 199_d\ 246_d\ 61_d\ ] \end{array},$$

and constant c_i is given by $c_2 = c_3 = c_4 = c_7 = 1$ and $c_1 = c_5 = c_6 = 0$. □

9. Finally, we check SAC for BOX-1 and BOX-0.

	00000001	00000010	00000100	00001000	00010000	00100000	01000000	10000000
h_7	0	-16	-8	-24	-32	-8	16	8
h_6	24	-16	8	-8	8	-24	16	-32
h_5	8	16	24	24	24	-8	-16	-8
h_4	24	-8	-16	-8	32	0	24	16
h_3	-32	16	24	-16	8	-8	16	-16
h_2	24	-16	32	24	-16	0	0	-8
h_1	-8	0	24	-16	8	-8	8	-24
h_0	-8	16	24	-8	-8	0	16	0
s_7	-8	16	-8	-16	24	24	-16	-8
s_6	-8	8	-8	-16	0	-8	-16	-32
s_5	24	-32	0	16	24	-8	16	-8
s_4	-32	0	16	24	-8	16	-8	-16
s_3	24	8	-32	0	0	16	16	8
s_2	8	24	0	-16	0	-24	-16	-16
s_1	24	0	-16	0	-24	-16	-16	8
s_0	0	-16	0	-24	-16	16	8	-8

Since an affine transformation rearranges additive correlation values, the Avalanche transform of h_i is possibly non-identical to that of s_i. However, for $\mathbf{w} \in \mathbb{F}_2^8$ with binary Hamming weight one, the maximum absolute correlation value of $(h_i * h_i)(\mathbf{w})$ is equal to that of $(s_j * s_j)(\mathbf{w})$ for $0 \le i, j \le 7$, and the frequency of occurrences of each possible values of both BOX-1 and BOX-0 are very similar. Therefore, BOX-1 and BOX-0 have almost the same level of performance in correlation aspect.

4.2 Using All Other Irreducible Polynomials of Degree 8

Analysis result of BOX-1, especially the items from 4 to 7 in the above list, and Theorem 1, shows that BOX-1 is equivalent to the original S-box of Rijndael in many aspects.

The effect of replacing the irreducible polynomial in Rijndael has been enough studied previously. Any replacement of irreducible polynomial in Rijndael cipher with different one can create a new cipher, but it is equivalent to the original in all aspects. Barkan and Biham [1] concluded that the arbitrary choice for the irreducible polynomial to be replaced works the same always, and hence, there is no advantage to changing the original irreducible polynomial with any other. Careless conclusion from the above information would lead to a guess that the remaining S-boxes, BOX-2, ... , BOX-29, using each of the remaining irreducible polynomials of degree 8, respectively, would have the similar properties. That is, every BOX-i for $2 \leq i \leq 29$ might be a balanced bijection with the same spectral properties (the same Hadamard and correlation transform profile) and whose coordinate functions would be all affine equivalent to that of Rijndael S-box. To our surprise, it turned out that this is not the case. Careful examination of the proposed scheme described in Section 3 will reveal that our scheme is completely different from simply changing the irreducible polynomial in Rijndael cipher. Instead, it is a method of constructing only a new S-box from the given one, and the whole cipher runs over the field defined by the same irreducible polynomial.

For example, we examine BOX-2, which is constructed using the irreducible polynomial $g_2(z) = z^8 + z^5 + z^3 + z^1 + 1$ in the conversion process. Again we use the parallel notations with Section 3, but in this case, we use the field $\mathcal{E}$ defined by $g_2(z)$. BOX-2 is shown in Table 4. The polynomial functions for BOX-2 are denoted by $u_i(x)$, their Hadamard transform profiles and SAC table are given in Table 5 and Table 6, respectively.

In summary, BOX-2 is completely different from BOX-1 or BOX-0:

1. BOX-2 is not bijective and no coordinate function is balanced. Therefore, it is worse against the linear attack than BOX-0.

Table 4. BOX-2 (in hexadecimal)

	0	1	2	3	4	5	6	7	8	9	a	b	c	d	e	f
0	63	12	31	1d	f9	50	e6	22	4f	2f	2e	e8	18	f1	03	08
1	4a	eb	84	c2	b9	90	34	d4	02	b6	61	6c	ea	29	46	2b
2	cd	d3	c7	f2	2f	34	9e	d4	c3	14	b3	56	7b	9d	d0	58
3	ff	d4	7e	82	85	55	90	88	21	ba	af	23	b2	aa	ba	49
4	1e	ac	27	2f	94	cb	0c	eb	7f	c3	9f	b1	53	2b	19	d2
5	78	2e	dd	ca	c3	18	a3	51	12	31	22	6e	2d	59	87	da
6	4a	ec	f2	a7	a8	1e	1b	33	5e	60	94	f5	07	f4	6d	ac
7	9b	01	64	55	93	d9	80	1c	2b	de	98	78	42	eb	65	c5
8	3f	56	f3	dc	e1	18	f0	db	59	e7	ab	cc	fa	3d	89	18
9	a8	3c	62	8b	70	55	7c	7a	0d	aa	c7	4c	9e	d4	bf	00
a	e7	48	50	7c	48	9b	89	72	cb	c4	a5	40	05	b1	00	fc
b	4a	b4	ac	85	bb	62	98	22	6d	b4	e4	b7	ac	30	d0	70
c	ce	09	bb	e8	ef	11	e6	f8	3a	14	ac	7c	75	29	c1	79
d	1b	ff	9c	31	49	7b	5a	57	cb	b6	d0	3e	b9	48	47	c8
e	1d	02	eb	7d	d7	df	31	3f	72	9c	a3	91	b5	75	c9	08
f	38	06	a4	b9	2d	f6	20	99	3a	9b	5e	6e	7e	36	58	14

Table 5. Hadamard transform profile (frequency distribution) of BOX-2

Absolute HT value	0	4	8	12	16	20	24	28	32	36	40	44	48	52	Total
u_7	27	59	45	28	21	30	25	7	5	4	2	0	3	0	256
u_6	26	45	46	42	31	22	17	13	6	4	1	2	1	0	256
u_5	22	55	42	38	32	23	18	8	10	3	4	1	0	0	256
u_4	25	45	38	33	42	31	15	17	5	2	3	0	0	0	256
u_3	23	46	44	46	34	25	16	7	5	3	4	1	2	0	256
u_2	33	53	38	32	33	22	15	15	5	6	3	0	1	0	256
u_1	22	55	40	39	35	21	21	10	6	1	3	1	1	1	256
u_0	30	44	47	41	29	20	15	16	4	6	2	1	1	0	256

Table 6. Check for SAC of BOX-2

	10000000	01000000	00100000	00010000	00001000	00000100	00000010	00000001
u_7	-8	8	-8	-24	0	-8	8	-24
u_6	16	24	-24	-24	0	-24	0	8
u_5	40	24	8	-8	-8	0	-8	-24
u_4	24	-32	16	0	8	0	56	8
u_3	24	16	-24	8	-8	8	24	-32
u_2	-8	8	-8	-8	-24	-8	24	0
u_1	0	8	8	32	16	-8	16	16
u_0	40	16	24	-16	0	16	-8	0

2. BOX-2 has worse spectrum in transform domain than BOX-0.
3. The Hadamard transform profiles of the eight component functions of BOX-2 are all distinct.
4. All coordinate functions of BOX-2 are *pairwise inequivalent* as boolean functions, which is one of the desirable characteristics of an S-box.
5. *None* of the component functions of BOX-2 has a simple algebraic expression over $\mathbb{F}_{2^n}$ with the multiplication performed modulo *any* irreducible polynomial, while *all* coordinates of BOX-0 do have the simplest equations such as (7) with the current Rijndael irreducible polynomial. Therefore, BOX-2 is better against the interpolation attack [6] than the original S-box, BOX-0.

We have experimentally checked all the remaining 27 S-boxes which are constructed from Rijndael S-box using the remaining 27 irreducible polynomials of degree 8, respectively. We have verified that all these share almost the same properties listed above with BOX-2.

5 Concluding Remarks

We proposed a simple scheme which produces a new S-box from the given S-box, which are based on operations over $\mathbb{F}_{2^n}$. The essential steps of the construction are (i) to determine the trace-represented polynomial functions of the given S-box over $\mathbb{F}_{2^n}$ with the multiplication performed modulo some other irreducible polynomial than the one originally used, (ii) to replace the coefficients in the trace-represented polynomial functions with the corresponding powers of the original primitive element, and finally, (iii) to evaluate new polynomials in $\mathbb{F}_{2^n}$ with the multiplication now performed modulo the original irreducible polynomial.

We have applied the scheme to Rijndael S-box, BOX-0, and constructed 29 different S-boxes, denoted by BOX-1, BOX-2, ... , BOX-29. All 29 S-boxes have much improved algebraic expressions over $\mathbb{F}_{2^n}$ with the multiplication performed modulo the original irreducible polynomial $g_0(z)$ (compare with (7)). Only BOX-1 has almost the same cryptographic properties as BOX-0. It is because only BOX-1 is equivalent to BOX-0 as boolean functions. Only BOX-0 and BOX-1 have the property that the algebraic expressions over $\mathbb{F}_{2^n}$ with the multiplication performed modulo some appropriate irreducible polynomial turned out to consist of a single trace function. No other S-boxes have such a simple algebraic expression.

Some theoretical developments that would be interesting are the following:

Q1 When and why the resulting S-box is a bijection or not a bijection?
Q2 When and why the resulting S-box has the same or different spectral properties as the original S-box?
Q3 Restricting to the case of Rijndael S-box, why is only BOX-1 similar to the original S-box? This is very surprising considering that $g_1(z)$ is an arbitrary choice among 29 irreducible polynomials of degree 8 over $\mathbb{F}_2$.
Q4 What are the distinctive properties of $g_1(z) = z^8 + z^4 + z^3 + z^2 + 1$ relative to $g_0(z) = z^8 + z^4 + z^3 + z^1 + 1$ compared with all other 28 irreducible polynomials of degree 8 over $\mathbb{F}_2$?

References

1. E. Barkan and E. Biham, "In how many ways can you write Rijndael?," In: Y. Zheng (Ed.), *ASIACRYPT 2002*, LNCS vol. 2501, Springer-Verlag, 2002, pp. 160–175.
2. E. Biham and A. Shamir, "Differential cryptanalysis of DES-like cryptosystems," *Journal of Cryptology*, vol. 4, pp. 3–72, 1991.
3. J. Fuller and W. Millan, "Linear redundancy in S-boxes," In: T. Johansson (Ed.), *Fast Software Encryption 2003*, LNCS vol. 2887, Springer-Verlag, 2003, pp. 74–86.
4. G. Gong and S.W. Golomb, "Transform domain analysis of DES," *IEEE Transactions on Information Theory*, vol. 45, no. 6, pp. 2065–2073, Sep., 1999.
5. S.W. Golomb and G. Gong, *Signal Design for Good Correlation: for wireless communication, cryptography, and radar.* Cambridge University Press, 2005.
6. T. Jakobsen and L.R. Knudsen, "The interpolation attack on block ciphers," In: E. Biham (Ed.), *Fast Software Encryption '97*, LNCS vol. 1267, Springer-Verlag, 1997, pp. 28–40.
7. L. Jing-mei, W. Bao-dian, C. Xiang-guo, and W. Xin-mei, "Cryptanalysis of Rijndael S-box and improvement," *Applied Mathematics and Computation*, vol. 170, pp. 958–975, 2005.
8. R. Lidl and H. Niederreiter, *Introduction to Finite Fields and Their Applications.* Cambridge University Press, 1986.

9. M. Matsui, "Linear cryptanalysis method for DES cipher," In: T. Helleseth (Ed.), *Advances in Cryptology: Eurocrypt '93*, LNCS vol. 765, Springer-Verlag, 1993, pp. 386–397.
10. C.E. Shannon, "Communication theory of secrecy systems," *Bell Systems Technical Journal*, vol. 28, pp. 656–715, 1949.
11. T. Siegenthaler, "Correlation-immunity of nonlinear combining functions for cryptographic applications," *IEEE Transactions on Information Theory*, vol. 30, no. 5, pp. 776–780, Sep., 1984.
12. A.F. Webster and S.E. Tavares, "On the design of S-box," In: H.C. Williams (Ed.), *Advances in Cryptology: Crypto '85*, LNCS vol. 218, Springer-Verlag, 1986, pp. 523–534.
13. A.M. Youssef and S.E. Tavares, "Affine equivalence in the AES round function," *Discrete Applied Mathematics*, vol. 148, pp. 161–170, 2005.
14. J. Daemen and V. Rijmen, *AES proposal: Rijndael*
15. J. Daemen and V. Rijmen, *The Design of Rijndael: AES—The Advanced Encryption Standard*, Springer-Verlag, 2002.
16. FIPS-197: Advanced Encryption Standard (AES), Nov. 2001, `http://csrc.nist.gov/publications/fips`
17. Block Cipher ARIA, `http://www.nsri.re.kr/ARIA/doc/ARIA-specification.pdf`
18. CRYPTEC, `http://www.ipa.go.jp/` (in Japanese).
19. NESSIE (The New European Schemes for Signatures, Integrity and Encryption), `http://www.cryptonessie.org`

Nonlinear Complexity of Binary Sequences and Connections with Lempel-Ziv Compression

Konstantinos Limniotis, Nicholas Kolokotronis, and Nicholas Kalouptsidis

Department of Informatics and Telecommunications
National and Kapodistrian University of Athens
TYPA Buildings, University Campus, 15784 Athens, Greece
{klimn, nkolok, kalou}@di.uoa.gr

Abstract. The nonlinear complexity of binary sequences is studied in this paper. A new recursive algorithm is presented, which produces the minimal nonlinear feedback shift register of a given sequence. Further, a connection between the nonlinear complexity and the compression capability of a sequence is established. A lower bound for the Lempel-Ziv compression ratio that a given sequence can achieve is proved, which depends on its nonlinear complexity.

Keywords: Cryptography, Lempel-Ziv compression, nonlinear complexity, nonlinear feedback shift registers, sequences.

1 Introduction

Binary sequences have a significant role in many applications, amongst others error control coding, spread spectrum communications and cryptography [3,5,14]. In particular, the security of cryptographic systems is strongly contingent on the unpredictability or pseudorandomness of the key streams [14]. Depending on the cryptographic system, a sequence is required to admit many properties in order to be considered as pseudorandom. The *nonlinear complexity* $c(y)$ of a sequence y, also called *maximum order complexity* or simply *complexity*, is an important cryptographic measure; it is defined as the length of the shortest feedback shift register (FSR) that generates y. For linear feedback shift registers (LFSRs), the corresponding complexity measure is referred to as *linear complexity* or *linear span* of y. The computation of the minimal LFSR that generates y is efficiently solved by the Berlekamp-Massey algorithm (BMA) [1,11]. Linear complexity has been widely studied in the literature using many different approaches [6,8,12,13,18].

On the contrary, the general case of nonlinear complexity has not been studied that extensively. In [4] a directed acyclic graph is used to exhibit the complexity profile of any sequence with values in arbitrary field. In [2] an approximate propability distribution for the nonlinear complexity of a random binary sequence is derived. Recent results are provided in [16], where the minimal nonlinear FSR that generates a given sequence is computed via an algorithmic approach. In [17] the special case of a quadratic feedback function of the FSR is treated.

G. Gong et al. (Eds.): SETA 2006, LNCS 4086, pp. 168–179, 2006.

The degree to which a given sequence can be compressed constitutes another important cryptographic measure. Clearly, a sequence can not be considered as pseudorandom if it can be significantly compressed. In this direction, a complexity measure related to the number of cumulatively distinct patterns in the sequence is proposed in [9]. Generalization of this procedure resulted in the prominent Lempel-Ziv compression algorithm, namely versions LZ77 and LZ78 proposed in [21] and [22] respectively. They are both asymptotically optimal, since the compression ratio approaches the source entropy for all finite-alphabet stationary ergodic sources [20,22]. However, the compression ratio for a finite sequence can be far from optimal.

The relationship between several of the currently established cryptographic criteria still remains an open problem. In this paper we focus on the connection between the nonlinear and Lempel-Ziv complexity, motivated by a statement of Niederreiter indicating this connection as an interesting open problem [15]. For any periodic sequence, we establish the dependence of the minimum achievable compression ratio on its nonlinear complexity. Furthermore, a new recursive algorithm producing the minimal FSR of any binary sequence is developed, thus generalizing the Berlekamp-Massey algorithm to the nonlinear case. This algorithm differs from the one proposed in [16] since it recursively computes the minimal FSR for any subsequence by utilizing special Boolean algebra arguments.

The paper is organized as follows: in Section 2 the basic terminology and definitions are introduced. Properties on the nonlinear complexity of sequences over any field are presented in Section 3. Based on these properties, a recursive algorithm that computes the minimal FSR of any binary sequence is derived in Section 4. The connection between the nonlinear complexity and Lempel-Ziv compression ratio is established in Section 5. Finally, concluding remarks are given in Section 6.

2 Preliminaries

Let $\mathbb{F}_2$ denote the binary field. A boolean function f with n variables is a mapping $f : \mathbb{F}_2^n \to \mathbb{F}_2$. The complement of a binary variable x is denoted by $x' = x \oplus 1$, where $\oplus$ represents the addition modulo 2. Let $x_1, x_2, \ldots, x_n$ be binary variables. Then, a product that contains each of variables $x_i, i = 1, 2, \ldots, n$, in either complemented or uncomplemented form is referred to as *minterm* [7]. Clearly, there are 2^n minterms. The minterm corresponding to the n-tuple $c = (c_1, c_2, \ldots, c_n) \in \mathbb{F}_2^n$ is uniquely determined by the property that it evaluates to 1 if the i-th variable of the minterm is replaced by c_i. For example, the minterm of the 5-tuple 00101 is $x_1'x_2'x_3x_4'x_5$, since $0' \cdot 0' \cdot 1 \cdot 0' \cdot 1 = 1$ and, clearly, no other minterm satisfies this property.

There are several ways to represent a boolean function. The *algebraic normal form (ANF)* of f is identified via the relation:

$$f(x_1, x_2, \ldots x_n) = \sum_{j \in \mathbb{F}_2^n} a_j x_1^{j_1} x_2^{j_2} \cdots x_n^{j_n}, \qquad a_j \in \mathbb{F}_2 \tag{1}$$

where $j = (j_1, j_2, \ldots, j_n)$ and the summation is taken modulo 2. A more general representation of a boolean function, the so-called *exclusive-or sum-of-products (ESOP)*, occurs if the variables in (1) are in either complemented or uncomplemented form [19].

Let $y = y_0\, y_1\, y_2 \ldots$ be a sequence over $\mathbb{F}_2$ and denote by y_i^j, with $i \le j$, the tuple $(y_i, y_{i+1}, \ldots, y_j)$. If y has finite length N, then clearly $y^N \triangleq y_0^{N-1}$ denotes the whole sequence. If there exist $t_0 \ge 0$ and $T > 0$ such that $y_i = y_{i+T}$ for all $i \ge t_0$ then sequence is called *ultimately periodic*. More precisely, if $t_0 = 0$ the sequence is simply *periodic*. The least integers t_0, T with this property are called *preperiod* and *period* respectively. Any ultimately periodic sequence can be generated by a feedback shift register, satisfying a recurring relation of the form

$$y_{i+n} = h(y_{i+n-1}, \ldots, y_i), \qquad i \ge 0$$

where the n-tuple $(y_{i+n-1}, \ldots, y_i)$ is the *state* of the FSR at time i, and $n > 0$ determines the number of stages of the FSR [3,5,10]. The function h is called the *feedback function* of the FSR. If $n = c(y^N)$, then the FSR and h are called the *minimal* FSR and *minimal nonlinear polynomial* of sequence y respectively. Clearly, the minimal FSR of any sequence is not necessarily unique. In the sequel, for any feedback function h of a FSR we assume that the constant term of its ANF is zero. For any $m > 0$, the $(N - m + 1) \times m$ state matrix $\boldsymbol{S}^m$ of y^N equals

$$\boldsymbol{S}^m(y^N) = \begin{pmatrix} y_{m-1} & y_{m-2} & \cdots & y_1 & y_0 \\ y_m & y_{m-1} & \cdots & y_2 & y_1 \\ \vdots & \vdots & \vdots & \vdots & \vdots \\ y_{N-1} & y_{N-2} & \cdots & y_{N-m+1} & y_{N-m} \end{pmatrix} . \tag{2}$$

Clearly, if an m-stage FSR produces y^N, the rows of $\boldsymbol{S}^m(y^N)$ coincide with the states of the FSR. Furthermore, $\boldsymbol{S}^m(y^N)$ is a Toeplitz matrix. The j-th row of $\boldsymbol{S}^m$ is denoted by $S_j^m(y^N)$, $j = 1, 2, \ldots, N - m + 1$.

3 Properties of Nonlinear Complexity

In this section we present some of the properties characterizing the nonlinear complexity of finite-length sequences.

Proposition 1 ([4]). *Let L be the length of the longest tuple in y^N that occurs at least twice with different successors. Then, $c(y^N) = L + 1$.*

In fact, Proposition 1 is valid in the case of a nonzero constant term in the feedback function of the FSR is nonzero. If we are confined to a zero constant term, then

$$c(y^N) = \max\{L + 1, M + 1\}$$

where L is the integer referred to in Proposition 1, and M is the length of the longest run of zeros in y^N followed by 1. Clearly, $M \le L + 1$.

Proposition 2. *Consider a minimal FSR of y^{N-1} with length m, which does not produce y^N. Then $c(y^N) = m$ if and only if*

$$y_{N-m-1}^{N-2} = (y_{N-m-1}, y_{N-m}, \ldots, y_{N-2})$$

appears only once within y^{N-1}.

Proof. Let $c(y^N) = m$ and assume that there exists $0 \leq i < N - m - 1$ such that $y_i^{i+m-1} = y_{N-m-1}^{N-2}$. Clearly, since the FSR does not produce the N-th element of y^N, we have $y_{i+m} \neq y_{N-1}$, and Proposition 1 gives $c(y^N) \geq m + 1$ – a contradiction.

Conversely, let y_{N-m-1}^{N-2} appear only once. Note that since $c(y^{N-1}) = m$, it clearly holds $c(y^N) \geq m$. Let us suppose that $c(y^N) = m + k$, $k \geq 1$. Then, from Proposition 1 there exist at least two identical tuples within y^N, each of length $m + k - 1$, with different successors. That is there exist $0 \leq i_1 < i_2 \leq N-m-k$ such that $y_{i_1}^{i_1+m+k-2} = y_{i_2}^{i_2+m+k-2}$ and $y_{i_1+m+k-1} \neq y_{i_2+m+k-1}$. Since $c(y^{N-1}) = m$, such pair of $(m + k - 1)$-tuples is not present in y^{N-1}, leading to $i_2 = N - m - k$. Hence, the tuple determined by the last $m + k - 1$ elements of y^{N-1} is present twice within y^{N-1}, contradicting our hypothesis. □

Proposition 2 and the structure of $\boldsymbol{S}^m(y^N)$ lead to the following result.

Corollary 3. *Let y^N be a finite-length sequence with $c(y^{N-1}) = m$. Then $c(y^N) > m$ if and only if there exists $i < N-m$ such that it holds $S_i^m(y^N) = S_{N-m}^m(y^N)$ and $S_{i+1}^m(y^N) \neq S_{N-m+1}^m(y^N)$.*

The above result indicates that the state matrix $\boldsymbol{S}^m$ can be utilized for determining the increment $j = c(y^N) - c(y^{N-1})$ or *jump* in the complexity. The exact value of j is given by the following theorem.

Theorem 4. *Let y^N be a finite-length sequence with $c(y^{N-1}) = m$ and $c(y^N) > m$. Further, let $K = \{k_1, k_2, \ldots, k_\ell\}$ be the set of integers with*

$$S_{k_i}^m(y^N) = S_{N-m}^m(y^N), \qquad i - 1, 2, \ldots, \ell$$

where $1 \leq k_1 < k_2 < \ldots < k_\ell < N - m$. Then $c(y^N) = m + j$, where j is the largest integer such that for all $0 \leq i < j$ we have

$$S_{N-m-i}^m(y^{N-1}) = S_{k_\ell-i}^m(y^{N-1}) \ . \tag{3}$$

Proof. Corollary 3 implies that K is a non-empty set and $S_{N-m+1}^m(y^N) \neq S_{k_\ell+1}^m(y^N)$. It is easy to verify that for all $i < j$, we have

$$S_{N-m-i}^{m+i}(y^N) = S_{k_\ell-i}^{m+i}(y^N) \quad \text{and} \quad S_{N-m-i+1}^{m+i}(y^N) \neq S_{k_\ell-i+1}^{m+i}(y^N) \ .$$

Hence, Corollary 3 gives $c(y^N) \geq m + j$. We shall prove that all the rows of $\boldsymbol{S}^{m+j}(y^N)$ are pairwise distinct, thus obtaining $c(y^N) = m+j$ from Proposition 1. The claim is straightforward for $\boldsymbol{S}^{m+j}(y^{N-1})$ from the definition of j and the fact that $c(y^{N-1}) = m$. Furthermore, the last $m + 1$ elements of y^N are present only once within y^N since $c(y^N) > m$. Therefore, none of the rows of $\boldsymbol{S}^{m+j}(y^{N-1})$ coincides with the last row $S_{N-m-j+1}^{m+j}(y^N)$ containing these $m+1$ elements, thus concluding our proof. □

If $c(y^{i-1}) = m$ and $c(y^i) > m$ for some integer i, then Corollary 3 implies that there exist $t_{0_S} \geq 0$ and $T_S > 0$ such that $S_\ell^m(y^{i-1}) = S_{\ell+T_S}^m(y^{i-1})$ for all $t_{0_S} < \ell \leq i - m - T_S$. Thus there exist $1 \leq \lambda \leq T_S$ and $k \geq 1$ satisfying $t_{0_S} + kT_S + \lambda = i - m$. Clearly, the jump j in the complexity according to Theorem 4 is given by $j = (k-1)T_S + \lambda$, or equivalently

$$j = i - m - T_S - t_{0_S} \ . \tag{4}$$

Proposition 5. *Let y^N be a finite-length sequence with $c(y^{N-1}) = m$ and $c(y^N) = m + j$ for some $j \geq 1$. If y^N is expanded arbitrarily by j elements, then the complexity of the expanded sequence y^{N+j} remains $m + j$.*

Proof. From the proof of Theorem 4, the matrix $\boldsymbol{S}^{m+j}(y^N)$ has pairwise distinct rows and the last $(m+1)$-tuple of y^N, that is y_{N-1-m}^{N-1}, is present only once within y^N. Let $\hat{y}_{N-1-m}^{N-1} = (y_{N-1}, y_{N-2}, \ldots, y_{N-1-m})$ be the reversed tuple of y_{N-1-m}^{N-1}. Then, among the rows of $\boldsymbol{S}^{m+j}(y^N)$, only the last row contains $\hat{y}_{N-1-m}^{N-1}$. Clearly, $\hat{y}_{N-1-m}^{N-1}$ also lies in the last j rows of $\boldsymbol{S}^{m+j}(y^{N+j})$ due to its Toeplitz structure. Hence, the last j rows of $\boldsymbol{S}^{m+j}(y^{N+j})$ are pairwise distinct and, moreover, they are distinct from any of the previous rows. Subsequently, all the rows of $\boldsymbol{S}^{m+j}(y^{N+j})$ are pairwise distinct, leading to $c(y^{N+j}) = m + j$. □

The above analysis is independent from the underlying field of the sequence. In the sequel we restrict our attention on binary sequences.

4 An Algorithm for Nonlinear Shift Register Synthesis

Let us consider a binary sequence y^N such that $c(y^{i-1}) = m$ for some $i < N$ and let h be the minimal nonlinear polynomial of y^{i-1}. Furthermore, let us suppose that the next bit y_{i-1} satisfies

$$h(y_{i-2}, y_{i-3}, \ldots, y_{i-m-1}) = y_{i-1} \tag{5}$$

Then, the same FSR produces y^i and clearly $c(y^i) = c(y^{i-1}) = m$. On the other hand, if (5) does not hold, we say that a *discrepancy* occurs. In this case, the minimal FSR of y^i is not known. Due to Proposition 2, if y_{i-m-1}^{i-2} appears only once within y^{i-1} the complexity does not increase; otherwise $c(y^i) > m$. Next, these two cases where a discrepancy occurs are treated separately, in order to obtain a minimal FSR of y^i for each case.

Case 1: $c(y^i) = c(y^{i-1})$. Consider the function $h' = h + f^\delta$ where f^δ is a function on m variables that equals 1 when evaluated at the last row of $\boldsymbol{S}^m(y^{i-1})$ and 0 otherwise. That is f^δ is the minterm corresponding to the last row of $\boldsymbol{S}^m(y^{i-1})$. Clearly, h' suffices to generate y^i and thus is a minimal nonlinear polynomial of y^i.

Algorithm 1 Computation of the minimal FSR of a sequence y^N

1: $r := 1$ {Start of initialization procedure}

2: **while** $y_r = y_0$ **do**

3:    $r := r + 1$

4: **end while**

5: **if** $y_r = 0$ **then**

6:    $m := r$

7: **else**

8:    $m := r + 1$

9: **end if**

10: $h := 0$ {End of initialization procedure}

11: **for** $i := r + 1$ to $N - 1$ **do**

12:    $\ell = (y_{i-1} \, y_{i-2} \, \cdots \, y_{i-m})$

13:    **if** $\ell = (0 \, 0 \, \ldots \, 0)$ **then**

14:       $m := m + 1$

15:       $\ell = (y_{i-1} \, y_{i-2} \, \cdots \, y_{i-m})$

16:    **end if**

17:    **if** $h(\ell) \neq y_i$ **then**

18:       **if** ℓ is appeared once in y^i **then**

19:          Set $f^\delta(x)$ equal to the minterm that corresponds to ℓ

20:          $h := h + f^\delta$

21:       **else**

22:          Compute the jump j of complexity

23:          $m := m + j$

24:          $\ell = (y_{i-1} \, y_{i-2} \, \cdots \, y_{i-m})$

25:          Set f^δ equal to the minterm that corresponds to ℓ

26:          $h := h + f^\delta$

27:       **end if**

28:    **end if**

29: **end for**

30: $c(y^N) := m$

31: The feedback polynomial of a minimal FSR of y^N is h

Fig. 1. Algorithm for nonlinear shift register synthesis

Case 2: $c(y^i) > c(y^{i-1})$. Recall that the exact jump j in the complexity is given by (4). Furthermore, note that the FSR of length $m + j$ with feedback function h also generates y^{i-1} if its initial load is the first $m + j$ elements of y^N. By using the same arguments as above, it is easy to verify that the FSR with feedback function $h' = h + f^\delta$ generates y^i, where f^δ is the minterm corresponding to the last row of $\boldsymbol{S}^{m+j}(y^{i-1})$. Hence, h' is a minimal nonlinear polynomial of y^i.

Combining the above cases, we conclude that if a feedback function h generates y^{i-1} but not y^i, there always exists a function f^δ such that $h' = h + f^\delta$ produces y^i. Any f^δ with this property is determined via its unique association with a minterm. Hence, the previous analysis settles the basis to construct a recursive algorithm for computing the minimal nonlinear polynomial of any binary sequence y^N. This algorithm also determines a minimal FSR of y^i, for any

current state	i	y_i	d	t_{0_S}, T_S	f^δ	next state
100	3	1	1	0,0	$x_1x_2'x_3'$	110
110	4	0	0	0,0	0	011
011	5	1	1	0,0	$x_1'x_2x_3$	101
101	6	1	1	0,0	$x_1x_2'x_3$	110
110	7	1	1	1,3	$x_1x_2x_3'x_4$	1110
1110	8	0	0	0,0	0	0111
0111	9	0	1	0,0	$x_1'x_2x_3x_4$	0011
0011	10	1	1	0,0	$x_1'x_2'x_3x_4$	1001
1001	11	0	1	0,0	$x_1x_2'x_3'x_4$	0100
0100	12	1	1	0,0	$x_1'x_2x_3'x_4'$	1010
1010	13	1	0	0,0	0	1101
1101	14	1	0	3,7	0	1110
1110	15	0	0	3,7	0	0111
0111	16	1	1	3,7	$x_1'x_2x_3x_4x_5'x_6x_7'$	1011101
1011101	17	1	0	0,0	0	1101110

Fig. 2. An example of the computation of the minimal FSR of y^{18}

$i \leq N$, and is depicted in Figure 1. Line 22 of the algorithm computes the jump j in the complexity according to (4). Note that t_{0_S} and T_S are estimated in line 18 while seeking for repetitions of tuples within the sequence; clearly, this step has linear computational complexity. Furthermore, according to Proposition 5, if $c(y^i) - c(y^{i-1}) = j$ then the next j bits of the sequence do not cause jump in the complexity and, thus no need arises to search for repetitions at the next j iterations. Finally, note that the minimal nonlinear polynomial h computed by the algorithm is given in ESOP representation. An illustrative example of the algorithm is given next.

Example 6. Consider the sequence $y^{18} = 001101110010111011$. The initialization procedure sets $m = 3$ and $r = 2$. Details are given in Figure 2, where d equals $y_i + h(y_{i-1}, y_{i-2}, \ldots, y_{i-m})$ and m is the complexity of y^i. Thus, if $d = 1$ a discrepancy occurs. Note also that for $i = 8$ and $i = 17$ there is no need to check the values of t_{0_S}, T_S due to Proposition 5. The complexity of y^{18} equals 7. The minimal nonlinear polynomial of $y^k, k = 4, 5, \ldots 18$ is given by the function consisting of the sum of all the functions f^δ appearing in Figure 2, for all $i < k$.

The computational complexity of the algorithm mainly rests with line 17 where the ESOP boolean function is evaluated. For each $n \leq N$ the ESOP of the minimal nonlinear polynomial of y^n has less than n terms, each consisting of at most n variables. Hence, the computational complexity of this step is at most $O(n^2)$. Therefore, in the worst case, the total computational complexity for a given sequence of length N is $O(N^3)$ due to the recursive structure of the algorithm.

5 Connections with Lempel-Ziv Compression Ratio

The LZ78 is a dictionary-based compression algorithm. A sequence y^N with values lying in a finite alphabet is partitioned into pairwise distinct words $s_1, s_2, \ldots$ (with a possible exception for the last word) such that each word s_i has the property that its prefix $s_i^{\ell(s_i)-1}$ is a previous word s_j for some $j < i$, where $\ell(s_i)$ is the length of s_i and s_i^k, $k \leq \ell(s_i)$, denotes the first k elements of s_i [22].

The compression of y^N is achieved by transforming each word s_i into a new codeword $\tilde{s}_i$ consisting of two parts: the first part is the binary representation of j where s_j is the uniquely defined word by $s_j = s_i^{\ell(s_i)-1}$ and the second part is simply the last symbol of s_i. Hence, if $c_{LZ}(y^N)$ is the number of distinct words produced via the incremental parsing procedure of LZ78, then the compression ratio of y^N equals

$$\rho_{y^N} = \frac{1}{N} \sum_{i=1}^{c_{LZ}(y^N)} \lceil \log_2(2i) \rceil \tag{6}$$

since the length of the codeword corresponding to s_i equals $\lceil \log_2(i) \rceil + 1$ [22]. The decoding procedure is straightforward. From (6) it is clear that as ρ_{y^N} approaches zero, better compression is achieved. Thus, sequences with large $c(y^N)$ and ρ_{y^N} are desirable for cryptographic purposes.

If a sequence y_i is partitioned into k words, then the average length of the words is denoted by $\overline{x}_i$ and the average length of the codewords is denoted by $\overline{w}_i$.

Lemma 7. *Consider two binary sequences y_1, y_2 parsed via the Lempel-Ziv incremental parsing into k_1, k_2 words respectively, with $k_1 > k_2$. Then, $\rho_{y_1} < \rho_{y_2}$ if and only if $\overline{x}_1 > \frac{\overline{w}_1}{\overline{w}_2} \overline{x}_2$.*

Proof. It is straightforward since $\rho_{y_i} = \frac{\overline{w}_i}{\overline{x}_i}$, $i = 1, 2$. □

Lemma 8. *With the above notation, if $k_1 > k_2$ then $\overline{w}_1 > \overline{w}_2$. Furthermore, if $\rho_1 < \rho_2$ then $\overline{x}_1 > \overline{x}_2$.*

Proof. Assume that $k_1 = k_2 + 1$. Then $\overline{w}_1 = \frac{1}{k_2+1} \sum_{i=1}^{k_2+1} \lceil \log_2(2i) \rceil$ and $\overline{w}_2 = \frac{1}{k_2} \sum_{i=1}^{k_2} \lceil \log_2(2i) \rceil$. Let us suppose that $\overline{w}_1 < \overline{w}_2$. Then we have

$$k_2 \sum_{i=1}^{k_2+1} \lceil \log_2(2i) \rceil < (k_2 + 1) \sum_{i=1}^{k_2} \lceil \log_2(2i) \rceil \Rightarrow$$

$$k_2 \lceil \log_2(2(k_2 + 1)) \rceil < \sum_{i=1}^{k_2} \lceil \log_2(2i) \rceil$$

leading to contradiction. Subsequently, $\overline{w}_1 > \overline{w}_2$ whenever $k_1 > k_2$ holds. Furthermore, Lemma 7 implies that, if $\rho_1 < \rho_2$ then $\overline{x}_1 > \overline{x}_2$. □

It is clear from the description of LZ78 that any sequence of the form

$$y^N = (y_{j_1} \underbrace{y_{j_1}\, y_{j_2}}_{2} \underbrace{y_{j_1}\, y_{j_2}\, y_{j_3}}_{3} \cdots \underbrace{y_{j_1}\, y_{j_2}\, y_{j_3} \cdots y_{j_s}}_{s}) \tag{7}$$

where $N = \frac{1}{2}s(s+1)$, is partitioned via the LZ78 into the minimum possible number of words amongst all the sequences of the same length.

Definition 9. *Any binary sequence y lying in the following set*

$$LZ^s_{opt} = \{y^N : N = \tfrac{1}{2}s(s+1),\ c_{LZ}(y^N) = s,\ s = 1, 2, 3, \ldots\}$$

is called s-optimal sequence and is denoted by y^s_{opt}.

Lemma 10. *Consider a binary sequence y^N and let s be the smallest integer such that $N \leq \frac{1}{2}s(s+1)$. Then, $\rho_{y^N} \geq \rho_{y^s_{opt}}$.*

Proof. Let k equal the number of words resulting from the parsing of y^N by LZ78. The corresponding compression ratios are given by

$$\rho_{y^N} = \frac{1}{N}\sum_{i=1}^{k}\lceil \log_2(2i)\rceil \quad \text{and} \quad \rho_{y^s_{opt}} = \frac{2}{s(s+1)}\sum_{i=1}^{s}\lceil \log_2(2i)\rceil \ .$$

It clearly holds $k \geq s$ and, since $N \leq \frac{1}{2}s(s+1)$, the claim follows. □

Lemma 11. *Let y_1, y_2 be s-optimal and $(s+1)$-optimal sequences respectively, where $s \geq 3$. Then $\rho_{y_1} > \rho_{y_2}$.*

Proof. The proof is provided in the appendix. □

Next the basic Theorem is presented, which illustrates the impact that complexity has on the Lempel-Ziv compression ratio.

Theorem 12. *Consider a periodic binary sequence y with period N and $c(y) = m \geq 3$. If ρ_y is the compression ratio that y^N achieves via the Lempel-Ziv compression algorithm, then it holds*

$$\rho_y > \frac{1}{m(2m-1)}\sum_{i=1}^{2m-1}\lceil \log_2(2i)\rceil \tag{8}$$

Proof. The lower bound given by (8) is the compression ratio of p-optimal sequences for $p = 2m-1$. Clearly, according to Lemmas 10 and 11, any sequence with period less than $\frac{p(p+1)}{2}$ can not achieve lower compression ratio. Let us suppose that there exists a periodic sequence y' with period $N > \frac{p(p+1)}{2}$ and complexity m, such that y'^N achieves better compression ratio than the one given in (8). Clearly, the Lempel-Ziv partitioning of y'^N results in $k > p$ words. Thus, according to Lemma 8, the average length of the parsed words is greater than $\frac{p+1}{2}$. But since $p = 2m-1$, it holds $\frac{p+1}{2} = \frac{2m}{2} = m$. Hence, the average length of the parsed words is greater than m. Subsequently, there exist at least two identical m-tuples in y'^N, contradicting the fact that y' has complexity m and period N. Since the same holds for any p-optimal sequence, the claim follows. □

According to Lemma 11, the lower bound given by (8) decreases as the complexity of a sequence increases, i.e. for any sequence with complexity m and compression ratio close to the above bound, it is always possible to find a sequence with complexity $m' > m$ achieving a lower compression ratio. Recalling that random sequences may attain a high compression ratio, this cryptographic measure should be used to filter out sequences with high complexity and compression ratio below some threshold.

6 Conclusions

This paper studies the nonlinear complexity of binary sequences and its connection with the Lempel-Ziv compression ratio. It provides a recursive algorithm, based on Boolean algebra arguments, which computes the minimal nonlinear shift register that generates a sequence. Furthermore, a lower bound is established for the compression ratio of sequences with given nonlinear complexity, illustrating a novel connection between these significant cryptographic criteria. Experimental results have shown that the above bound could be further improved; ongoing research is performed towards this direction.

References

1. Berlekamp, E.R.: Algebraic coding theory. New York: McGraw-Hill (1968).
2. Erdmann, D., Murphy, S.: An approximate distribution for the maximum order complexity. Des. Codes and Cryptography **10** (1997) 325–339.
3. Golomb, S.W.: Shift Register Sequences. Holden-Day, San Francisco (1967).
4. Jansen, C.J., Boekee, D.E.: The shortest feedback shift register that can generate a given sequence. Proc. Advances in Cryptology-CRYPTO '89 (1990) 90–99.
5. Kalouptsidis, N.: Signal Processing Systems. Telecommunications and Signal Processing Series, John Wiley & Sons (1996)
6. Key, E.L.: An analysis of the structure and complexity of nonlinear binary sequence generators. IEEE Trans. Inform. Theory **22** (1976) 732–736.
7. Kohavi, Z.: Switching and finite automata theory. McGraw-Hill Book Company (1978)
8. Kolokotronis, N., Kalouptsidis, N.: On the linear complexity of nonlinearly filtered PN-sequences. IEEE Trans. Inform. Theory **49** (2003) 3047–3059.
9. Lempel, A., Ziv, J.: On the complexity of finite sequences. IEEE Trans. Inform. Theory **22** (1976) 75–81.
10. Lidl, R., Niederreiter, H.: Finite Fields. vol. 20 of Encyclopedia of Mathematics and its Applications. Cambridge University Press 2nd ed. (1996)
11. Massey, J.L.: Shift register synthesis and BCH decoding. IEEE Trans. Inform. Theory **15** (1969) 122–127.
12. Massey, J.L., Serconek, S.: A Fourier transform approach to the linear complexity of nonlinearly filtered sequences. Advances in Cryptology - CRYPTO '94, Lecture Notes in Computer Science, **839** 332–340.
13. Massey, J.L., Serconek, S.: Linear complexity of periodic sequences: a general theory. in Proc. Advances in Cryptology - CRYPTO '96, Lecture Notes in Computer Science 358–371.

14. Menezes, A.J., van Oorschot, P.C., Vanstone, S.A.: Handbook of Applied Cryptography. CRC Press (1996).
15. Niederreiter, H.: Some computable complexity measures for binary sequences. C. Ding, T. Helleseth, and H. Niederreiter, eds., in: Sequences and Their Applications, Discrete Mathematics and Theoretical Computer Science, Springer-Verlag (1999) 67–78.
16. Rizomiliotis, P., Kalouptsidis, N.: Results on the nonlinear span of binary sequences. IEEE Trans. Inform. Theory **51** (2005) 1555–1563.
17. Rizomiliotis, P., Kolokotronis, N., Kalouptsidis, N.: On the quadratic span of binary sequences. IEEE Trans. Inform. Theory **51** (2005) 1840–1848.
18. Rueppel, R.A.: Analysis and design of stream ciphers. Berlin, Germany: Springer-Verlang (1986).
19. Stergiou, S., Voudouris, D., Papakonstantinou, G.: Multiple-value exclusive-or sum-of-products minimization algorithms. IEICE Transactions on Fundamentals. E.87-A **5** (2004) 1226–1234.
20. Wyner, A.D., Ziv, J.: The sliding-window Lempel-Ziv algorithm is asymptotically optimal. Proceedings of the IEEE **82** (1994) 872–877.
21. Ziv, J., Lempel, A.: A universal algorithm for sequential data compression. IEEE Trans. Inform. Theory **23** (1977) 337–343.
22. Ziv, J., Lempel, A.: Compression of individual sequences via variable-rate coding. IEEE Trans. Inform. Theory **24** (1978) 530–536.

Appendix: Proof of Lemma 11

Note that $\overline{x}_1 = \frac{s+1}{2}$ and $\overline{x}_2 = \frac{s+2}{2}$. Then, according to Lemma 7, it holds $\rho_{y_2} < \rho_{y_1}$ if and only if

$$\frac{s+2}{2} > \frac{\frac{\sum_{i=1}^{s+1} \lceil \log_2(2i) \rceil}{s+1}}{\frac{\sum_{i=1}^{s} \lceil \log_2(2i) \rceil}{s}} \frac{s+1}{2}$$

which leads to the following relation

$$\frac{s+2}{2} > \frac{s}{2}\left(1 + \frac{\lceil \log_2(2(s+1)) \rceil}{\sum_{i=1}^{s} \lceil \log_2(2i) \rceil}\right) . \quad (9)$$

From (9) it is readily derived that

$$\frac{2}{s} > \frac{\lceil \log_2(2(s+1)) \rceil}{\sum_{i=1}^{s} \lceil \log_2(2i) \rceil} \Leftrightarrow s \lceil \log_2(2(s+1)) \rceil < 2 \sum_{i=1}^{s} \lceil \log_2(2i) \rceil . \quad (10)$$

Let us first consider the case where $s = 2^n$ for some $n \geq 2$. Then, (10) becomes

$$2 \sum_{i=1}^{2^n} \lceil \log_2(2i) \rceil > 2^n \lceil \log_2(2(2^n + 1)) \rceil . \quad (11)$$

Next we prove the validity of (11). Indeed, (11) is equivalently written as follows

$$2\sum_{i=1}^{2^n}\lceil \log_2(2i)\rceil > 2^n(1+\lceil \log_2(2^n+1)\rceil)$$

$$\Leftrightarrow 2\sum_{i=1}^{2^n}\lceil \log_2(2i)\rceil > 2^n(1+n+1)$$

$$\Leftrightarrow \sum_{i=1}^{2^n}\lceil \log_2(2i)\rceil > 2^n + n2^{n-1} \ . \tag{12}$$

Furthermore it holds

$$\sum_{i=1}^{2^n}\lceil \log_2(2i)\rceil = \sum_{i=1}^{2^n}\lceil 1+\log_2(i)\rceil = 2^n + \sum_{i=1}^{2^n}\lceil \log_2(i)\rceil \ . \tag{13}$$

Relation (13) implies that (12) is equivalent to

$$\sum_{i=1}^{2^n}\lceil \log_2(i)\rceil > n2^{n-1} \ . \tag{14}$$

It is easily proved by induction that $\sum_{i=1}^{2^n}\lceil \log_2(i)\rceil = \sum_{i=0}^{n} i2^{i-1}$. This clearly proves the validity of (14).

Let us now consider the case that s is not a power of 2. We shall proceed by induction. The claim holds for $s = 3$. Let us suppose that it holds for $s = k$, namely

$$k\lceil \log_2(2(k+1))\rceil < 2\sum_{i=1}^{k}\lceil \log_2(2i)\rceil \ . \tag{15}$$

It is next proved that the claim holds for $s = k+1$, where $k+1$ is not a power of 2. In this case, it suffices to show that

$$(k+1)\lceil \log_2(2(k+2))\rceil < 2\sum_{i=1}^{k+1}\lceil \log_2(2i)\rceil \ .$$

Since $2\sum_{i=1}^{k+1}\lceil \log_2(2i)\rceil = 2\sum_{i=1}^{k}\lceil \log_2(2i)\rceil + 2\lceil \log_2(2(k+1))\rceil$, (15) indicates that is suffices to show that

$$k\lceil \log_2(2(k+1))\rceil + 2\lceil \log_2(2(k+1))\rceil > (k+1)\lceil \log_2(2(k+2))\rceil$$

which is equivalently written as

$$\frac{k+2}{k+1} > \frac{\lceil \log_2(2(k+2))\rceil}{\lceil \log_2(2(k+1))\rceil} \ .$$

But since $k+1$ is not a power of 2, it holds

$$\lceil \log_2(2(k+2))\rceil = \lceil \log_2(2(k+1))\rceil$$

and, thus, the claim follows.

On Lempel-Ziv Complexity of Sequences

Ali Doğanaksoy[1,2,4] and Faruk Göloğlu[2,3]

[1] Department of Mathematics, Middle East Technical University
Ankara, Turkey
aldoks@metu.edu.tr
[2] Institute of Applied Mathematics, Middle East Technical University
Ankara, Turkey
[3] Dept. of Computer Technology and Information Systems
Bilkent University, Ankara, Turkey
gologlu@bilkent.edu.tr
[4] TUBITAK-UEKAE, Gebze, Turkey

Abstract. We derive recurrences for counting the number $a(n,r)$ of sequences of length n with Lempel-Ziv complexity r, which has important applications, for instance testing randomness of binary sequences. We also give algorithms to compute these recurrences. We employed these algorithms to compute $a(n,r)$ and expected value, EP_n, of number of patterns of a sequence of length n, for relatively large n. We offer a randomness test based on the algorithms to be used for testing randomness of binary sequences. We give outputs of the algorithms for some n. We also provide results of the proposed test applied to the outputs of contestant stream ciphers of ECRYPT's eSTREAM.

Keywords: Lempel-Ziv complexity, randomness, χ^2-statistics.

1 Introduction

There are several complexity measures to test the randomness of a sequence. Linear complexity, for example, is one of these measures. Lempel-Ziv complexity of a sequence was defined by Lempel and Ziv in 1976 [1]. This measure counts the number of different patterns in a sequence when scanned from left to right. For instance Lempel-Ziv complexity of $s = 101001010010111110$ is 8, because when scanned from left to right, different patterns observed in s are $1|0|10|01|010|0101|11|110|$.

Lempel-Ziv complexity is the basis of LZ77 compression algorithm [2]. It is also an important measure used in cryptography. For instance, it was used to test the randomness of the output of a symmetric cipher [3]. One expects a 'random' sequence of length n has a close Lempel-Ziv complexity to the expected value of Lempel-Ziv complexity of a sequence of length n. However, the expected value of Lempel-Ziv complexity for arbitrary n is unknown. For limiting behaviour of this value, the reader is referred to Jacquet and Szpankowski [4] and Kirschenhofer et. al. [6]. Some cryptographic applications of Lempel-Ziv complexity are given in [5].

G. Gong et al. (Eds.): SETA 2006, LNCS 4086, pp. 180–189, 2006.

Some sequences end with a pattern that was observed before (one simplest example is: $s = 0|0$), which we call *open*; and remaining sequences (i.e., that end without same pattern appearing twice) are called *closed*.

In this paper we derive a recurrence for $a(n,r)$, the number of sequences of length n with Lempel-Ziv complexity r; and a recurrence for $c(n,r)$, the number of closed sequences of length n with Lempel-Ziv complexity r. By using these recurrences and with the help of a computer, we compute $a(n,r)$ for as large n as possible.

A test based on Lempel-Ziv complexity was used in the NIST test suite, to test the randomness of sequences. However the test had some weaknesses. First of all, the test could only be applied to data of a specified length: 10^6 bits. Moreover, the test used empirical data generated by SHA-1 (under randomness assumptions) for estimating the expected value of Lempel-Ziv complexity of sequences of length 10^6 bits. Apparently, the data generated by SHA-1 led to not-so-good an estimate, hence, for instance, first 10^6 bits of the binary expansion of e failed the randomness test. Using asymptotic formulae for an estimate will not work either, since the sequences, as we will see in the forthcoming sections, are distributed tightly around the mean. Recently, apparently because of the spelt out reasons, Lempel-Ziv test had been excluded from the NIST test suite. Inclusion of a Lempel-Ziv complexity based randomness test in a statistical test suite is important concerning completeness. In the last section, we offer a new and stronger variant of this test, which employs the results we found and present in this paper. The data we use are neither empirical nor derived from asymptotic formulae, but are exact results; thanks to the recurrences (1),(2), hence avoid the errors present in the previous test.

2 Preliminaries

Lempel-Ziv complexity was first defined in [1]. We include the definitions here. For the sequel, juxtaposition denotes concatenation of strings.

Let $p = p_1 p_2 \cdots p_k$ and $s = s_1 s_2 \cdots s_k \cdots s_n$ be binary strings. p is a *prefix* of s if $p_i = s_i$ for $1 \le i \le k$. If $k < n$, then p is said to be a *proper prefix* of s.

Let again $s = s_1 s_2 \cdots s_n$ be a binary string of length n. $\sigma_1 | \cdots | \sigma_r$ is called the *Lempel-Ziv partition* of s, if

- for $1 \le i < r$, σ_i is different from σ_j for $0 \le j < i$, satisfying
- $s = \sigma_1 \sigma_2 \cdots \sigma_r$, and
- for $1 \le i \le r$, every proper prefix of σ_i is equal to σ_j for some $0 \le j < i$.

where σ_i are binary strings (*patterns*) and σ_0 is defined to be the empty string.

Lempel-Ziv complexity of s is then defined to be the number of patterns, r, in the Lempel-Ziv partition of s.

Note that σ_r may or may not satisfy $\sigma_r = \sigma_i$ for some $1 \le i < r$. If $\sigma_r = \sigma_i$ for some $1 \le i < r$, then we call s an *open* sequence. s is called *closed* otherwise.

Lempel-Ziv partition of:

- an open sequence s is denoted by $s = \sigma_1 | \cdots | \sigma_r$,
- a closed sequence s is denoted by $s = \sigma_1 | \cdots | \sigma_r |$.

Succint background for statistical tests (especially for randomness) can be found in [3].

3 The Recurrences

Let $A(n,r)$ denote the set of binary strings of length n with Lempel-Ziv complexity r. For any $s = s_1 \cdots s_n \in A(n,r)$ and $s_{n+1} \in \{0,1\}$, it is evident that $ss_{n+1} \in A(n+1,r) \cup A(n+1,r+1)$. In fact $s0 \in A(n+1,r) \iff s1 \in A(n+1,r)$. We define

$$C(n,r) = \{s \in A(n,r) : s0 \in A(n+1,r+1)\} .$$

Note that $C(n,r)$ is the set of *closed* sequences. One has

$$a(n,r) = 2c(n-1,r-1) + 2\left[a(n-1,r) - c(n-1,r)\right] , \tag{1}$$

where $a(n,r) = |A(n,r)|$ and $c(n,r) = |C(n,r)|$.

Given $s = s_1 \cdots s_n \in C(n,r)$, let $\sigma_1|\ldots|\sigma_r|$ be the Lempel-Ziv partition of s. We define the mapping $\delta^0_{n,r} : C(n,r) \to C(n+r+1,r+1)$ by setting $\delta^0_{n,r}(s) = 00\sigma_1 0\sigma_2 \cdots 0\sigma_r$ for $s = \sigma_1 \cdots \sigma_r \in C(n,r)$. $\delta^1_{n,r}$ is defined in a similar way. Let $C_0(n,r) = \mathrm{Im}(\delta^0_{n-r,r-1})$, $C_1(n,r) = \mathrm{Im}(\delta^1_{n-r,r-1})$, and $C_*(n,r) = C_0(n,r) \cup C_1(n,r)$. It follows that $c_*(n,r) = c_0(n,r) + c_1(n,r) = 2c(n-r,r-1)$, where $c_*(n,r) = |C_*(n,r)|$, $c_0(n,r) = |C_0(n,r)|$, $c_1(n,r) = |C_1(n,r)|$, and $\mathrm{Im}(f)$ denotes the image of the map f.

Any $s = \sigma_1|\cdots|\sigma_r| \in C(n,r) \setminus C_*(n,r)$ has a unique substring $\alpha = \alpha_1|\cdots|\alpha_p| \in C_0(a,p)$, and a unique substring $\beta = \beta_1|\cdots|\beta_q| \in C_1(b,q)$ such that $a+b=n$ and $p+q=r$.

For any pair (p,q) of positive integers, we consider the subset $\Xi^{p,q}$ of the symmetric group S^{p+q} given by:

$$\Xi^{p,q} = \left\{\sigma \in S^{p+q} : i < j \le p \text{ or } p+1 \le i < j \Rightarrow \sigma(i) < \sigma(j)\right\} .$$

For $\alpha = \alpha_1|\cdots|\alpha_p| \in C_0(a,p), \beta = \beta_1|\cdots|\beta_q| \in C_1(b,q)$ and $\pi \in \Xi^{p,q}, \pi(\alpha,\beta)$ stands for $\pi(\alpha_1,\ldots,\alpha_p,\beta_1,\ldots,\beta_q)$.

Any triple (π,α,β), where $\pi \in \Xi^{p,q}, \alpha \in C_0(a,p), \beta \in C_1(b,q)$, corresponds to a unique string in $C(n,r) \setminus C_*(n,r)$, namely to $\pi(\alpha,\beta)$. Conversely given any $\sigma \in C(n,r) \setminus C_*(n,r)$, there exist a unique triple (π,α,β), such that $\pi(\alpha,\beta) = \sigma$.

Given a,b,p and q, the number of all possible triples (π,α,β) with $\pi \in \Xi^{p,q}, \alpha \in C_0(a,p), \beta \in C_1(b,q)$ is

$$\binom{p+q}{p} c_0(a,p)c_1(b,q) .$$

It follows that

$$\begin{aligned} c(n,r) - c_*(n,r) &= \sum_{a+b=n} \sum_{\substack{p+q=r \\ p,q\ge 1}} \binom{p+q}{p} c_0(a,p)c_1(b,q) \\ &= \sum_{a+b=n} \sum_{\substack{p+q=r \\ p,q\ge 1}} \binom{p+q}{p} c(a-p,p-1)c(b-q,q-1) \end{aligned}$$

$$\begin{aligned}\Rightarrow c(n,r) &= 2c(n-r,r-1)+\\ &\sum_{a+b=n}\sum_{\substack{p+q=r\\ p,q\geq 1}}\binom{p+q}{p}c(a-p,p-1)c(b-q,q-1)\\ &= 2c(n-r,r-1)+ \qquad (2)\\ &\sum_{0\leq a\leq n}\sum_{1\leq p<r}\underbrace{\binom{r}{p}c(a-p,p-1)c(n-a-r+p,r-p-1)}_{\tau(n,r,a,p)}\end{aligned}$$

We can give upper and lower bounds for r, since not all r are possible given any n. Indeed, observing $s = 0|00|000|\cdots$ has minimum complexity, and

$$s = 0|1|00|01|10|11|000|001|010|011|\cdots$$

has maximum complexity among all sequences of length n, we limit r by:

$$\left\lceil \frac{-1+\sqrt{1+8n}}{2} \right\rceil \leq r \leq \left\lceil \frac{2^{t+2}+n-2t-4}{t+1} \right\rceil \qquad (3)$$

where $t = \max\{i \in \mathbb{N} : (i-1)2^{i+1}+2 \leq n\}$. Note here that r is bounded by $r < k\frac{n}{\log n}$, for some $k \in \mathbb{N}$. Indeed, $t < \log n$ for all $n \geq 2$. Also

$$\left\lceil \frac{2^{t+2}+n-2t-4}{t+1} \right\rceil = \left\lceil \frac{2^{t+2}-2(t+2)}{t+1} + \frac{n}{t+1} \right\rceil$$

increases when t increases, hence

$$r \leq \left\lceil \frac{2^{t+2}+n-2t-4}{t+1} \right\rceil \leq \left\lceil \frac{4\cdot 2^{\log n}+n-2\log n-4}{\log n+1} \right\rceil < 5\frac{n}{\log n}.$$

4 Algorithms and Their Complexities

(1) implies computing $c(n,r)$ for all $k \leq n$, and knowing $a(1,1) = 2$, is enough to compute $a(n,r)$ for any $n \geq 2$. Therefore we use (2) to compute $c(n,r)$, the result of which is used by another algorithm to compute $a(n,r)$. However, it is inefficient to compute larger values (e.g., computing $a(2000,r)$ for all r takes two hours on a standard PC with our implementation). We use the recurrence (2) in the following algorithm.

```
Compute-c(n,r)(N)
c(1,1) ← 2
for n ← 2 to N
   do for r ← r_l(n) to r_u(n)
         do c(n,r) ← 2c(n − r, r − 1)
            for a ← 0 to n
               do for p ← 1 to r
                     do c(n,r) ← c(n,r) + τ(n,r,a,p)
```

After we compute all $c(n,r)$ for $n < N$, we use the following algorithm which is based on the recurrence (1) to compute $a(n,r)$.

COMPUTE-A(N,R)$(N, c(i,j))$

$a(1,1) \leftarrow 2$
for $n \leftarrow 2$ **to** N
    do for $r \leftarrow r_l(n)$ **to** $r_u(n)$
        do $a(n,r) \leftarrow 2c(n-1,r-1) + 2\,[a(n-1,r) - c(n-1,r)]$

In the algorithms, r_l and r_u are computed by the inequalities (3).

We have the following observations for the complexity of the algorithms. For any (n,r) pair, $c(n,r) < 2^n$, hence an (at most) n-bit integer. Since $r \leq k \cdot n/\log n$, and complexity of multiplication of two n bit integers is $\mathcal{O}(n \log n)$ we have :

Proposition 1. *Complexity of the algorithm:*

- COMPUTE-C(N,R) *is* $\mathcal{O}(n^5/\log n)$, *and*
- COMPUTE-A(N,R) *is* $\mathcal{O}(n^2/\log n)$ *(after computing* $c(n,r)$*).*

5 Computing $a(n,r)$ for Large n

Tables 1 and 2 in Appendix A display the results for $n = 100$ and $n = 250$. Note that without using the recurrences (1) and (2), time complexity to find these results is $\mathcal{O}(n2^{n-1})$, impractical for today's computers for $n = 100$ or $n = 250$.

Expected values EP_n of number of patterns of a sequence of length n, for $n = 100$ and $n = 250$ are $EP_{100} = 29.04319$ and $P_{250} = 57.93485$.

Table 4 in Appendix C displays the EP_n values for some $n \leq 1000$.

6 An Application: A Randomness Test for Binary Sequences

We design a randomness test for binary sequences which employs the algorithms as follows.

Given a sequence of length n bits. First divide the sequence into $M = \lfloor \frac{n}{k} \rfloor$ non-overlapping blocks of length k bits, omitting if necessary last few bits. Apply Lempel-Ziv partitioning procedure to each of these M blocks to get the number of Lempel-Ziv partitions π_i for $1 \leq i \leq M$. From now on we choose $k = 1024$. Set:

$$r_1 = |\{i \;:\; \pi_i \leq 174,\; 1 \leq i \leq M\}|,$$
$$r_2 = |\{i \;:\; \pi_i = 175,\; 1 \leq i \leq M\}|,$$
$$r_3 = |\{i \;:\; \pi_i = 176,\; 1 \leq i \leq M\}|,$$
$$r_4 = |\{i \;:\; \pi_i = 177,\; 1 \leq i \leq M\}|,$$
$$r_5 = |\{i \;:\; \pi_i \geq 178,\; 1 \leq i \leq M\}|.$$

We obviously have $\sum_{i=1}^{5} r_i = M$. The numbers 174 through 178 are chosen to align $EP_{1024} = 176.09949$ to the center.

Define the random variable X to be the number of partitions of a random sequence of fixed length k bits. Employing the algorithm described in Section 4, we obtain the following probabilities for $k = 1024$.

$$\begin{aligned} p_1 &= Pr(X \leq 174) = 0.05262, \\ p_2 &= Pr(X = 175) = 0.19987, \\ p_3 &= Pr(X = 176) = 0.39720, \\ p_4 &= Pr(X = 177) = 0.29107, \\ p_5 &= Pr(X \geq 178) = 0.05924. \end{aligned}$$

Then apply the χ^2-statistic to the observed data:

$$X(obs) = \sum_{i=1}^{5} \frac{(r_i - Mp_i)^2}{Mp_i}$$

to get the χ^2 random variable $X(obs)$ with degree of freedom 4. Then, the P-value of the test is:

$$\frac{\int_{X(obs)}^{\infty} e^{-u/2} u \, du}{\Gamma(2)\, 2^2} = \frac{1}{2}(X(obs) + 2)e^{-X(obs)/2}.$$

A condition that can be safely used with χ^2-approximation is:

$$M \cdot \mathsf{min}\{p_i : 1 \leq i \leq 5\} = M \cdot 0.05262 \geq 5.$$

Hence, if k is chosen to be 1024, then n should satisfy $n \geq 100000$ approximately. Note that the test can be applied for any k with respective p_i 's and 'bins' are aligned around EP_k and of course provided that computation of $a(k, l)$ is feasible.

If the P-value of the observed data is less than some threshold (e.g., 0.01), one can conclude that the given sequence is not random. The test applied to the outputs of stream ciphers contesting in ECRYPT's eSTREAM can be found in Appendix B.

7 Conclusion and Future Work

We give two recurrences for the number of sequences of length n with Lempel-Ziv complexity r. We also give the the algorithms and the output of the computer programs that we run to calculate $a(n, r)$ for relatively large values.

We also offer a randomness test that can be applied to the output of ciphers.

The recurrence (2) is quite hard to simplify, but can be used to improve the limiting behaviour of the expected value of $a(n, r)$.

Acknowledgments

The authors would like to thank to anonymous refeeres for their comments, which improved the presentation of the paper. The authors also would like to thank Meltem Sönmez Turan and Çağdaş Çalık for making the outputs of eSTREAM contestant stream ciphers available.

References

1. Lempel, A., Ziv, J.: On the complexity of finite sequences. IEEE Transactions on Information Theory **IT-22** (1976) 75–81
2. Ziv, J., Lempel, A.: A universal algorithm for sequential data compression. IEEE Transactions on Information Theory **IT-23** (1977) 337–343
3. Soto, J.: Statistical testing of random number generators. In: Proceedings of the 22nd National Information Systems Security Conference, Crystal City, Virginia (1999)
4. Jacquet, P., Szpankowski, W.: Asymptotic behavior of the Lempel-Ziv parsing scheme and digital search trees. Theoretical Computer Science **144** (1995)
5. Mund, S.: Ziv-Lempel complexity for periodic sequences and its cryptographic application. In: Advances in cryptology – EUROCRYPT 91 (Brighton, 1991). Volume 547 of Lecture Notes in Comput. Sci. Springer, Berlin (1991) 114–126
6. Kirschenhofer, P., Prodinger, H., and Szpankowski, W.: Digital Search Trees Again Revisited: The Internal Path Length Perspective, SIAM Journal on Computing **23** (1994) 598–616

A Tables for $n = 100$ and $n = 250$

Table 1. $a(100, r)$ and their probabilities

$\mathbf{r}$	$\mathbf{a(100, r)}$	$\mathbf{a(100, r)/2^{100}}$
14	122880	0.000000000000000
15	96129024	0.000000000000000
16	1754408140	0.000000000000000
17	169010698649	0.000000000000000
18	12282745099264	0.000000000000000
19	726896570696704	0.000000000000006
20	35864704163873996	0.000000000000283
21	1555171539525474304	0.000000000012268
22	629504083451115732992	0.000000000496591
23	23657061862581861351424	0.000000018662131
24	796717339700675605430272	0.000000628499162
25	25016712354109852183691264	0.000019734706353
26	701956405285233154502688768	0.000553745965299
27	14929637765344244033503887360	0.011777407562191
28	190072463603886098540862111744	0.149940735696149
29	785071700104053917078962307072	0.619312372007483
30	276807820976750678936983175168	0.218362868227975
31	41183640732091617843871744	0.000032488164108

Table 2. $a(250, r)$

r	a(250, r)
22	12582912
23	172462440448
24	207405092700160
25	100022234734919680
26	29027442465801502720
27	5970493862438356647936
28	947059437548499752058880
29	124084577391675511972954112
30	14104448150286646440414281728
31	1429659188269782925153552039936
32	131701470381268947695969402486784
33	11234825836624304676748166609502208
34	902346385748231250614173057894580224
35	68848617082812392433571189369104498688
36	5026197932887293151555523266808542920704
37	353853624800555379505246051484079264628736
38	24168383146155367527845519053853996700663808
39	1608511050085914176405326626207802044763340800
40	104662286330519094422345952269389211024618422272
41	6671326955511762120610779318782504320898951020544
42	417414764650712462333990517379047167232803455631360
43	25695057332640828405359742259152343883668370141216768
44	1557264023287411624909081480426191697539573325326450688
45	92701547946190343870914352507209237010079274453119795200
46	5404237040271065934800659259750349232918183084398749941760
47	308137037472343306269203510492671163205021043338817023508480
48	17174851395953502738636183446022389293687597007698530091925504
49	931658807304593772659970661068671319161079379311405098319478784
50	48481739469168604779398196043362721869737926460656763302282264576
51	2362999038927091779739893669395333742467402505543139386088962916352
52	104713381850515314827585466063284598037259318649923004999242356883456
53	4060841943120707511367011625729606171901794317261167326529027544449024
54	130173461076947361401256777713070992922594587328928980706197194351050752
55	3144432233774945014197088996141663864291615518236615683726382172345466880
56	49270065837659893327338857961415416600618856787734271093560018969747783680
57	389584816963822529432810975258074767638122905768577642480619865183753338880
58	990152575211374388394386838415206581167171380538108666357249216731099955200
59	373709741073491221158081813947703844668617084164776240113666220254822400000
60	3255419797444187375980631332217791151748569890952982076721561862144000000
61	886666544515752607846923627683910176444275554890872258560000000000

B Test Results

Table 3. Results of LZ randomness test applied to eSTREAM contestants with parameters $M = 800, k = 1024$ and threshold < 0.01

eSTREAM stream cipher	**P – value**
ABC-v2	0.215686
ACHTERBAHN	0.856026
CryptMT	0.281958
DECIM	0.435354
DICING	0.391681
Dragon	0.784314
Edon80	0.958401
F-FCSR-8	0.559503
FUBUKI	0.805604
Frogbit	0.247524
Grain	0.092822
HC-256	0.189772
Hermes8	0.548511
LEX	0.192730
MAG	0.172511
MICKEY-128	0.951844
MICKEY	0.958706
Mir-1	0.624140
POMARANCH	0.864929
Phelix	0.422482
Polar-Bear	0.032209
ProVEST-4	0.902847
Py	0.518629
Rabbit	0.654306
SFINKS	0.327318
Salsa20	0.325591
TRIVIUM	0.624686
TSC-3	0.943600
WG	0.836510
Yamb	0.514665
ZK-Crypt	0.590525

C Table of Expected Values

Table 4. Expected values EP_n for some $n \leq 1000$

n	EP_n
968	168.285154708125871909
969	168.425325208575350399
970	168.565472359678417715
971	168.705595531148748041
972	168.845694563342602216
973	168.985769897582357720
974	169.125822268928967939
975	169.265852191319223625
976	169.405859611321451745
977	169.545843978226506736
978	169.685804672692994605
979	169.825741477831057654
980	169.965654751385158620
981	170.105545179628954902
982	170.245413297909854937
983	170.385259128160740312
984	170.525082193178254703
985	170.664881890971218212
986	170.804657952890038172
987	170.944410653561969008
988	171.084140625407537432
989	171.223848418188401220
990	171.363534125420147438
991	171.503197345751651805
992	171.642837501466368323
993	171.782454278665110691
994	171.922047870128312202
995	172.061618849379214303
996	172.201167772807546273
997	172.340694800677496380
998	172.480199609262570074
999	172.619681652361296729
1000	172.759140578329111086

Computing the k-Error N-Adic Complexity of a Sequence of Period p^n*

Lihua Dong, Yupu Hu, and Yong Zeng

The Key Lab. of Computer Networks and Information Security
the Ministry of Education, Xidian University, Xi'an,
ShaanXi Province, 710071, P.R. China
lih_dong@hotmail.com, yphu@mail.xidian.edu.cn,
yzeng@mail.xidian.edu.cn

Abstract. Cryptographically strong sequences should have a large N-adic complexity to thwart the known feedback with carry shift register (FCSR) synthesis algorithms. At the same time the change of a few terms should not cause a significant decrease of the N-adic complexity. This requirement leads to the concept of the k-error N-adic complexity. In this paper, an algorithm for upper bounding the k-error N-adic complexity of the sequence with period $T = p^n$, and p is just a prime, is proposed by extending the 2-adic complexity synthesis algorithm of Wilfried Meidl, and the Stamp-Martin algorithm. This algorithm is the first concrete construction of the algorithm for calculating the k-error N-adic complexity. Using the algorithm proposed, the upper bound of the k-error N-adic complexity can be obtained in n steps.

1 Introduction

The notion of feedback with carry shift registers (FCSRs), introduced by Klapper and Goresky [1], has received a great amount of attention in the cryptography[2], [3], [4], [5], [6], [7], [8], [9]. Some basic properties of FCSR sequences have been discussed, see [10] for a recent survey. Additionally, Wilfried Meidl [11] presented an FCSR analog of the (extended) Games-Chan algorithm for up bounding the 2-adic complexity of a periodic binary sequence with period $T = 2^n$ or p^n, where p is an odd prime and 2 is a primitive element modulo p^2.

It is well known that the linear complexity of a periodic sequence is unstable under small perturbations [12], [13]. This is also true for the case of the N-adic complexity. For example, let $S = (1, 0, 0, ..., 0)^\infty$ or $(0, 1, 1, ..., 1)^\infty$ with period T. Then the N-adic complexity $\lambda_N(S)$ of the sequence S is $\log_N(N^T - 1)$. However, after changing 1 bit within every period, the N-adic complexity becomes 0. Hence it is interesting to investigate the properties of the k-error N-adic complexity of periodic sequences.

The area of k-error N-adic complexity for the case of $N = 2$ was first formally studied by Wang [14], and a lower bound of it was given by Hu [15]. The definition of the k-error N-adic complexity of sequences is described as follows:

* This work was supported in part by the National Science Foundation(No. 60273084) and Doctoral Foundation(No.20020701013) in China.

G. Gong et al. (Eds.): SETA 2006, LNCS 4086, pp. 190–198, 2006.

Definition 1. *Let S be a sequence with period T, then the k-error N-adic complexity is defined as*

$$\lambda_{k,N}(S) = \min_{per(t)=T, d(S,t)\le k} \lambda_N(t).$$

Remark 1. The minimum is extended over all T-periodic sequences $t = t_0, t_1, \cdots, t_{T-1}, \cdots$, for which the Hamming distance of the vectors $(s_0, s_1, \cdots, s_{T-1})$ and $(t_0, t_1, \cdots, t_{T-1})$ is at most k. In this case we write $d(S,t) \le k$. The k-error N-adic complexity defined above is similar to that of k-error linear complexity. In other words, $\lambda_{k,N}(S)$ is the least N-adic complexity $\lambda_N(t)$ among all T-periodic sequences t that are obtained by changing up to k terms among the first T terms of S and continuing these changes periodically with period T.

There are no known efficient algorithms in the literature for calculating the k-error N-adic complexity of a periodic sequence. In this paper, we will construct an algorithm for up bounding the k-error N-adic complexity of a periodic sequence with period p^n, p is an prime, based on the 2-adic complexity synthesis algorithm in [11].

In Section 2, the necessary background is established. The algorithm for computing the k-error N-adic complexity is described in Section 3.

2 Preliminary

An FCSR is determined by coefficients $q_1, q_2, \cdots, q_c$, and an initial memory m_{c-1}, with $q_i \in \{0, 1, \cdots, N-1\}$ for $i = 1, 2, \cdots, c$, and $m_{c-1} \in Z$, which can iteratively generate an FCSR-sequence S with initial state $\{s_0, s_1, \cdots, s_{c-1}\}$ in the following way, for $n = c, c+1, \cdots$, and $s_i \in \{0, 1, \cdots, N-1\}$ for $i = 0, 1, 2, \cdots$:

- Form the integer sum $\sigma_n = \sum_{k=1}^{c} q_k s_{n-k} + m_{n-1}$,
- Shift the contents one step to the right, outputting the rightmost digit s_{n-c},
- Put $s_n = \sigma_n \, modN$,
- Replace the memory integer m_{n-1} with $m_n = (\sigma_n - s_n)/N = \lfloor \sigma_n/N \rfloor$.

The integer $q = q_c N^c + q_{c-1} N^{c-1} + \cdots + q_1 N - 1$ is called the connection integer of the FCSR. There is a useful polynomial $f(x) = \sum_{i=0}^{T-1} s_i x^i$ that associates a sequence S with its N-adic interpretation. In this case the corresponding N-adic number is given as

$$\alpha = f(N)N^0 + f(N)N^T + f(N)N^{2T} + \cdots = \frac{-f(N)}{N^T - 1}$$

Let us write $\alpha = -r/q$ as a fraction reduced to lowest terms and q is odd. Then the eventual period T of the associated sequence with $\alpha = -r/q$ equals $ord_q(N)$, where $ord_q(N)$ is the minimal integer t such that $N^t \equiv 1 \pmod{q}$.

Then $q = (N^T - 1)/\gcd(N^T - 1, f(N))$ is the connection integer of the smallest FCSR, and $\log_N(q)$ is the N-adic complexity of the sequence S. Since the N-adic complexity $\lambda_N(S)$ measures the size of the smallest FCSR that can generate S, it is of comparable significance as the linear complexity of the periodic sequence S.

Suppose that the period T of a sequence S is a power of a prime p, i.e. $T = p^n, n \geq 1$. The integer $N^{p^n} - 1$ can be written as the product $N^{p^n} - 1 = \prod_{m=1}^{n} F_m^{(p)}$ with $F_m^{(p)} = \frac{N^{p^m}-1}{N^{p^{m-1}}-1}$. We will need a key result, which may be proved similarly to [11].

Lemma 1. *Let $S^T = (s_0, s_1, \ldots, s_{T-1})$ be a T-tuple, with $s_i \in \{0, 1, \cdots, N-1\}$ for $i = 0, 1, 2, \ldots, T-1, \ldots, T = p^n, n \geq 1$. Let $f(x)$ be the polynomial $f(x) = \sum_{i=0}^{T-1} s_i x^i$, and let A_j be the p^{n-1}-tuple consisting of the string beginning at $s_{(j-1)p^{n-1}}$, i.e. $A_j = (s_{(j-1)p^{n-1}}, \cdots, s_{jp^{n-1}-1}), j = 1, 2, \ldots, p$. Then*

(i) *$F_n^{(p)}$ divides $f(N)$ if and only if $A_1 = A_2 = \cdots = A_p$, and*
(ii) *$F_m^{(p)}, 1 \leq m < n$, divides $f(N)$ if and only if it divides*

$$A_1(N) + A_2(N) + \cdots + A_p(N),$$

where

$$A_j(x) = \sum_{t=0}^{p^{n-1}-1} s_{(j-1)p^{n-1}+t} x^t, \quad j = 1, \ldots, p.$$

Given the first period $S^T = (s_0, s_1, \ldots, s_{T-1}), T = p^n$, of a sequence S, the following Algorithm 1 is a simple extension of the algorithm that has given in [11]. In the algorithm,

- If we have the equation $A_1 = A_2 = \cdots = A_p$, then the N-adic complexity does not increase and we apply the procedure to A_1;
- Otherwise we increase the N-adic complexity by $p^{n-1}(p-1)$ and apply the procedure to $A_1(N) + A_2(N) + \cdots + A_p(N)$.

Since $C = A_1(N) + A_2(N) + \cdots + A_p(N)$ has to be not larger than $p(N^{p^{n-1}} - 1)$, the N-adic expansion of it may have up to $p^{n-1} + \lceil \log_N(p) \rceil$ digits. In this case we can write C in the form $a + bN^{p^{n-1}}$ with $0 \leq a < N^{p^{n-1}}$ and $1 \leq b < p$. Now,

$$C = a + bN^{p^{n-1}} = a + b(F_m^{(p)} \prod_{u=1, u\neq m}^{n-1} F_u^{(p)} + 1)$$

$$= a + b + bF_m^{(p)} \prod_{u=1, u\neq m}^{n-1} F_u^{(p)}$$

Thus $F_m^{(p)}$ divides C if and only if it divides $a + b, 1 \leq m \leq n$.

Note that the N-adic complexity only increases at a step unless $A_1 = A_2 = \cdots = A_p$. The following algorithm determines for which m the integer $F_m^{(p)}$

divides $f(N) = \sum_{i=0}^{T-1} s_i N^i, 1 \leq m \leq n$. Since in general the integer $F_m^{(p)}$ is not a prime, hence the algorithm yields an upper bound for $q = (N^T - 1)/\gcd(N^T - 1, f(N))$, and thus for the N-adic complexity $\lambda_N(s) = \log_N(q)$.

Algorithm 1: *N-adic complexity synthesis algorithm*

$A = S, l = p^n, \delta = 1, \lambda_N(S) = 0,$
while $n > 0,$
  $l = l/p,$
  $A_j = (a_{(j-1)l}, a_{(j-1)l+1}, \cdots, a_{jl-1}), j = 1, 2, \cdots, p,$
  if $A_1 = A_2 = \cdots = A_p,$
    $A = A_1,$
  else
    $\delta = \delta F_n^{(p)},$
    $\lambda_N(S) = \lambda_N(S) + p^{n-1}(p-1),$
    $A = A_1 \oplus A_2 \oplus \cdots \oplus A_p,$
    if $B = (a_{p^{n-1}}, a_{p^{n-1}+1}, \cdots, a_{p^{n-1}+\lceil \log_N(p) \rceil - 1}) \neq 0,$
    $A = (a_0, a_1, \cdots, a_{p^{n-1}-1}),$
    $A = A \oplus B,$
  $n = n - 1,$
end while

Remark 2. Here the algebraic operation $\oplus$ is the N-adic addition, which is performed by carrying overflow to higher order terms.

3 Extended Stamp-Martin Algorithm for Solving the k-Error N-Adic Complexity Synthesis Problem

In the algorithm above, we found that the estimate of the N-adic complexity increases unless $A_1 = A_2 = \cdots = A_p$. Thus to find the k-error N-adic complexity, that is, to find the least N-adic complexity among all T-periodic sequences t that are obtained by changing up to k terms among the first T terms of the sequence S, the principal goal of the algorithm is to change as few terms among the first T terms of S as possible to make the equation $A_1 = A_2 = \cdots = A_p$ hold. In the following we give an algorithm for estimating the upper bound for the k-error N-adic complexity of a periodic sequence with period $T = p^n$ that is an analog of the Stamp-Martin algorithm [16] for computing the k-error linear complexity of a periodic sequence with the same period. Here p is a prime. Similar to Algorithm 1, we denote $\oplus$ as the N-adic addition. The vector of $\cos t[a_i, h]$'s is intended to measure the " $\cos t$"-in terms of the least number of terms changes required in the original sequence S-of changing the current element a_i into h without disturbing the results $A_1 = A_2 = \cdots = A_p$ of any previous steps.

3.1 The Synthesis Algorithm for Computing the k-Error N-Adic Complexity of a Periodic Sequence with Period $T = p^n$, Here p Is a Prime

Algorithm 2: *k-error N-adic complexity synthesis algorithm*

$A \leftarrow S$, $l \leftarrow p^n$, $\cos t[a_i, a_i] \leftarrow 0$, for all $h \in 0, 1, \cdots, N-1$ and $h \neq a_i$, $\cos t[a_i, h] \leftarrow 1$, $i = 0, 1, 2, \cdots, l-1$, $\cos t[c_j, m] \leftarrow 1$, $j = 0, 1, 2, \cdots, l$, $m \neq c_j$, $\cos t[c_j, c_j] \leftarrow 0, c_0 \leftarrow 0$. $\delta \leftarrow 1$, $\lambda_N(S) \leftarrow 0$,

1. If $l = 1$, then stop; else, $l \leftarrow l/p$, $A_j = (a_{(j-1)l}, a_{(j-1)l+1}, \cdots, a_{jl-1})$, $j = 1, 2, \cdots, p$, $T_{ih} = \sum_{j=0}^{p-1} [a_{i+jl}, h]$, $h = 0, 1, \cdots, N-1$, $T_i = \min_{0 \leq h \leq N-1} \{T_{ih}\}$, $i = 0, 1, 2, \cdots, l-1$, $T = \sum_{i=0}^{l-1} T_i$, turn to 2.

2. If $T \leq k$, then $k \leftarrow k - T$, $\cos t[a_i, h] \leftarrow T_{ih} - T_i$, $h = 0, 1, \cdots, N-1$, and $i = 0, 1, 2, \cdots, l-1$, turn to 3; else, $A \leftarrow A_1 \oplus A_2 \oplus \cdots \oplus A_p$, and the corresponding overflows are denoted as integer vector $(c_0, c_1, c_2, \cdots, c_{l-1}, c_l), c_0 = 0$, $0 \leq c_i \leq p-1, 1 \leq i \leq l$, $\lambda_N(S) \leftarrow \lambda_N(S) + l(p-1)$, $\delta \leftarrow \delta F_2^{(p)}$. Let temp $= d_0 + d_1 + \cdots + d_p$, then

$$\cos t[c_{i+1}, g] \leftarrow \min_{\text{temp}=(g-c_{i+1})N} \left\{ \sum_{k=0}^{p-1} \cos t[a_{i+kl}, a_{i+kl} + d_k], \cos t[c_i, c_i + d_p] \right\},$$

$$\cos t[a_i, h] \leftarrow \min_{\text{temp}=h-a_i} \left\{ \sum_{k=0}^{p-1} \cos t[a_{i+kl}, a_{i+kl} + d_k], \cos t[c_i, c_i + d_p] \right\},$$

$i = 0, 1, 2, \cdots, l-1$, $h \in \{0, 1, \cdots, N-1\}$, and $g \in \{0, 1, \cdots, N-1\}$. If $c_l \neq 0$, we denote B as $(a_l, a_{l+1}, \cdots, a_{l+\lceil \log_N(p) \rceil - 1})$, A as $(a_0, a_1, \cdots, a_{l-1})$, U as $A \oplus B$, and $V = (v_0, v_1, \cdots, v_{l-1})$ as the corresponding overflows, $v_0 = 0$. Then

$$\cos t[u_i, h] \leftarrow \min_{d_0 - d_1 N = h - u_i} \{\cos t[a_i, a_i + d_0], \cos t[a_{i+1}, a_{i+1} + d_1]\},$$

$a_{i+1} \leftarrow a_{i+1} + d_1$, $i = 0, 1, 2, \cdots, l-2$, $h \in 0, 1, \cdots, N-1$, here $d_1 \in \{-1, 0, 1\}$ and $a_i + d_0 \in \{0, 1, \cdots, N-1\}$. Then

   - If $a_{i+1} + d_1 = N$, then $\cos t[a_{i+1}, a_{i+1} + d_1] \leftarrow \cos t[a_{i+1}, 0]$;
   - Else if $a_{i+1} + d_1 = -1$, and we have $a_{i+k} = 0$ for all $k = 2, \cdots, w$, but $a_{i+w+1} \neq 0$, then

$$\cos t[a_{i+1}, a_{i+1} + d_1] \leftarrow \sum_{k=1}^{w} \cos t[a_{i+k}, N-1] + \cos t[a_{i+w+1}, a_{i+w+1} - 1].$$

   $U \rightarrow A$, turn to 1.
3. For $j = 1, 2, \cdots, l$, if $T_{jh} = T_j$, then $a_j = h$. $A \leftarrow A_1$, turn to 1.

Remark 3. The output δ is the connection integer of an FCSR that can generate the sequence $t = t_0, t_1, \cdots, t_{T-1}, \cdots$ for which the Hamming distance of the vectors $(s_0, s_1, \cdots, s_{T-1})$ and $(t_0, t_1, \cdots, t_{T-1})$ is at most k. Since in general the integer $F_n^{(p)}$ is not a prime, δ might not be the connection integer of the smallest FCSR that can generate t. Thus, the k-error N-adic complexity $\lambda_{k,N}(S)$ satisfies $\lambda_{k,N}(S) \leq \log_N(\delta)$. The coefficients of the FCSR correspond to the coefficients of the N-adic expansion of $\delta + 1 = q_c N^c + q_{c-1} N^{c-1} + \cdots + q_1 N$. If $\delta + 1 = N^{\lambda_{k,N}(S)+1}$, trivially we have $c = \lambda_{k,N}(S) + 1$. Else we have $c = \lfloor \log_N(\delta + 1) \rfloor = \lambda_{k,N}(S)$. Thus the output $\lambda_{k,N}(S)$ satisfies $\lambda_{k,N}(S) < \log_N(\delta) < \lambda_{k,N}(S) + 1$.

3.2 The Validity of the k-Error N-Adic Complexity Synthesis Algorithm

Theorem 1. *Let S be a sequence with period $T = p^n$. Here p is a prime, and $0 \leq k \leq p^n$. Then the integer $\lambda_N(S)$ that has been obtained by Algorithm 2 above is an upper bound for the k-error N-adic complexity of the sequence S.*

Proof. The outline of the proof follows: first two paragraphs will tell us two switches ($k > 0$ and $k = 0$) to be proved, finally, an epagoge is given to prove that $\cos t[i, h]$ correctly records the cost of changing a_i without disturbing the results $A_1 = A_2 = \cdots = A_p$ of any previous steps.

When $k = 0$, Algorithm 2 just reduces to Algorithm 1.

When $k > 0$, to obtain the k-error N-adic complexity, we are allowed to make k (or fewer) changes in S in order to reduce the N-adic complexity as much as possible. But as with Algorithm 1, the N-adic complexity increases unless $A_1 = A_2 = \cdots = A_p$. Notice that if the equation doesn't hold in step m of Algorithm 2, and we can change up to k terms among the first T terms of S to make it hold, we do so, by which we can avoid adding the $(p-1)p^{n-m}$ into $\lambda_N(S)$, and the total of all remaining possible additions is only p^{n-m}. This is the basic logic of the algorithm.

Now, suppose we have computed to step m, and $\cos t[i, h]$ correctly records the cost of changing a_i into h. If not all of the terms $a_{i+(j-1)l}, j = 1, 2, \cdots, p$, are equal, then we need change all of which into a same element, say h, to make them equal. Thus the total cost to make them equal is just the minimal cost that makes these $a_{i+(j-1)l}, j = 1, 2, \cdots, p$, equal to h, that is $T_i = \min_{0 \leq h \leq N-1} \{T_{ih}\}$. Thus the variable $T = \sum_{i=0}^{l-1} T_i$ correctly records the total cost of making $A_1 = A_2 = \cdots = A_p$.

If $T \leq k$,

- If not all of the terms $a_{i+(j-1)l}, j = 1, 2, \cdots, p$, are equal (in step m), and $T_{ih} = T_i$, then we need change all of the them into h, since such changes can minimize the corresponding total cost that makes these $a_{i+(j-1)l}, j = 1, 2, \cdots, p$, equal. Note that at the end of this step, we have $A \leftarrow A_1$. If we need change a_i (which has been made equal to d in step m) into h in step $(m + 1)$, to keep all of the terms $a_{i+(j-1)l}, j - 1, 2, \cdots, p$, are equal in step

m, we have to change them into h in step m, which has a net cost of $T_{ih} - T_i$ and hence $\cos t[i, h]$ is computed correctly in this case.

- If all of the terms $a_{i+(j-1)l}, j = 1, 2, \cdots, p$, are equal, then $T_i = 0$. Note that at the end of this step, we have $A \leftarrow A_1$. If we need change a_i (which has been made equal to d in step m) into h in step $(m+1)$, to keep all of the terms $a_{i+(j-1)l}, j = 1, 2, \cdots, p$, are equal in step m, we have to change all of them into h in step m, which has a net cost of $T_{ih} = T_{ih} - T_i$, and hence $\cos t[i, h]$ is computed correctly in this case.

If $T > k$, we haven't the means to make $A_1 = A_2 = \cdots = A_p$. However at the end of this step, we have $A \leftarrow A_1 \oplus A_2 \oplus \cdots \oplus A_p$. If the corresponding overflows are denoted as integer vector $(c_0, c_1, c_2, \cdots, c_{l-1}, c_l)$, c_0 is 0, $0 \leq c_i \leq p-1, 1 \leq i \leq l$, then,

- If we need change c_{i+1} into g, we have to change a_{i+jl}, $j = 0, 1, \cdots, p-1$, into $a_{i+jl} + d_j$, $j = 0, 1, \cdots, p-1$, and c_i into $c_i + d_p$, in step m, here $\sum_{j=0}^{p} d_j = (g - c_{j+1})N$. Then the cost for changing c_{i+1} into g in step $(m+1)$ is just

$$\cos t[c_{i+1}, g] = \min_{\text{temp}=(g-c_{i+1})N} \left\{ \sum_{k=0}^{p-1} \cos t[a_{i+kl}, a_{i+kl} + d_k], \cos t[c_i, c_i + d_p] \right\},$$

here temp $= d_0 + d_1 + \cdots + d_p$, and hence $\cos t[c_i, g]$ is computed correctly.

- If we need change a_i into h in step $(m+1)$, we have to change a_{i+jl}, $j = 0, 1, \cdots, p-1$, into $a_{i+jl} + d_j$, $j = 0, 1, \cdots, p-1$, and c_i into $c_i + d_p$, in step m, here $\sum_{j=0}^{p} d_j = h - a_i$. Then the cost for changing a_i into h in step $(m+1)$ is just

$$\cos t[a_i, h] = \min_{\text{temp}=h-a_i} \left\{ \sum_{k=0}^{p-1} \cos t[a_{i+kl}, a_{i+kl} + d_k], \cos t[c_i, c_i + d_p] \right\},$$

here temp $= d_0 + d_1 + \cdots + d_p$, and hence $\cos t[a_i, h]$ is computed correctly in this case.

- If $c_l \neq 0$, then $B = (a_l, a_{l+1}, \cdots, a_{l+\lceil \log_N(p) \rceil - 1}) \neq 0$, from Theorem 1, if we denote A as vector $(a_0, a_1, \cdots, a_{l-1})$, then at the end of this step we have $A \leftarrow A \oplus B$. In the following, we denote $U = A \oplus B$, $A = (a_0, a_1, \cdots, a_{l-1})$, and the corresponding overflows as vector $V = (v_0, v_1, \cdots, v_{l-1})$, $v_0 = 0$. If we need change u_i into h and keep the values of others in step $(m+1)$, we have to change a_i into $a_i + d_0$, and a_{i+1} into $a_{i+1} + d_1$, in step m, here $d_0 - d_1 N = h - u_i$, $d_1 \in \{-1, 0, 1\}$, $a_i + d_0 \in \{0, 1, \cdots, N-1\}$. Then the cost for changing u_i into h in step $(m+1)$ is just

$$\cos t[u_i, h] = \min_{d_0 - d_1 N = h - u_i} \{\cos t[a_i, a_i + d_0], \cos t[a_{i+1}, a_{i+1} + d_1]\},$$

and hence $\cos t[a_i, h]$ is computed correctly in this case. And to keep the value of u_{i+1}, we have $a_{i+1} + d_1 \rightarrow a_{i+1}$. However,

- $a_{i+1} + d_1$ may equal to N, then it is clear that

$$\cos t[a_{i+1}, a_{i+1} + d_1] = \cos t[a_{i+1}, 0];$$

- $a_{i+1} + d_1$ may equal to -1, and if $a_i + k = 0$ for all $k = 2, \cdots, w$, but $a_{i+w+1} \neq 0$, then we have

$$\cos t[a_{i+1}, a_{i+1} + d_1] = \sum_{k=1}^{w} \cos t[a_{i+k}, N-1] + \cos t[a_{i+w+1}, a_{i+w+1} - 1].$$

Finally, when $n = 0$, there remains only one term a_0. Since for the all-0 sequence and the all-1 sequence, the N-adic complexity is 0. Thus the algorithm is terminated when the vector (0) or the vector (1) is encountered.

Remark 4. Algorithm 2 given yields an upper bound for the k-error N-adic complexity of a given p^n-periodic sequence S in n steps. In each step we just have to compute the cost for p^{n-1} terms c_i and p^{n-1} terms a_i respectively, and have to add p N-adic integers of the length at most p^{n-1} (plus one supplementary addition if we have an overflow). The time complexity of Algorithm 2 is $O(p^n)$, while the obvious algorithm of computing the N-adic complexity of every sequence obtained by modifying up to k bits of S via Algorithm 1, has complexity $O(2^k \cdot p^n)$.

4 Conclusion

We have exhibited an efficient algorithm which upper-bounds the k-error N-adic complexity of a periodic sequence of period $T = p^n$. The algorithm given in this paper is an extension of the Stamp-Martin algorithm. It remains a challenging open problem to design an algorithm, which efficiently computes the k-error N-adic complexity of sequences of arbitrary period.

References

1. A. Klapper, M. Goresky, 2-adic shift registers, in: R. Anderson (Ed.), Fast Software Encryption, LNCS, Vol. 809, Springer-Verlag, New York, 1994, pp. 174-178.
2. M. Goresky, A. Klapper, Feedback registers based on ramified extensions of the 2-adic numbers, Advances in Cryptology-Eurocrypt'94, LNCS, vol, 950, Springer-Verlag, Berlin, 1995, pp. 215-222.
3. M. Goresky, A. Klapper, Cryptanalysis based on 2-adic rational approximation, Advances in Cryptology-Crypt'95, LNCS, vol. 963, Springer-Verlag, Berlin, 1995, pp. 262-273.
4. M. Goresky, A. Klapper, Large periods nearly de Bruijn FCSR sequences, Advances in Cryptology-Eurocrypt'95, LNCS, vol. 921, Springer-Verlag, Berlin, 1995, pp. 263-273.
5. François Arnault, Thierry P.Berger, F-FCSR: Design of a new class of stream cipher, Fast Software Encryption, LNCS, 3557, Springer-Verlag, Berlin, 2005, pp.83-97.
6. François Arnault, Thierry P.Berger, Design and properties of a new pseudorandom generator based on a filtered FCSR automaton, IEEE Transactions on Computers, 54(11), 2005, pp.1374-1383.

7. A. Klapper, Jingzhong Xu, Register synthesis for algebraic feedback shift registers based on nonprimes, Designs, Codes and Cryptography, vol.31, 2004, pp.227-250.
8. A. Klapper, M. Goresky, Feedback shift registers, 2-adic span, and combiners with memory, J. Cryptology, vol.10, 1997, pp.111-147.
9. François Arnault, Thiery P. Berger, and Abdelkadar Necer, Feedback with carry shift registers synthesis with the Euclidean algorithm, IEEE Trans. Information Theory, 50(5), 2004, pp.910-917.
10. A. Klapper, A survey of feedback with carry shift registers, Sequence and Its Application, SETA 2004, LNCS, 3486, Springer-Verlag, Berlin, 2005, pp.56-71.
11. Wilfried Meidl, Extended Games-Chan algorithm for the 2-adic complexity of FCSR-sequences, Theoretical Computer Science, vol.290, Elsevier Science B.V., 2003, pp.2045-2051.
12. Niederreiter, H.: Periodic sequences with large k-error linear complexity. IEEE Trans.Inform. Theor. Vol.49, 2003, pp.501-505.
13. Wei Shimin, Dong Qingkuan, Xiao Guozhen, An efficient algorithm for the k-error linear complexity of periodic sequences, Journal of Xidian University (in Chinese), Vol. 28, No. 4, Aug. 2001, pp.421-424.
14. WangLei, CaiMian and Xiao Guozhen, On stability of 2-adic complexity of periodic sequence, Journal of Xidian University (in Chinese), vol.27, 2000, pp.348-350.
15. Honggang Hu, Dengguo Feng, On the 2-Adic complexity and the k-error 2-adic complexity of periodic binary sequences, Sequence and Its Application, SETA 2004, LNCS, vol.3486, Springer-Verlag, Berlin, 2005, pp.185-196.
16. M. Stamp and C.F. Martin, An algorithm for the k-error linear complexity of binary sequences of period 2^n, IEEE Trans. Inform. Theory, 39(1993), pp.1398-1401.

On the Expected Value of the Joint 2-Adic Complexity of Periodic Binary Multisequences*

Honggang Hu[1], Lei Hu[2], and Dengguo Feng[1]

[1] State Key Laboratory of Information Security (Institute of Software, Chinese Academy of Sciences), Beijing, 100080, China
hghu@ustc.edu, feng@is.iscas.ac.cn

[2] State Key Laboratory of Information Security (Graduate School of Chinese Academy of Sciences), Beijing, 100049, China
hu@is.ac.cn

Abstract. Recently people show some interest in the word-based stream ciphers. The theory of such stream ciphers requires the study of the complexity of multisequences. The 2-adic complexity is the FCSR analog of the linear complexity, and it is very useful in the study of the security of stream ciphers. The improved version of 2-adic complexity—the symmetric 2-adic complexity was presented in 2004 which is a better measure for the cryptographic strength of binary sequences. In this paper, we derive the expected value of the joint 2-adic complexity of periodic binary multisequences. A nontrivial lower bound for the expected value of the joint symmetric 2-adic complexity of periodic binary multisequences is also given.

1 Introduction

By adding a memory to linear feedback shift register, Klapper and Goresky introduced feedback with carry shift register (FCSR) in [1] (see also [2,3,4,6]). Based on the new feedback architecture they proposed the concept of the 2-adic complexity which is very useful in the study of the security of stream ciphers. An FCSR is determined by r coefficients $q_1, q_2, ..., q_r$, where $q_i \in \{0, 1\}, i = 1, 2, ..., r$, and an initial memory m_{r-1}. If the contents of the register at any time are $(a_{r-1}, a_{r-2}, ..., a_1, a_0)$ and the memory is m, then the operation of the shift register is defined as follows:

1. Take the integer sum $\sigma = \sum_{k=1}^{r} q_k a_{r-k} + m$;
2. Shift the contents one step to the right, while outputting the rightmost bit a_0;
3. Put $a_r \equiv \sigma \bmod 2$ into the leftmost cell of the shift register;
4. Replace m with $m = (\sigma - a_r)/2$;

The integer $q = -1 + q_1 2 + q_2 2^2 + ... + q_r 2^r$ is called the connection integer of the FCSR.

* This work was supported in part by the National Natural Science Foundation of China (No. 90604011 and 90604036) and the National Grand Fundamental Research 973 Program of China (No. 2004CB318004).

G. Gong et al. (Eds.): SETA 2006, LNCS 4086, pp. 199–208, 2006.

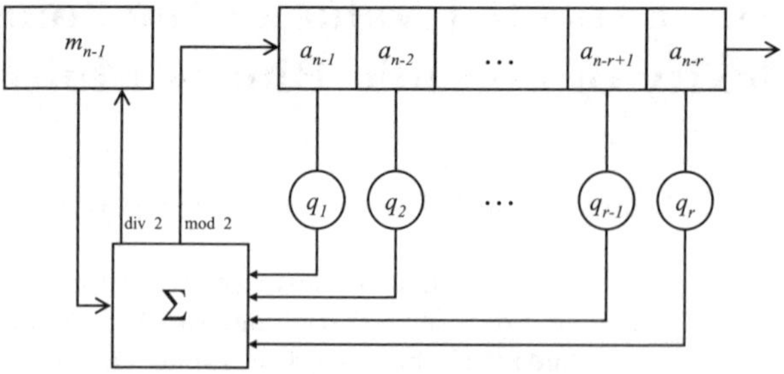

Fig. 1. Feedback with carry shift register

In [1,2,3,4,6], Klapper and Goresky discussed some basic properties of FCSR sequences, such as their periods, rational expressions, exponential representations, rational approximation algorithms and their randomness. The summation cipher proposed by Rueppel [7] is insecure under the attack of the rational approximation algorithm (an analog of the Berlekamp-Massey algorithm, see [6]). In [8], Meidl presented an FCSR analog of (extended) Games-Chan algorithm which can efficiently give an upper bound for the 2-adic complexity of a p^n-periodic binary sequence.

Any infinite binary sequence $S = \{s_i\}_{i=0}^{\infty}$ can be identified with the element $\alpha = \sum_{i=0}^{\infty} s_i 2^i$ in the ring Z_2 of 2-adic numbers. For a comprehensive survey of p-adic numbers the reader is referred to [9]. The sequence S is eventually periodic if and only if the 2-adic number α is rational, i.e., there exist integers p, q such that $\alpha = -p/q \in Z_2$. In particular, if S is strictly periodic with minimal period T, then

$$\alpha = \sum_{i=0}^{\infty} s_i 2^i = -\frac{\sum_{i=0}^{T-1} s_i 2^i}{2^T - 1} = -\frac{p}{q},$$

where $0 \le p \le q$. $p = q$ if and only if S is the all-1 sequence. We call $-p/q$ the rational expression of S. If gcd(p, q)=1, $-p/q$ is called the reduced rational expression of S. In that case, $T = ord_q(2)$, where $ord_q(2)$ is the minimal integer t such that $2^t \equiv 1 \pmod{q}$, and q is the connection integer of the smallest FCSR [6], i.e. the FCSR with minimal number r of coefficients q_i which can generate the binary sequence S. From now on we only consider strictly periodic sequences, and we just call them periodic sequences for simplicity.

Definition 1. [6] *Let S be a periodic binary sequence with reduced rational expression $-p/q$, then the 2-adic complexity $\Phi(S)$ of S is the real number $log_2 q$.*

Remark 1. If S is the all-0 sequence or the all-1 sequence, then $\Phi(S) = 0$.

The improved version of 2-adic complexity—the symmetric 2-adic complexity is given below, which is a better measure for the cryptographic strength of binary sequences.

Definition 2. [5] *Let S be a periodic binary sequence with reduced rational expression $-p/q$. Let $\widehat{S}$ be its inverse sequence with reduced rational expression*

$-p'/q'$. Then the symmetric 2-adic complexity $\overline{\Phi}(S)$ of S is the real number $min(log_2 q, log_2 q')$.

Using the rational approximation algorithm, only a knowledge of $\lceil 2\overline{\Phi}(S)\rceil+2$ bits is sufficient to reproduce the sequence S (see [6], Theorem 10.2). Therefore any binary sequence with low symmetric 2-complexity is insecure for cryptographic applications.

The linear complexity of a periodic sequence is the least order of a linear recurrence relation that the sequence satisfies [11]. In [7], Rueppel conjectured that periodic sequences have expected linear complexity close to the period T. In [12], Meidl and Niederreiter determined the expected value of the linear complexity of T-periodic sequences explicitly by the generalized discrete Fourier transform and confirmed Rueppel's conjecture. Moreover, they also determined the expected value of the joint linear complexity of T-periodic multisequences explicitly in [13]. Hu and Feng determined the expected value of the 2-adic complexity of periodic binary sequences, and gave a nontrivial lower bound for the expected value of the symmetric 2-adic complexity of periodic binary sequences in [5]. In this paper, we derive the expected value of the joint 2-adic complexity of periodic binary multisequences, and give a nontrivial lower bound for the expected value of the joint symmetric 2-adic complexity of periodic binary multisequences.

The underlying stochastic model is that every binary sequence with period T has the same probability $1/2^T$ in the following sections.

2 The Expected Value of the Joint 2-Adic Complexity of Periodic Binary Multisequences

From now on, we call q a connection integer of a sequence if the sequence can be generated by an FCSR with connection integer q.

Let $S = \{s_i\}_{i=0}^{\infty}$ be a periodic binary sequence with period T. We can describe S by the notation $S = (s_0, s_1, ..., s_{T-1})^{\infty}$ because S is completely determined by its first T terms. We define $S^T(x)$ to be the polynomial

$$S_T(x) = s_0 + s_1 x + ... + s_{T-1}x^{T-1}.$$

Suppose that $2^T - 1 = p_1^{e_1} p_2^{e_2} ... p_h^{e_h}$, where p_i are prime numbers with $p_1 < p_2 < ... < p_h, e_i \geq 1, i = 1, 2, ..., h$. Let Γ be the set of binary sequences with period T except the all-1 sequence. If $S \in \Gamma$, then we have

$$\alpha = -\frac{S_T(2)}{2^T - 1} = -\frac{a}{p_{i_1}^{f_{i_1}} p_{i_2}^{f_{i_2}} ... p_{i_t}^{f_{i_t}}}, \tag{1}$$

where $1 \leq t \leq h, 1 \leq i_1 < i_2 < ... < i_t \leq h$, $1 \leq f_{i_j} \leq e_{i_j}$, $0 \leq a < \prod_{j=1}^{t} p_{i_j}^{f_{i_j}}$. $p_{i_1}^{f_{i_1}} p_{i_2}^{f_{i_2}} ... p_{i_t}^{f_{i_t}}$ is a connection integer of S. Moreover, if $gcd(a, \prod_{j=1}^{t} p_{i_j}^{f_{i_j}}) = 1$, $p_{i_1}^{f_{i_1}} p_{i_2}^{f_{i_2}} ... p_{i_t}^{f_{i_t}}$ is the minimal connection integer of S.

By (1), there are $p_{i_1}^{f_{i_1}} p_{i_2}^{f_{i_2}} ... p_{i_t}^{f_{i_t}}$ binary sequences of period T with $p_{i_1}^{f_{i_1}} p_{i_2}^{f_{i_2}} ... p_{i_t}^{f_{i_t}}$ as a connection integer except for the all-1 sequence. In particular, there are $\prod_{j=1}^{t} \phi(p_{i_j}^{f_{i_j}})$ binary sequences of period T with $p_{i_1}^{f_{i_1}} p_{i_2}^{f_{i_2}} ... p_{i_t}^{f_{i_t}}$ as the minimal connection integer, where ϕ is the Euler function.

For any $1 \leq t \leq h, 1 \leq i_1 < i_2 < ... < i_t \leq h$, $1 \leq f_{i_j} \leq e_{i_j}$, let

$$\Gamma_{f_{i_1}, f_{i_2}, ..., f_{i_t}} = \{(a_{i_1}, a_{i_2}, ..., a_{i_t}) | 0 \leq a_{i_j} < p_{i_j}^{f_{i_j}}, j = 1, 2, ..., t\},$$

and let $\Omega_{f_{i_1}, f_{i_2}, ..., f_{i_t}}$ be the set of periodic binary sequences of period T with $p_{i_1}^{f_{i_1}} p_{i_2}^{f_{i_2}} ... p_{i_t}^{f_{i_t}}$ as a connection integer except the all-1 sequence.

We define the mapping

$$\rho : \Gamma_{f_{i_1}, f_{i_2}, ..., f_{i_t}} \mapsto \Omega_{f_{i_1}, f_{i_2}, ..., f_{i_t}}$$

as follows: for any $(a_{i_1}, a_{i_2}, ..., a_{i_t}) \in \Gamma_{f_{i_1}, f_{i_2}, ..., f_{i_t}}$, by Chinese Remainder Theorem [10], there exists unique a such that $0 \leq a < \prod_{j=1}^{t} p_{i_j}^{f_{i_j}}$ and $a \equiv a_{i_j} (mod\, p_{i_j}^{f_{i_j}})$, $j = 1, 2, ..., t$. Let S be the sequence with rational expression

$$-\frac{a}{p_{i_1}^{f_{i_1}} p_{i_2}^{f_{i_2}} ... p_{i_t}^{f_{i_t}}}.$$

Put $\rho((a_{i_1}, a_{i_2}, ..., a_{i_t})) = S$.

We have the following lemma.

Lemma 1. *With the notations as above, ρ is a one-to-one correspondence between $\Gamma_{f_{i_1}, f_{i_2}, ..., f_{i_t}}$ and $\Omega_{f_{i_1}, f_{i_2}, ..., f_{i_t}}$. Moreover, for any $(a_{i_1}, a_{i_2}, ..., a_{i_t}) \in \Gamma_{f_{i_1}, f_{i_2}, ..., f_{i_t}}$, $p_{i_1}^{f_{i_1}} p_{i_2}^{f_{i_2}} ... p_{i_t}^{f_{i_t}}$ is the minimal connection integer of $\rho((a_{i_1}, a_{i_2}, ..., a_{i_t}))$ if and only if $gcd(a_{i_j}, p_{i_j}^{f_{i_j}}) = 1, j = 1, 2, ..., t$.*

Now we consider the multisequence case.

Lemma 2. *For any $m \geq 1$, let $S_1, S_2, ..., S_m$ be m binary sequences with period T, and their reduced rational expressions are $-p_1/q_1, -p_2/q_2, ..., -p_m/q_m$ respectively. Then $q = lcm(q_1, q_2, ..., q_m)$ is the smallest integer such that there exists an FCSR with connection integer q which can generate $S_1, S_2, ..., S_m$ simultaneously.*

Remark 2. $q = lcm(q_1, q_2, ..., q_m) \leq 2^T - 1$.

With the notations as in Lemma 2, we call $\log_2 lcm(q_1, q_2, ..., q_m)$ the 2-adic complexity of the m sequences $S_1, S_2, ..., S_m$, and denote it by $\Phi(S_1, S_2, ..., S_m)$.

For any $f_1, f_2, ..., f_h$ satisfying $1 \leq f_{i_j} \leq e_{i_j}, 1 \leq i_1 < i_2 < ... < i_t \leq h, 1 \leq t \leq h, f_j = 0, j \neq i_1, ..., i_t$, by Lemmas 1 and 2, the number $N_m(f_1, f_2, ..., f_h)$ of sequences $S_1, S_2, ..., S_m \in \Gamma$ such that $\Phi(S_1, S_2, ..., S_m) = \sum_{i=1}^{h} f_i \log_2 p_i$ satisfies

$$N_m(f_1, f_2, ..., f_h) = \prod_{k=1}^{t} [(p_{i_k}^{f_{i_k}})^m - (p_{i_k}^{f_{i_k}-1})^m] = \prod_{k=1}^{h} \psi(p_k, m, f_k), \qquad (2)$$

where $\psi(p_k, m, f_k) = p_k^{mf_k} - p_k^{m(f_k-1)}$ if $f_k \geq 1$; $\psi(p_k, m, f_k) = 1$ if $f_k = 0$.

Lemma 3. *For any real number* $x \neq 0, 1$,

$$\sum_{n=1}^{e} nx^{n-1} = \frac{1-(e+1)x^e + ex^{e+1}}{(x-1)^2}.$$

Lemma 4. *Suppose that* $T \geq 2$ *and* $2^T - 1 = p_1^{e_1} p_2^{e_2} ... p_h^{e_h}$, *where* p_i *are prime numbers with* $p_1 < p_2 < ... < p_h, e_i \geq 1, i = 1, 2, ..., h$. *For any* $m \geq 1$, *we have*

$$\sum_{S_1,S_2,...,S_m \in \Gamma} \Phi(S_1, S_2, ..., S_m)$$
$$= (2^T-1)^m \log_2(2^T-1) - (2^T-1)^m \sum_{i=1}^{h} \frac{(1-p_i^{-me_i})\log_2 p_i}{p_i^m - 1}.$$

Proof. We have

$$\sum_{S_1,S_2,...,S_m \in \Gamma} \Phi(S_1, S_2, ..., S_m)$$
$$= \sum_{f_1=0}^{e_1} \sum_{f_2=0}^{e_2} \cdots \sum_{f_h=0}^{e_h} N_m(f_1, f_2, ..., f_h)(f_1 \log_2 p_1 + f_2 \log_2 p_2 + ... + f_h \log_2 p_h).$$

By (2),

$$\sum_{S_1,S_2,...,S_m \in \Gamma} \Phi(S_1, S_2, ..., S_m)$$
$$= \sum_{f_1=0}^{e_1} \sum_{f_2=0}^{e_2} \cdots \sum_{f_h=0}^{e_h} \prod_{k=1}^{h} \psi(p_k, m, f_k)(f_1 \log_2 p_1 + f_2 \log_2 p_2 + ... + f_h \log_2 p_h)$$
$$= \sum_{i=1}^{h} \sum_{f_i=0}^{e_i} \psi(p_i, m, f_i) f_i \log_2 p_i \prod_{k=1, k\neq i}^{h} \sum_{f_k=0}^{e_k} \psi(p_k, m, f_k)$$
$$= \sum_{i=1}^{h} p_1^{me_1} ... p_{i-1}^{me_{i-1}} p_{i+1}^{me_{i+1}} ... p_h^{me_h} \sum_{f_i=0}^{e_i} \psi(p_i, m, f_i) f_i \log_2 p_i$$
$$= \sum_{i=1}^{h} p_1^{me_1} ... p_{i-1}^{me_{i-1}} p_{i+1}^{me_{i+1}} ... p_h^{me_h} (p_i^m - 1) \log_2 p_i \sum_{f_i=1}^{e_i} f_i p_i^{m(f_i-1)}.$$

By Lemma 3,

$$\sum_{S_1,S_2,...,S_m \in \Gamma} \Phi(S_1, S_2, ..., S_m)$$
$$= (2^T-1)^m \sum_{i=1}^{h} \log_2 p_i \frac{e_i p_i^{m(e_i+1)} - (e_i+1)p_i^{me_i} + 1}{(p_i^m - 1)p_i^{me_i}}$$

$$= (2^T-1)^m \sum_{i=1}^{h} \left[e_i \log_2 p_i - \frac{(1-p_i^{-me_i}) \log_2 p_i}{p_i^m - 1} \right]$$

$$= (2^T-1)^m \log_2(2^T-1) - (2^T-1)^m \sum_{i=1}^{h} \frac{(1-p_i^{-me_i}) \log_2 p_i}{p_i^m - 1}.$$

Let $E_{T,n}$ denote the expected value of the joint 2-adic complexity of n periodic binary sequences with period T. We have the following theorem.

Theorem 1. *Suppose that $T \geq 2$ and $2^T - 1 = p_1^{e_1} p_2^{e_2} ... p_h^{e_h}$, where p_i are prime numbers with $p_1 < p_2 < ... < p_h, e_i \geq 1, i = 1, 2, ..., h$. Then the expected value $E_{T,n}$ of the joint 2-adic complexity of n binary sequences of period T satisfies*

$$E_{T,n} = \log_2(2^T - 1) - \sum_{i=1}^{h} \sum_{k=1}^{e_i} (\frac{2^T - 1}{2^T p_i^k} + \frac{1}{2^T})^n \log_2 p_i.$$

Proof. We denote the all-1 sequence by $\overrightarrow{1}$. Then we have

$$\begin{aligned}
& 2^{nT} \cdot E_{T,n} \\
= & \sum_{S_1, S_2, ..., S_n \in \Gamma \cup \{\overrightarrow{1}\}} \Phi(S_1, S_2, ..., S_n) \\
= & \sum_{m=1}^{n} \binom{n}{n-m} \sum_{S_1, S_2, ..., S_m \in \Gamma} \Phi(S_1, S_2, ..., S_m) \\
= & \sum_{m=1}^{n} \binom{n}{m} \left[(2^T-1)^m \log_2(2^T-1) - (2^T-1)^m \sum_{i=1}^{h} \frac{(1-p_i^{-me_i}) \log_2 p_i}{p_i^m - 1} \right] \\
= & (2^{Tn} - 1) \log_2(2^T - 1) - \sum_{i=1}^{h} \left[\sum_{k=1}^{e_i} (\frac{2^T-1}{p_i^k} + 1)^n - e_i \right] \log_2 p_i \\
= & 2^{Tn} \log_2(2^T-1) - \sum_{i=1}^{h} \sum_{k=1}^{e_i} (\frac{2^T-1}{p_i^k} + 1)^n \log_2 p_i.
\end{aligned}$$

Thus

$$E_{T,n} = \log_2(2^T - 1) - \sum_{i=1}^{h} \sum_{k=1}^{e_i} (\frac{2^T - 1}{2^T p_i^k} + \frac{1}{2^T})^n \log_2 p_i.$$

Corollary 1. *If T is as above, then the expected value $E_{T,n}$ of the joint 2-adic complexity of binary sequences of period T satisfies*

$$E_{T,n} > \log_2(2^T - 1) - \sum_{i=1}^{h} \frac{2^n}{p_i^n - 1} \log_2 p_i.$$

Proof. With the notations as in Theorem 1, because $p_i^k < 2^T$, we have

$$\sum_{i=1}^{h}\sum_{k=1}^{e_i}(\frac{2^T-1}{2^T p_i^k}+\frac{1}{2^T})^n \log_2 p_i < \sum_{i=1}^{h}\sum_{k=1}^{e_i}(\frac{2}{p_i^k})^n \log_2 p_i$$
$$= \sum_{i=1}^{h}\frac{2^n(1-p_i^{-ne_i})}{p_i^n-1}\log_2 p_i$$
$$< \sum_{i=1}^{h}\frac{2^n}{p_i^n-1}\log_2 p_i.$$

By Theorem 1, the result follows.

In particular, if 2^T-1 is prime, we have the following corollary.

Corollary 2. *If 2^T-1 is prime, then the expected value E_T^n of the joint 2-adic complexity of binary sequences of period T is given by*

$$E_{T,n} = [1-\frac{1}{2^{n(T-1)}}]\log_2(2^T-1).$$

Proof. With the notations as in Theorem 1, because 2^T-1 is prime, we have $h=1, e_1=1$, and $p_1=2^T-1$. The result follows from Theorem 1.

3 The Expected Value of the Joint Symmetric 2-Adic Complexity of Periodic Binary Multisequences

For any binary sequence $S=(s_0,s_1,...,s_{T-1})^\infty$ of period T, let $\widehat{S}$ be the inverse sequence of S, i.e., $\widehat{S}=(s_{T-1},s_{T-2},...,s_1,s_0)^\infty$. For any n binary sequences $S_1,S_2,...,S_n$ of period T, their joint symmetric 2-adic complexity is defined to be $min(\Phi(S_1,S_2,...,S_n),\Phi(\widehat{S_1},\widehat{S_2},...,\widehat{S_n}))$, and we denote it by $\overline{\Phi}(S_1,S_2,...,S_n)$. One can check that $\overline{\Phi}(S_1,S_2,...,S_n)\geq\overline{\Phi}(S_i), i=1,2,\cdots,n$.

Lemma 5. [5] *There are at least $2\phi(2^T-1)-(2^T-1)$ binary sequences S of period T such that the symmetric 2-adic complexity of S is $\overline{\Phi}(S)=\log_2(2^T-1)$.*

Let $\overline{E}_{T,n}$ denote the expected value of the joint symmetric 2-adic complexity of n periodic binary sequences with period T. For any $i, 1\leq i\leq n$, if the symmetric 2-adic complexity of S_i is $\overline{\Phi}(S_i)=\log_2(2^T-1)$, then $\overline{\Phi}(S_1,S_2,...,S_n) = \log_2(2^T-1)$.

Let $\mathcal{A}=\{S\in\Gamma \mid \overline{\Phi}(S)=\log_2(2^T-1)\}$, and $\mathcal{B}=\{S\in\Gamma \mid gcd(S_T(2),2^T-1)>1\}$. Then the cardinality of $\mathcal{A}$ is at least $2\phi(2^T-1)-(2^T-1)$, and the cardinality of $\mathcal{B}$ is $2^T-1-\phi(2^T-1)$. Moreover, $\mathcal{A}\cap\mathcal{B}=\varnothing$, where $\varnothing$ is the empty set. Put

$$\mathcal{H}_1=\{(S_1,S_2,...,S_n)\mid S_1\in\mathcal{A}, S_j\in\Gamma\cup\{\overrightarrow{1}\}, j=2,3,\cdots,n\},$$
$$\mathcal{H}_n=\{(S_1,S_2,...,S_n)\mid S_j\in\mathcal{B}, j=1,2,\cdots,n-1, S_n\in\mathcal{A}\}.$$

For any $2 \leq i \leq n-1$, put $\mathcal{H}_i = \{(S_1, S_2, ..., S_n) \mid S_j \in \mathcal{B}, j = 1, 2, \cdots, i-1, S_i \in \mathcal{A}, S_j \in \Gamma \cup \{\overrightarrow{1}\}, j = i+1, \cdots, n\}$. Then for any $1 \leq i < j \leq n$, we have $\mathcal{H}_i \cap \mathcal{H}_j = \varnothing$.

Theorem 2. *Suppose that $T \geq 2$ and $2^T - 1 = p_1^{e_1} p_2^{e_2} ... p_h^{e_h}$, where p_i are prime numbers with $p_1 < p_2 < ... < p_h, e_i \geq 1, i = 1, 2, ..., h$. Then the expected value $\overline{E}_{T,n}$ of the joint symmetric 2-adic complexity of n binary sequences of period T satisfies*

$$\overline{E}_{T,n} \geq \sum_{k=0}^{n-1} (\frac{2^T - 1}{2^T})^{k+1} \left[1 - (1 - \frac{1}{p_1}) \cdots (1 - \frac{1}{p_h})\right]^k \cdot \left[2(1 - \frac{1}{p_1}) \cdots (1 - \frac{1}{p_h}) - 1\right] \log_2(2^T - 1).$$

Proof. Let $N = 2^T - 1, a = \phi(2^T - 1)$. By the analysis above, we have

$$\begin{aligned} 2^{nT} \cdot \overline{E}_{T,n} &\geq [(2a - N)2^{(n-1)T} + (N - a)(2a - N)2^{(n-2)T} + \\ &\quad \cdots + (N - a)^{n-1}(2a - N)] \log_2(2^T - 1) \\ &= \sum_{k=0}^{n-1} (N - a)^k (2a - N) 2^{(n-1-k)T} \log_2(2^T - 1). \end{aligned}$$

Thus

$$\begin{aligned} \overline{E}_{T,n} &\geq \sum_{k=0}^{n-1} (N - a)^k (2a - N) 2^{(-1-k)T} \log_2(2^T - 1) \\ &= \sum_{k=0}^{n-1} (\frac{2^T - 1}{2^T})^{k+1} \left[1 - (1 - \frac{1}{p_1}) \cdots (1 - \frac{1}{p_h})\right]^k \\ &\quad \cdot \left[2(1 - \frac{1}{p_1}) \cdots (1 - \frac{1}{p_h}) - 1\right] \log_2(2^T - 1). \end{aligned}$$

Remark 3. Because $p_1 < p_2 < ... < p_h$, if T is large, we have

$$\begin{aligned} \overline{E}_{T,n} &> \sum_{k=0}^{n-1} (\frac{2^T - 1}{2^T})^{k+1} \left[1 - (1 - \frac{1}{p_h})^h\right]^k \cdot \left[2(1 - \frac{1}{p_1})^h - 1\right] \log_2(2^T - 1) \\ &\approx \sum_{k=0}^{n-1} \left[1 - (1 - \frac{1}{p_h})^h\right]^k \cdot \left[2(1 - \frac{1}{p_1})^h - 1\right] \log_2(2^T - 1). \end{aligned}$$

However, if

$$(1 - \frac{1}{p_1}) \cdots (1 - \frac{1}{p_h}) \leq \frac{1}{2},$$

the result of Theorem 2 is meaningless.

In particular, if $2^T - 1$ is prime, we can get the exact value.

Proposition 1. *If $2^T - 1$ is prime, then the expected value $\overline{E}_{T,n}$ of the joint symmetric 2-adic complexity of binary sequences of period T is given by*

$$\overline{E}_{T,n} = [1 - \frac{1}{2^{n(T-1)}}] \log_2(2^T - 1).$$

Proof. Because $2^T - 1$ is prime, if S isn't the all-0 sequence or the all-1 sequence, the symmetric 2-adic complexity of S is $\log_2(2^T - 1)$. Hence the result follows.

4 Concluding Remarks

The linear complexity of the all-1 sequence is 1, but the 2-adic complexity of the all-1 sequence is 0. This difference makes the computation of the expected value of the joint 2-adic complexity of periodic binary multisequences more difficult than that of the linear complexity case.

For the expected value of the joint symmetric 2-adic complexity, we haven't found a method for the exact value yet. We give a nontrivial lower bound, and the method is just the generalization of that in [5]. In order to derive the exact value for any T, we need to know the factorization of $S_T(2)$ and $\widehat{S}_T(2)$ simultaneously for any binary sequence S with period T. However, in our opinion, it is a difficult problem.

Acknowledgment

The authors wish to express their deep gratitude to the anonymous reviewers for many helpful comments and suggestions.

References

1. A. Klapper, M. Goresky, 2-adic shift registers, in: R. Anderson (Ed.), Fast Software Encryption, Lecture Notes in Computer Science, Vol. 809, Springer, New York, 1994, pp. 174-178.
2. M. Goresky, A. Klapper, Feedback registers based on ramified extensions of the 2-adic numbers, Advances in Cryptology-Eurocrypt'94, LNCS, vol, 950, Springer-Verlag, Berlin, 1995, pp. 215-222
3. M. Goresky, A. Klapper, Large periods nearly de Bruijn FCSR sequences, Advances in Cryptology-Eurocrypt'95, LNCS, vol. 921, Springer-Verlag, Berlin, 1995, pp. 263-273
4. M. Goresky, A. Klapper, Cryptanalysis based on 2-adic rational approximation, Advances in Cryptology-Crypt'95, LNCS, vol. 963, Springer-Verlag, Berlin, 1995, pp. 262-273
5. H. Hu, D. Feng, On the 2-adic complexity and the k-error 2-adic complexity of periodic binary sequences, Proceedings of SETA'04, Lecture Notes in Computer Science 3486, Springer-Verlag, pp. 185-196, 2005
6. A. Klapper, M. Goresky, Feedback shift registers, 2-adic span, and combiners with memory, J. Cryptology, vol. 10, pp. 111-147, 1997.

7. R. A. Rueppel, Analysis and Design of Stream Cipher. Berlin, Germany: Springer-Verlag, 1986
8. W. Meidl, Extended Games-Chan algorithm for the 2-adic complexity of FCSR-sequences,Theoretical Computer Science, Volume 290, 2003, Pages 2045-2051
9. N. Koblitz, p-Adic Numbers, p-Adic Analysis, and Zeta Functions, Graduate Texts in Mathematics, Vol. 58, Springer, New York, 1984.
10. K. Ireland and M. Rosen, A Classical Introduction to Modern Number Theory (second edition), in GTM. New York: Springer Verlag, 1990, vol. 84
11. James L. Massey, Shift-register synthesis and BCH decoding, IEEE Trans. Info. Theory, vol. IT-15, pp. 122-127, January 1969.
12. W. Meidl, H. Niederreiter, On the expected value of the linear complexity and the k-error linear complexity of periodic sequences, IEEE Trans. Inform. Theory 48 (2002), pp. 2817-2825.
13. W. Meidl, H. Niederreiter, The expected value of the joint linear complexity of periodic multisequences, Journal of Complexity 19 (2003), pp. 61-72.

On the Classification of Periodic Binary Sequences into Nonlinear Complexity Classes

George Petrides[1,⋆] and Johannes Mykkeltveit[2]

[1] School of Mathematics, University of Manchester
P.O. Box 88, Sackville Street, Manchester M60 1QD, UK
g.petrides@maths.manchester.ac.uk
[2] IRIS, International Research Institute of Stavanger
Thormohlensgt. 55, N-5008 Bergen, Norway
Johannes.Mykkeltveit@irisresearch.no

Abstract. In this paper we investigate the notion of nonlinear complexity, or maximal order complexity as it was first defined in 1989 [4]. Our main purpose is to begin classification of periodic binary sequences into nonlinear complexity classes. Previous work on the subject also includes approximation of the size of each class, found in [2]. Once the classification is completed, we can use it to show how to perform checks for short cycles in large nonlinear feedback shift registers using our proposed algorithm.

Keywords: Nonlinear complexity, nonlinear feedback shift register, short cycles.

1 Introduction

The classical complexity measure assessing the cryptographic strength of binary sequences used in stream ciphers is the *linear complexity*. It can be calculated using the well known Berlekamp-Massey algorithm [6] and is used in statistical tests for the randomness of sequences.

In this paper we investigate the generalised notion of *nonlinear complexity* (or maximum order complexity as first introduced in [4]) which can be calculated using, for example, the directed acyclic word graph [4]. In particular, we try to classify periodic binary sequences of given period into nonlinear complexity classes.

In [2] an approximate number of sequences in each class was calculated and used in finding the approximate probability distribution of nonlinear complexity. Our results, though incomplete, give the exact number for the cases considered. The cases not yet dealt with are left for future work, with the idea and method of approach having been established.

A complete classification will be useful in the implementation of an algorithm checking for short cycles in large *nonlinear feedback shift registers* (NLFSRs, see [3] for definition). In [5] it is claimed that for a given large NLFSR it is hard to check whether short cycles have been embedded by the given method: brute

⋆ Supported by the Marie Curie Fellowship HPMT-CT-2001-00260.

G. Gong et al. (Eds.): SETA 2006, LNCS 4086, pp. 209–222, 2006.

force is inefficient due to largeness, and so are the two algorithms which are given for such a check, based on algebraic approach. Our proposed algorithm checks for short cycles, regardless of whether they are embedded or not.

Another use for this classification (once complete) can be found in the construction of statistical tests for the randomness of sequences, as shown for example in [2], thus further upgrading the level of interest of nonlinear complexity from just theoretical to also practical. For instance, we could say that a sequence is considered random if it belongs to a large nonlinear complexity class.

We begin in Section 2 with the definition of nonlinear complexity (a name preferred to maximum order complexity as it is more easily seen as the counterpart to linear complexity). In Section 3 we attempt to classify the sequences into nonlinear complexity classes, before presenting the algorithm checking for short cycles in Section 4. We finish with our conclusions in Section 5.

2 Preliminaries

In this section we give the definition of nonlinear complexity and a brief discussion of how to proceed in the next section.

Definition 1. *A sequence s is periodic if there exists a positive integer r such that $s_{i+r} = s_i$, for $i = 0, 1, \dots$, and aperiodic otherwise. The smallest such positive integer r is called the period of s and denoted by $p(s)$.*

Definition 2. *The nonlinear complexity* $\mathrm{C}(s)$ *of a periodic sequence s is the least integer k such that all k-vectors $(s_q, s_{q+1}, \dots, s_{q+k-1})$, $q = 0, 1, \dots, p(s) - 1$ are different. Indices are reduced modulo $p(s)$.* $\mathrm{C}(s)$ *is defined to be 1 if $p(s) = 1$.*

We will denote by $nlin(k, e)$ the number of binary sequences of nonlinear complexity k and period e.

Definition 3. *A binary necklace of length l is an equivalence class of binary strings of length l under rotation. It is periodic if the strings it contains are periodic, and aperiodic otherwise. In the periodic case we have that the period e of the strings divides l and $e/l > 1$.*

In order to classify the binary sequences of period e into nonlinear complexity classes we have to consider a representative from each binary aperiodic necklace of length e. This can be deduced from the fact that all members of a binary necklace have the same nonlinear complexity [4]. Therefore, throughout the paper, s will denote the repeating part of the binary periodic sequence $(s_0 s_1 s_2 \dots s_{e-1})^\infty$ of period $p(s) = e$. Also, all indices will be reduced modulo e.

Proposition 1. [4] *For any integer* e*, we have that $nlin(k, e) = 0$, where $1 \leq k < \lceil \log_2(e) \rceil$.*

The total number of binary aperiodic necklaces of length e is well known [1,7] to be equal to the number of irreducible binary polynomials of degree e. Hence

$$\sum_{k=\lceil \log_2(e) \rceil}^{e} nlin(k,e) = \frac{1}{e} \sum_{d|e} \mu\left(\frac{e}{d}\right) 2^d \quad , \tag{1}$$

where μ denotes the Möbius function.

Table 1 in the Appendix tabulates the first values of $nlin(e-\gamma, e)$, as found by exhaustive search. The sum of each row is described by (1). Our aim is to find a general formula for each individual entry of the table.

3 Determining $nlin(e-\gamma, e)$

As discussed in the previous section, consider $s = s_0 s_1 s_2 \ldots s_{e-1}$. For s to have complexity $\mathrm{C}(s) = e - \gamma$, where $0 \leq \gamma \leq e - \lceil \log_2(e) \rceil$, it would mean, by definition, that all $(e-\gamma)$-vectors $(s_q, s_{q+1}, \ldots, s_{q-1-\gamma})$, where $0 \leq q \leq e-1$, are different, and also that at least one pair of the $(e-1-\gamma)$-vectors

$$\begin{array}{ll}
s_0 s_1 \ldots s_{e-2-\gamma} & (\mathcal{S}_0) \\
s_1 s_2 \ldots s_{e-1-\gamma} & (\mathcal{S}_1) \\
\vdots & \vdots \\
s_i s_{i+1} \ldots s_{i-2-\gamma} & (\mathcal{S}_i) \\
\vdots & \vdots \\
s_{e-1} s_0 \ldots s_{e-3-\gamma} & (\mathcal{S}_{e-1})
\end{array}$$

is the same (otherwise, if none are equal we would have $\mathrm{C}(s) \leq e-1-\gamma$, and if a triplet or more are equal, we would have that at least two of the $(e-\gamma)$-vectors are equal and so $\mathrm{C}(s) \geq e+1-\gamma$).

Without loss of generality, we consider the $e-1$ cases of $(\mathcal{S}_0)$ being the same as $(\mathcal{S}_i)$, $1 \leq i \leq e-1$:

$$\begin{array}{lllll}
(\mathcal{S}_0) = (\mathcal{S}_1) & \Rightarrow s_0 = s_1, & s_1 = s_2, & \ldots, & s_{e-2-\gamma} = s_{e-1-\gamma} \\
(\mathcal{S}_0) = (\mathcal{S}_2) & \Rightarrow s_0 = s_2, & s_1 = s_3, & \ldots, & s_{e-2-\gamma} = s_{e-\gamma} \\
\vdots & & & \vdots & \\
(\mathcal{S}_0) = (\mathcal{S}_i) & \Rightarrow s_0 = s_i, & s_1 = s_{i+1}, & \ldots, & s_{e-2-\gamma} = s_{i-2-\gamma} \\
\vdots & & & \vdots & \\
(\mathcal{S}_0) = (\mathcal{S}_{e-1}) & \Rightarrow s_0 = s_{e-1}, & s_1 = s_0, & \ldots, & s_{e-2-\gamma} = s_{e-3-\gamma}
\end{array}$$

Now, suppose $(\mathcal{S}_0) = (\mathcal{S}_i)$ for some $1 \leq i \leq e-1$. To ensure that the $(e-\gamma)$-vectors $(s_0 s_1 \ldots s_{e-1-\gamma})$ and $(s_i s_{i+1} \ldots s_{i-1-\gamma})$ are different, we must also have that $s_{e-1-\gamma} \neq s_{i-1-\gamma}$.

Similarly, for the $(e-\gamma)$-vectors $(s_{e-1}, s_0, s_1, \ldots, s_{e-2-\gamma})$ and $(s_{i-1}, s_i, s_{i+1}, \ldots, s_{i-2-\gamma})$ to be different, we need $s_{e-1} \neq s_{i-1}$.

When $\gamma = 0$, there is only one inequality, namely $s_{e-1} \neq s_{i-1}$. However, as $s_j = s_{j+i}$, where $0 \leq j \leq e-2$, and $i-1 < e-1$ we have

$$s_{i-1} = s_{i+i-1} = s_{2i-1} = \ldots = s_{ki-1} \quad ,$$

for some $k \in \mathbb{N}$. Now, if we let $gcd(e,i) = d$ then, when $k = \frac{e}{d}$ we have

$$s_{i-1} = s_{\frac{e}{d}i-1} = s_{\frac{i}{d}e-1} = s_{e-1} \ ,$$

a contradiction. We have thus provided an alternative proof of the following proposition.

Proposition 2. [4] *$nlin(e,e) = 0$.*

In the sequel we will consider fixed i and γ such that $1 \leq i \leq e-1$ and $1 \leq \gamma \leq e - \lceil \log_2(e) \rceil$.

For $\gamma > 1$, no relations between s_{e-t} and s_{i-t}, $\gamma \geq t \geq 2$, are imposed and thus we can have all possible combinations of equalities and inequalities. The only constraint is that we have to have an even number of inequalities, otherwise we would reach a contradiction, as in the case $\gamma = 0$. However, as will be seen in Proposition 6, this constraint is a necessary but not sufficient condition to avoid such a contradiction.

The total number of possibilities for the choice of relations is

$$\sum_{k=0}^{\lfloor \frac{\gamma-1}{2} \rfloor} \binom{\gamma-1}{2k} = 2^{\gamma-2} \ .$$

Our approach will be to view the relations between s_t and s_{i+t}, $0 \leq t \leq e-1$ in each recursion as a binary vector v of length e, where a 0 would denote an equality and a 1 an inequality. It is obvious that these vectors are of even parity (that is they have an even number of 1's) and we have

$$s_{i+t} = s_t \oplus v_t \ .$$

We will call such a vector v the *relating* vector of the recursion $R_v(e,i)$. The set of all $2^{\gamma-2}$ possible relating vectors for a given e and γ will be denoted by $\mathcal{V}_{e,\gamma}$. If v is of the form

$$v = 0 \ldots 01 \underbrace{0 \ldots 0}_{\gamma-1} 1 \ ,$$

which means we only have equalities, then $R_v(e,i)$ is the same as $R(e,i,\gamma)$ of Definition 4. Note that with this notation the case $\gamma = 1$ is now also covered.

Definition 4. *By recursion $R(e,i,\gamma)$ we mean the following list of relations between the digits of a sequence s of length e:*

$$s_j = s_{i+j}, \ s_{e-1-\gamma} \neq s_{i-1-\gamma}, \ s_{e-1} \neq s_{i-1} \ ,$$

where $j \in \{0, \ldots, e-2\} \setminus \{e-1-\gamma\}$.

Definition 5. *Two sequences of length e are cyclically inequivalent if they do not belong to the same binary necklace.*

Definition 6. *$S(R_v(e,i))$ is the set of cyclically inequivalent sequences which satisfy recursion $R_v(e,i)$ and $|S(R_v(e,i))|$ denotes its cardinality.*

Definition 7. *$S(R_v(e,i),\delta)$ is the set of cyclically inequivalent sequences of nonlinear complexity δ which satisfy recursion $R_v(e,i)$ and $|S(R_v(e,i),\delta)|$ denotes its cardinality.*

Using the definition above, we can now relate $nlin(e-\gamma,e)$ and recursions $R_v(e,i)$ in the following obvious theorem.

Theorem 1. $nlin(e-\gamma,e) = \sum_{v\in\mathcal{V}_{e,\gamma}} \sum_{i=0}^{e-1} |S(R_v(e,i),e-\gamma)|$.

Proposition 3. *Let $v_1,\ldots,v_\gamma$ be the relating vectors of recursions $R(e,i,1)$, $\ldots,R(e,i,\gamma)$ respectively. Then, the relating vectors in $\mathcal{V}_{e,\gamma}$ are obtained from all possible sums of v_γ with an even number of vectors v_z, where $z\in\{1\ldots\gamma-1\}$.*

Proof. All vectors $v_1,\ldots,v_\gamma$ have a 1 in the $(e-1)^{th}$ position, and since there's an odd number of them in each sum, the resulting vector will also have a 1 in the $(e-1)^{th}$ position. Also, the $(e-\gamma-1)^{th}$ position will also be a 1 due to v_γ. The remaining positions will depend on which v_z, where $z\in\{1\ldots\gamma-1\}$, are in the sum: the $(e-z-1)^{th}$ positions will be a 1 and the rest will be 0. □

Theorem 2. *Consider the recursion $R_v(e,i)$ and suppose that $v=v^0\oplus v^1\oplus\ldots\oplus v^l\in\mathcal{V}_{e,\gamma}$, where v^0 and v^k are the relating vectors of the recursions $R(e,i,\gamma_0=\gamma)$ and $R(e,i,\gamma_k)$ respectively, $1\le k\le l$, l is even and $\gamma_k<\gamma$ $\forall k$. Let s^0, s^k be sequences satisfying these recursions respectively. Then the sequence $s=s^0\oplus s^1\oplus\ldots\oplus s^l\in S(R_v(e,i))$.*

Proof. We have that $s^m=s_0^m\ldots s_{e-1}^m$, $v^m=v_0^m\ldots v_{e-1}^m$ and $s_{n+i}^m=s_n^m\oplus v_n^m$, where $0\le m\le l$, $0\le n\le e-1$. Therefore,

$$s_{n+i}=\bigoplus_{m=0}^{l} s_{n+i}^m=\bigoplus_{m=0}^{l}(s_n^m\oplus v_n^m)=\bigoplus_{m=0}^{l} s_n^m\bigoplus_{m=0}^{l} v_n^m=s_n\oplus v_n .$$

□

Proposition 3 together with Theorem 2 above suggests that we only need to investigate recursions $R(e,i,\gamma)$ and sums of sequences satisfying them.

3.1 The Recursions $R(e,i,\gamma)$

In this subsection we study the properties of the recursions $R(e,i,\gamma)$. In the way we defined them, we have not imposed any conditions to exclude the case of a third $(e-\gamma-1)$-vector being equal to the other two. In that case, since we are working in binary, at least two of the $(e-\gamma)$-vectors would be the same and therefore the nonlinear complexity of the sequence greater than $(e-\gamma)$. Hence the following two results are immediate.

Proposition 4. $S(R(e,i,\gamma),e-\gamma)\subseteq S(R(e,i,\gamma))$.

Corollary 1. $|S(R(e,i,\gamma))| = \sum_{d=1}^{\gamma} |S(R(e,i,\gamma), e-d)|$.

Definition 8. *Given m such that $0 \le m \le e-1$, the recursion $R^m(e,i,\gamma)$ is the following list of relations between the digits of a sequence s of length e:*

$$s_j = s_{i+j},\ s_{m-1-\gamma} \ne s_{i-1-\gamma+m},\ s_{m-1} \ne s_{i-1+m} \ ,$$

where $j \in \{0,\dots,e-1\} \setminus \{m-1-\gamma, m-1\}$.

The recursion $R^m(e,i,\gamma)$ corresponds to the case $(\mathcal{S}_m) = (\mathcal{S}_{m+i})$ (as compared to $(\mathcal{S}_0) = (\mathcal{S}_i)$ for recursion $R(e,i,\gamma)$). It has as relating vector the m-digits cyclic shift of the relating vector for $R(e,i,\gamma)$.

Proposition 5. *Let sequence $s = s_0 s_1 \dots s_{e-2} s_{e-1} \in S(R(e,i,\gamma))$. Then its m-digits right cyclic shift $s' = s_{e-m} s_{e-m+1} \dots s_{e-m-2} s_{e-m-1} \in S(R^m(e,i,\gamma))$.*

Proof. Obvious. □

It is a consequence of Proposition 5 that recursions $R(e,i,\gamma)$ and $R^m(e,i,\gamma)$, where $0 \le m \le e-1$, share the same properties.

Definition 9. *We will call two recursions equivalent (denoted by $\sim$) if the sequences they define belong to the same binary necklace.*

Lemma 1. $R(e,i,\gamma) \sim R(e,i,e-\gamma)$.

Proof. Let sequence $s = s_0 s_1 \dots s_{e-2} s_{e-1} \in S(R(e,i,\gamma))$. By Proposition 5, its γ-digits right cyclic shift $s' = s_{e-\gamma} s_{e-\gamma+1} \dots s_{e-\gamma-2} s_{e-\gamma-1} \in S(R^\gamma(e,i,\gamma))$, where the recursion $R^\gamma(e,i,\gamma)$ is the following list of relations between the digits of a sequence s of length e:

$$s_j = s_{i+j},\ s_{\gamma-1-\gamma} = s_{e-1} \ne s_{i-1-\gamma+\gamma} = s_{i-1},\ s_{\gamma-1} \ne s_{i-1+\gamma} \ ,$$

where $j \in \{0,\dots,e-1\} \setminus \{e-1, \gamma-1\}$. This is exactly $R(e,i,e-\gamma)$. □

Lemma 2. $R(e,i,\gamma) \sim R(e,e-i,\gamma)$.

Proof. Let sequence $s = s_0 s_1 \dots s_{e-2} s_{e-1} \in S(R(e,i,\gamma))$. By Proposition 5, its $(e-i)$-digits right cyclic shift $s' = s_i s_{i+1} \dots s_{i-2} s_{i-1} \in S(R^{e-i}(e,i,\gamma))$, where the recursion $R^{e-i}(e,i,\gamma)$ is the following list of relations between the digits of a sequence s of length e:

$$s_j = s_{i+j},\ s_{i-1-\gamma+e-i} = s_{e-1-\gamma} \ne s_{e-i-1-\gamma},\ s_{i-1+e-i} = s_{e-1} \ne s_{e-i-1} \ ,$$

where $j \in \{0,\dots,e-1\} \setminus \{e-1-\gamma, e-1\}$. This is exactly $R(e,e-i,\gamma)$. □

Combining the results of the lemmas above we come to the following corollary:

Corollary 2

$$R(e,i,\gamma) = R^i(e,e-i,\gamma) = R^{e-\gamma}(e,i,e-\gamma) = R^{i-\gamma}(e,e-i,e-\gamma) \ .$$

By Proposition 2 and Corollary 2 we deduce that in the sequel, for each e we have to look only at the distinct recursions $R(e,i,\gamma)$ for $1 \le i,\ \gamma \le \lfloor \frac{e}{2} \rfloor$. Proposition 6 below limits them further to those such that $\gcd(e,i) \mid \gamma$.

Given a positive integer n, we will denote by $Div(n)$ the set of divisors of n and by $Div^*(n)$ the set $Div(n) \setminus \{n\}$.

Proposition 6. *If $gcd(e,i) \notin Div(\gamma)$, then $|S(R(e,i,\gamma))| = 0$.*

Proof. $R(e,i,\gamma)$ is given by

$$s_j = s_{i+j},\ s_{e-1-\gamma} \neq s_{i-1-\gamma},\ s_{e-1} \neq s_{i-1}\ ,$$

where $j \in \{0,\dots,e-2\} \setminus \{e-1-\gamma\}$. Let $gcd(e,i) = g$. Then, there exist $x, y \in \mathbb{N}$ such that $i = xg$ and $e = yg$.

First we observe that for $i \neq \gamma$, by definition of $R(e,i,\gamma)$ we have

$$s_{i-\gamma-1} = s_{i+i-\gamma-1} = s_{2i-\gamma-1} = \dots = s_{k_1 i-\gamma-1}\ ,$$

for some $k_1 \in \mathbb{N}$. In the case $i = \gamma$ we just have $s_{e-1} \neq s_{\gamma-1}$.

Also,

$$s_{i-1} = s_{i+i-1} = s_{2i-1} = \dots = s_{k_2 i-1}\ ,$$

for some $k_2 \in \mathbb{N}$. Now, when $k_1 = \frac{e}{g}$ we obtain

$$s_{i-\gamma-1} = s_{\frac{e}{g}i-\gamma-1} = s_{\frac{i}{g}e-\gamma-1} = s_{e-\gamma-1}$$

and when $k_2 = \frac{e}{g}$

$$s_{i-1} = s_{\frac{e}{g}i-1} = s_{\frac{i}{g}e-1} = s_{e-1}\ ,$$

both contradicting the definition of $R(e,i,\gamma)$.

So, what we need to avoid the contradiction is to find $k_1, k_2 < \frac{e}{g}$ such that

$$k_1 i - \gamma - 1 \equiv e - 1 \text{ and } k_2 i - 1 \equiv e - \gamma - 1\ .$$

In that case we would have

$$s_{i-\gamma-1} = \dots = s_{e-1} \neq s_{i-1} = \dots = s_{e-\gamma-1}\ ,$$

satisfying both inequalities.

Now, if $k_1 i - \gamma - 1 \equiv e - 1$, then we would have

$$\begin{aligned} k_1 i \equiv \gamma \bmod e &\Rightarrow k_1 i = k_3 e + \gamma \\ &\Rightarrow k_1 x g = k_3 y g + \gamma \\ &\Rightarrow g(k_1 x - k_3 y) = \gamma\ , \end{aligned}$$

for some $k_3 \in \mathbb{N}$.

Similarly, if $k_2 i - 1 \equiv e - \gamma - 1$, then

$$\begin{aligned} k_2 i \equiv e - \gamma \bmod e &\Rightarrow k_2 i = k_4 e - \gamma \\ &\Rightarrow k_2 x g = k_4 y g - \gamma \\ &\Rightarrow g(k_4 y - k_2 x) = \gamma\ , \end{aligned}$$

for some $k_4 \in \mathbb{N}$.

Since all of $g, k_1x - k_3y$ and $k_4y - k_2x \in \mathbb{N}$, the only possible way these equalities can hold is when g is a divisor of γ. Otherwise, for $g \notin Div(\gamma)$, we have that $s_{e-\gamma-1} = s_{i-\gamma-1}$ and $s_{e-1} = s_{i-1}$, contradicting the definition of $R(e,i,\gamma)$ and as a result $|S(R(e,i,\gamma)| = 0$. □

Proposition 7. *Let* $\gcd(e,i) = g$. *If* $g \in Div(\gamma)$, *then*

$$|S(R(e,i,\gamma))| = \begin{cases} 2^g & \text{if} \quad e \neq 2\gamma \\ 2^{g-1} & \text{if} \quad e = 2\gamma \end{cases} .$$

Proof. **I.** Let $\gcd(e,i) = g$. $R(e,i,\gamma)$ is given by

$$s_j = s_{i+j},\ s_{e-1-\gamma} \neq s_{i-1-\gamma},\ s_{e-1} \neq s_{i-1} \ ,$$

where $j \in \{0,\dots,e-2\} \setminus \{e-1-\gamma\}$. This gives

$$\begin{array}{ccccc} s_0 & = \dots = & s_{ki} & = \dots = & s_{e-i} \ , \\ s_1 & = \dots = & s_{ki+1} & = \dots = & s_{e-i+1} \ , \\ \vdots & & \vdots & & \vdots \\ s_{g-2} & = \dots = & s_{ki+g-2} & = \dots = & s_{e-i+g-2} \ , \\ \multicolumn{5}{l}{\text{where } k \in \mathbb{Z}^*, \text{ and also}} \\ s_{i-1} & = \dots = & s_{e-\gamma-1} \neq s_{i-\gamma-1} & = \dots = & s_{e-1} \ , \end{array} \tag{2}$$

as seen in the proof of Proposition 6. The above form of representing the recursion will be called its *structure*. There are 2 possibilities for each line and so $|S(R(e,i,\gamma))| = 2^g$.

II. $e = 2\gamma$ means that $1 \leq i \leq \gamma$. We use the same reasoning as in **I**. However, due to Corollary 2, for each sequence we obtain we will also obtain its γ-digits cyclic shift. Therefore, $|S(R(e,i,\gamma))| = 2^{g-1}$. □

Proposition 7 tells us that when $e = 2\gamma$ we have two repeating $(e-1-\gamma)$-vectors: $(\mathcal{S}_0) = (\mathcal{S}_i)$ and $(\mathcal{S}_\gamma) = (\mathcal{S}_{\gamma+i})$. In addition, when $g = 1$, $S(R(e,i,\gamma))$ contains the self-complementary sequences.

Conjecture 1. For all γ, i and i' such that $0 \leq \gamma, i, i' \leq \lfloor \frac{e}{2} \rfloor$, $i \neq i'$ and for all m such that $0 \leq m \leq e-1$, we have that

$$S(R(e,i,\gamma)) \cap S(R^m(e,i',\gamma)) = \emptyset \ .$$

Theorem 3. *Let* $\gcd(e,i) = g$. *If* $g \in Div(\gamma)$, *then*

$$|S(R(e,i,\gamma), e-\gamma)| = \begin{cases} \sum_{d|g} \mu(d) 2^{g/d} & \text{if} \ \ g = i = \gamma < \frac{e}{2} \\ \sum_{d|g} \left(\mu(d) 2^{g/d}\right) - 2^{g-1} & \text{if} \ \ g = i = \gamma = \frac{e}{2} \\ 0 & \text{if} \ \ g < i = \gamma \\ 2^g & \text{if} \ \ g, \gamma < i \text{ and } \gamma < \frac{e}{2} \\ 2^{g-1} & \text{if} \ \ i \notin Div(\gamma) \text{ and } \gamma = \frac{e}{2} \end{cases} ,$$

where μ *denotes the Möbius function.*

Remark 1. The present version of Theorem 3 does not cover the following cases, left for future work: $g = i < \gamma$, $g < i \in Div^*(\gamma)$ and $g < i < \gamma$ with $i \notin Div^*(\gamma)$.

Proof. **I.** Let $e > 2\gamma$ and $\gcd(e, \gamma) = \gamma$. By Proposition 7 $R(e, \gamma, \gamma)$ gives

$$\begin{array}{lllll} s_0 & = \ldots = & s_{k_1\gamma} & = \ldots = & s_{e-\gamma} \ , \\ s_1 & = \ldots = & s_{k_1\gamma+1} & = \ldots = & s_{e-\gamma+1} \ , \\ \vdots & & \vdots & & \vdots \\ s_{\gamma-2} & = \ldots = & s_{k_1\gamma+\gamma-2} & = \ldots = & s_{e-2} \ , \\ s_{\gamma-1} & = \ldots = & s_{e-\gamma-1} & \neq & s_{e-1} \ , \quad \text{where } k_1 \in \mathbb{Z}^* \ . \end{array} \tag{3}$$

Now, consider $d \in Div^*(\gamma)$. We have that $\gcd(e, d) = d$ and $R(e, d, d)$ is given by

$$\begin{array}{lllllll} s_0 & = \ldots = & s_{k_2 d} & = \ldots = & s_{e-\gamma} & = \ldots = & s_{e-d} \ , \\ s_1 & = \ldots = & s_{k_2 d+1} & = \ldots = & s_{e-\gamma+1} & = \ldots = & s_{e-d+1} \ , \\ \vdots & & \vdots & & \vdots & & \vdots \\ s_{d-2} & = \ldots = & s_{k_2 d+d-2} & = \ldots = & s_{\gamma-2} & = \ldots = & s_{e-2} \ , \\ s_{d-1} & = \ldots = & s_{e-\gamma-1} & = \ldots = & s_{e-d-1} & \neq & s_{e-1} \ , \quad \text{where } k_2 \in \mathbb{Z}^* \ . \end{array} \tag{4}$$

We can see that the set $\{k_2 d + q\}$ contains the set $\{k_1\gamma + q + pd\}$, where $0 \le q \le d-1$, $0 \le p \le \frac{\gamma}{d} - 1$ and $k_1, k_2 \in \mathbb{Z}^*$. Therefore, the sequences satisfying $R(e, d, d)$ also satisfy $R(e, \gamma, \gamma)$ (splitting each line of (4) in $\frac{\gamma}{d}$ parts, we obtain (3)). Since $|S(R(e, d, d), e - \epsilon)| = 0$ when $\epsilon > d$ and $Div^*(d) \subset Div^*(\gamma)$, we are only interested in $S(R(e, d, d), e - d)$ for each $d \in Div^*(\gamma)$.

Hence, by Corollary 1 and Proposition 7,

$$|S(R(e, \gamma, \gamma), e - \gamma)| = 2^\gamma - \sum_{d \in Div^*(\gamma)} |S(R(e, d, d), e - d)| \ . \tag{5}$$

Rearranging gives

$$2^\gamma = \sum_{d \in Div(\gamma)} |S(R(e, d, d), e - d)| \ . \tag{6}$$

Now, applying the Möbius Inversion Formula (see e.g. [1]) to (6), we obtain

$$|S(R(e, \gamma, \gamma), e - \gamma)| = \sum_{d \in Div(\gamma)} \mu(d) 2^{\gamma/d} \ , \tag{7}$$

where μ denotes the Möbius function.

II. Let $e = 2\gamma$. We start by following the same arguments as the proof above and so, by Proposition 7, we have

$$|S(R(e, \gamma, \gamma), e - \gamma)| = 2^{\gamma-1} - \sum_{d \in Div^*(\gamma)} |S(R(e, d, d), e - d)| \ . \tag{8}$$

Now, since $d < \gamma = \frac{e}{2}$, we know from (7) that

$$|S(R(e,d,d), e-d)| = \sum_{d' \in Div(d)} \mu(d') 2^{d/d'} .$$

By the Möbius Inversion Formula

$$\sum_{d \in Div(\gamma)} \sum_{d' \in Div(d)} \mu(d') 2^{d/d'} = 2^{\gamma} ,$$

and therefore (8) becomes

$$\begin{aligned} |S(R(e,\gamma,\gamma), e-\gamma)| &= 2^{\gamma-1} - 2^{\gamma} + \sum_{d \in Div(\gamma)} \left(\mu(d) 2^{\gamma/d}\right) \\ &= \sum_{d \in Div(\gamma)} \left(\mu(d) 2^{\gamma/d}\right) - 2^{\gamma-1} . \end{aligned}$$

III. Let $\gcd(e,\gamma) = g \in Div^*(\gamma)$. $R(e,\gamma,\gamma)$ gives

$$\begin{array}{lllll} s_0 & = \ldots = & s_{k_1\gamma} & = \ldots = & s_{e-\gamma} , \\ s_1 & = \ldots = & s_{k_1\gamma+1} & = \ldots = & s_{e-\gamma+1} , \\ \vdots & & \vdots & & \vdots \\ s_{g-2} & = \ldots = & s_{k_1\gamma+g-2} & = \ldots = & s_{e-\gamma+g-2} , \\ s_{\gamma-1} & = \ldots = & s_{k_1\gamma+\gamma-1} & = \ldots = & s_{e-\gamma-1} \neq s_{e-1} , \text{where } k_1 \in \mathbb{Z}^* . \end{array} \quad (9)$$

Since $\gamma > g$, we have that $\gamma = gx$, for some $x > 1$. Therefore, $s_{k_1\gamma+q} = s_{k_1 gx+q} = s_{k_2 g+q}$, where $0 \leq q \leq g-2$ and $k_1, k_2 \in \mathbb{Z}^*$. Similarly $s_{k_1\gamma+\gamma-1} = s_{k_2 g+gx-1}$. Hence, each line of (9) is just a rearrangement of a line obtained by recursion $R(e,g,g)$ for which $|S(R(e,g,g), e-\epsilon)| = 0$, when $\epsilon > g$.

IV. Let $\gcd(e,i) = g$. We have $g < i$ and $i > \gamma$. $R(e,i,\gamma)$ gives (2). None of the combinations yield a different recursion and so by Proposition 7, $|S(R(e,i,\gamma), e-\gamma)| = 2^g$.

V. Let $\gcd(e,i) = g < i$. The case we have is $\gamma > i \notin Div(\gamma)$. $R(e,i,\gamma)$ gives (2) and by Proposition 7 we get $|S(R(e,i,\gamma), e-\gamma)| = 2^{g-1}$.

□

The next theorem follows from Theorems 1 and 3.

Theorem 4. *$nlin(e-1,e) = \phi(e)$, where $\phi(n)$ is Euler's totient function.*

3.2 The Recursions $R_v(e,i)$.

We have already studied recursions $R_v(e,i) = R(e,i,\gamma)$, that is when the relating vector v has only two $1's$, in the previous subsection. By Proposition 3 and Theorem 2, studying the rest of the cases is equivalent to investigating the properties of sums of sequences satisfying recursions $R(e,i,\gamma)$. Doing this is further work.

4 An Algorithm Checking for Short Cycles in Large Nonlinear Feedback Shift Registers

In this section we present our proposed algorithm that checks for short cycles in large NLFSRs. A *cycle* is the periodic part of a sequence generated by a NLFSR. Every NLFSR is described by its nolinear recursion.

Definition 10. *A recursion* $s_n = f(s_0, s_1, \ldots, s_{n-1})$ *is irreducible if all sequences it generates have complexity* n*, and reducible otherwise. We say it is of order* n*.*

Jansen [4] defined the *maximum order complexity* of s as the length of the shortest Feedback Shift Register that generates s. We relate to this in the following:

Definition 11. *The minimum recursion* $\mathrm{MinR}(s)$ *for* s *is the recursion of order* $\mathrm{C}(s)$ *and with the fewest binary operations that generates* s*.*

Finally, we let $\tau(e)$ denote the total number of cycles of period $\leq e$.

4.1 Algorithm Check_Sh

The algorithm takes the following two items of input data:

1. An array L of all sequences of period $\leq e$, e a given parameter:
$$L = ((0, s_0), (1, s_0), (0, \overline{s_0}), (01, s_0 \oplus s_1), \ldots,$$
$$\left(s_0^q s_1^q \ldots s_{C(s^q)-1}^q, MinR\left(s^q\right)\right), \ldots,$$
$$\left(s_0^{\tau(e)} s_1^{\tau(e)} \ldots s_{C\left(s^{\tau(e)}\right)-1}^{\tau(e)}, MinR\left(s^{\tau(e)}\right)\right)) .$$
where $\overline{s_a}$ is the complement of s_a, that is $\overline{s_a} = s_a \oplus 1$.
2. The recursion $s_n = f(s_0, s_1, \cdots, s_{n-1})$ to be checked for short cycles.

Each element in L is a pair, the left member being a binary vector of length $\mathrm{C}(s)$ contained in the sequence s, and the right member being the minimum recursion for s. Obviously such a pair defines s uniquely. For example, the first member in the list is the all zero sequence, satisfying $s_{m+1} = s_m$ with $s_0 = 0$. The second is the all one sequence, the third is $0101\ldots$, and the fourth $011011\ldots$, which satisfies $s_{m+2} = s_m \oplus s_{m+1}$ with $s_0 = 0$ and $s_1 = 1$.

The main function of the algorithm is the following:

Set $q = 1$. **While** $q \leq \tau(e)$ **do**

1. Read the qth member of L, $L(q) = \left(s_0^q s_1^q \ldots s_{C(s^q)-1}^q, MinR\left(s^q\right)\right)$.
2. Generate the whole period of the sequence $s_0^q s_1^q \ldots s_{e(s^q)-1}^q$.
3. **If** $s_{n+h}^q = f\left(s_h^q, s_{h+1}^q, \ldots, s_{n+h-1}^q\right),\ h = 0, 1, \ldots, e\left(s^q\right) - 1$, **then quit Check_Sh** with the result that $f(s_0, s_1, \ldots, s_{n-1})$ is reducible. It generates the qth member in L (we assume that n is greater than $\mathrm{C}(s^q)$ for all (s^q) in L) . **Otherwise** increase q by one.

End Check_Sh with the result that $f(s_0, s_1, \ldots, s_{n-1})$ does not generate cycles of period $\leq e$.

4.2 Implementing the Algorithm

By definition, the size of L is $\tau(e)$. However, when checking recursion $s_n = f(s_0, s_1, \cdots, s_{n-1})$ for short cycles, we only want to go through those sequences s in L that have complexity $\mathrm{C}(s) < n$. Therefore the "effective" size of L is

$$|L| = \sum_{e=2}^{e_{max}} \sum_{k=\lceil \log_2(e) \rceil}^{n-1} nlin(k, e) \ ,$$

where e_{max} is the largest period considered in L. Since our purpose is to check for short cycles, we will be looking at a reasonable, from a computational complexity aspect, value for e_{max} (≈ 30).

Definition 12. *The repeativity of a cycle s, denoted by $R(s)$, is defined as the number of* $(\mathrm{C}(s)-1)$*-vectors of s which are repeated.*

Theorem 5. *The probability that recursion $s_n = f(s) = F(s_1, s_2, \ldots, s_{n-1}) \oplus s_0$ generates a sequence of period $e(s)$, complexity $C(s)$ and repeativity $R(s)$ is:*

$$\mathcal{P}(s) = \begin{cases} 2^{-e(s)} & \text{if} \quad C(s) < n \\ 2^{R(s)-e(s)} & \text{if} \quad C(s) = n \\ 0 & \text{if} \quad C(s) > n \end{cases} .$$

Proof. There are $2^{2^{n-1}}$ different F's. In the case $C(s) < n$, s determines $F(s_1, s_2, \ldots, s_{n-1})$ for e digits of $(s_1, s_2, \ldots, s_{n-1})$, leaving $2^{n-1} - e$ digits to be chosen arbitrarily. In other words, the fraction of all possible F's which generate s is $\frac{2^{n-1}-e}{2^{n-1}} = 2^{-e}$ as required.

In the case $C(s) = n$, $R(s)$ $(n-1)$-vectors occur twice in s. Therefore, if we denote by $D(s)$ the number of different $(n-1)-vectors$ occurring in s, we have $e(s) = D(s) + R(s)$. This gives

$$\mathcal{P}(s) = \frac{2^{2^{n-1}-D(s)}}{2^{2^{n-1}}} = \frac{2^{2^{n-1}+R(s)-e(s)}}{2^{2^{n-1}}} = 2^{R(s)-e(s)} \ .$$

Finally, $f(s)$ cannot generate a sequence s with $C(s) > n$. □

To optimise the algorithm, we should reorder the sequences in L in a way that $\mathcal{P}(s)$ does not increase. Furthermore, this version of the algorithm is not optimal if $f(s_0, s_1, \cdots, s_{n-1})$ has a symmetry. For example, if $s_n = f(s_0, s_1, \ldots, s_{n-1})$ is mapped onto itself if we replace s_q with $\overline{s_q}$ (self-complementary), if we replace s_q with s_{n-q} (reversible) or a combination of both, where $0 \leq q \leq n$. For example, $s_2 = \overline{s_1} \oplus s_0$ is reversible, and $s_2 = \overline{s_0}$ is both self-complementary and reversible. These symmetries define equivalence classes of sequences. In these cases a more optimal algorithm would be to include in L one representative for each equivalence class.

Nevertheless, the biggest difficulty in generating L lies in obtaining $\mathrm{MinR}(s)$ for each sequence s. The corresponding register synthesis with the fewest number of terms (not fewest binary operations as we want) proposed in [4] is of order $2e^2 \log_2(e)$, where e is the length of s.

It is a problem for further research to try to reformulate the algorithm in such a way that we do not need MinR. We may for instance let the algorithm which generates L be reentrant and let **Check_Sh** call it each time it needs new bits of the short sequence it is working on.

5 Conclusions

In this paper we have commenced the classification of periodic binary sequences into nonlinear complexity classes. Not all cases have been covered but our results and methodology suggest a way forward. Finally, we have presented an algorithm that performs checks for short cycles in large Nonlinear Feedback Shift Registers. Its implementation and efficiency depend on the above mentioned classification.

Acknowledgements. The first author would like to thank the Marie Curie Fellowship Scheme of the European Union and Professor Tor Helleseth for making his visit to the University of Bergen in Norway possible and his stay enjoyable.

References

1. E.R. Berlekamp: Algebraic Coding Theory, McGraw-Hill Book Company (1968).
2. D. Erdmann and S. Murphy: An Approximate Distribution for the Maximum Order Complexity, Designs, Codes and Cryptography, Vol. 10 (1997), 325–339, Springer.
3. S.W. Golomb: Shift Register Sequences, Revised edition, Aegean Park Press (1982).
4. C.J.A. Jansen: Investigations on Nonlinear Streamcipher Systems: Construction and Evaluation Methods, PhD Thesis, Technical University of Delft (1989).
5. L.V. Ly and W. Schindler: How to Embed Short Cycles into Large Nonlinear Feedback-Shift Registers, Revised Selected Papers of the 4th International Conference in Security in Communication Networks 2004, Lecture Notes in Computer Science, Vol. 3352 (2004), 367–380, Springer.
6. J.L. Massey: Shift-register synthesis and BCH decoding, IEEE Transactions on Information Theory, Vol. IT-15 (1969), 81–92.
7. The On-Line Encyclopedia of Integer Sequences, http://www.research.att.com/~njas/sequences/?Anum=A001037.

Appendix

Table 1. The first values of $nlin(e - \gamma, e)$

2	1																		
3	2	0																	
4	2	1	0																
5	4	2	0	0															
6	2	4	3	0	0														
7	6	4	4	4	0	0													
8	4	6	8	10	2	0	0												
9	6	4	14	18	14	0	0	0											
10	4	10	12	16	40	17	0	0	0										
11	10	8	14	30	46	64	14	0	0	0									
12	4	6	16	38	54	104	100	13	0	0	0								
13	12	10	18	38	70	134	194	142	12	0	0	0							
14	6	16	20	36	74	132	303	366	188	20	0	0	0						
15	8	6	26	40	106	170	324	558	644	268	32	0	0	0					
16	8	14	20	52	92	176	348	732	1110	1122	390	16	0	0	0				
17	16	14	26	54	104	214	410	806	1448	2162	1918	538	0	0	0	0			
18	6	16	28	46	94	236	416	816	1803	2966	4105	3297	703	0	0	0	0		
19	18	16	30	62	120	246	482	958	1880	3560	5960	7914	5506	842	0	0	0	0	
20	8	14	20	68	140	248	494	952	1916	4176	7340	12070	14800	9046	1085	0	0	0	0
21	12	10	42	64	122	242	618	1096	2158	4312	8302	15132	23836	27942	14660	1310	0	0	0
22	10	28	36	68	140	274	550	1094	2188	4330	9583	17380	31069	47571	51611	23160	1465	0	0
23	22	20	38	78	152	310	614	1230	2450	4890	9700	19056	35894	63504	94132	94660	36428	1544	0
24	8	14	36	70	126	306	558	1354	2452	4894	9774	21236	39740	74422	128876	185014	172188	56232	1570
25	20	18	34	70	200	344	680	1364	2720	5404	10816	21560	42510	82526	153554	261740	361356	310652	84640
26	12	34	44	84	172	338	680	1356	2716	5408	10826	21596	47102	88711	171375	315939	528631	701641	555775
27	18	16	54	88	170	338	682	1364	3220	5938	11860	23686	47302	94080	184402	354656	647200	1064850	1358056
e/γ	1	2	3	4	5	6	7	8	9	10	11	12	13	14	15	16	17	18	19

Sequences of Period $2^N - 2$

Rainer Göttfert

Infineon Technologies AG
Am Campeon 1–12
85579 Neubiberg, Germany
rainer.goettfert@infineon.com

Abstract. We derive a formula for the minimal polynomial of the termwise product of binary sequences of least periods $2^N - 2$. The obtained results are important in the analysis of keystream generators based on binary nonlinear N-stage feedback shift registers producing sequences of period $2^N - 2$. Sequences of period $2^N - 1$ are also considered.

Keywords: Periodic sequences, minimal polynomial, nonlinear feedback shift registers.

1 Introduction

A fundamental problem in the theory of stream ciphers is the determination of the linear complexity of keystreams. In many practical stream ciphers binary feedback shift registers and Boolean combining functions serve as building blocks. Oftentimes linear feedback shift registers (LFSRs) with primitive characteristic polynomials are employed. When started in any nonzero initial state, these shift registers will produce so-called m-sequences, which are sequences of least period $2^N - 1$, where N is the length of the shift register (see e.g. Golomb [9] or Gong and Golomb [10]). The downside of an LFSR is that its next-state function, which maps the state of the register at time t into the state at time $t + 1$, is—as the name suggests—linear. This fact makes systems relying on LFSRs potentially vulnerable to algebraic attacks (see [2], [3], [18]).

An approach to counteract algebraic attacks at the root (and to lower hardware costs at the same time) is to use nonlinear feedback shift registers (NLFSRs) instead. Here the next state function is a nonlinear mapping from $\mathbb{F}_2^N$ into itself, where N is the length of the shift register. In order to facilitate the analysis of combined nonlinear feedback shift register sequences, the underlying NLFSRs should have a simple cycle structure. The simple cycle structure of the shift register is often accompanied with a simple algebraic structure of the minimal polynomial of the produced shift register sequence.

The nonlinear counterparts of LFSRs producing m-sequences are N-stage NLFSRs which produce binary sequences of least period $2^N - 1$ for every nonzero initial state. These shift registers were called *primitive* in [6]. A stream cipher deploying several primitive NLFSRs and a suitable Boolean combining function was recently proposed by Gammel, Göttfert, and Kniffler [7].

G. Gong et al. (Eds.): SETA 2006, LNCS 4086, pp. 223–236, 2006.

We mention some prior work: The minimal polynomial of a sequence ζ derived from combining the nonzero output sequences $\sigma_1, \ldots, \sigma_d$ of several primitive binary NLFSRs of pairwise relatively prime lengths, using an arbitrary Boolean combining function $F : \mathbb{F}_2^d \to \mathbb{F}_2$, is uniquely determined and its degree, the linear complexity of ζ, can be expressed by the formula

$$L(\zeta) = F^*(L(\sigma_1), \ldots, L(\sigma_d)). \tag{1}$$

Here $L(\sigma_i)$ is the linear complexity of the sequence σ_i for $1 \leq i \leq d$. The function F^* is formally identical with the algebraic normal form of F but is regarded as an element of $\mathbb{Z}[x_1, \ldots, x_d]$. This result is implicitly already contained in Rueppel and Staffelbach [24, Corollary 6]. Formula (1) is also an implication of Golić [8, Theorem 5]. A formula for the minimal polynomial of the sequence ζ was derived by Gammel and Göttfert in [6, Theorem 3]. The case that the lengths of the primitive NLFSRs are distinct—but no longer pairwise relatively prime—was dealt with in [7].

In this paper we investigate the case of binary N-stage NLFSRs capable to produce sequences of least period $2^N - 2$. Like in the $2^N - 1$ case, the output sequences of such shift registers have almost ideal k-tuple distribution for all $1 \leq k \leq N$. Furthermore the elements 0 and 1 are equidistributed over a full portion of the period.

We can distinguish two types of N-stage binary shift registers producing sequences of least period $2^N - 2$. The first type fixes the all-zero state of the register and the all-one state. The shift register has three different cycles. There is a one-to-one correspondence between these shift registers and binary span N De Bruijn sequences. This correspondence shows that there are

$$2^{2^{N-1}-N}$$

shift registers of the first type (see De Bruijn [1]). The second type of shift register has two different cycles. The short cycle contains the two states $(0, 1, 0, \ldots)$ and $(1, 0, 1, \ldots)$. The long cycle contains all the remaining $2^N - 2$ states. The number of these shift registers is given by

$$\frac{1}{3}\left(2^{N-2} + (-1)^{N-1}\right) 2^{2^{N-1}-2N+2},$$

as was shown by Fredricksen [5].

At the present time there is no algorithm known to produce the mentioned NLFSRs (with a reasonably sparse feedback function) efficiently. Using a heuristic search method, the authors of [7] are currently able to produce such shift registers up to the length 35. These lengths are sufficient for the design of stream ciphers with 128-bit security.

Over the finite field $\mathbb{F}_2$ the effect of a Boolean combining function on individual sequences reduces to two simpler problems: termwise addition of sequences and termwise multiplication of sequences. Since termwise addition of periodic sequences is comparatively easy to analyze, we focus our attention to the termwise product of sequences. Although we are primarily interested in output sequences of binary N-stage NLFSRs, we prove the results for arbitrary binary sequences of least periods $2^N - 1$ and $2^N - 2$. It suffices to treat the product of two sequences, as we can then proceed by induction to obtain results on the product of any finite number of sequences.

2 The Case Per(σ) = $2^M - 1$ and Per(τ) = $2^N - 1$

Proposition 1. *Let $\sigma = (s_n)_{n=0}^{\infty}$ be a periodic binary sequence of least period $2^M - 1$. Then the minimal polynomial of σ is the product of distinct binary irreducible polynomials $\neq x$ whose degrees divide M.*

Proof. The minimal polynomial m_σ of σ divides the polynomial $x^p - 1$, where $p = 2^M - 1$. The polynomial $x^p - 1$ is the product of all binary polynomials f with $\deg(f)$ divides M and $f(0) \neq 0$. □

We recall some results of Selmer [25, Chap. 4]. Let f and g be nonconstant polynomials over $\mathbb{F}_q$ without multiple roots and with nonzero constant terms. Then $f \vee g$ is defined to be the monic polynomial whose roots are the distinct elements of the form $\alpha\beta$, where α is a root of f and β is a root of g. The polynomial $f \vee g$ is again a polynomial over $\mathbb{F}_q$. This follows from the fact that all conjugates over $\mathbb{F}_q$ of a root of $f \vee g$ are roots of $f \vee g$. The following lemma is due to Selmer.

Lemma 1. *Let f and g be nonconstant polynomials over $\mathbb{F}_q$ without multiple roots and with nonzero constant terms. The polynomial $f \vee g \in \mathbb{F}_q[x]$ is irreducible if and only if the polynomials f and g are both irreducible and of pairwise relatively prime degrees. In this case, $\deg(f \vee g) = \deg(f)\deg(g)$. If $\sigma = (s_n)_{n=0}^{\infty}$ and $\tau = (t_n)_{n=0}^{\infty}$ are periodic sequences of elements of $\mathbb{F}_q$ with irreducible minimal polynomials f and g of pairwise relatively prime degrees, then $f \vee g$ is the minimal polynomial of $\sigma\tau = (s_n t_n)_{n=0}^{\infty}$.*

Proof. See Selmer [25, Chap. 4]. □

Theorem 1. *Let $\sigma = (s_n)_{n=0}^{\infty}$ and $\tau = (t_n)_{n=0}^{\infty}$ be binary periodic sequences of least periods $2^M - 1$ and $2^N - 1$, respectively. Let the canonical factorizations over $\mathbb{F}_2$ of the minimal polynomials of σ and τ be given by $m_\sigma = \prod_{i=1}^{r} f_i$ and $m_\tau = \prod_{j=1}^{s} g_j$, respectively. If $\gcd(M, N) = 1$, then the minimal polynomial of the product sequence $\sigma\tau = (s_n t_n)_{n=0}^{\infty}$ is given by*

$$m_{\sigma\tau} = \prod_{i=1}^{r}\prod_{j=1}^{s}(f_i \vee g_j). \tag{2}$$

In fact, (2) is the canonical factorization of the minimal polynomial of $\sigma\tau$.

Proof. The proof is a straightforward application of Proposition 3A in Appendix A and of Lemma 1. The hypothesis $\gcd(M, N) = 1$ guarantees that the rs irreducible polynomials $f_i \vee g_j$, $1 \leq i \leq r$, $1 \leq j \leq s$, are distinct. For more details see [6, Lemma 7]. □

3 The Case $\mathrm{Per}(\sigma) = 2^M - 2$ and $\mathrm{Per}(\tau) = 2^N - 1$

Proposition 2. *Let $\sigma = (s_n)_{n=0}^{\infty}$ be a binary periodic sequence of least period $2^M - 2$. The canonical factorization of m_σ, the minimal polynomial of σ, in $\mathbb{F}_2[x]$ has the form $m_\sigma = \prod_{i=1}^{r} f_i^{e_i}$, where the f_i are distinct irreducible binary polynomials with $\deg(f_i)$ divides $M - 1$ and $f_i(0) \neq 0$. The exponents e_i satisfy $1 \leq e_i \leq 2$. At least one exponent e_i is 2.*

Proof. Consider the characteristic polynomial of σ:

$$x^{2^M - 2} - 1 = (x^2)^{2^{M-1}-1} - 1 = \prod f(x^2) = \prod f(x)^2,$$

where the two products are extended over all irreducible binary polynomials f with $f(0) \neq 0$ whose degrees divide $M - 1$. Since the minimal polynomial of σ divides any characteristic polynomial of σ, it must have the form $m_\sigma = \prod_{i=1}^{r} f_i^{e_i}$ with $e_i \in \{1, 2\}$ for all i. It remains to show that at least one e_i is equal to 2. Assume to the contrary that all e_i are 1. Then $\mathrm{ord}(m_\sigma) = \mathrm{lcm}(\mathrm{ord}(f_1), \ldots, \mathrm{ord}(f_r))$ (see [16, p. 84f]). Since $\mathrm{ord}(f_i)$ divides $2^{M-1} - 1$ for $1 \leq i \leq r$, it follows that $\mathrm{ord}(m_\sigma)$ divides $2^{M-1} - 1$. However, the order of the minimal polynomial of σ is equal to the least period of σ which is $2^M - 2$, a contradiction. □

Theorem 2. *Let $\sigma = (s_n)_{n=0}^{\infty}$ be a binary periodic sequence of least period $2^M - 2$, and let $\tau = (t_n)_{n=0}^{\infty}$ be binary periodic sequences of least period $2^N - 1$. Let the canonical factorizations over $\mathbb{F}_2$ of the minimal polynomials of σ and τ be given by $m_\sigma = \prod_{i=1}^{r} f_i^{e_i}$ and $m_\tau = \prod_{j=1}^{s} g_j$, respectively. If $\gcd(M - 1, N) = 1$, then the minimal polynomial of the product sequence $\sigma\tau = (s_n t_n)_{n=0}^{\infty}$ is given by*

$$m_{\sigma\tau} = \prod_{i=1}^{r} \prod_{j=1}^{s} (f_i \vee g_j)^{e_i}. \tag{3}$$

In fact, the formula represents the canonical factorization of the minimal polynomial of $\sigma\tau$.

Before we can proof the theorem we need some auxiliary results. Let f be a polynomial over $\mathbb{F}_2$ with $\deg(f) = d \geq 1$. We define $M(f)$ to be the set of all periodic binary sequences with minimal polynomial f. We use $S(f)$ to denote the set of all binary sequences with characteristic polynomial f. Under termwise operations on sequences, $S(f)$ is a vector space over $\mathbb{F}_2$ of dimension d. Clearly, $M(f) \subseteq S(f)$. If f is irreducible then $S(f) = M(f) \cup \{\mathbf{0}\}$, where $\mathbf{0} = (0, 0, \ldots)$ is the zero sequence. In the the following, (n) denotes the binary sequence $(n)_{n=0}^{\infty} = (0, 1, 0, \ldots)$. Likewise, $(n + 1)$ denotes the binary sequence $(n + 1)_{n=0}^{\infty} = (1, 0, 1, \ldots)$.

Lemma 2. *Let f be a nonzero binary polynomial without multiple roots and with a nonzero constant term. If $\sigma = (s_n)_{n=0}^{\infty}$ is a binary periodic sequence with minimal polynomial f, then $(n)\sigma = (ns_n)_{n=0}^{\infty}$ and $(n+1)\sigma = ((n+1)s_n)_{n=0}^{\infty}$ are binary* *periodic sequences with minimal polynomial f^2. Furthermore, if $h/f \in \mathbb{F}_2(x)$ is the rational generating function of σ in its reduced form (see Appendix A), then the rational generating function of $(n)\sigma$ and $(n+1)\sigma$ in their reduced forms are given by*

$$\frac{hf + x(hf)'}{f^2} \quad \text{and} \quad \frac{x(hf)'}{f^2},$$

respectively, where $(hf)'$ stands for the formal derivative of the polynomial hf.

Proof. Consider the generating function of σ (see Appendix A):

$$s_0x^{-1} + s_1x^{-2} + s_2x^{-3} + \cdots = \frac{h}{f}.$$

We differentiate both sides and multiply the result by x. This yields

$$s_0x^{-1} + s_2x^{-3} + s_4x^{-5} + \cdots = \frac{x(h'f - hf')}{f^2}.$$

By adding the first equation to the second we obtain

$$s_1x^{-2} + s_3x^{-4} + s_5x^{-6} + \cdots = \frac{hf + x(hf)'}{f^2}, \tag{4}$$

which is the generating function of $(n)\sigma$. The numerator of the fraction on the right-hand side of (4) has the form $gf + xhf'$, where $g = h + xh'$. Thus the fraction is in reduced form if and only if $\gcd(f, xhf') = 1$. The latter is true since, by hypothesis, we have $f(0) \neq 0$, $\sigma \in M(f)$, and f has no multiple roots. These facts imply that f is not divisible by x, $\gcd(f, h) = 1$, and $\gcd(f, f') = 1$. It follows that $(n)\sigma \in M(f^2)$. The claims concerning the sequence $(n+1)\sigma$ are proved similarly. □

There are two fundamental linear operators on the $\mathbb{F}_2$-vector space $V = \mathbb{F}_2^\infty$. The shift operator T defined by $T\sigma = (s_{n+1})_{n=0}^\infty$ and the decimation operator D defined by $D\sigma = (s_{2n})_{n=0}^\infty$ for all $\sigma = (s_n)_{n=0}^\infty$ in V. If f is any nonzero polynomial over $\mathbb{F}_2$, then $f(T)$ is again a linear operator on V. We can write $S(f) = \{\sigma \in V : f(T)\sigma = \mathbf{0}\}$.

One readily checks that $S(f)$ is closed under the actions of T and D. That is, $T\sigma \in S(f)$ and $D\sigma \in S(f)$ whenever $\sigma \in S(f)$. Thus, T and D are linear operators on the finite dimensional $\mathbb{F}_2$-vector space $S(f)$. Moreover, T is an automorphism of $S(f)$ if and only if the polynomial f is not divisible by x, and D is an automorphism of $S(f)$ if and only if f has no multiple roots. For more information on the decimation operator, we refer to Niederreiter [22], [23].

Lemma 3. *Let f be a binary irreducible polynomial not equal to x, and let $\sigma, \tau \in S(f)$. The following statements are equivalent:* (i) $\sigma = \tau$, (ii) $T\sigma = T\tau$, (iii) $D\sigma = D\tau$, (iv) $DT\sigma = DT\tau$.

Proof. This follows from the fact that T and D are automorphisms of $S(f)$ if f is irreducible with $f(0) \neq 0$. □

Lemma 4. *Let f be a binary irreducible polynomial not equal to x. For each $\sigma \in M(f^2)$ there are uniquely determined sequences $\sigma_0 \in S(f)$ and $\sigma_1 \in M(f)$ such that $\sigma = \sigma_0 + (n)\sigma_1$, and there are uniquely determined sequences $\tau_0 \in S(f)$ and $\tau_1 \in M(f)$ such that $\sigma = \tau_0 + (n+1)\tau_1$.*

Proof. If σ_0 is any sequence of $S(f)$ and σ_1 is any sequence of $M(f)$, then $\sigma_0 + (n)\sigma_1 \in M(f^2)$. This follows immediately from Lemma 2 and Proposition 2A. We prove the first assertion in the lemma by showing that the mapping

$$\begin{aligned} G : S(f) \times M(f) \to \ & M(f^2) \\ (\sigma_0, \sigma_1) \ & \mapsto \sigma_0 + (n)\sigma_1 \end{aligned}$$

is one-to-one and onto.

Suppose we have $G(\sigma_0, \sigma_1) = G(\tilde{\sigma}_0, \tilde{\sigma}_1)$, that is

$$\sigma_0 + (n)\sigma_1 = \tilde{\sigma}_0 + (n)\tilde{\sigma}_1, \tag{5}$$

with $\sigma_0, \tilde{\sigma}_0 \in S(f)$ and $\sigma_1, \tilde{\sigma}_1 \in M(f)$. Applying the decimation operator D to both sides of (5), we get $D\sigma_0 = D\tilde{\sigma}_0$, which is equivalent to $\sigma_0 = \tilde{\sigma}_0$, by Lemma 3. Therefore, from (5), we obtain $(n)\sigma_1 = (n)\tilde{\sigma}_1$. Applying the shift operator T to this equation, we get $(n+1)T\sigma_1 = (n+1)T\tilde{\sigma}_1$. Another application of the decimation operator yields $DT\sigma_1 = DT\tilde{\sigma}_1$, which is equivalent to $\sigma_1 = \tilde{\sigma}_1$, by Lemma 3. Thus the mapping G is one-to-one.

To show that G is onto it now suffices to show that the sets $S(f) \times M(f)$ and $M(f^2)$ have the same cardinality. In fact, if $\deg(f) = d$, then $\#S(f) = 2^d$ and $\#M(f) = 2^d - 1$, so that $S(f) \times M(f)$ has cardinality $2^d(2^d - 1)$. On the other hand, $M(f^2) = S(f^2) \setminus S(f)$, so that $\#M(f^2) = \#S(f^2) - \#S(f) = 2^{2d} - 2^d$. Hence G is onto and the proof of the first assertion in the lemma is complete. The second assertion can be proved analogously by considering the mapping $H : S(f) \times M(f) \to M(f^2)$ defined by $(\tau_0, \tau_1) \mapsto \tau_0 + (n+1)\tau_1$. □

Lemma 5. *Let f and g be binary irreducible polynomials of relatively prime degrees and with $f(0)g(0) \neq 0$. If $\sigma \in M(f^2)$ and $\tau \in M(g)$, then $\sigma\tau \in M((f \vee g)^2)$.*

Proof. According to Lemma 4 we can write $\sigma = \sigma_0 + (n)\sigma_1$ with uniquely determined sequences $\sigma_0 \in S(f)$ and $\sigma_1 \in M(f)$. It follows that $\sigma\tau = \sigma_0\tau + (n)\sigma_1\tau$. By Lemma 1, we have $\sigma_1\tau \in M(f \vee g)$. By Lemma 2, we get $(n)\sigma_1\tau \in M((f \vee g)^2)$. Since $\sigma_0\tau \in S(f \vee g)$, we conclude, using Proposition 2A, that $\sigma\tau \in M((f \vee g)^2)$. □

Proof of Theorem 2. Since $m_\sigma = \prod_{i=1}^r f_i^{e_i}$ and $m_\tau = \prod_{j=1}^s g_j$ are the canonical factorizations of the minimal polynomials of σ and τ, respectively, it follows from Proposition 3A that σ and τ possess unique representations $\sigma = \sum_{i=1}^r \sigma_i$ and $\tau = \sum_{j=1}^s \tau_j$ with $\sigma_i \in M(f_i^{e_i})$ for $1 \leq i \leq r$, and $\tau_j \in M(g_j)$ for $1 \leq j \leq s$. It follows that

$$\sigma\tau = \sum_{i=1}^r \sum_{j=1}^s \sigma_i \tau_j.$$

By Proposition 2, $\deg(f_i)$ divides $M-1$ for all i, and by Proposition 1, $\deg(g_j)$ divides N for all j. By hypothesis we have $\gcd(M-1, N) = 1$. Consider the sequence $\sigma_i\tau_j$. If $e_i = 1$, then Lemma 1 implies that $\sigma_i\tau_j \in M(f_i \vee g_j)$. If $e_i = 2$, then Lemma 5 implies that $\sigma_i\tau_j \in M((f_i \vee g_j)^2)$. It can be shown that the rs irreducible polynomials $f_i \vee g_j$, $1 \le i \le r$, $1 \le j \le s$, are distinct (compare the proof of [6, Lemma 7]). Another application of Proposition 3A yields

$$m_{\sigma\tau} = \prod_{i=1}^{r}\prod_{j=1}^{s}(f_i \vee g_j)^{e_i}.$$

As the polynomials $f_i \vee g_j$ are irreducible, this is the canonical factorization of $m_{\sigma\tau}$ over $\mathbb{F}_2$. □

4 The Case $\mathrm{Per}(\sigma) = 2^M - 2$ and $\mathrm{Per}(\tau) = 2^N - 2$

In this section we deal with the most interesting case in which the minimal polynomials of both sequences have multiple roots. Notice that Herlestam [15] and Golić [8] achieved their respective results on the linear complexity of the product of periodic sequences (with elements in a finite field) under the assumption that at most one sequence may have a minimal polynomial with multiple roots. The case of both sequences having minimal polynomials with multiple roots was treated by Göttfert and Niederreiter [12], [13], [14].

In their investigations binomial coefficients of the form $\binom{a+b-2}{a-1}$ play a crucial role, where a is the multiplicity of a root of the first minimal polynomial, and b the multiplicity of a root of the second minimal polynomial. In the binary case, if the binomial coefficient is odd, then the corresponding root product will in general contribute to the linear complexity of the product sequence with the amount $a+b-1$. If the binomial coefficient is even, there might be no contribution at all.

The minimal polynomial of a binary sequence of least period $2^N - 2$ will in general have many roots of multiplicity 2. As the binomial coefficient $\binom{a+b-2}{a-1}$ is even for $a = b = 2$, the results found in [12], [13], and [14] are of no assistance for the matter of this section. We need two more lemmas.

Lemma 6. *Let f and g be two irreducible polynomials over $\mathbb{F}_2$ of relatively prime degrees and with $f(0)g(0) \neq 0$. Let $\sigma, \tilde{\sigma} \in M(f)$ and $\tau, \tilde{\tau} \in M(g)$. Then $\sigma\tau = \tilde{\sigma}\tilde{\tau}$ if and only if $\sigma = \tilde{\sigma}$ and $\tau = \tilde{\tau}$.*

Proof. Clearly, $\sigma = \tilde{\sigma}$ and $\tau = \tilde{\tau}$ implies $\sigma\tau = \tilde{\sigma}\tilde{\tau}$. To show the converse, let $\deg(f) = M$ and $\deg(g) = N$. Recall that $S(f)$ and $S(g)$ are vector spaces over $\mathbb{F}_2$ with $\dim(S(f)) = M$ and $\dim(S(g)) = N$, respectively. As $\sigma \neq \mathbf{0}$, we can extend $\{\sigma\}$ to a basis $\mathcal{B} = \{\sigma, \sigma_2, \ldots, \sigma_M\}$ of $S(f)$. Similarly, let $\mathcal{C} = \{\tau, \tau_2, \ldots, \tau_N\}$ be a basis of $S(g)$. It is known from Zierler and Mills [26] that $S(f \vee g)$ is the uniquely determined subspace of V spanned by all product sequences $\eta\theta$ with $\eta \in S(f)$ and $\theta \in S(g)$. Now, $\dim(S(f \vee g)) = \deg(f \vee g) = MN$, by Lemma 1. It follows that

$$\mathcal{D} = \{\sigma_i \tau_j : 1 \le i \le M,\ 1 \le j \le N\}$$

is a basis of $S(f \vee g)$, where we set $\sigma_1 = \sigma$ and $\tau_1 = \tau$.

Let the unique representations of $\tilde{\sigma} \in M(f) \subseteq S(f)$ and $\tilde{\tau} \in M(g) \subseteq S(g)$ as linear combinations relative to bases $\mathcal{B}$ and $\mathcal{C}$ be given by

$$\tilde{\sigma} = \sum_{i=1}^{M} a_i \sigma_i \qquad \text{and} \qquad \tilde{\tau} = \sum_{j=1}^{N} b_j \tau_j,$$

respectively. The unique representation of $\tilde{\sigma}\tilde{\tau} \in M(f \vee g) \subseteq S(f \vee g)$ as a linear combination relative to the basis $\mathcal{D}$ is

$$\tilde{\sigma}\tilde{\tau} = \sum_{i=1}^{M} \sum_{j=1}^{N} a_i b_j \sigma_i \tau_j. \tag{6}$$

The unique representation of $\sigma\tau \in M(f \vee g) \subseteq S(f \vee g)$ as a linear combination relative to the basis $\mathcal{D}$ is

$$\sigma\tau = \sum_{i=1}^{M} \sum_{j=1}^{N} c_{ij} \sigma_i \tau_j, \tag{7}$$

where $c_{11} = 1$ and all other coordinates $c_{ij} = 0$. Comparing coordinates in (6) and (7), we obtain $a_1 = b_1 = 1$, and $a_i = 0$ for $2 \le i \le M$, and $b_j = 0$ for $2 \le j \le N$. Thus, $\tilde{\sigma} = \sigma_1 = \sigma$ and $\tilde{\tau} = \tau_1 = \tau$. □

Lemma 7. *Let f and g be binary irreducible polynomials of relatively prime degrees with $f(0)g(0) \neq 0$. For any $\sigma_0 \in S(f)$, $\sigma_1 \in M(f)$, $\tau_0 \in S(g)$, and $\tau_1 \in M(g)$, the sequence*

$$\sigma_0\tau_0 + (n)\sigma_1\tau_0 + (n+1)\sigma_0\tau_1 \tag{8}$$

is either the zero sequence or has minimal polynomial $(f \vee g)^2$. The sequence is the zero sequence if and only if $(\sigma_0, \tau_0) = (\mathbf{0}, \mathbf{0})$ or $(\sigma_0, \tau_0) = (\sigma_1, \tau_1)$.

Proof. If $(\sigma_0, \tau_0) = (\mathbf{0}, \mathbf{0})$ or $(\sigma_0, \tau_0) = (\sigma_1, \tau_1)$, then one readily verifies that the sequence in (8) is the zero sequence $\mathbf{0}$.

We now show that if $(\sigma_0, \tau_0) \neq (\mathbf{0}, \mathbf{0})$ and $(\sigma_0, \tau_0) \neq (\sigma_1, \tau_1)$, then the sequence in (8) has minimal polynomial h^2, where $h = f \vee g$ is irreducible by Lemma 1. We have to consider three cases.

In the first case, $\sigma_0 = \mathbf{0}$ and $\tau_0 \neq \mathbf{0}$. The sequence in (8) reduces to $(n)\sigma_1\tau_0$. The sequence $\sigma_1\tau_0$ has minimal polynomial h by Lemma 1, so that $(n)\sigma_1\tau_0$ has minimal polynomial h^2 by Lemma 2.

In the second case, $\sigma_0 \neq \mathbf{0}$ and $\tau_0 = \mathbf{0}$. The sequence in (8) now is equal to $(n+1)\sigma_0\tau_1$ which again has minimal polynomial h^2 according to Lemma 2.

In the last case, $\sigma_0 \neq \mathbf{0}$ and $\tau_0 \neq \mathbf{0}$. We now have $\sigma_0, \sigma_1 \in M(f)$ and $\tau_0, \tau_1 \in M(g)$, so that all three product sequences $\sigma_0\tau_0$, $\sigma_1\tau_0$, and $\sigma_0\tau_1$ appearing in (8) have the minimal polynomial h. Let the uniquely determined rational generating functions of the sequences $\sigma_0\tau_0$, $\sigma_1\tau_0$, and $\sigma_0\tau_1$ be given by

$$\frac{a}{h}, \qquad \frac{b}{h}, \qquad \text{and} \qquad \frac{c}{h}, \tag{9}$$

respectively. Applying Lemma 2, we get the rational generating function of the sequence in (8):

$$\frac{(a+b+xb'+xc')h + xh'(b+c)}{h^2}. \tag{10}$$

It remains to show that the fraction is in reduced form.

By hypothesis, we have $f(0)g(0) \neq 0$, which implies $h(0) \neq 0$. As h is irreducible we have $\gcd(h, h') = 1$. Thus the numerator of the fraction in (10) is divisible by h only if h divides $b + c$. However, since $\deg(b) < \deg(h)$ and $\deg(c) < \deg(h)$, this is only possible if $b + c$ is the zero polynomial, that is, if $b = c$. The latter, however, is equivalent to $\sigma_1\tau_0 = \sigma_0\tau_1$ which, by Lemma 6, is equivalent to $\sigma_0 = \sigma_1$ and $\tau_0 = \tau_1$, a contradiction. □

Let σ and τ be a binary sequences of least periods $2^M - 2$ and $2^N - 2$, respectively. According to Proposition 2, the minimal polynomial of σ has the form

$$m_\sigma = f_1^2 \cdots f_a^2 f_{a+1} \cdots f_r \tag{11}$$

with irreducible polynomials f_i whose degrees divide $M - 1$. Of course, we may have $r = a$ in which case all factors have multiplicity 2. Similarly,

$$m_\tau = g_1^2 \cdots g_b^2 g_{b+1} \cdots g_s \tag{12}$$

with irreducible polynomials g_j whose degrees divide $N - 1$. Again, we may have $s = b$. An application of Proposition 3A and a subsequent application of Lemma 4 yields that σ can be represented in the form

$$\sigma = \sum_{i=1}^{a} [\sigma_i^{(0)} + (n)\sigma_i^{(1)}] + \sum_{i=a+1}^{r} \sigma_i \tag{13}$$

with uniquely determined sequences $\sigma_i^{(0)} \in S(f_i)$ and $\sigma_i^{(1)} \in M(f_i)$ for $1 \leq i \leq a$, and $\sigma_i \in M(f_i)$ for $a + 1 \leq i \leq r$. Similarly, τ has a unique representation of the form

$$\tau = \sum_{j=1}^{b} [\tau_j^{(0)} + (n+1)\tau_j^{(1)}] + \sum_{j=b+1}^{s} \tau_j \tag{14}$$

with $\tau_j^{(0)} \in S(g_j)$ and $\tau_j^{(1)} \in M(g_j)$ for $1 \leq j \leq b$, and $\tau_j \in M(g_j)$ for $b + 1 \leq j \leq s$.

Theorem 3. *Let $\sigma = (s_n)_{n=0}^\infty$ and $\tau = (t_n)_{n=0}^\infty$ be binary periodic sequences of least periods $2^M - 2$ and $2^N - 2$, respectively. Let the canonical factorizations of the minimal polynomials of σ and τ over $\mathbb{F}_2$ be given by (11) and (12). Assume that $\gcd(M - 1, N - 1) = 1$. The canonical factorization of the minimal polynomial of the product sequence $\sigma\tau = (s_n t_n)_{n=0}^\infty$ over $\mathbb{F}_2$ is given by*

$$m_{\sigma\tau} = \frac{\prod_{i=1}^{r}\prod_{j=1}^{s}(f_i \vee g_j)^2}{\prod_{i=a+1}^{r}\prod_{j=b+1}^{s}(f_i \vee g_j) \prod_{(i,j)\in I_0\times J_0}(f_i \vee g_j)^2 \prod_{(i,j)\in I_1\times J_1}(f_i \vee g_j)^2}. \tag{15}$$

The index sets I_0, I_1, J_0, and J_1 are uniquely determined by the representations (13) *and* (14):

$$I_0 = \{i\colon 1 \le i \le a,\ \sigma_i^{(0)} = \mathbf{0}\}, \qquad I_1 = \{i\colon 1 \le i \le a,\ \sigma_i^{(0)} = \sigma_i^{(1)}\},$$
$$J_0 = \{j\colon 1 \le j \le b,\ \tau_j^{(0)} = \mathbf{0}\}, \qquad J_1 = \{j\colon 1 \le j \le b,\ \tau_j^{(0)} = \tau_j^{(1)}\}.$$

As usual, an empty product—which may occur in the denominator of the big fraction—has the value 1.

Proof. Consider equation (11). By Proposition 3A, we can write σ in the form $\sigma = \sigma^{[2]} + \sigma^{[1]}$ with uniquely determined sequences $\sigma^{[2]} \in M(f_1^2 \cdots f_a^2)$ and $\sigma^{[1]} \in M(f_{a+1} \cdots f_r)$. (In the case that $r = a$, $M(f_{a+1} \cdots f_r) = M(1)$ and $\sigma^{[1]}$ is the zero sequence.) Similarly, $\tau = \tau^{[2]} + \tau^{[1]}$ with uniquely determined sequences $\tau^{[2]} \in M(g_1^2 \cdots g_b^2)$ and $\tau^{[1]} \in M(g_{b+1} \cdots g_s)$. It follows that

$$\sigma\tau = \sigma^{[2]}\tau^{[2]} + \sigma^{[2]}\tau^{[1]} + \sigma^{[1]}\tau^{[2]} + \sigma^{[1]}\tau^{[1]}. \tag{16}$$

We already know the minimal polynomial of the second, third, and fourth product in (16). By Theorem 2, we have:

$$m_{\sigma^{[2]}\tau^{[1]}} = \prod_{i=1}^{a} \prod_{j=b+1}^{s} (f_i \vee g_j)^2, \tag{17}$$

$$m_{\sigma^{[1]}\tau^{[2]}} = \prod_{i=a+1}^{r} \prod_{j=1}^{b} (f_i \vee g_j)^2. \tag{18}$$

By Theorem 1, we have:

$$m_{\sigma^{[1]}\tau^{[1]}} = \prod_{i=a+1}^{r} \prod_{j=b+1}^{s} (f_i \vee g_j). \tag{19}$$

It remains to determine the minimal polynomial of the first product in (16). We have

$$\begin{aligned}\sigma^{[2]}\tau^{[2]} &= \left(\sum_{i=1}^{a}[\sigma_i^{(0)} + (n)\sigma_i^{(1)}]\right)\left(\sum_{j=1}^{b}[\tau_j^{(0)} + (n+1)\tau_j^{(1)}]\right) \\ &= \sum_{i=1}^{a}\sum_{j=1}^{b}[\sigma_i^{(0)}\tau_j^{(0)} + (n)\sigma_i^{(1)}\tau_j^{(0)} + (n+1)\sigma_i^{(0)}\tau_j^{(1)}].\end{aligned}$$

By Lemma 7, the term in the brackets is the zero sequence if and only if $(i,j) \in I_0 \times J_0$ or $(i,j) \in I_1 \times J_1$. Otherwise it is a sequence with minimal polynomial $(f_i \vee g_j)^2$. It follows that

$$m_{\sigma^{[2]}\tau^{[2]}} = \prod_{(i,j)\in C\setminus K} (f_i \vee g_j)^2, \tag{20}$$

where $C = \{(i,j) : 1 \le i \le a,\ 1 \le j \le b\}$ and $K = (I_0 \times J_0) \cup (I_1 \times J_1)$.

Since the irreducible polynomials $f_i \vee g_j$, $1 \le i \le r$, $1 \le j \le s$, are again distinct (a consequence of $\gcd(M-1, N-1) = 1$), the four polynomials in (17), (18), (19), and (20) are pairwise relatively prime. Therefore, it follows from equation (16) and Proposition 3A that the minimal polynomial of $\sigma\tau$ is the product of those four polynomials. The product of the four polynomials, however, coincides with the polynomial in (15). □

Consider equation (11). The squarefree part of m_σ is defined as $\langle m_\sigma \rangle = f_1 \cdots f_r$. It can be computed by $\langle m_\sigma \rangle = m_\sigma / \sqrt{\gcd(m_\sigma, m'_\sigma)}$. In the next theorem, $m_{D\sigma}$ denotes the minimal polynomial of the sequence $D\sigma$.

In order to obtain the index sets I_0, I_1, J_0, and J_1 appearing in Theorem 3, it is not necessary to actually know the representations (13) and (14) of the sequences σ and τ.

Theorem 4. *Under the provisions of Theorem 3, we have*

$$S_0 = \frac{\langle m_\sigma \rangle}{m_{D\sigma}} = \prod_{i \in I_0} f_i, \qquad S_1 = \frac{\langle m_\sigma \rangle}{m_{DT\sigma}} = \prod_{i \in I_1} f_i,$$

$$T_0 = \frac{\langle m_\tau \rangle}{m_{DT\tau}} = \prod_{j \in J_0} g_j, \qquad T_1 = \frac{\langle m_\tau \rangle}{m_{D\tau}} = \prod_{j \in J_1} g_j.$$

Proof. We prove the first formula. The other three are proved similarly. Consider the representation (13). We apply the decimation operator to both sides. This yields

$$D\sigma = \sum_{i=1}^{a} D\sigma_i^{(0)} + \sum_{i=a+1}^{r} D\sigma_i.$$

The minimal polynomial $m_{D\sigma}$ of the sequence $D\sigma$ is the product of all f_i, $1 \le i \le r$, for which $\sigma_i^{(0)} \ne \mathbf{0}$. It therefore follows that

$$S_0 = \frac{\langle m_\sigma \rangle}{m_{D\sigma}} = \prod_{i \in I_0} f_i.$$

□

Corollary 1. *Under the provisions of Theorem 3, the linear complexity $L(\sigma\tau)$ of the product sequence $\sigma\tau$ is given by*

$$\begin{aligned} L(\sigma\tau) = & L(\sigma)L(\tau) - 2\deg\left(\sqrt{\gcd(m_\sigma, m'_\sigma)}\right)\deg\left(\sqrt{\gcd(m_\tau, m'_\tau)}\right) \\ & - 2\deg(S_0)\deg(T_0) - 2\deg(S_1)\deg(T_1), \end{aligned}$$

where S_0, T_0, S_1, and T_1 are the polynomials specified in Theorem 4.

Proof. The assertion follows from Theorem 3 and Theorem 4. □

Let us assume that the sequences σ and τ are chosen at random from the respective sets of all binary periodic sequences of periods $2^M - 2$ and $2^N - 2$. By

Dai and Yang [4, Lemma 1], most irreducible factors appearing in the canonical factorization of the minimal polynomial m_σ and m_τ, respectively, will have multiplicity 2 with high probability. One can also show that for such randomly chosen sequences σ and τ with high probability the corresponding index sets I_0, I_1, J_0, and J_1 will have small cardinalities.

If all irreducible factors of m_σ and m_τ have multiplicity 2, if at least one of the two sets I_0 and J_0, and at least one of the two sets I_1 and J_1 is empty, then the formula in the corollary reduces to $L(\sigma\tau) = \frac{1}{2}L(\sigma)L(\tau)$.

The expected value for the linear complexity of a randomly chosen periodic sequence is close to the period length (see Dai and Yang [4], and Meidl and Niederreiter [17]). Experimental results show that the majority of the NLFSRs under consideration have output sequences of maximum linear complexity $L(\sigma) = 2^M - 2$ respectively $L(\tau) = 2^N - 2$. As a consequence, the linear complexity of $\sigma\tau$ can be equal to the least period of $\sigma\tau$ which is $\frac{1}{2}\left(2^M - 2\right)\left(2^N - 2\right)$.

Acknowledgements

The author would like to thank the anonymous reviewers for their valuable comments and suggestions and the careful proofreading.

References

1. N. G. de Bruijn: A combinatorial problem, *Indag. Math.* **8**, 461–467 (1946).
2. N. T. Courtois: Fast algebraic attacks on stream ciphers with linear feedback, *Advances in Cryptology – CRYPTO 2003* (D. Boneh, ed.), Lecture Notes in Computer Science, vol. 2729, pp. 176–194, Springer-Verlag, Berlin, 2003.
3. N. T. Courtois and W. Meier: Algebraic attacks on stream ciphers with linear feedback, *Advances in Cryptology — EUROCRYPT 2003* (E. Biham, ed.), Lecture Notes in Computer Science, vol. 2656, pp. 345–359, Springer-Verlag, Berlin, 2003.
4. Z.-D. Dai and J.-H. Yang: Linear complexity of periodically repeated random sequences, *Advances in Cryptology — EUROCRYPT '91* (D. W. Davies, ed.), Lecture Notes in Computer Science, vol. 547, pp. 168–175, Springer-Verlag, Berlin, 1991.
5. H. Fredricksen: A survey of full length nonlinear shift register cycle algorithms, *SIAM Rev.* **24**, 195–221 (1982).
6. B. M. Gammel and R. Göttfert: Linear filtering of nonlinear shift register sequences, *Proc. of The Intern. Workshop on Coding and Cryptography — WCC 2005* (Ø. Ytrehus, ed.), Lecture Notes in Computer Science, vol. 3969, pp. 354–370, Springer-Verlag, Berlin, 2006.
7. B. M. Gammel, R. Göttfert, and O. Kniffler: ACHTERBAHN-80 and ACHTERBAHN-128/80, eSTREAM, ECRYPT Stream Cipher Project, Juni 2006. `http://www.ecrypt.eu.org/stream/papers.html`
8. J. Dj. Golić: On the linear complexity of functions of periodic GF(q) sequences, *IEEE Trans. Inform. Theory* **35**, 69–75 (1989).
9. S. W. Golomb: *Shift Register Sequences*, Aegean Park Press, Laguna Hills, Cal., 1982.

10. S. W. Golomb and G. Gong: *Signal Design for Good Correlation: For Wireless Communication, Cryptography, and Radar*, Cambridge Univ. Press, 2005.
11. R. Göttfert: *Produkte von Schieberegisterfolgen*, Ph.D. Thesis, Univ. of Vienna, 1993.
12. R. Göttfert and H. Niederreiter: On the linear complexity of products of shift-register sequences, *Advances in Cryptology — EUROCRYPT '93* (T. Helleseth, ed.), Lecture Notes in Computer Science, vol. 765, pp. 151–158, Springer-Verlag, Berlin, 1994.
13. R. Göttfert and H. Niederreiter: A general lower bound for the linear complexity of the product of shift-register sequences, *Advances in Cryptology — EUROCRYPT '94* (A. De Santis, ed.), Lecture Notes in Computer Science, vol. 950, pp. 223–229, Springer-Verlag, Berlin, 1995.
14. R. Göttfert and H. Niederreiter: On the minimal polynomial of the product of linear recurring sequences, *Finite Fields Appl.* **1**, 204–218 (1995).
15. T. Herlestam: On functions of linear shift register sequences, *Advances in Cryptology — EUROCRYPT '85* (F. Pichler, ed.), Lecture Notes in Computer Science, vol. 219, pp. 119–129, Springer-Verlag, Berlin, 1986.
16. R. Lidl and H. Niederreiter: *Finite Fields*, Encyclopedia of Mathematics and Its Applications, vol. 20, Addison-Wesley, Reading, Mass., 1983. (Now Cambridge Univ. Press.)
17. W. Meidl and H. Niederreiter: On the expected value of the linear complexity and the k-error linear complexity of periodic sequences, *IEEE Trans. Inform. Theory* **48**, 2817–2825 (2002).
18. W. Meier, E. Pasalic, and C. Carlet: Algebraic attacks and decomposition of Boolean functions, *Advances in Cryptology — EUROCRYPT 2004* (C. Cachin and J. Camenisch, eds.), Lecture Notes in Computer Science, vol. 3027, pp. 474–491, Springer-Verlag, Berlin, 2004.
19. H. Niederreiter: Cryptology – The mathematical theory of data security, *Prospects of Mathematical Science* (T. Mitsui, K. Nagasaka, and T. Kano, eds.), pp. 189–209, World Sci. Pub., Singapore, 1988.
20. H. Niederreiter: Distribution properties of feedback shift register sequences, *Problems Control Inform. Theory* **15**, 19–34 (1986).
21. H. Niederreiter: Sequences with almost perfect linear complexity profile, *Advances in Cryptology — EUROCRYPT '87* (D. Chaum and W.L. Price, eds.), Lecture Notes in Computer Science, vol. 304, pp. 37–51, Springer-Verlag, Berlin, 1988.
22. H. Niederreiter: Some new cryptosystems based on feedback shift register sequences, *Math. J. Okayama Univ.* **30**, 121–149 (1988).
23. H. Niederreiter: A simple and general approach to the decimation of feedback shift-register sequences, *Problems Control Inform. Theory* **17**, 327-331 (1988).
24. R. A. Rueppel and O. J. Staffelbach: Products of linear recurring sequences with maximum complexity, *IEEE Trans. Inform. Theory* **IT-33**, 124–131 (1987).
25. E. S. Selmer: *Linear Recurrence Relations over Finite Fields*, Department of Mathematics, Univ. of Bergen, 1966.
26. N. Zierler and H. W. Mills: Products of linear recurring sequences, *J. Algebra* **27**, 147–157 (1973).

Appendix A

In the following three propositions, $\mathbb{F}_q$ denotes the finite field of order q. Throughout the entire article, however, only sequences with elements in the binary field $\mathbb{F}_2$ occur.

Proposition 1A. Let $\sigma = (s_n)_{n=0}^{\infty}$ be a sequence of elements of $\mathbb{F}_q$, and let g be a monic polynomial over $\mathbb{F}_q$ with $g(0) \neq 0$. Then σ is a periodic sequence in $V = \mathbb{F}_q^{\infty}$ with characteristic polynomial g if and only if

$$\sum_{n=0}^{\infty} s_n x^{-n-1} = \frac{f(x)}{g(x)}$$

with $f \in \mathbb{F}_q[x]$ and $\deg(f) < \deg(g)$.

Proof. See Niederreiter [19], [21]. □

Proposition 2A. Let $\sigma = (s_n)_{n=0}^{\infty}$ be a sequence of elements of $\mathbb{F}_q$, and let m be a monic polynomial over $\mathbb{F}_q$ with $m(0) \neq 0$. Then σ is a periodic sequence with minimal polynomial m if and only if

$$\sum_{n=0}^{\infty} s_n x^{-n-1} = \frac{h(x)}{m(x)}$$

with $h \in \mathbb{F}_q[x]$, $\deg(h) < \deg(m)$, and $\gcd(h, m) = 1$.

Proof. This follows from Proposition 1A and the definition of the minimal polynomial, see [19], [21]. □

Proposition 3A. Let $\sigma_1, \ldots, \sigma_r$ be periodic sequences in $V = \mathbb{F}_q^{\infty}$ with minimal polynomials $m_i \in \mathbb{F}_q[x]$, $1 \leq i \leq r$. If the polynomials $m_1, \ldots, m_r$ are pairwise relatively prime, then the minimal polynomial of the sum $\sigma = \sigma_1 + \cdots + \sigma_r$ is equal to the product $m_1 \cdots m_r$. Conversely, let σ be a periodic sequence in V whose minimal polynomial $m \in \mathbb{F}_q[x]$ is the product of pairwise relatively prime monic polynomials $m_1, \ldots, m_r \in \mathbb{F}_q[x]$. Then, for each $i = 1, \ldots, r$, there exists a uniquely determined periodic sequence σ_i with minimal polynomial $m_i \in \mathbb{F}_q[x]$ such that $\sigma = \sigma_1 + \cdots + \sigma_r$.

Proof. A proof of the first part of the proposition can be found in [16, p. 426]. A proof of the second part can be found in [11, Korollar 2.5] and [6, Lemma 6]. □

A New Algorithm to Compute Remote Terms in Special Types of Characteristic Sequences

Kenneth J. Giuliani[1] and Guang Gong[2]

[1] Dept. of Mathematical and Computational Sciences
University of Toronto at Mississauga
Mississauga, ON, Canada, L5L 1C6
kgiulian@utm.utoronto.ca

[2] Dept. of Electrical and Computer Engineering
University of Waterloo
Waterloo, ON, Canada, N2L 3G1
ggong@calliope.uwaterloo.ca

Abstract. This paper proposes a new algorithm, called the Diagonal Double-Add (DDA) algorithm, to compute the k-th term of special kinds of characteristic sequences. We show that this algorithm is faster than Fiduccia's algorithm, the current standard for computation of general sequences, for fourth- and fifth-order sequences.

1 Introduction

Linear feedback shift register (LFSR) sequences have found an important role in public-key cryptography. One of the first to use them was Niederreiter [11,12,13] who proposed several cryptosystems based on LFSR's. More recently, cryptosystems such as LUC [10,16], GH [4], XTR [7], and a fifth-order system [14,2] have been based upon linear recurrence sequences. As cryptography places a high priority on efficiency of computation, algorithms to compute sequence terms became very important.

Several algorithms to compute sequences terms, such as by Miller and Spencer-Brown [9], Shortt [15], Shortt and Wilson [19], Gries and Levin [5], Urbanek [17], and Fiduccia [1], have been proposed to compute LFSR sequences. Of these, Fiduccia's appears to be the most efficient.

In this paper, we propose a new algorithm, called the Diagonal-Double-Add (DDA) algorithm to compute remote terms for a special type of sequences. These sequences have found use in cryptography as the basis for XTR [7] and a special fifth-order cryptosystem [14,2]. We examine its computational cost and show that it is more efficient than Fiduccia's algorithm for fourth- and fifth-order sequences.

This paper is organized as follows. In Section 2, we give background on the special type of sequence we are working with. In Section 3, we introduce the DDA algorithm and analyze its computational cost in Section 4. We note here that the analysis of the DDA is very long and tedious. For this reason, we give the reader the computational cost of each step and a strong indication of how

G. Gong et al. (Eds.): SETA 2006, LNCS 4086, pp. 237–247, 2006.

this cost is derived, but leave the complete details for the full paper. We then present and analyze Fiduccia's algorithm in Section 5 and compare the two in Section 6.

2 Preliminaries

The sequences we will be looking at were first proposed as third-order sequences for the XTR [7] cryptosystem. They were also employed in a related fifth-order cryptosystem [14,2]. These sequences were later generalized for use in cryptography for any order [3].

Let $p \equiv 2 \pmod 3$ be a prime and $q = p^2$. We denote by $GF(q)$ the finite field of order q.

Let $f(x)$ be an irreducible polynomial of degree n over $GF(q)$ and α a root of $f(x)$ in $GF(q^n)$. Then the roots of $f(x)$ are $\alpha_i = \alpha^{q^i}$ for $i = 0, \ldots, n-1$. Note that these roots all have the same order in $GF(q^n)$.

Suppose α has order dividing both $(q^n - 1)/(q-1) = q^{n-1} + q^{n-2} + \cdots q + 1$ and $(p^{2n} - 1)/(p^n - 1) = p^n + 1$. Then we may represent $f(x)$ as

$$f(x) = x^n - a_1 x^{n-1} + a_2 x^{n-2} - \cdots + (-1)^{n-1} a_{n-1} x + (-1)^n \tag{1}$$

Note that the constant term is $(-1)^n \alpha^{q^{n-1}+q^{n-2}+\cdots+q+1} = (-1)^n$ by the assumption on the order of α.

Since the order of α also divides $p^n + 1$, we have that $\alpha^{-1} = \alpha^{p^n}$. It then follows that

$$\begin{aligned} a_j &= \sum_{0 \le i_1 < i_2 < \ldots < i_j \le n-1} \alpha^{q^{i_1} + \cdots + q^{i_j}} \\ &= \sum_{0 \le i_1 < i_2 < \ldots < i_{n-j} \le n-1} \alpha^{-(q^{i_1} + \cdots + q^{i_{n-j}})} \\ &= \sum_{0 \le i_1 < i_2 < \ldots < i_{n-j} \le n-1} \alpha^{p^n (q^{i_1} + \cdots + q^{i_{n-j}})} \\ &= \left(\sum_{0 \le i_1 < i_2 < \ldots < i_{n-j} \le n-1} \alpha^{q^{i_1} + \cdots + q^{i_{n-j}}} \right)^{p^n} \\ &= a_{n-j}^{p^n} \end{aligned} \tag{2}$$

for all $j = 1, \ldots, n-1$. Since $a_j \in GF(p^2)$, we must have $a_j = a_{n-j}$ if n is even and $a_j = a_{n-j}^p$ if n is odd.

Consider the recurrence relation of order n over $GF(q)$

$$s_{k+n} = a_1 s_{k+n-1} - a_2 s_{k+n-2} + \cdots + (-1)^n a_{n-1} s_{k+1} + (-1)^{n+1} s_k. \tag{3}$$

The sequence $\{s_i\}$ of elements in $GF(q)$ obtained from (3) with fixed initial conditions

$$s_i = Tr(\alpha^i) = \alpha_0^i + \alpha_1^i + \cdots + \alpha_{n-1}^i \tag{4}$$

for $i = 0, \ldots, n-1$ is called the n-th order characteristic sequence over $GF(q)$ generated by α. The period Q of this sequence is equal to the order of α and defining $s_{-i} = s_{Q-i}$, we get that (4) holds for s_i for all $i \in \mathbb{Z}$. We also have that

$$\begin{aligned} s_{-i} &= \alpha^{-i} + \alpha^{-iq} + \cdots + \alpha^{iq^{n-1}} \\ &= \alpha^{p^n i} + \alpha^{p^n iq} + \cdots + \alpha^{p^n q^{n-1}} \\ &= (\alpha^i + \alpha^{iq} + \cdots + \alpha^{q^{n-1}})^{p^n} \\ &= s_i^{p^n} \end{aligned}$$

Thus, we have $s_{-i} = s_i$ if n is even and $s_{-i} = s_i^p$ if n is odd.

For any integer k, let

$$f_k(x) = x^n - a_{1,k}x^{n-1} + \cdots + (-1)^n a_{n-1,k}x + (-1)^{n+1}$$

be the polynomial whose roots are α_i^k for all $i = 0, \ldots, n-1$. Using a similar argument as in (2), we see that for all $i = 1, \ldots, n-1$, $a_{i,k} = a_{n-i,k}$ if n is even and $a_{i,k} = a_{n-i,k}^p$ if n is odd. We also have an analogue to (3), namely

$$s_{kn+l} = a_{1,k}s_{k(n-1)+l} - \cdots + (-1)^n a_{n-1,k}s_{k+l} + (-1)^{n+1}s_l \tag{5}$$

for all integers k and l.

The s_k and $a_{i,k}$ terms are related by the Newton Formula (see [8]).

Theorem 1 (Newton's Formula). *For a characteristic sequence $\{s_i\}_{i\in\mathbb{Z}}$ and any integers k and i with $1 \le i \le n-1$,*

$$s_{i,k} = a_{1,k}s_{(i-1)k} - \cdots + (-1)^i a_{i-1,k}s_k + (-1)^{i+1} i a_{i,k} \tag{6}$$

and hence

$$a_{i,k} = i^{-1}((-1)^{i+1}s_{ik} + (-1)^i a_{1,k}s_{(i-1)k} + \ldots + a_{i-1,k}s_k) \tag{7}$$

The key use for this theorem is that given the sequence terms $s_0, s_k, s_{2k}, \ldots, s_{ik}$, we can efficiently calculate the Newton coefficients $a_{1,k}, a_{2,k}, \ldots, a_{i,k}$ and vice versa.

3 The Diagonal Double-Add (DDA) Algorithm

We now introduce a new algorithm called the Diagonal Double-Add (DDA) algorithm to calculate the k-th term of a characteristic sequence. The following exposition is for odd n. The case for even n is analogous. Let $v = \frac{n-1}{2}$. For each integer j, define the $(n-1)/2 \times n$ array $\hat{S}_j$ as

$$\hat{S}_j = \begin{bmatrix} s_{-v} & s_{-v+1} & \cdots & s_0 & \cdots & s_v \\ s_{j-v} & s_{j-v+1} & \cdots & s_j & \cdots & s_{j+v} \\ s_{2j-v} & s_{2j-v+1} & \cdots & s_{2j} & \cdots & s_{2j+v} \\ \vdots & \vdots & \ddots & \vdots & \ddots & \vdots \\ s_{vj-v} & s_{vj-v+1} & \cdots & s_{vj} & \cdots & s_{vj+v} \end{bmatrix} \tag{8}$$

The (m, l)-entry, that is the entry in the m-th row and l-th column, is s_{jm+l} where m is indexed from 0 to v and l is indexed from $-v$ to v.

Our goal is to calculate either $\hat{S}_{2j}$ or $\hat{S}_{2j+1}$ from $\hat{S}_j$. This will enable us to piece together a double-and-add type of algorithm to compute $\hat{S}_k$ given an integer k. The element s_k can then be read off from $\hat{S}_k$.

To calculate $\hat{S}_{2j}$ from $\hat{S}_j$, we simply append rows to the bottom of $\hat{S}_j$. Given the Newton coefficients $a_{1,j}, \ldots, a_{n-1,j}$, we can use the variant of (5)

$$s_{jm+l} = a_{1,j} s_{j(m-1)+l} - a_{2,j} s_{j(m-2)+l} + \cdots + (-1)^{n-1} s_{j(m-n)+l} \tag{9}$$

to progressively compute the sequence terms s_{jm+l} for $m = v+1, \ldots, 2v$ and $l = -v, \ldots, v$. We then keep the rows of even index to form $\hat{S}_{2j}$. We note here that since either $s_{-i} = s_i$ if n is even and $s_{-i} = s_i^p$ if n is odd, all the terms needed to compute these new terms are already in $\hat{S}_j$. Note also that the Newton coefficients can be computed from the elements in $\hat{S}_j$.

To calculate $\hat{S}_{2j+1}$, we need to compute terms of the form $s_{(2j+1)u+w} = s_{2ju+u+w}$ for $u = 0, \ldots, v$ and $w = -v, \ldots, v$. Terms with $2u \le v$ and $u + w \le v$ already exist in $\hat{S}_j$. Terms with $2u \le v$ and $u + w > v$ can be computed using (3) from the terms in the same row. When $2u > v$ and $-v \le u + w \le v$, these terms can be computed using (9) as occurred when computing $\hat{S}_{2j}$. However, if $2u > v$ and $u + w > v$, we require another recurrence.

We can compute the Newton coefficients $a_{1,j+1}, \ldots, a_{n-1,j+1}$ from $\hat{S}_j$. We can then subsequently and progressively compute the terms s_{jm+l} where $m, l > v$ by using the recurrence

$$s_{jm+l} = a_{1,j+1} s_{j(m-1)+l-1} - a_{2,j+1} s_{j(m-2)+l-2} + \cdots + (-1)^{n-1} s_{j(m-n)+l-n} \tag{10}$$

Pictorially, the new terms are calculated diagonally about $\hat{S}_j$ instead of vertically as in the double step.

We now formally state the DDA algorithm.

Algorithm 1. DDA

INPUT: A positive integer k
OUTPUT: $\hat{S}_k$

1. Let w and k_i for $i = 0, \ldots, w$ be such that $k = \sum_{i=0}^{w} k_i 2^i$.
2. $B \leftarrow \hat{S}_1$, $j \leftarrow 1$.
3. For i from $w-2$ down to 0 do
   3.1 Compute the Newton coefficients $a_{1,j}, \ldots, a_{(n-1)/2,j}$.
   3.2 If $k_i = 0$, then $B \leftarrow S_{2j}$, $j \leftarrow 2j$.
   3.3 If $k_i = 1$
      3.3.1 Compute the Newton coefficients $a_{1,j+1}, \ldots, a_{(n-1)/2,j+1}$.
      3.3.2 $B \leftarrow \hat{S}_{2j+1}$, $j \leftarrow 2j+1$.
4. Output $B = \hat{S}_k$.

---

4 The Computational Cost of the DDA

In this section, we examine the computational cost of the DDA. We start by examining the cost of simple operations.

4.1 Measuring Operations

We will measure the cost of the DDA in terms of the number of multiplications in $GF(p)$ it uses. Additions and subtractions and are not as costly as multiplications and so will not be counted. We will include terms only which involve k and disregard those that depend only on n. We shall also assume that k has roughly the same number of 0's as 1's in its binary representation.

The following lemma due to Lenstra and Verheul [7], details the costs of basic operations in $GF(q)$.

Lemma 1 (Lenstra, Verheul). *Suppose* $p \equiv 2 \pmod 3$. *Let* $x, y, z \in GF(p^2)$ *and* $c \in GF(p)$.

- *The* p*-th power* x^p *is for free.*
- *The squaring* x^2 *requires* 2 *multiplications in* $GF(p)$.
- *The multiplication* xy *requires* 3 *multiplications in* $GF(p)$.
- *The joint multiplication* $xz + yz^p$ *requires* 4 *multiplications in* $GF(p)$.
- *The scalar multiplication* cx *requires* 2 *multiplications in* $GF(p)$.

This lemma, and the fact that $a_{i,j} = a_{n-i,j}^{p^n}$ tells us that computing a single sequence term from (3), (5), (9), or (10) requires $2(n-1)$ and $3n/2$ multiplications in $GF(p)$ if n is odd and even respectively.

From (7), we see that the i-th Newton coefficients $a_{i,j}$ where $1 \leq i \leq \lfloor n/2 \rfloor$ requires $3(i-1)$ multiplications in $GF(p)$ in addition to the multiplication by i^{-1}. When i is a power of 2, division by i can essentially be done by a shift of the bits in the representation the elements. Hence it may considered free of cost. Otherwise, $i^{-1} \pmod p$ can be precomputed and requires 2 multiplications in $GF(p)$. Calculating the total cost is now straightforward by summing over i. It is listed in Table 1.

Table 1. Cost of Computing the Newton Coefficients $a_{1,j}, \ldots, a_{\lfloor n/2 \rfloor,j}$

n	# of Multiplications in $GF(p)$
n even	$(3n^2 + 2n - 8 - 16\log n)/8$
n odd	$(3n^2 - n + 1 - 16\log n)/8$

4.2 The Computational Cost of the DDA

We now calculate the total cost of the DDA. Please note that some parts of this analysis are excessively tedious. When this is the case, we will omit some of the

Table 2. Cost of Step 3.2 in the DDA

n	# of Multiplications in $GF(p)$
n even	$3n^3/4$
n odd	$n^3 - 2n^2 + n$

Table 3. Cost of Adding New Rows in Step 2.3.2 of the DDA

n	# odd rows	# even rows	total cost (in mults)
$n \equiv 1 \pmod 4$	$(n-1)/4$	$(n-1)/4$	$(2n^3 - 5n^2 + 4n - 1)/2$
$n \equiv 2 \pmod 4$	$(n-2)/4$	$(n+2)/4$	$(6n^3 - 3n^2 + 6n)/8$
$n \equiv 3 \pmod 4$	$(n-3)/4$	$(n+1)/4$	$(2n^3 - 5n^2 + 6n - 3)/2$
$n \equiv 0 \pmod 4$	$n/4$	$n/4$	$(6n^3 - 3n^2)/8$

Table 4. Cost of Adding Side Terms in Step 3.3.2 of the DDA

n	row index	# multiplications in $GF(p)$
$n \equiv 1 \pmod 4$	even	$(n^3 + n^2 - 5n + 3)/16$
$n \equiv 2 \pmod 4$	even	$(3n^3 - 12n)/64$
$n \equiv 3 \pmod 4$	even	$(n^3 - 3n^2 - n + 3)/16$
$n \equiv 4 \pmod 4$	even	$(3n^3 + 12n^2)/64$
$n \equiv 1 \pmod 4$	odd	$(n^3 - 11n^2 + 27n - 17 + 8(-1)^{(n-5)/4})/32$
$n \equiv 2 \pmod 4$	odd	$(3n^3 + 12n^2 - 12n + 24n(-1)^{(n+2)/4})/128$
$n \equiv 3 \pmod 4$	odd	$(n^3 + n^2 - 9n + 7 + 8(-1)^{(n+1)/4})/32$
$n \equiv 4 \pmod 4$	odd	$(3n^3 - 24n^2 + 24n + (-1)^{(n-4)/4})/128$

details and give the total computational cost. A complete detailed analysis will appear in the full paper.

The cost of the Newton coefficients in Steps 3.1 and 3.3.1 are listed in Table 1.

We calculate $(n-1)/2$ and $n/2$ new rows in Step 3.2 for the case of n odd and even respectively. Each row has n new terms, each of which is derived from an application of (9). Thus, the total cost of this step is the cost of computing $n(n-1)/2$ and $n^2/2$ new terms using (9). It is listed in Table 2.

The only remaining step to be analyzed is Step 3.3.2. There are $\lfloor n/2 \rfloor$ new rows added to the bottom of $\hat{S}_j$. The new rows of even index have n terms. However, the odd-indexed rows are only needed to calculate the even-indexed terms. Because the right-most term is calculated using (10), odd-indexed rows need to contain only $n-1$ terms. The total cost of the new rows is listed in Table 3.

Table 5. Cost of DDA Algorithm

n	# of Multiplications in $GF(p)$
$n \equiv 1 \pmod 4$	$(67n^3 - 117n^2 + 65n - 207 - 192\log n + 8(n-1)(-1)^{(n-5)/4})/64 \log k$
$n \equiv 2 \pmod 4$	$(201n^3 - 108n^2 + 156n - 384 - 768\log n + 24n(-1)^{(n+2)/4})/256 \log k$
$n \equiv 3 \pmod 4$	$(67n^3 - 113n^2 + 69n - 215 - 192\log n + 8(n-1)(-1)^{(n+1)/4})/64 \log k$
$n \equiv 0 \pmod 4$	$(201n^3 - 84n^2 + 144n - 384 - 768\log n + 24n(-1)^{(n-4)/4})/256 \log k$

Step 3.3.2 must also compute terms of the form s_{jm+l} with $m \leq v$ and $l > v$. We shall refer to these terms as side terms. Side terms in even-indexed rows are needed to form $\hat{S}_{2j+1}$. Side terms in odd-indexed rows are needed for use in (10) to compute terms in the rows below it. A proper analysis of the cost of these type of terms is extremely tedious. It will be shown in the full paper. The total cost of these terms is listed in Table 4.

Putting these all together, we get the total cost of the DDA as listed in Table 5.

5 Fiduccia's Algorithm

To evaluate the efficiency of the DDA algorithm, we will compare it to the current standard for efficient computation of linear recurrences, Fiduccia's algorithm [1]. In this section, we describe Fiduccia's algorithm in detail and then analyze its computational cost. Let us start with some background.

The companion matrix C of the linear recurrence (3) is defined as

$$C = \begin{bmatrix} 0 & 1 & 0 & \cdots & 0 \\ 0 & 0 & 1 & \cdots & 0 \\ \vdots & \vdots & \vdots & \ddots & \vdots \\ 0 & 0 & 0 & \cdots & 1 \\ (-1)^{n+1} & (-1)^n a_{n-1} & (-1)^{n-1} a_{n-2} & \cdots & a_1 \end{bmatrix}$$

The characteristic polynomial of C is $f(x)$ as in (1).

We define the sequence vector $\boldsymbol{s}_j$ as

$$\boldsymbol{s}_j = \begin{bmatrix} s_j \\ s_{j+1} \\ \vdots \\ s_{j+n-1} \end{bmatrix}$$

Observe that $C\boldsymbol{s}_j = \boldsymbol{s}_{j+1}$, whence for any positive integer k, $C^k \boldsymbol{s}_j = \boldsymbol{s}_{j+k}$.

Fiduccia makes use of the Cayley-Hamilton theorem.

Theorem 2 (Cayley-Hamilton). *Let $\lambda(x)$ be the characteristic polynomial of an $n \times n$ matrix M. Then $\lambda(M) = 0_n$ where 0_n is the $n \times n$ zero matrix.*

This theorem tells us that $C^k = r(C)$ where $r(x)$ is any polynomial such that $r(x) \equiv x^k \pmod{f(x)}$. Thus, Fiduccia's algorithm computes $r(x)$ in this way, then uses $r(C)$ to get $\boldsymbol{s}_k$. We now state it formally.

Algorithm 2. Fiduccia

INPUT: A positive integer k
OUTPUT: $\boldsymbol{s}_k$

1. $d(x) \leftarrow x$.
2. For i from $w-2$ down to 0 do
   2.1 $d(x) \leftarrow d(x) \times d(x) \bmod f(x)$.
   2.2 If $k_i = 1$, $d(x) \leftarrow d(x) \times x \bmod f(x)$.
3. $D \leftarrow d(C)$ and $\boldsymbol{s}_k \leftarrow D\boldsymbol{s}_0$.

Remark 1. Fiduccia also gave a speedup of Step 3 which only calculated one column of $d(C)$. However, this will not affect the asymptotic complexity of the algorithm so we will not list it here.

Step 2.1 requires a polynomial squaring with reduction modulo $f(x)$. The squaring can be done by using Karatsuba [6] multiplication for a total cost of n^2+n multiplications in $GF(p)$. The reduction modulo $f(x)$ would cost a total of $3n^2-6n+3$ multiplications.

Step 2.2 requires a multiplication by x with reduction modulo $f(x)$. The multiplication has no cost since we merely shift coefficients. The reduction costs $3(n-1)$ multiplications.

Putting this all together, the total cost of Fiduccia's algorithm is $(8n^2-7n+3)/2\log k$ multiplications in $GF(p)$.

6 The Efficiency of the DDA

As Fiduccia's algorithm has a highest term of $n^2 \log k$ while the DDA has a highest term $n^3 \log k$, Fiduccia's algorithm is faster asymptotically for large values of n and for all types of sequences.

The question of which is more efficient for small values of n requires further examination. Table 6 lists the values obtained by substituting $n = 2,3,4,5,6,7$ into the cost of Fiduccia's algorithm and Table 5 for the DDA.

Table 6 shows that for characteristic sequences where $n \geq 6$, Fiduccia is more efficient. For the cases where $n = 2,3$, the DDA is faster. We note here that in these cases, the DDA is essentially the same as given in the LUC cryptosystem [10,16], for $n = 2$ and the XTR cryptosystem [7] for $n = 3$. However, in these two cases no Newton coefficients need to be computed since $a_{1,k} = s_k$. The real power of the DDA occurs when $n \geq 4$ since this is where non-trivial Newton coefficients are computed.

For the cases of $n = 4,5$, the cost is very close between the two algorithms. However, there are some speedups available for the DDA. We examine them in the next two subsections.

Table 6. Comparison of Fiduccia's Algorithm and the DDA for Small Values of n (in $\log k$ operations)

n	Fiduccia	DDA
2	10.5 mult	6 mult
3	27 mult	12 mult
4	51.5 mult	52.5 mult
5	84 mult	84.5 mult
6	124.5 mult	183 mult
7	173 mult	274.5 mult

For some terms during the course of the DDA, (5) can be made more efficient than the standard count by making several observations. We examine these individually for each n.

6.1 Fourth-Order Sequences

For fourth-order sequences, since $a_{1,k} = s_k$ and $s_0 = 4$, we can rewrite (5) for the following terms.

$$s_{3k} = a_{1,k}s_{2k} - a_{2,k}s_k + a_{1,k}s_0 - s_k = s_k(s_{2k} - a_{2,k} + 3)$$

$$s_{3k+3} = a_{1,k+1}s_{2k+2} - a_{2,k+1}s_{k+1} + a_{1,k+1}s_0 - s_{k+1} = s_{k+1}(s_{2k+2} - a_{2,k+1}) + 3s_{k+1}$$

The cost of 6 multiplications in $GF(p)$ changes to 3 for s_{3k} and s_{3k+3}. Note also that the cost of computing the Newton coefficients $a_{2,k}$ and $a_{2,k+1}$ changes from 3 multiplications to 2 multiplications in $GF(p)$ each.

Thus, the total cost of the DDA becomes $48 \log k$ multiplications.

6.2 Fifth-Order Sequences

For fifth-order sequences, (5) requires 8 multiplications in $GF(p)$. Again since $a_{1,k} = s_k$ and $s_0 = 5$, we can rewrite (5) for the following terms.

$$s_{3k} = s_k(s_{2k} - a_{2,k}) + 5a_{2,k}^p - (s_k^p)^2 + s_{2k}^p$$

$$s_{3k+3} = s_{k+1}(s_{2k+2} - a_{2,k+1}) + 5a_{2,k+1}^p - (s_{k+1}^p)^2 + s_{2k+2}^p$$

$$s_{4k} = s_k(s_{3k} + a_{2,k}^p) - a_{2,k}s_{2k} - 4s_k^p$$

$$s_{4k+4} = s_{k+1}(s_{3k+3} + a_{2,k+1}^p) - a_{2,k+1}s_{2k+2} - 4s_{k+1}^p$$

This reduces from a cost of 8 multiplications in $GF(p)$ to 4 for s_{3k} and s_{3k+3} and 6 for s_{4k} and s_{4k+4}. Note also that the cost of computing the Newton coefficients $a_{2,k}$ and $a_{2,k+1}$ changes from 3 multiplications to 2 multiplications in $GF(p)$ each.

Thus the total cost of the DDA becomes $74 \log k$ multiplications. It should be noted that Quoos and Mjølsnes [14] also gave an algorithm for computing fifth-order sequences of this type. However, the computational cost of their algorithm was found to be $102 \log k$ multiplications in $GF(p)$.

6.3 Summary

The computational cost for Fiduccia's algorithm and the DDA algorithm with improvements are listed in Table 7 for $n = 4, 5$.

Table 7. Updated Comparison of Fiduccia and the DDA for $n = 4, 5$ (in $\log k$ operations)

n	Fiduccia	DDA
4	51.5 mult	48 mult
5	84 mult	74 mult

Hence, the DDA is the faster algorithm for $n = 2, 3, 4, 5$ while Fiduccia's algorithm is faster for $n \geq 6$.

References

1. Fiduccia, C.M.: An Efficient Formula for Linear Recurrences. SIAM J. Comput. **14** (1985) 106–112.
2. Giuliani, K., Gong, G.: Efficient Key Agreement and Signature Schemes Using Compact Representations in $GF(p^{10})$. In: Proceedings of the 2004 IEEE International Symposium on Information Theory - ISIT 2004. Chicago (2004) 13–13.
3. Giuliani, K., Gong, G.: New LFSR-Based Cryptosystems and the Trace Discrete Log Problem (Trace-DLP). In: Sequence and Their Applications – SETA 2004. Lecture Notes In Computer Science, Vol. 3486. Springer-Verlag, Berlin Heidelberg New York (2005) 298–312.
4. Gong, G., Harn, L.: Public-Key Cryptosystems Based on Cubic Finite Field Extensions. IEEE Trans. IT. **24** (1999) 2601–2605.
5. Gries, D., Levin, D.: Computing Fibonacci Numbers (and Similarly Defined Functions) in Log Time. Information Processing Letters **11** (1980) 68–69.
6. Karatsuba, A., Ofman, Y.: Mulplication of Many-Digital Numbers by Automatic Computers. Physics-Daklady **7** (1963) 595–596.
7. Lenstra, A., Verheul, E.: The XTR Public Key System. In: Advances in Cryptology – Crypto 2000. Lecture Notes In Computer Science, Vol. 1880. Springer-Verlag, Berlin Heidelberg New York (2000) 1–19.
8. Lidl, N., Niederreiter, H.: Finite Fields. Addison-Wesley, Reading (1983).
9. Miller, J. C. P., Spencer-Brown, D. J.: An Algorithm For Evaluation of Remote Terms in a Linear Recurrence Sequence. Computer Journal **9** (1966/67) 188–190.
10. Müller, W. B., Nobauer, R.: Cryptanalysis of the Dickson scheme. In: Advances in Cryptology – Eurocrypt 1985. Lecture Notes In Computer Science, Vol. 219. Springer-Verlag, Berlin Heidelberg New York (1986) 50–61.

11. Niederreiter, H.: A Public-Key Cryptosystem Based on Shift-Register Sequences. In: Advances in Cryptology – Eurocrypt 1985. Lecture Notes In Computer Science, Vol. 219. Springer-Verlag, Berlin Heidelberg New York (1986) 35–39.
12. Niederreiter, H.: Some New Cryptosystems Based on Feedback Shift Register Sequences. Math. J. Okayama Univ. **30** (1988) 121-149.
13. Niederreiter, H.: Finite Fields and Cryptology. In: Finite Fields, Coding Theory, and Advances in Communications and Computing. M. Dekker, New York (1993) 359–373.
14. Quoos, L., Mjølsnes, S.-F.: Public Key Systems Based on Finite Field Extensions of Degree Five. Presented at Fq7 conference (2003).
15. Shortt, J.: An Iterative Algorithm to Calculate Fibonacci Numbers in $O(\log n)$ Arithmetic Operations. Information Processing Letters **7** (1978) 299–303.
16. Smith, P., Skinner, C.: A Public-Key Cryptosystem and a Digital Signature System Based on the Lucas Function Analogue to Discrete Logarithms. In: Advances in Cryptology – Asiacrypt '94. Lecture Notes In Computer Science, Vol. 917. Springer-Verlag, Berlin Heidelberg New York (1994) 357–364.
17. Urbanek, F. J.: An $(O(\log n)$ Algorithm for Computing the nth Element of a Solution of a Difference Equation. Information Processing Letters **11** (1980) 66–67.
18. Ward, M.: The Algebra of Recurring Series. Annals of Math **32** (1931) 1–9.
19. Wilson, T. C., Shortt, J.: An $O(\log n)$ Algorithm for Computing General Order-k Fibonacci Numbers. Information Processing Letters **10** (1980) 68–75.

Implementation of Multi-continued Fraction Algorithm and Application to Multi-sequence Linear Synthesis*

Quanlong Wang[1,2], Kunpeng Wang[2], and Zongduo Dai[2]

[1] School of Mathematical Sciences, Peking University, Beijing 100871, P.R. China
quanlongwang@yahoo.com.cn

[2] State Key Laboratory of Information Security (Graduate School of Chinese Academy of Sciences), Beijing 100049, P.R. China
{kpwang, daizongduo}@is.ac.cn

Abstract. In this paper, we present a method of implementing the multi-continued fraction algorithm on a class of infinite multi-sequences. As applications of our implementing method, we get the linear complexity and minimal polynomial profiles of some non-periodic multi-sequences.

1 Introduction

It is well known that Berlekamp-Massey algorithm[1,2] (BMA) can be used for the problem of single sequence linear synthesis, and its generalization, the generalized Berlekamp-Massey algorithm (GBMA) [3,4], can be used for the same problem of multi-sequences. However, it is a never ending job when acting the GBMA on a non-periodic multi-sequence S, because GBMA is an iterative algorithm and may only return the linear complexity and minimal polynomial of the length n prefix of S at the n-th step.

Recently, a multi-continued fraction algorithm (m-CFA), as a generalization of simple continued fraction algorithm for formal Laurent series, was introduced for multi-formal Laurent series over any given field [5,6]. It is known [5,6] that the m-CFA always provides optimal rational approximation to multi-formal Laurent series, while it is proved [14] that both Jacobi-Perron Algorithm (JPA) [15] and modified Jacobi-Perron Algorithm (MJPA) [17], which are also algorithms for the problem of rational approximation to multi-formal Laurent series, do not guarantee providing optimal rational approximation. In addition, any multi-sequence can be identified with a multi-formal Laurent series [6] and the multi-sequence linear synthesis problem is essentially the optimal rational approximation problem of the multi-formal Laurent series (identified with it). The linear complexity and minimal polynomial profiles of any given multi-sequence can then be obtained immediately from its multi-continued fraction expansion. Thus, it is not

* This work was supported in part by the National Science Foundation of China (NSFC) under Grants No.90604011.

G. Gong et al. (Eds.): SETA 2006, LNCS 4086, pp. 248–258, 2006.

surprising that m-CFA is a powerful tool in dealing with the problems related to linear complexities of multi-sequences. In fact, there are already advances in problems based on the technique of m-CFA, such as on d-perfect multi-sequence conjecture [8,9,10], asymptotic behavior of normalized linear complexities of multi-sequences [11], and the conjecture [8,13] of the expected value of linear complexity of multi-binary sequences and the relation [12] between the GBMA and the m-CFA, etc.. In view of the usefulness of m-CFA, it is natural to consider how to implement m-CFA in computer programs.

In this paper, we present a method of implementing the m-CFA on a class of multi-formal Laurent series whose components are algebraic functions of degree greater than 1 over the rational function field over any given finite field. We have run our implementing on some multi-formal Laurent series of the form $(\gamma, \gamma^2)^T$ (here and later, T means transpose), where γ is a special kind of algebraic functions of degree 3 over the rational function field over the binary field, the experiment results show that their multi-continued fraction expansions are all periodic. As applications, the linear complexity and minimal polynomial profiles for the multi-infinite sequences corresponding to $(\gamma, \gamma^2)^T$ are obtained. It is worth pointing out that these multi-sequences are all non-periodic, since these γ are irrational.

2 m-CFA and Linear Complexity

In this section, we recall the multi-continued fraction algorithm which is given in [5,6] and its relation with the linear complexity of multi-sequence. For this purpose, we need to recall some concepts and give some notations.

Let F_q be the finite field with q elements, and $F_q(z)$ be the rational function field of the indeterminate z over the field F_q. Let $S = (\underline{s}_1, \cdots, \underline{s}_m)^T$ be an m-dimensional multi-sequence, where $\underline{s}_i = \{s_{i,t}\}_{t\geq 0}$ is an infinite sequence over F_q. For any positive integer n, let $S^{(n)} = (\underline{s}_1^{(n)}, \cdots, \underline{s}_m^{(n)})^T$ be the length n prefix of S, where $\underline{s}_i^{(n)} = \{s_{i,t}\}_{t=0}^{n-1}$. Then the problem of multi-sequence shift-register synthesis is to determine a minimum l and a $\sigma(x)$, where

$$\sigma(x) = \sigma_l + \sigma_{l-1}x + \cdots + \sigma_1 x^{l-1} + x^l \in F_q,$$

such that the following equation holds:

$$s_{i,j} + \sigma_1 s_{i,j-1} + \cdots + \sigma_l s_{i,j-l} = 0,$$

for $j = l, l+1, \cdots, n-1$ and $i = 1, 2, \cdots, m$. Namely, it is to find a LFSR of shortest length capable of generating these m sequences. This minimum l and the polynomial $\sigma(x)$, denoted by $L_n(S)$ and $f_n(s)$, is called the linear complexity of $S^{(n)}$ and the minimal polynomial of $S^{(n)}$, respectively.

For a formal Laurent series $\alpha = \sum_{t\geq b} a_t z^{-t} \in F_q((z^{-1}))(a_t \in F_q)$, its discrete valuation $v(\alpha)$ is defined by

$$v(\alpha) = \begin{cases} +\infty, & if\ \alpha = 0, \\ b, & if\ a_b \neq 0. \end{cases}$$

Each $\alpha \in F_q((z^{-1}))$ can be written as a sum $\alpha = \lfloor\alpha\rfloor + \{\alpha\}$, where $\lfloor\alpha\rfloor = \sum_{d\geq -i\geq 0} a_i z^{-i}$ is its polynomial part and $\{\alpha\} = \sum_{1\leq i<\infty} a_i z^{-i}$ is its remaining part. Let $F_q((z^{-1}))^m$ and $F_q[z]^m$ denote the set of all m-tuples over $F_q((z^{-1}))$ and $F_q[z]$ respectively. For the m-dimensional multi-sequence $S = (\underline{s}_1, \cdots, \underline{s}_m)^T$, we identify it with the m-tuple Laurent series

$\underline{r} = (r_1(z), r_2(z), \cdots, r_m(z))^T \in F_q((z^{-1}))^m$, where $r_j(z) = \sum_{t\geq 0} s_{j,t} z^{-(t+1)}$ for $1 \leq j \leq m$. The m-tuple $(\lfloor r_1(z)\rfloor, \lfloor r_2(z)\rfloor, \cdots, \lfloor r_m(z)\rfloor)^T \in F_q[z]^m$ is called the polynomial part of $\underline{r}$ and denoted as $\lfloor\underline{r}\rfloor$, and the m-tuple

$\{\underline{r}\} = (\{r_1(z)\}, \{r_2(z)\}, \cdots, \{r_m(z)\})^T \in F_q((z^{-1}))^m$ is called the remaining part of $\underline{r}$.

Multi-continued Fraction Algorithm (m-CFA, in Short):

Given a multi-Laurent series

$$\underline{r} = (\sum_{t\geq 0} s_{1,t} z^{-(t+1)}, \cdots, \sum_{t\geq 0} s_{j,t} z^{-(t+1)}, \cdots, \sum_{t\geq 0} s_{m,t} z^{-(t+1)})^T.$$

Initially, let $v_{0,1} = \cdots = v_{0,m} = 0$; $\underline{a}_0 = \underline{0}$; $\underline{\beta}_0 = (\beta_{0,1}, \cdots, \beta_{0,m})^T = \underline{r}$. Repeat the following rounds successively for $k \geq 1$, and the k-th round consists of the following five steps:

(1) $v_k = min\{v_{k-1,j} + v(\beta_{k-1,j}) | 1 \leq j \leq m\}$.
(2) $h_k = min\{j | v_{k-1,j} + v(\beta_{k-1,j}) = v_k, 1 \leq j \leq m\}$.
(3) $v_{k,j} = v_{k-1,j}$ if $j \neq h_k$ and $v_{k,h_k} = v_k$.
(4) $\underline{\rho}_k = (\rho_{k,1}, \cdots, \rho_{k,j}, \cdots, \rho_{k,m})^T \in F((z^{-1}))^m$, where $\rho_{k,j} = \frac{\beta_{k-1,j}}{\beta_{k-1,h_k}}$ if $j \neq h_k$ and $\rho_{k,h_k} = \frac{1}{\beta_{k-1,h_k}}$.
(5) $\underline{a}_k = \lfloor\underline{\rho}_k\rfloor = (a_{k,1}(z), \cdots, a_{k,m}(z))^T \in F_q[z]^m$ and $\underline{\beta}_k = \underline{\rho}_k - \underline{a}_k \in F_q((z^{-1}))^m$, and set $\underline{\beta}_k = (\beta_{k-1,1}, \cdots, \beta_{k-1,m})^T$.

As results, we get the following expansion:

$$C(S) = [\underline{a}_0 = \underline{0}, h_1, \underline{a}_1, \cdots, h_k, \underline{a}_k, \cdots], \quad 1 \leq k < \infty,$$

which is called the multi-continued fraction expansion (m-CFE, in short) of S. Denote $\underline{a}_k = (a_{k,1}(z), \cdots, a_{k,j}(z), \cdots, a_{k,m}(z))^T$, $a_{k,j}(z) \in F_q[z]$. It is known that deg $(a_{k,h_k}(z)) \geq 1$. Associated with $C(S)$, we define the following parameters:

$$t_k = \deg(a_{k,hk}(z)) \geq 1, \quad t_0 = 0,$$
$$d_k = \sum_{1\leq i\leq k} t_i, \quad d_0 = 0,$$
$$v_k = \sum_{h_i = h_k, 1\leq i\leq k} t_i, \quad v_0 = 0,$$
$$n_k = d_{k-1} + v_k,$$
$$l(k,j) = max\{0, i \mid 1 \leq i \leq k, h_i = j\}, \quad \forall\, k \geq 0.$$

Then, based on the function $l(k,j)$, we define a set of polynomials $\{q_k(z)\}_{k\geq 0}$ inductively :

$$q_0(z) = 1,$$
$$q^*_{l(k,j)-1}(z) = \begin{cases} q_{l(k,j)-1}(z), if & l(k,j) \geq 1, \\ 0 \qquad\qquad\quad if & l(k,j) = 0, \end{cases} k \geq 1,$$
$$q_k(z) = q^*_{l(k-1,h_k)-1}(z) + \sum_{1\leq j\leq m} a_{k,j}(z) q^*_{l(k,j)-1}(z).$$

Similar to the one-dimensional case [7], the minimal polynomial and linear complexity of the length n prefix $S^{(n)}$ of the m-dimensional multi-sequence S, denoted by $f_n(S)$ and $L_n(S)$ respectively, can be read out immediately from the above parameters, as shown below.

Proposition 1. *[6]*
For any n such that $n_k \leq n < n_{k+1}$, $k \geq 0$, we have $f_n(S) = q_k(z)$ and $L_n(S) = d_k$.

3 Method of Implementing m-CFA

In this section, we present a method of implementing the m-CFA on multi-formal Laurent series over $F_q(z)$ whose components are algebraic functions of degree greater than 1.

Proposition 2. *[15] Let $n \geq 2$, and*

$$f(Y) = Y^n + k_1(z)d(z)Y^{n-1} + k_2(z)d(z)Y^{n-2} + \cdots + k_{n-1}(z)d(z)Y - d(z) \in F_q[z][Y], \tag{1}$$

where $d(z), k_i(z) \in F_q[z]$ for $1 \leq i \leq n-1$, $k_{n-1}(z)d(z) \neq 0$, and $\deg k_{n-1}(z) > \deg d(z) + max\{\deg k_i(z) \mid 0 \leq i \leq n-2\} (k_0 = 1)$. Then

(1) $f(Y)$ is irreducible in $F_q(z)[Y]$.
(2) There exists a unique root, denoted by γ, of $f(Y)$ in $F_q((z^{-1}))$, i.e., $f(\gamma) = 0$.
(3) $v(\gamma) = \deg k_{n-1}(z)$.

In the sequel we always let γ denote the unique root of any given polynomial $f(Y)$ which is of the form as in (1).

Below we discuss how to implement the m-CFA on the (n-1)-tuple Laurent series $\underline{r} = (\gamma, \gamma^2, \cdots, \gamma^{n-1})^T$.

In implementing the k-th round of the m-CFA, we have to solve the following problems:

(1) Compute $v(\beta_{k-1,j})$, for $1 \leq j \leq m$.
(2) Compute $a_{k,j}(z)$, for $1 \leq j \leq m$.
(3) Decide whether the m-CFE has just finished a period (Later we will give the definition of periodicity of a multi-continued fraction).

3.1 Criterion for Periodicity Decision

The m-CFA is an iterative algorithm. For a general multi-formal Laurent series S, the result obtained after finite number of rounds in acting the m-CFA on S gives only partial information about the m-CFE $C(S)$. However, it gives complete information whenever the m-CFE $C(S)$ is ultimately periodic. In implementing m-CFA, it is desired to know whether the m-CFE $C(S)$ is periodic. When $C(S)$ is periodic, one should determine whether it enters into a period at each round. However, the periodicity of $C(S)$ is not guaranteed by the condition $(h_{\lambda+U}, a_{\lambda+U}) = (h_\lambda, a_\lambda)$ for some $\lambda \geq 0$ and $U \geq 1$. Fortunately, a sufficient condition for periodic m-CFE $C(S)$ and a criterion for the fact that the process of the m-CFA acting on a S is entering in a period are found. We give the relevant results as below.

Definition 1. *Let $C(S) = [\underline{a}_0 = \underline{0}, h_1, \underline{a}_1, \cdots, h_k, \underline{a}_k, \cdots]$ be the m-CFE of S . $C(S)$ is called (λ, U)-periodic for some integers $\lambda \geq 1$ and $U \geq 1$ if it satisfies the following conditions:*

$$(h_{\lambda+k},\ \underline{a}_{\lambda+k}) = (h_{\lambda+k+U},\ \underline{a}_{\lambda+k+U}),\ \forall k \geq 0. \tag{2}$$

We denote $[\underline{a}_0 = \underline{0}, h_1, \underline{a}_1, \cdots, h_k, \underline{a}_k, \cdots]$ simply by $[\underline{0}, h_1, \underline{a}_1, \cdots, h_{\lambda-1}, \underline{a}_{\lambda-1}, \cdots, \overline{h_\lambda, \underline{a}_\lambda, h_{\lambda+1}, \underline{a}_{\lambda+1}, \cdots,\ h_{\lambda+U-1}, \underline{a}_{\lambda+U-1}}\,]$ if $C(S)$ is (λ, U)-periodic.

Proposition 3. *Let $C(S) = [\underline{a}_0 = \underline{0}, h_1, \underline{a}_1, h_2, \underline{a}_2, \cdots, h_k, \underline{a}_k, \cdots]$ be the m-CFE of S, which is obtained by the actiing m-CFA on S. If there exist integers $\lambda \geq 1$ and $U \geq 1$ such that $\underline{\beta}_{\lambda-1} = \underline{\beta}_{\lambda-1+U}$ and $v_{\lambda-1+U,j} - v_{\lambda-1,j} = v_{\lambda-1+U,i} - v_{\lambda-1,i}$ hold true for $1 \leq i \leq j \leq m$, then $C(S)$ is (λ, U)-periodic.*

Proof. Induction on k.

Let $\Delta_{k-1} = Diag.(z^{-v_{k,1}}, z^{-v_{k,2}}, \cdots, z^{-v_{k,m}})$ be a diagonal matrix. Let Iv be the indexed valuation defined in [6]. From [6] we have

$$\begin{aligned}(h_{\lambda+U},\ v_{\lambda+U}) &= Iv(\Delta_{\lambda-2+U}\beta_{\lambda-1+U}) \\ &= Iv(z^l\Delta_{\lambda-2}\beta_{\lambda-1}) \\ &= (h_\lambda,\ v_\lambda + v(z^l)) \\ &= (h_\lambda,\ v_\lambda - l).\end{aligned}$$

So $h_{\lambda+U} = h_\lambda$, together with $\beta_{\lambda-1} = \beta_{\lambda-1+U}$, we have $\rho_\lambda = \rho_{\lambda+U}$. It follows that $\underline{a}_\lambda = \underline{a}_{\lambda+U}$, $\beta_\lambda = \beta_{\lambda+U}$. Thus for $k = \lambda$, we have proved that $(h_k,\ \underline{a}_k) = (h_{k+U},\ \underline{a}_{k+U})$.

Next, since $z^l\Delta_{\lambda-2} = \Delta_{\lambda-2+U}$ and $v_{\lambda+U} = v_\lambda - l$, by the definition of Δ_k, we have $z^l\Delta_{\lambda-1} = \Delta_{\lambda-1+U}$.

So, we can do the above procedure inductively, and the proposition follows.

3.2 Method of Computing Valuation and Polynomial Part

Since $\underline{\beta}_0 = (\gamma, \gamma^2, \cdots, \gamma^{n-1})^T$, it is easy to see by induction that $\beta_{k-1,j}$ belongs to the field $F_q(z)(\gamma)$ for $k \geq 1$, $1 \leq j \leq m$. By Proposition 2, $f(Y)$ in 1 is irreducible over $F_q(z)$, so each element $\alpha \in F_q(z)(\gamma)$ can be uniquely expressed by $\alpha = c_1\gamma^{n-1} + c_2\gamma^{n-2} + \cdots + c_n$, where $c_i \in F_q(z)$. We call this representation a normalized form of α.

Given $f_1(Y) = c_1Y^{n-1} + c_2Y^{n-2} + \cdots + c_n \in F_q(z)[Y]$. In order to compute $v(f_1(\gamma))$ in a computer program, it is convenient to find $\mu \in F_q(z)$ such that $v(f_1(\gamma)) = v(f_1(\mu))$. Similarly for computing $\lfloor f_1(\gamma) \rfloor$, we find $\mu \in F_q(z)$ such that $\lfloor f_1(\gamma) \rfloor = \lfloor f_1(\mu) \rfloor$.

Therefore, we can solve the problem of computing $v(\beta_{k-1,j})$ and $a_{k,j}(z)$ by setting up three subprograms(**PROCEDURES**): **Normalization(), Valuation()** and **Polynomial-Part()**. For any $\alpha \in F_q(z)(\gamma)$, **Normalization**($\alpha$) returns the normalized form of α; for any $\alpha \in F_q(z)(\gamma)$ in normalized form, **Valuation**(α) and **Polynomial-Part**(α) returns the valuation and polynomial part of α respectively. These procedures are mainly based on the following:

Proposition 4. *[18](Newton approximation) Let* $O=\{\alpha | v(\alpha) \geq 0, \alpha \in F_q((z^{-1}))\}$. *Suppose* $g(Y) \in O[Y]$, $a_0 \in O$ *and* $\infty \neq v(g(a_0)/g'(a_0)^2) > 0$, *then the sequence induced by the following iterative relation*

$$a_{i+1} = a_i - g(a_i)/g'(a_i), \ i = 0, 1, 2, \cdots,$$

satisfies

(1) $\lim_{i\to\infty} a_i = a \in O$, *and* $g(a) = 0$.

(2) $v(a - a_i) \geq 2^i v(g(a_0)/g'(a_0)^2)$.

Now we return to equation 1. It can be transformed into

$$g(Y) = \frac{1}{k_{n-1}(z)d(z)}Y^n + \frac{k_1(z)}{k_{n-1}(z)}Y^{n-1} + \cdots + Y - \frac{1}{k_{n-1}(z)} = 0. \quad (3)$$

Take a_0 to be 0. It is easy to check that $g(Y)$ in 3 satisfies the condition in Proposition 4. Thus, by applying the Newton approximation to $g(Y)$, we get a sequence $\{a_i\}_{i=0}^{\infty}$ satisfying $a_i \in F_q(z)$, $\lim_{i\to\infty} a_i = \gamma$ and $v(\gamma - a_i) \geq 2^i v(1/k_{n-1}(z))$.

Proposition 5. *Let* $\{a_i\}_{i=0}^{\infty}$ *be defined as in Proposition 4 with* $g(Y)$ *in (3). Suppose* $f_1(Y) = c_1Y^{n-1} + c_2Y^{n-2} + \cdots + c_n \in F_q(z)[Y]$ *is a nonconstant polynomial in* Y. *Then there exists an integer* N *such that* $v(f_1(\gamma)) = v(f_1(a_i))$, *for all* $i \geq N$.

Proof. Let X be a undetermined element different from Y. Then

$$f_1(X) - f_1(Y) = (X - Y)h(X, Y),$$

where $0 \neq h(X,Y) = \sum c_{jk}X^jY^k, 0 \leq j,k \leq n-2, c_{jk} \in F_q(z)$. Since $v(\gamma - a_i) \geq 2^i v(1/k_{n-1}(z))$, then $v(\gamma) = v(a_i)$ for i large enough. Thus,

$$\begin{aligned} v(f_1(a_i) - f_1(\gamma)) &= v((a_i - \gamma)h(a_i, \gamma)) \\ &= v(a_i - \gamma) + v(h(a_i, \gamma)) \\ &\geq 2^i v(1/k_{n-1}(z)) + min\{v(c_{jk}) + (j + k)v(\gamma)\} \end{aligned}$$

for i large enough. Therefore $\lim_{i\to\infty} v(f_1(a_i) - f_1(\gamma)) = +\infty$. Since $f_1(\gamma) \neq 0$, we have $v(f_1(\gamma)) < +\infty$. So there must exist an integer N such that $v(f_1(a_i) - f_1(\gamma)) > v(f_1(\gamma)), i \geq N$; that is $v(f_1(a_i)) = v(f_1(\gamma))$ for $i \geq N$.

Remark 1. In order to compute $v(f_1(\gamma))$, we compare the value of $2^i v(1/k_{n-1}(z)) + v(h(a_i, \gamma))$ with $v(f_1(a_i))$ for each i from $i = 1$ to $i = N$ such that N is the least integer satisfying $2^N v(1/k_{n-1}(z)) + v(h(a_N, \gamma)) > v(f_1(a_N))$. Then $v(f_1(a_N)) = v(f_1(\gamma))$ by Proposition 5. In fact, such an N can be soon found since $2^i v(1/k_{n-1}(z)) + v(h(a_i, \gamma))$ increases exponentially in respect to i while $v(f_1(a_i))$ is bounded.

Proposition 6. *Let $\{a_i\}_{i=0}^{\infty}$ be defined as in Proposition 4 with $g(Y)$ in 3. Suppose $f_1(Y) = c_1Y^{n-1} + c_2Y^{n-2} + \cdots + c_n \in F_q(z)[Y]$ is a nonconstant polynomial in Y. Then there exists an integer N such that $\lfloor f_1(\gamma)\rfloor = \lfloor f_1(a_i)\rfloor$, for all $i \geq N$.*

Proof. From the proof of Proposition 5 we know, $\lim_{i\to\infty} v(f_1(a_i) - f_1(\gamma)) = +\infty$. So there must exist an integer N such that $v(f_1(a_i) - f_1(\gamma)) > 0, i \geq N$. That is $\lfloor f_1(\gamma)\rfloor = \lfloor f_1(a_i)\rfloor$ for $i \geq N$.

Remark 2. In order to compute $\lfloor f_1(\gamma)\rfloor$, we compare the value of $2^i v(1/k_{n-1}(z)) + v(h(a_i, \gamma))$ with 0 for each i from $i = 1$ to $i = N$ such that N is the least integer satisfying $2^N v(1/k_{n-1}(z)) + v(h(a_N, \gamma)) > 0$. Then $\lfloor f_1(\gamma)\rfloor = \lfloor f_1(a_N)\rfloor$ by Proposition 6. In fact, such an N can be soon found since $2^i v(1/k_{n-1}(z)) + v(h(a_i, \gamma))$ increases exponentially in respect to i.

Let $f(Y)$ be defined as in 1. Then we set up the three subprograms as follows.

- **PROCEDURE Normalization**(α) :
  On input $\alpha \in F_q(z)(\gamma)$, it must has the form $\alpha = \frac{g_1(\gamma)}{g_2(\gamma)}$, where $g_1(Y)$, $g_2(Y) \in F_q(z)[Y]$. By Euclid division algorithm, we can get $h_2(Y) \in F_q(z)[Y]$ such that $h_2(\gamma)g_2(\gamma) \equiv 1 \pmod{f(\gamma)}$. Again by the Euclid division algorithm, we get $c_1\gamma^{n-1} + c_2\gamma^{n-2} + \cdots + c_n \equiv h_2(\gamma)g_1(\gamma) \pmod{f(\gamma)}$ for $c_i \in F_q(z), i = 1, 2, \cdots, n$. Then $c_1\gamma^{n-1} + c_2\gamma^{n-2} + \cdots + c_n$ is the normalized form of α.
  Return $c_1\gamma^{n-1} + c_2\gamma^{n-2} + \cdots + c_n$.
- **PROCEDURE Valuation**(α) :
  On input $\alpha \in F_q(z)(\gamma)$ in normalized form. By Proposition 5 we find a_N such that $v(\alpha) = v(c_1a_N^{n-1} + c_2a_N^{n-2} + \cdots + c_n)$. Since c_i, $a_N \in F_q(z)$, there exist $h_1(z)$, $h_2(z) \in F_q[z]$ such that $c_1a_N^{n-1} + c_2a_N^{n-2} + \cdots + c_n = \frac{h_1(z)}{h_2(z)}$, $gcd(h_1(z), h_2(z)) = 1$. Therefore, $v(\alpha) = v(c_1a_N^{n-1} + c_2a_N^{n-2} + \cdots + c_n) = deg(h_2(z)) - deg(h_1(z))$.
  Return $deg(h_2(z)) - deg(h_1(z))$.
- **PROCEDURE Polynomial-Part**(α) :
  On input $\alpha \in F_q(z)(\gamma)$ in normalized form. we can get $F_N(\alpha) = f_1(\gamma) = c_1\gamma^{n-1} + c_2\gamma^{n-2} + \cdots + c_n$. By Proposition 6 we find a_N such that $\lfloor c_1\gamma^{n-1} + c_2\gamma^{n-2} + \cdots + c_n\rfloor = \lfloor c_1a_N^{n-1} + c_2a_N^{n-2} + \cdots + c_n\rfloor$. Since $c_i, a_N \in F_q(z)$, there exist $h_1(z), h_2(z) \in F_q[z]$ such that $c_1a_N^{n-1} + c_2a_N^{n-2} + \cdots + c_n = \frac{h_1(z)}{h_2(z)}$, $gcd(h_1(z), h_2(z)) = 1$. Let $q(z)$ be the quotient of $h_1(z)$ divided by $h_2(z)$ in $F_q[z]$, $q_1(z) \equiv q(z) \pmod{f(\gamma)}$, then $\lfloor\alpha\rfloor = q_1(z)$.
  Return $q_1(z)$.

3.3 Program for the Implementation of m-CFA

In this subsection, we give a program to implement the m-CFA on a multi-formal Laurent series $\underline{r} = (\gamma, \gamma^2, \cdots, \gamma^{n-1})^T$, where γ is the unique root of $f(Y)$ in $F_q((z^{-1}))$, and $f(Y)$ is defined as in (1). Let $m = n-1$ in the following program.

Program:
Input: $v_{0,1} = \cdots = v_{0,m} = 0$; $\underline{a}_0 = \underline{0}$; $\underline{\beta}_0 = \underline{r} = (\gamma, \gamma^2, \cdots, \gamma^{n-1})^T$.

Output: $C(\underline{r})$.

(1) **for** $k = 1$ **to** l (a chosen positive integer large enough) **do**
  (1.1) $\beta_{k-1,j} \leftarrow$ **Normalization**$(\beta_{k-1,j})$, $1 \le j \le m$
  (1.2) $v(\beta_{k-1,j}) \leftarrow$ **Valuation**$(\beta_{k-1,j})$, $1 \le j \le m$
  (1.3) $v_k \leftarrow min\{v_{k-1,j} + v(\beta_{k-1,j}) | 1 \le j \le m\}$
  (1.4) $h_k \leftarrow min\{j | v_{k-1,j} + v(\beta_{k-1,j}) = v_k, 1 \le j \le m\}$
  (1.5) $v_{k,j} \leftarrow v_{k-1,j}$ if $j \ne h_k$
  $v_{k,h_k} \leftarrow v_k$
  (1.6) $\rho_{k,j} \leftarrow \frac{\beta_{k-1,j}}{\beta_{k-1,h_k}}$ if $j \ne h_k$
  $\rho_{k,h_k} \leftarrow \frac{1}{\beta_{k-1,h_k}}$
  $\underline{\rho}_k \leftarrow (\rho_{k,1}, \cdots, \rho_{k,j}, \cdots, \rho_{k,m})^T$
  (1.7) $\rho_{k,j} \leftarrow$**Normalization**$(\rho_{k,j})$, $1 \le j \le m$
  (1.8) $a_{k,j}(z) \leftarrow$**Polynomial-Part**$(\rho_{k,j})$, $1 \le j \le m$
  $\underline{a}_k \leftarrow (a_{k,1}(z), \cdots, a_{k,m}(z))^T$
  $\underline{\beta}_k = (\beta_{k-1,1}, \cdots, \beta_{k-1,m})^T \leftarrow (\underline{\rho}_k - \underline{a}_k)$
  (1.9) If there exist an integer $1 \le \lambda \le k = \lambda - 1 + U$ such that $\underline{\beta}_{\lambda-1} = \underline{\beta}_k$, and $v_{k,j} - v_{\lambda-1,j} = v_{k,i} - v_{\lambda-1,i}, for 1 \le i \le j \le m$, then $C(\underline{r}) \leftarrow [\underline{0}, h_1, \underline{a}_1, \cdots, h_{\lambda-1}, \underline{a}_{\lambda-1}, \overline{h_\lambda, \underline{a}_\lambda, h_{\lambda+1}, \underline{a}_{\lambda+1}, \cdots, h_{\lambda+U-1}, \underline{a}_{\lambda+U-1}}]$
  return $C(\underline{r})$
  end Program
  (1.10) $C(\underline{r}) \leftarrow [\underline{a}_0 = \underline{0}, h_1, \underline{a}_1, \cdots, h_k, \underline{a}_k]$
(2) **return** $C(\underline{r})$
(3) **end** Program

Remark 3. Here we restrict $\underline{\beta}_0$ to a special form $\underline{\beta}_0 = (\gamma, \gamma^2, \cdots, \gamma^{n-1})^T$, because this is a typical form that is discussed in some references such as [15] and [16]. In fact, if $\underline{\beta}_0 = (f_1(\gamma), f_2(\gamma), \cdots, f_{n-1}(\gamma))^T$, where $f_i(\gamma) \in F_q(z)(\gamma)$ for all i, our program still works.

4 Application of the Implementing of m-CFA

In this section we give some examples of applying our implementing m-CFA to the multi-formal Laurent series $(\gamma, \gamma^2)^T$, where γ is a root of an irreducible polynomial $f(Y)$ of degree 3 over $F_2(z)$ with the form as shown in the equation (1).

Denote by S the multi-sequences corresponding to $(\gamma, \gamma^2)^T$. It is worth pointing out that these S are all non-periodic, since γ is irrational. The experiment results show that their multi-continued fraction expansions are all periodic. As applications, the linear complexity and minimal polynomial profiles for these S are obtained.

- **Example 1:** Let $\underline{r} = \begin{pmatrix} \gamma \\ \gamma^2 \end{pmatrix}$, where γ is the root of $Y^3 + k(z)Y - 1 = 0$, $k(z) \in F_2[z]$, $\deg k(z) > 0$. The m-CF expansion of $\underline{r}$ is as follows.

$$C(\underline{r}) = \left[\begin{pmatrix} 0 \\ 0 \end{pmatrix}, \overline{1, \begin{pmatrix} k(z) \\ 0 \end{pmatrix}, 2, \begin{pmatrix} 0 \\ k(z) \end{pmatrix}} \right].$$

Let $\tau = deg(k(z))$. From Proposition 1 we get $d_k = k\tau$, $v_k = \lfloor \frac{k+1}{2} \rfloor \tau$, $n_k = (k - 1 + \lfloor \frac{k+1}{2} \rfloor)\tau$, $q_0(z) = 1$, $q_1(z) = k(z)$, $q_2(z) = k(z)^2$ and $q_k(z) = k(z)q_{k-1}(z) + q_{k-3}(z)$ for all $k \geq 3$. Hence, we get

$$(f_n(S), L_n(S)) = (q_k(z), k\tau), \quad \forall\, (k - 1 + \lfloor \frac{k+1}{2} \rfloor)\tau \leq n < (k + 1 + \lfloor \frac{k}{2} \rfloor)\tau.$$

- **Example 2:** Let $\underline{r} = \begin{pmatrix} \gamma \\ \gamma^2 \end{pmatrix}$, where γ is the root of $Y^3 + z^2Y^2 + z^4Y - z = 0$. The m-CFE of $\underline{r}$ is as follows.

$$C(\underline{r}) = \left[\begin{pmatrix} 0 \\ 0 \end{pmatrix}, \overline{1, \begin{pmatrix} z^3 \\ 0 \end{pmatrix}, 2, \begin{pmatrix} z \\ z^3 \end{pmatrix}, 2, \begin{pmatrix} 0 \\ z^2 \end{pmatrix}, 1, \begin{pmatrix} z^2 \\ 0 \end{pmatrix}} \right].$$

Then we have the following table:

k	v_k	d_k	n_k
$4i+1$	$5i+3$	$10i+3$	$15i+3$
$4i+2$	$5i+3$	$10i+6$	$15i+6$
$4i+3$	$5i+5$	$10i+8$	$15i+11$
$4i+4$	$5i+5$	$10i+10$	$15i+13$

For $i \geq 0$, we have

$$l(4i+j, h) = \begin{cases} 0 & if\ h = 2, j = 1, i = 0, \\ 4i + j & if\ h = 1, j = 1, 4, \\ 4i - 1 & if\ h = 2, j = 1, i > 0, \\ 4i + 3 & if\ h = 2, j = 4, \\ 4i + j & if\ h = 2, j = 2, 3, \\ 4i + 1 & if\ h = 1, j = 2, 3; \end{cases}$$

then $q_1(z) = z^3, q_2(z) = z + z^6, q_3(z) = z^8, q_4(z) = 1 + z^{10}$ and for $i \geq 1$,

$$q_{4i+j}(z) = \begin{cases} q_{4i-1}(z) + z^3 q_{4i}(z) & if\ j = 1, \\ q_{4i-2}(z) + z q_{4i}(z) + z^3 q_{4i+1}(z) & if\ j = 2, \\ q_{4i+1}(z) + z^2 q_{4i+2}(z) & if\ j = 3, \\ q_{4i}(z) + z^2 q_{4i+3}(z) & if\ j = 4. \end{cases}$$

Based on Proposition 1, for $i \geq 1$ we get

$$(f_n(S), L_n(S)) = \begin{cases} (q_{4i+1}(z), 10i+3) & if\ 15i+3 \leq n < 15i+6, \\ (q_{4i+2}(z), 10i+6) & if\ 15i+6 \leq n < 15i+11, \\ (q_{4i+3}(z), 10i+8) & if\ 15i+11 \leq n < 15i+13, \\ (q_{4i+4}(z), 10i+10) & if\ 15i+13 \leq n < 15(i+1)+3. \end{cases}$$

Remark 4. The above computations of $C(\underline{r})$ were done on a Pentium IV processor using Mathematica.

References

1. Berlekamp E R.: Algebraic Coding Theory, McGraw-Hill, New York, 1968
2. Massey J L.: Shift-register synthesis and BCH decoding, IEEE Trans. Inform. Theory, IT-15, no. 1, 1969. 122-127
3. Sakata S.: Extension of Berlekamp-Massey algorithm to N dimension, Inform. and Comput. 84(1990). 207-239
4. Feng G L and Tzeng K K.: A generalization of the Berlekamp-Massey algorithm for multisequence shift-register synthesis with applications to decoding cyclic codes, IEEE Trans. Inform. Theory, vol. 37, 1991. 1274-1287
5. Zongduo Dai, Kunpeng Wang and Dingfeng Ye: m-Continued Fraction Expansions of Multi-Laurent Series, Advances in Mathematics (China), vol.33, No.2, 2004. 246-248
6. Zongduo Dai, Kunpeng Wang and Dingfeng Ye: Multi-continued fraction algorithm on multi-formal Laurent series, Acta Arithmetica, 122.1(2006). 1-16
7. Harald Niederreiter, Sequences with Almost Perfect Linear Complexity Profile, Advances in Cryptology-EUROCRYPT'87, Lecture Notes in Computer Science, vol.304, Springer-Verlag, 1988, 37-51
8. Chaoping Xing: Multi-sequences with Almost Perfect Linear Complexity Profile and Function Fields over Finite Fields, Journal of Complexity 16, 2000. 661-675
9. Xiutao Feng, Quanlong Wang and Zongduo Dai: Multi-sequences with d-perfect property, Proceedings of the 2004 **IEEE** International Symposium on Information Theory (ISIT'04), June 2004. 86-86
10. Xiutao Feng, Quanlong Wang and Zongduo Dai. Multi-Sequences with d-Perfect Property, Journal of Complexity 21, 2005. 230-242
11. Zongduo Dai, Kyoki Imamura and Junhui Yang: Asymptotic Behavior of Normalized Linear Complexity of Multi-Sequences, Proceedings Extended Abstracts) of 2004 International Conference on Sequences and Their Applications (SETA'04), Oct. 24-28, 2004. 29-33
12. Zongduo Dai and Xiutao Feng: Multi- Continued Fraction Algorithm and Generalized B-M Algorithm over F_2, Proceedings Extended Abstracts) of 2004 International Conference on Sequences and Their Applications (SETA'04), Oct. 24-28, 2004. 113-117
13. Xiutao Feng and Zongduo Dai: The Expected value of the Normalized linear complexity of 2-dimensional binary sequences, Proceedings Extended Abstracts) of 2004 International Conference on Sequences and Their Applications (SETA'04), Oct. 24-28, 2004. 24-28
14. Quanlong Wang, Zongduo Dai: The Proof of the Non-optimality of JPA and MJPA for Multi-formal Laurent Series, Jounal of the Graduate School of the Chinese Academy of Sciences, vol.22, No.1, 2005. 51-58(in Chinese)

15. Keqin Feng, Feirong Wang: The Jacobi-Perron Algorithm on Function Fields, Algebra Colloq., No.1, 1994. 149-158
16. Ito S, Fujii J, Higashino H, Yasutomi S-I: On Simultaneous Approximation to (α, α^2) with $\alpha^3 + k\alpha - 1 = 0$, Jounal of Number Theory, 99, 2003. 255-283
17. Inoue K, Nakada H: The modified Jacobi-Perron Algorithm over $F_q(X)^d$, Tokyo Journal of Mathematics, 26(2), 2003. 447-470
18. Weiss E: Algebraic Number Theory, McGraw Hill, 1963.

The Hausdorff Dimension of the Set of r-Perfect M-Multisequences

Michael Vielhaber* and Mónica del Pilar Canales Ch.*

Instituto de Matemáticas, Universidad Austral de Chile, Casilla 567, Valdivia, Chile
{vielhaber, monicadelpilar}@gmail.com

Abstract. We introduce a stochastic infinite state machine (Markov chain) BDM, the "Battery–Discharge–Model", which keeps track of all linear complexities of all $q^{M \cdot n}$ prefixes of length n of M-multisequences over $\mathbb{F}_q$.

We then use a finite subset of the BDM, dealing with those multisequences which are r-perfect. The largest eigenvalue λ of its transition matrix then yields the Hausdorff dimension of the set of r-perfect multisequences as

$$D_H = 1 + \frac{\log_q(\lambda)}{M}.$$

Also, we give a general formula for 1-perfect multisequences, for any M and q.

Keywords: Linear complexity, multisequence, Battery Discharge Model, isometry, Hausdorff dimension, perfect linear complexity profile.

1 Introduction

For a multisequence $a \in \left(\mathbb{F}_q^M\right)^\infty$ with M symbols of the finite field $\mathbb{F}_q$ in parallel, its linear complexity $L_a(n)$ is the least length of an LFSR able to produce all M prefix rows of length n with appropriate initial contents.

Since $L_a(n) \approx \lceil n \cdot \frac{M}{M+1} \rceil$ typically (and exactly for $q \to \infty$), we define the *linear complexity deviation* $d(n) := L_a(n) - \lceil n \cdot \frac{M}{M+1} \rceil$ and expect $d \approx 0$ for all n.

In this paper, we first recall the multi Strict Continued Fraction Algorithm (mSCFA) by Dai and Feng and introduce our Battery–Discharge–Model (BDM). The BDM is a stochastic infinite state machine which keeps track of *all* linear complexity deviations of *all* multisequences in $\left(\mathbb{F}_q^M\right)^\infty$.

r-perfectness for $M > 1$ was introduced first by Xing [6]. However, we define two *different* (for $M > 1$) notions of r-perfectness (distinct to Xing's model as well) and the finite portion of the BDM which corresponds to dealing only with r-perfect multisequences.

We then introduce the concept of Hausdorff dimension and its connection to the largest eigenvalue of the transition matrix of the finite subset of the BDM.

We finish with explicit numerical values for $r = 1, \ldots, 5$, $M = 1, \ldots, 4$, and $q = 2, 3, 4, 5, 8, 16$. We also give a general formula for $r = 1$, for any M and q.

* Supported by Project FONDECYT 2004, No. 1040975 of CONICYT, Chile.

G. Gong et al. (Eds.): SETA 2006, LNCS 4086, pp. 259–270, 2006.

2 Diophantine Approximation of Multisequences

We start with the multi–strict continued fraction algorithm (mSCFA) by Dai and Feng [2]. The mSCFA calculates a best simultaneous approximation to a set of M formal power series $G_m = \sum_{t=1}^{\infty} a_{m,t}x^{-t} \in \mathbb{F}_q[[x^{-1}]],\ 1 \le m \le M$. It computes a sequence $(u_m^{(m,n)}/v^{(m,n)})$ of approximations in $\mathbb{F}_q(x)$, in the order $(m,n) = (M,0),(1,1),(2,1),\ldots,(M,1)(1,2,),(2,2),\ldots$ with

$$G_m = \sum_{t\in\mathbb{N}} a_{m,t}\cdot x^{-t} = \frac{u_m^{(m,n)}(x)}{v^{(m,n)}(x)} + o(x^{-n}), \quad \forall\, 1 \le m \le M, n \in \mathbb{N}_0.$$

We will denote the degree of $v^{(m,n)}(x)$ by $\deg(m,n) \in \mathbb{N}_0$ instead of d as in [2] (we will use d differently). Then the multisequence has the linear complexity profile $(\deg(M,n))_{n\in\mathbb{N}_0} = (L_{(G_m,1\le m\le M)}(n))_{n\in\mathbb{N}_0}$.

The mSCFA also uses M auxiliary degrees $w_1,\ldots,w_M \in \mathbb{N}_0$. The update of these values depends on a so–called "discrepancy" $\delta(m,n) \in \mathbb{F}_q$. $\delta(m,n)$ is zero if the current approximation predicts correctly the value $a_{m,n}$, and $\delta(m,n)$ is nonzero otherwise. Furthermore, the polynomials $u_m(x)$ and $v(x)$ are updated, crucial for the mSCFA, but of no importance for our concern.

Algorithm 1. `mSCFA`

```
deg := 0; w_m := 0, 1 ≤ m ≤ M
FOR n := 1, 2, ...
    FOR m := 1, ..., M
        compute δ(m,n)  //discrepancy
        IF δ(m,n) = 0: {} // do nothing, [2, Thm. 2, Case 2a]
        IF δ(m,n) ≠ 0 AND n − deg −w_m ≤ 0 : {} // [2, Thm. 2, Case 2c]
        IF δ(m,n) ≠ 0 AND n − deg −w_m > 0: // [2, Thm. 2, Case 2b]
            deg_copy := deg
            deg := n − w_m
            w_m := n − deg_copy
    ENDFOR
ENDFOR
```

The linear complexity grows like $\deg(M,n) \approx \left\lceil n\cdot\frac{M}{M+1}\right\rceil$ (exactly, if always $\delta(m,n) \ne 0$), and the $w_m \approx \left\lfloor \frac{n}{M+1}\right\rfloor$. We therefore extract the *deviation* from this average behaviour as

$$d := \deg - \left\lceil n\cdot\frac{M}{M+1}\right\rceil, \tag{1}$$

the degree deviation, which we call the "*drain*" value, and

$$b_m := \left\lfloor n\cdot\frac{1}{M+1}\right\rfloor - w_m, \quad 1 \le m \le M, \tag{2}$$

the deviation of the auxiliary degrees, which we call the "*battery charges.*"

We establish the behaviour of d and b_m in two steps. First we treat the change of d, b_m when increasing n to $n+1$ (keeping deg, w_m fixed for the moment):

$$-\left\lceil (n+1) \cdot \frac{M}{M+1} \right\rceil = \begin{cases} -1 - \left\lceil n \cdot \frac{M}{M+1} \right\rceil, & n \not\equiv M \bmod M+1, \\ -\left\lceil n \cdot \frac{M}{M+1} \right\rceil, & n \equiv M \bmod M+1, \end{cases} \tag{3}$$

and

$$\left\lfloor (n+1) \cdot \frac{1}{M+1} \right\rfloor = \begin{cases} \left\lfloor n \cdot \frac{1}{M+1} \right\rfloor, & n \not\equiv M \bmod M+1, \\ 1 + \left\lfloor n \cdot \frac{1}{M+1} \right\rfloor, & n \equiv M \bmod M+1. \end{cases} \tag{4}$$

Hence, by (3) we have to decrease d in all steps, except when $n \equiv M \to n \equiv 0 \bmod (M+1)$, and only here we increase all M battery values b_m, by (4).

With $d(M,0) = b_m(M,0) := 0, \forall m$, initially, we obtain the invariant

$$d(M,n) + \left(\sum_{m=1}^{M} b_m(M,n) \right) + n \bmod (M+1) = 0, \ \forall n \in \mathbb{N}_0. \tag{5}$$

Now, for n fixed, the M steps of the inner loop of the mSCFA change w_m and deg only in the case of $\delta(m,n) \neq 0$ and $n - \deg - w_m > 0$ that is

$$n - \deg - w_m > 0 \overset{(1;2)}{\Longleftrightarrow} n - \left(d + \left\lceil n \cdot \frac{M}{M+1} \right\rceil\right) - \left(\left\lfloor n \cdot \frac{1}{M+1} \right\rfloor - b_m\right) > 0 \Leftrightarrow b_m > d.$$

In the case $\delta \neq 0$ and $b_m > d$, the new values are (see mSCFA)

$$\deg^+ = n - w_m \quad \text{and} \quad w_m^+ = n - \deg \tag{6}$$

and thus in terms of the BDM variables:

$$d^+ \overset{(1;6)}{=} (n - w_m) - \left\lceil \frac{n \cdot M}{M+1} \right\rceil \overset{(2)}{=} \left\lfloor \frac{n}{M+1} \right\rfloor + b_m - \left\lfloor \frac{n}{M+1} \right\rfloor = b_m$$

and

$$b_m^+ \overset{(2;6)}{=} \left\lfloor \frac{n}{M+1} \right\rfloor - (n - \deg) \overset{(1)}{=} -\left\lceil \frac{n \cdot M}{M+1} \right\rceil + \left(d + \left\lceil \frac{n \cdot M}{M+1} \right\rceil\right) = d,$$

an interchange of the values d and b_m. We say in this case that "battery b_m *discharges* (the excess of charge) into the drain". A discharge does not affect the invariant (5), which is thus valid for every timestep (m,n).

3 mSCFA and BDM Induce an Isometry on $\left(\mathbb{F}_q^M\right)^\infty$

We want to combine the effects of *all* $q^{M \cdot n}$ prefixes of length n:

If we arrive in E cases out of the $q^{M \cdot n}$ prefix strings, in a certain configuration $(w_1, \ldots, w_M, \deg)$, then we want to have a mass (probability) of $E/q^{M \cdot n}$ on the corresponding state of the probabilistic analogue, $(b_1, \ldots, b_M, d)$, with d, b_m depending on deg, w_n according to (1), (2).

In the limit $n \to \infty$, we will then obtain d as a probability distribution over *all* multisequences $(a_{m,n}) \in \left(\mathbb{F}_q^M\right)^\infty$. Since we do not actually compute the discrepancy δ, we have to model the distinction between $\delta = 0$ and $\delta \neq 0$ probabilistically.

Proposition 1. *In any given position $(m, n), 1 \leq m \leq M, n \in \mathbb{N}$ of the formal power series, exactly one choice for the next symbol $a_{m,n}$ will yield a discrepancy $\delta = 0$, all other $q - 1$ symbols from $\mathbb{F}_q$ result in some $\delta \neq 0$.*

Proof. The current approximation $u_m^{(m,n)}(x)/v^{(m,n)}(x)$ determines exactly *one* approximating coefficient sequence for the m-th formal power series G_m. The (only) corresponding symbol belongs to $\delta = 0$. □

In fact, for every position (m, n), each discrepancy value $\delta \in \mathbb{F}_q$ occurs exactly once for some $a_{m,n} \in \mathbb{F}_q$, in other words (see [1][5] for $M = 1$):

Fact. *The mSCFA induces an isometry on $\left(\mathbb{F}_q^M\right)^\infty$.*

Hence, we can model $\delta = 0$ as occurring with probability $1/q$, and $\delta \neq 0$ as having probability $(q-1)/q$.

4 The Battery–Discharge–Model: A Stochastic Infinite State Machine

In this part we describe a stochastic infinite state machine, the *Battery–Discharge–Model.* Our model shall compute the measure (probability) of prefixes with certain linear complexity profiles. As before, let $M \in \mathbb{N}$ be the number of sequences over $\mathbb{F}_q$ to be approximated simultaneously. We consider M battery values $b_m \in \mathbb{Z}, 1 \leq m \leq M$ and a drain $d \in \mathbb{Z}$, which gives the linear complexity deviation.

The model is self–similar in time: Assume that the automaton is in the same state $(b_1, \ldots, b_M, d)$ (with mass 1) for two timesteps $n_1 \equiv n_2 \bmod M + 1$. Then for every $\tau \in \mathbb{N}$, the resulting probability distribution over all $q^{M \cdot \tau}$ prolongations of the two sequence prefixes at times $n_1 + \tau$ and $n_2 + \tau$, resp., is the same. However, we have to distinguish time mod $M + 1$ to be able to adjust $b_m := b_m + 1, 1 \leq m \leq M$ $(n \equiv 0 \bmod M + 1)$ or $d := d - 1$ $(n \not\equiv 0 \bmod M + 1)$.

The state set of our *Battery–Discharge-Model* is

$$S := \left\{ (b_1, \ldots, b_M; d, t) \in \mathbb{Z}^M \times \mathbb{Z} \times \{0, \ldots, M\} \colon \sum_{m=1}^{M} b_m + d + t = 0 \right\}$$

(battery values; drain, time mod $M + 1$) with initial state $(0, \ldots, 0; 0, 0) =: s_0$.

We attach the following three actions to the cases of [2, Thm. 2]:

D, battery discharge: A battery can discharge, provided its charge is higher than that of the drain, $b_m > d$, moving the excess charge to d, and it does so with probability $(q-1)/q$, action $a_m = D$ for battery b_m, when $\delta \neq 0$, corresponding to case 2b of [2, Thm. 2].

I, *inhibition*: A battery, although having a charge $b_m > d$ higher than the drain, does not discharge, since it is inhibited by $\delta = 0$ with probability $1/q$, modeling case 2a with $b_m > d$, action $a_m = I$.

N, *do nothing*: If $b_m \leq d$, the action is $a_m = N$, do nothing (case 2c and part of 2a).

The probabilistic version of the mSCFA is then:

```
Algorithm BDM
d := 0; b_m := 0, 1 ≤ m ≤ M
FOR n := 1,2,...
    IF n ≡ 0 mod M+1 : b_m := b_m + 1, 1 ≤ m ≤ M ELSE d := d − 1 ENDIF
    FOR m := 1,...,M
        IF b_m > d:
            WITH prob. (q − 1)/q:
                swap(b_m, d) // action D
            WITH prob. 1/q:
                {} // action I
        ELSE
            {} // action N
        ENDIF
    ENDFOR
ENDFOR
```

We combine the actions at the M batteries to a word $\underline{a} = a_1 \dots a_M \in \{D, I, N\}^M$, describing a transition between states from S.

Let d_m be the value of the drain *before* the action of battery b_m that is $d_{m+1} = \begin{cases} d, & m+1=1, \\ d_m, & a_m \in \{I, N\}, \\ b_m, & a_m = D. \end{cases}$ Then $s \xrightarrow{\underline{a}} s'$ is feasible, if $b_m > d_m$ for $a_m \in \{D, I\}$ and $b_m \leq d_m$ for $a_m = N$, and the probability then is

$$prob(\underline{a}) = \left(\frac{q-1}{q}\right)^{a_D} \cdot \left(\frac{1}{q}\right)^{a_I} \cdot \left(\frac{q}{q}\right)^{a_N} = \frac{(q-1)^{a_D}}{q^{a_D+a_I}},$$

where a_D, a_I, a_N are the number of occurrences in $\underline{a}$ of the respective symbol.

We define a probability or mass distribution $\mu_n(s)$ for timestep $n \in \mathbb{N}_0$ and state $s \in S$ as follows. Initially ($n = 0$), let $\mu_0(s_0) = 1$ and $\mu_0(s) = 0$ for $s \in S \backslash \{s_0\}$. Also, let $S_T = \{s \in S \mid s.t = T\}$, for $0 \leq T \leq M$.

With every step $n \in \mathbb{N}$, we update the mass distribution of the states in S_T with $T \equiv n \bmod M+1$. The total mass a state $s' \in S_T$ receives, is $\mu_{n+1}(s') = \sum_{s \xrightarrow{\underline{a}} s'} \mu_n(s) \cdot prob(\underline{a})$. After the first $M+1$ steps, we thus have $\sum_{s \in S_T} \mu_n(s) = 1$ for each $0 \leq T \leq M$, and so $\sum_{s \in S} \mu_n(s) = M+1$ for $n \geq M+1$.

To give an example, let $M = 3$ and $s = (0, 2, 1; -3, 0)$. We first decrement d by (3) (and increment t) to $(0, 2, 1; -4, 1)$, and then have 6 feasible transitions:

a_1 a_2 a_3	$prob(\underline{a})$	s'	a_1 a_2 a_3	$prob(\underline{a})$	s'
D D N	$(q-1)^2/q^2$	$(-4, 0, 1; 2, 1)$	I D N	$(q-1)/q^2$	$(0, -4, 1; 2, 1)$
D I D	$(q-1)^2/q^3$	$(-4, 2, 0; 1, 1)$	I I D	$(q-1)/q^3$	$(0, 2, -4; 1, 1)$
D I I	$(q-1)/q^3$	$(-4, 2, 1; 0, 1)$	I I I	$1/q^3$	$(0, 2, 1; -4, 1)$

For instance, the first transition of the second line consists of these actions: $(\underline{0}, 2, 1; -4, 1) \overset{D;\frac{q-1}{q}}{\longrightarrow} (-4, \underline{2}, 1; 0, 1) \overset{I;\frac{1}{q}}{\longrightarrow} (-4, 2, \underline{1}; 0, 1) \overset{D;\frac{q-1}{q}}{\longrightarrow} s' = (-4, 2, 0; 1, 1)$

In the next section, we will consider the notion of r-perfect multisequences, defined by the allowed states from S that may be touched.

5 r-Perfect Multisequences

The notion of r-perfect linear complexity profiles can be described by two equivalent conditions for $M = 1$: The r-perfect multisequences are those, whose linear complexity deviation d is bounded by

$$\frac{-r-t+1}{2} \leq d \leq \frac{r-t}{2} \tag{7}$$

forever (where $t \in \{0, 1\}$). Equivalently (for $M = 1$), these sequences are just those where all partial denominator degrees are at most r (jumps by no more than $|b_m - d| \leq r$).

For $M > 1$, we adjust the inequality (7) for d to

$$\frac{-r-\varepsilon_L}{2} \leq d \leq \frac{r-\varepsilon_H}{2}, \text{with } \varepsilon_L = \begin{cases} 1, t < M/2, \\ 0, t \geq M/2, \end{cases} \text{and } \varepsilon_H = \begin{cases} 0, t \leq M/2, \\ 1, t > M/2, \end{cases} \tag{8}$$

and call all multisequences, which only touch BDM states satisfying (8) "r-L–perfect" (L: linear complexity).

However, there may be *several* partial denominators, all of degree at most r, in one transition, whose combined effect is to move between states with d outside the range of r-L–perfectness. Hence, the two conditions are no longer equivalent for $M > 1$, and we define r-J–perfect as $|b_m - d| \leq r$ (J: jump height).

In both cases the interesting value is the Hausdorff dimension of the sets $\mathcal{A}_L(r; M, q)$ and $\mathcal{A}_J(r; M, q) \subset \left(\mathbb{F}_q^M\right)^\infty \equiv [0, 1] \subset \mathbb{R}$ of r-L–perfect resp. r-J–perfect M-multisequences over $\mathbb{F}_q$. Certainly, if some multisequence a is r-L-perfect, then a is also r-J–perfect, hence $D_H(\mathcal{A}_L) \leq D_H(\mathcal{A}_J)$.

Xing [6] has generalized the notion of perfectness [4] to $M > 1$ by defining r_X-perfect $\Leftrightarrow d \geq \lceil (M(n+1) - r_X)/(M+1) \rceil - \lceil Mn/(M+1) \rceil$. However, the same r should give a *higher* Hausdorff dimension for larger M (more choices possible), which is the case with both r-L and r-J, but not with r_X.

Definition 1. *Let $\mathcal{T}_L(r; M, q)$ be the transition matrix over the set of states satisfying* (8)*, and $\mathcal{T}_J(r; M, q)$ the matrix for states from which we may eventually return to s_0 via actions with all $a_m = D$ satisfying $|b_m - d_m| \leq r$.*

Let the largest eigenvalue of $\mathcal{T}_{L/J}(r; M, q)$ be $\lambda_{L/J}(r; M, q) \in \mathbb{R}$.

$\lambda_{L/J}(r; M, q) \in \mathbb{R}$ is strictly less than 1, since $\mathcal{T}_{L/J}(r; M, q)$ is substochastic, and we will obtain the Hausdorff dimension as

$$D_H(\mathcal{A}_{L/J}(r; M, q)) = \frac{\log(\lambda_{L/J}(r; M, q) \cdot q^M)}{\log q^M} = 1 + \frac{\log_q(\lambda_{L/J}(r; M, q))}{M},$$

as will be explained in more detail in the next section. All the λ are roots of polynomials in $\mathbb{Z}[q]$, since $\mathcal{T}_{L/J}(r; M, q)$ is a *finite* matrix with entries of the form $\sum_i (q-1)^{a_i}/q^{b_i}$.

For computational stability, it is usually preferable to use the $(M+1)$-st power of the transition matrix to return immediately to states with $t = 0$. The largest eigenvalue of that matrix is just λ^{M+1}.

6 Hausdorff Dimension

This section is taken essentially from [4], following the introduction of the Hausdorff dimension given in Chapter 2 of Falconer [3] for a subset $\mathcal{A}$ of the reals. Set

$$h^s_\varepsilon(\mathcal{A}) = \inf \sum_{i=1}^{\infty} |U_i|^s \quad \text{for } s \geq 0, \varepsilon > 0,$$

where the infimum runs over all covers $\mathcal{U} = \{U_1, U_2, \ldots\}$ of $\mathcal{A}$ with intervals U_i of length $|U_i| \leq \varepsilon$, and letting $\varepsilon \to 0$:

$$h^s(\mathcal{A}) := \lim_{\varepsilon \to 0+} h^s_\varepsilon(\mathcal{A}).$$

Then

$$h^s(\mathcal{A}) = \begin{cases} 0, & s > D_H(\mathcal{A}) \\ \infty, & s < D_H(\mathcal{A}) \end{cases}$$

for a certain real number $D_H(\mathcal{A})$ ($h^{D_H(\mathcal{A})}(\mathcal{A})$ may assume any value in $[0, \infty]$).

Definition 2. *The Hausdorff dimension of a set $\mathcal{A}$ is defined as*

$$D_H(\mathcal{A}) = \inf\{s | h^s(\mathcal{A}) = 0\} = \sup\{s | h^s(\mathcal{A}) = \infty\}.$$

Remark. The definition of $h^s_\varepsilon(\mathcal{A})$ and thus of $h^s(\mathcal{A})$ involves an infimum. Thus, an upper bound for the Hausdorff dimension is considerably easier to obtain than a lower bound. For the former, one essentially defines a sequence of covers $\mathcal{U}^{(k)} = \{U_1{}^{(k)}, U_2{}^{(k)}, \ldots\}$, where $|U_i{}^{(k)}| \leq \varepsilon_k$ and $\varepsilon_k \to 0$. If then $\sum_{i=1}^{\infty} |U_i^{(k)}|^s$ remains bounded for every cover of the sequence, the infimum cannot be infinity. Hence the candidate s actually is an upper bound.

On the contrary, if s is below the Hausdorff dimension, it will lead to a sum $\sum_{i=1}^{\infty} |U_i^{(k)}|^s = \infty$ for each and every cover, and so the infimum cannot be determined in this way. Here we have to apply an analog of the Mass Distribution Principle (see Theorem 4.2 in [3]). Other special techniques to get lower bounds are given in Chapter 4 of [3].

Lemma 1. [4] *Let ν be a mass distribution on some set $A \subseteq [0,1] \subset \mathbb{R}$. We assume that for a given s there exist two real numbers $c > 0$ and $\delta > 0$ such that*

$$\nu(U) \leq c \cdot |U|^s$$

for all intervals $U \subseteq [0,1]$ with $|U| \leq \delta$. Then $D_H(A) \geq s$.

Proof. Let $0 < \varepsilon \leq \delta$. Let $\mathcal{U} = \{U_i\}$ be any cover of A by intervals $U_i \subseteq [0,1]$ of length $|U_i| \leq \varepsilon \leq \delta$. Then

$$0 < \nu(A) = \nu\left(\bigcup_i U_i\right) \leq \sum_i \nu(U_i) \leq c \cdot \sum |U_i|^s,$$

hence

$$\sum_i |U_i|^s \geq \frac{\nu(A)}{c}.$$

It follows that the infimum over all $\mathcal{U}$ gives

$$h^s_\varepsilon(A) \geq \frac{\nu(A)}{c} \quad \text{for all } \varepsilon \leq \delta,$$

and so $h^s(A) \geq \nu(A)/c > 0$, hence $s \leq D_H(A)$. □

Definition 3. *An N-ary interval of degree $k, N \in \mathbb{N}, k \in \mathbb{N}_0$, is an interval of the form $[r \cdot N^{-k}, (r+1) \cdot N^{-k}), 0 \leq r \leq N^k - 2, r \in \mathbb{N}_0$, or $[1 - N^k, 1]$.*

Lemma 2. [4] *Consider a nonempty subset $A \subseteq [0,1] \subset \mathbb{R}$ of the reals and N-ary intervals with $N \geq 2$. Let there be a natural number $S \leq N$ such that for each $k \in \mathbb{N}_0$ we have: If an N-ary interval I of degree k has nonempty intersection with A, then exactly S of the N-ary subintervals of I of degree $k+1$ also have nonempty intersection with A. In this case*

$$D_H(A) \geq \frac{\log S}{\log N}.$$

Proof. Each interval $U \subset [0,1]$ with $|U| < 1$ satisfies an inequality $N^{-k-1} \leq |U| < N^{-k}$ for a certain $k \in \mathbb{N}_0$. Thus, U can intersect at most two N-ary intervals of degree k.

Define a mass distribution ν on A such that each of the S^k N-ary intervals of degree k (of length N^{-k}) that intersect A contains a mass of S^{-k}. The mass that is covered by U can thus be bounded by $\nu(U) \leq 2 \cdot S^{-k}$.

For $s := (\log S)/(\log N)$ we therefore obtain

$$\begin{aligned}\nu(U) &\leq 2 \cdot S^{-k} = 2 \cdot (N^{-k})^s = 2 \cdot N^s \cdot (N^{-k-1})^s \\ &\leq 2 \cdot N^s \cdot |U|^s \\ &\leq 2 \cdot N \cdot |U|^s,\end{aligned}$$

where we used that $0 \leq s \leq 1$. Now we can apply Lemma 1. □

Example 1. Let $N = 3$ and $S = 2$. This describes the Cantor set, and indeed $(\log 2)/(\log 3)$ is its Hausdorff dimension.

Definition 4. *The space* $\left(\mathbb{F}_q^M\right)^\infty$ *of all infinite multisequences can be mapped onto the unit interval* $[0,1]$ *by*

$$\iota := \iota_{q^M} \colon \left(\mathbb{F}_q^M\right)^\infty \ni (a_{i,j})_{i=1,j=1}^{M\ \ \infty} \mapsto \sum_{i=1}^{M}\sum_{j=1}^{\infty} \psi(a_{i,j}) q^{-i-M\cdot j+M} \in [0,1] \subset \mathbb{R},$$

where ψ *is a fixed bijection from* $\mathbb{F}_q$ *to* $\{0,1,\ldots,q-1\}$.

If $\mathcal{A}_{L/J}(r;M,q) \subset \left(\mathbb{F}_q^M\right)^\infty$ is the set of r-L/J–perfect multisequences, then we study the subset $\mathcal{B}_{L/J}(r;M,q) := \iota\left(\mathcal{A}_{L/J}(r;M,q)\right)$ of $[0,1]$.

Theorem 2. *For all* $r \in \mathbb{N}, M \in \mathbb{N}$ *and prime powers* q *we have*

$$D_H\left(\mathcal{B}_{L/J}(r;M,q)\right) = 1 + \frac{\log_q(\lambda_{L/J}(r;M,q))}{M},$$

where $\log_q$ *denotes the logarithm to the base* q *and* $\lambda_{L/J}(r;M,q)$ *is as in Definition* 1.

Proof. We first show an upper bound for the Hausdorff dimension. Let $\mathcal{A} := \mathcal{A}_{L/J}(r;M,q)$, $\mathcal{B} := \mathcal{B}_{L/J}(r;M,q)$, and $\lambda := \lambda_{L/J}(r;M,q)$. Since λ is the largest eigenvalue of the transition matrix, starting from one sequence (the empty one, ε) at $t = 0$, for every $h \in \mathbb{N}$ there exists a constant C_h such that for all $t \in \mathbb{N}_0$, there are at most $C_h \cdot \left(q^M\left(\lambda + \frac{1}{h}\right)\right)^t$ prefixes of length t in the set $\mathcal{A}$. Each initial multistring of length t and width M defines a cylinder set in $\left(\mathbb{F}_q^M\right)^\infty$ consisting of all infinite continuations of this string. The image of each such cylinder set under the map ι is a closed interval of length q^{-Mt} in $[0,1]$. Thus, $\mathcal{B}$ can be covered by $\left\lfloor C_h \cdot \left(q^M\left(\lambda + \frac{1}{h}\right)\right)^t\right\rfloor$ intervals of length q^{-Mt}. With $\varepsilon_t = q^{-Mt}$ it follows that

$$h^s_{\varepsilon_t}(\mathcal{B}) \le C_h \cdot \left(q^M\left(\lambda + \frac{1}{h}\right)\right)^t \cdot q^{-Mts} = C_h \cdot \left(\frac{q^M(\lambda + \frac{1}{h})}{q^{Ms}}\right)^t.$$

For any $s > 1 + \log_q(\lambda + \frac{1}{h})/M$ we have $q^{Ms} > q^M(\lambda + \frac{1}{h})$. Thus, letting $t \to \infty$ (hence $\varepsilon_t \to 0$), we get $h^s(\mathcal{B}) = 0$. By the definition of $D_H(\mathcal{B})$ it follows that $D_H(\mathcal{B}) \le s$. Since $s > 1 + \log_q(\lambda + \frac{1}{h})/M$ is arbitrary, we obtain $D_H(\mathcal{B}) \le 1 + \log_q(\lambda + \frac{1}{h})/M$ for all $h \in \mathbb{N}$ and thus

$$D_H(\mathcal{B}) \le 1 + \frac{\log_q(\lambda)}{M}.$$

Thus, the upper bound is shown.

To prove the lower bound, we define for $h \in \mathbb{N}$ the number X_h as the least multiple of $(M+1)$ such that at timestep X_h there are at least $(q^M(\lambda - \frac{1}{h}))^{X_h}$

prefix strings leading to state s_0. Let the exact number of such strings be S_h. Let also

$$\mathcal{A}(h) := \{a \in \mathcal{A} \mid s(n \cdot X_h) = s_0, \forall n \in \mathbb{N}\}, \qquad \mathcal{B}(h) := \iota(\mathcal{A}(h)),$$

be the set of multisequences which visit state s_0 every X_h timesteps. Since each prefix of length $n \cdot X_h$ repeats n times the process of duplication by S_h, there are S_h^n prefixes of length $n \cdot X_h$ in $\mathcal{A}(h)$. By the mapping $\iota(\mathcal{A}(h)) = \mathcal{B}(h)$, we thus obtain a subset of $[0,1]$ for which we can apply Lemma 2 with $S := S_h$ and $N := q^{M \cdot X_h}$. So we obtain

$$D_H(\mathcal{B}(h)) \geq \frac{\log_q(S_h)}{\log_q(q^{M \cdot X_h})} \geq \frac{MX_h + \log_q\left((\lambda - \frac{1}{h})^{X_h}\right)}{M \cdot X_h} = 1 + \frac{\log_q(\lambda - \frac{1}{h})}{M}.$$

The last inequality is valid for all $h \in \mathbb{N}$ and we have $\mathcal{A} \supseteq \mathcal{A}(h)$ and thus $\mathcal{B} \supseteq \mathcal{B}(h)$. Hence the Hausdorff dimension of $\mathcal{B}$ is bounded from below by

$$D_H(\mathcal{B}(h)) \geq 1 + \frac{\log_q(\lambda - \frac{1}{h})}{M}$$

for all $h \in \mathbb{N}$, and together with the upper bound we finally arrive at

$$D_H(\mathcal{A}_{L/J}(r; M, q)) := D_H(\mathcal{B}_{L/J}(r; M, q)) = 1 + \frac{\log_q(\lambda_{L/J}(r; M, q))}{M}.$$

□

7 Numerical Results

This section assembles the Hausdorff dimensions of the sets of r-L–perfect, resp. r-J–perfect M-multisequences over $\mathbb{F}_q$, for $r = 1, 2, 3, 4, 5$, $M = 1, 2, 3, 4$, and $q = 2, 3, 4, 5, 8, 16$.

Unlike the case $M = 1$, it seems to be hopeless to find a general formula (in r) of the characteristic polynomial (remember that $D_H(\mathcal{A}_{L/J}(r; 1, q)) = (1 + \log_q(\varphi^{(r)}))/2$, where $\varphi^{(r)}$ is the largest real root of $x^r - (q-1) \cdot \sum_{k=0}^{r-1} x^k = 0$, see [4]). However, for fixed $r = 1$, varying M, we have achieved the following result:

Theorem 3. *The Hausdorff dimension of the set of* 1*-L–perfect multisequences over* $\mathbb{F}_q^M$ *is* $D_H(\mathcal{A}_L(1; M, q)) = 1 + \log_q(\lambda_L(1; M, q))/M$ *with*

$$\lambda_L(1; M, q) = \sqrt[M+1]{\frac{(q-1)^M \cdot \prod_{k=0}^{M-1}\left(\sum_{i=0}^{k} q^i\right)}{q^{\binom{M+1}{2}}}} = \sqrt[M+1]{\prod_{k=1}^{M}\left(1 - \frac{1}{q^k}\right)}.$$

Proof. For 1-L–perfect multisequences, all involved states must have $d = 0$ by (8), hence all batteries must be either zero, or -1 (from a drain $0 \to -1$ after decrementing, with immediate discharge).

There are 2^M such states, with $b_m \in \{0,-1\}$ and $d = 0$. This subset of the BDM is isomorphic to the M-dimensional hypercube with corners labeled from $\{0,-1\}^M$ (corresponding to $b_1,\ldots,b_M$), every edge being directed towards the vertex with (one) more '-1's and with attached probability $(q-1)/q^a$, if the a-th zero from the beginning is replaced by -1. Furthermore, an additional edge goes from $(-1,\ldots,-1)$ to $(0,\ldots,0)$, with probability 1.

It may be seen that every state with T '-1's is from S_T, and in fact we may identify all states from the same S_T: There is only one state from S_M, and all states from S_{M-1} have exactly one transition (with probability $(q-1)/q$) to the single state in S_M, hence the states in S_{M-1} can all be identified. By induction, all states in S_{M-K} have K transitions with probabilities $(q-1)/q^k, 1 \le k \le K$, to the (now only) state in S_{M-K+1}.

Thus we have $M+1$ states, one for each T, with transition probability $(q-1)\cdot\sum_{k=1}^{M-T} q^{-k}, T = 0,\ldots,M-1$, and a transition from $(-1,\ldots,-1) \in S_M$ to $(0,\ldots,0) \in S_0$, with probability 1. The product of these probabilities gives the formula for $\lambda_L(1;M,q)$, and the Hausdorff dimension now follows as before. □

We finish with a table of Hausdorff dimensions (truncated, not rounded). Apparently, with $r \to \infty$ and/or $M \to \infty$, $D_H \to 1$. However, we always have $D_H < 1$, and the $\mathcal{A}_{L/J}$ have Haar measure 0 in $\left(\mathbb{F}_q^M\right)^\infty$.

Hausdorff Dimensions

$D_H(\mathcal{A}_L(r;M,2))$

r	M = 1	2	3	4
1	0.5000	0.7641	0.8660	0.9149
2	0.8471	0.9442	0.9646	0.9865
3	0.9395	0.9743	0.9905	0.9964
4	0.9733	0.9931	0.9974	0.9995
5	0.9876	0.9965	0.9993	0.9999

$D_H(\mathcal{A}_L(r;M,3))$

r	M = 1	2	3	4
1	0.8154	0.9206	0.9574	0.9739
2	0.9574	0.9891	0.9937	0.9984
3	0.9876	0.9964	0.9992	0.9999
4	0.9961	0.9995	0.9999	0.9999
5	0.9987	0.9998	0.9999	0.9999

$D_H(\mathcal{A}_L(r;M,4))$

r	M = 1	2	3	4
1	0.8962	0.9576	0.9778	0.9865
2	0.9806	0.9962	0.9979	0.9996
3	0.9955	0.9990	0.9998	0.9999
4	0.9989	0.9999	0.9999	0.9999
5	0.9997	0.9999	0.9999	0.9999

$D_H(\mathcal{A}_J(r;M,2))$

r	M = 1	2	3	4
1	0.5000	0.8166	0.9200	0.9622
2	0.8471	0.9591	0.9880	0.9965
3	0.9395	0.9889	0.9980	0.9996
4	0.9733	0.9968	0.9996	0.9999
5	0.9876	0.9991	0.9999	0.9999

$D_H(\mathcal{A}_J(r;M,3))$

r	M = 1	2	3	4
1	0.8154	0.9480	0.9830	0.9941
2	0.9574	0.9934	0.9990	0.9998
3	0.9876	0.9991	0.9999	0.9999
4	0.9961	0.9998	0.9999	0.9999
5	0.9987	0.9999	0.9999	0.9999

$D_H(\mathcal{A}_J(r;M,4))$

r	M = 1	2	3	4
1	0.8962	0.9760	0.9937	0.9983
2	0.9806	0.9981	0.9998	0.9999
3	0.9955	0.9998	0.9999	0.9999
4	0.9989	0.9999	0.9999	0.9999
5	0.9997	0.9999	0.9999	0.9999

$D_H(\mathcal{A}_L(r; M, 5))$

r	M = 1	2	3	4
1	0.9306	0.9726	0.9859	0.9915
2	0.9891	0.9982	0.9991	0.9998
3	0.9979	0.9996	0.9999	0.9999
4	0.9996	0.9999	0.9999	0.9999
5	0.9999	0.9999	0.9999	0.9999

$D_H(\mathcal{A}_J(r; M, 5))$

r	M = 1	2	3	4
1	0.9306	0.9864	0.9970	0.9993
2	0.9891	0.9992	0.9999	0.9999
3	0.9979	0.9999	0.9999	0.9999
4	0.9996	0.9999	0.9999	0.9999
5	0.9999	0.9999	0.9999	0.9999

$D_H(\mathcal{A}_L(r; M, 8))$

r	M = 1	2	3	4
1	0.9678	0.9880	0.9939	0.9963
2	0.9965	0.9996	0.9998	0.9999
3	0.9995	0.9998	0.9999	0.9999
4	0.9999	0.9999	0.9999	0.9999

$D_H(\mathcal{A}_J(r; M, 8))$

r	M = 1	2	3	4
1	0.9678	0.9956	0.9993	0.9999
2	0.9965	0.9999	0.9999	0.9999
3	0.9995	0.9999	0.9999	0.9999
4	0.9999	0.9999	0.9999	0.9999

$D_H(\mathcal{A}_L(r; M, 16))$

r	M = 1	2	3	4
1	0.9883	0.9958	0.9979	0.9987
2	0.9993	0.9999	0.9999	0.9999
3	0.9999	0.9999	0.9999	0.9999

$D_H(\mathcal{A}_J(r; M, 16))$

r	M = 1	2	3	4
1	0.9883	0.9991	0.9999	0.9999
2	0.9993	0.9999	0.9999	0.9999
3	0.9999	0.9999	0.9999	0.9999

8 Conclusion

We developed a model of multidimensional linear complexity, using a stochastic infinite state machine, the Battery-Discharge-Model "BDM", which is selfsimilar on the time axis, folding back time mod $(M + 1)$ onto itself.

We introduced two different notions of r-perfectness and defined the corresponding finite subset of the transition matrix of the BDM. Its largest eigenvalue λ gives the Hausdorff dimension of the set of r-perfect multisequences as $D_H = 1 + \log_q(\lambda)/M$. We finished with explicit numerical values and a general formula for the Hausdorff dimension of the set of 1-perfect multisequences.

References

1. M. del P. Canales Chacón, M. Vielhaber, *Structural and Computational Complexity of Isometries and their Shift Commutators*, Electronic Colloquium on Computational Complexity, ECCC **TR04–057**, 2004.
2. Z. Dai, X. Feng, *Multi–Continued Fraction Algorithm and Generalized B–M Algorithm over* $\mathbb{F}_2$, in: SETA '04, International Conference on Sequences and Their Applications, October 24 – 28, 2004, Seoul, Korea, LNCS **3486**, Springer, 2005.
3. K. Falconer, *Fractal Geometry — Mathematical Foundations and Applications*, Wiley, Chichester, 1990.
4. H. Niederreiter, M. Vielhaber, *Linear complexity profiles: Hausdorff dimensions for almost perfect profiles and measures for general profiles*, J. Cpx **13**, 353–383, 1997.
5. M. Vielhaber, *A Unified View on Sequence Complexity Measures as Isometries*, in: SETA '04, International Conference on Sequences and Their Applications, October 24 – 28, 2004, Seoul, Korea, LNCS **3486**, Springer, 2005.
6. C. Xing, *Multi–sequences with Almost Perfect Linear Complexity Profile and Function Fields over Finite Fields*, J. Cpx **16**, 661–675, 2000.

Lower Bounds on Sequence Complexity Via Generalised Vandermonde Determinants

Nicholas Kolokotronis, Konstantinos Limniotis, and Nicholas Kalouptsidis

Department of Informatics and Telecommunications
National and Kapodistrian University of Athens
TYPA Buildings, University Campus, 15784 Athens, Greece
{nkolok, klimn, kalou}@di.uoa.gr

Abstract. Binary sequences generated by nonlinearly filtering maximal length sequences with period $2^n - 1$ are studied in this paper. We focus on the particular class of equidistant filters and provide improved lower bounds on the linear complexity of the filtered sequences. This is achieved by first considering and proving properties of generalised Vandermonde determinants. Furthermore, it is shown that the methodology developed can be used for studying properties of any nonlinear filter.

Keywords: Binary sequences, filter functions, linear complexity, linear feedbak shift registers, symmetric functions, Vandermonde determinants.

1 Introduction

Binary sequences have been traditionally employed in many applications, ranging from spread spectrum communication systems to stream ciphers and cryptography in general. This is mainly due to the ease and efficiency of their implementation, most notably via a *linear feedback shift register* (LFSR) [4]. The properties such sequences are required to possess depend on the application and among others include long period, balance of ones and zeros, low out-of-phase autocorrelation spectra, as well as large *linear complexity*. The last property, defined as the length of the shortest LFSR that generates a given sequence, is an important measure for evaluating the cryptographic strength of the sequence against cryptanalytic attacks, such as the *Berlekamp-Massey algorithm* [13].

Sequences with large linear complexity are most commonly generated by applying appropriately chosen *nonlinear filters*, i.e. Boolean functions, to distinct phases of a maximal length sequence [16]. It is well-known that the maximum possible linear complexity attained by nonlinear filterings depends on the degree of the Boolean function used [7], [8]. The problem of determining the exact value of linear complexity attained by any filtering is still open; however, several classes of filters have been proposed that allow to derive lower bounds on its value. These constructions primarily study filters that consist of a single *equidistant* or *norm-phase* product of phases of a maximal length sequence, and extend

G. Gong et al. (Eds.): SETA 2006, LNCS 4086, pp. 271–284, 2006.

the results obtained to the sum of such products. In the former case, the distance d between any two successive phases is taken to be coprime to the period $N = 2^n - 1$ of the maximal length sequence [3], [9], [15], [16]. In the latter case, phases are properly chosen from the elements of cyclotomic cosets corresponding to a normal basis [2], [9], [10]. It is well-known that the problem of finding the linear complexity of a filtered maximal length sequence is equivalent to determining the degree of its minimal polynomial [5], [6], or the weight of its *discrete Fourier transform* (DFT) [14], [15]. When a filter function of degree k is applied, the best lower bound on the linear complexity of filterings derived so far is equal to $\binom{n}{k}$, and rely on proving that all Fourier coefficients of field elements whose exponent has weight k do not vanish. This procedure is formally known as the *root presence test* [16].

In this paper, we focus on the case of nonlinearly filtering a maximal length sequence with period $N = 2^n - 1$ by an equidistant filter of degree k. We extend the work in [9], [16] by deriving a simple root presence test for field elements whose exponent has weight $k - 1$. This is achieved by formulating the test in terms of *generalised Vandermonde determinants* [17], [18], and obtain the new improved lower bound $\binom{n}{k} + \binom{n}{k-1}$ in some cases. Moreover, we investigate simple variants of equidistant filters and prove that they also attain the lower bound $\binom{n}{k}$ on the linear complexity, based on the methodology developed. The paper is organised as follows. Section 2 gives the basic background and settles the notation. Properties of the generalised Vandermonde determinants are considered in Section 3, whereas the improved lower bounds and the new class of nonlinear filters are given in Section 4. Finally, Section 5 summarizes the conclusions.

2 Background

Let $x = \{x_j\}_{j\geq 0}$ be a maximal length sequence of period $N = 2^n - 1$ with elements over the finite field $\mathbb{F}_2$, and let $\mu(z)$ be its minimal polynomial, with $\deg(\mu) = n$. Then, sequence x is generated by a LFSR with feedback polynomial $\mu^*(z) = z^n \mu(1/z)$, the reciprocal of $\mu(z)$. Both polynomials are primitive and the roots of $\mu^*(z)$ are the inverses of the roots of $\mu(z)$ in the extension field $\mathbb{F}_{2^n}$ of $\mathbb{F}_2$ [11]. It is known that the linear complexity L_x of the maximal length sequence x equals n [4], [16]. Let $\alpha \in \mathbb{F}_{2^n}$ be a primitive element of $\mathbb{F}_{2^n}$ with $\mu^*(\alpha) = 0$. Then, sequence x is given by

$$x_j = \mathrm{tr}_1^n(\beta\alpha^{-j}) = \beta\alpha^{-j} + \left(\beta\alpha^{-j}\right)^2 + \cdots + \left(\beta\alpha^{-j}\right)^{2^{n-1}} \quad (1)$$

for some $\beta \in \mathbb{F}_{2^n}^* = \mathbb{F}_{2^n} \setminus \{0\}$, where $\mathrm{tr}_1^n(\cdot)$ is the *trace function* that maps elements of $\mathbb{F}_{2^n}$ onto $\mathbb{F}_2$. This representation of sequence x is referred to as the *trace representation*, and is uniquely associated with the discrete Fourier transform of x, since every binary sequence x of period $N = 2^n - 1$ can be expressed as $x_j = \sum_{i=0}^{N-1} \beta_i \alpha^{-ij}$, where $\beta_i \in \mathbb{F}_{2^n}$ [9], [14], [15].

Let $y = \{y_j\}_{j\geq 0}$ be the binary sequence that results from the nonlinear filtering of the maximal length sequence x by function $h : \mathbb{F}_2^n \to \mathbb{F}_2$, which maps elements of vector space $\mathbb{F}_2^n$ onto $\mathbb{F}_2$. Then, sequence y is given by $y_j = h(x_{j-t_1}, x_{j-t_2}, \ldots, x_{j-t_n})$, where the phases t_i belong to the residue class ring $\mathbb{Z}_N = \{0, 1, \ldots, N-1\}$ of the integers modulo N. The filter h can always be reduced to an equivalent form $y_j = \tilde{h}(x_{j-1}, x_{j-2}, \ldots, x_{j-n})$ of consecutive phases by applying the linear recurrence relation satisfied by x [16]. Let $\boldsymbol{z} = (z_1, z_2, \ldots, z_n)$ and $\boldsymbol{r} = (r_1, r_2, \ldots, r_n)$ be elements of the vector space $\mathbb{F}_2^n$. The Boolean function h is commonly expressed in its *algebraic normal form* (ANF) given by

$$h(\boldsymbol{z}) = \sum_{\boldsymbol{r}\in\mathbb{F}_2^n} a_{\boldsymbol{r}}\, z_1^{r_1} z_2^{r_2} \cdots z_n^{r_n}, \qquad a_{\boldsymbol{r}} \in \mathbb{F}_2 \ . \tag{2}$$

In the sequel, we assume that $a_{\mathbf{0}} = 0$. The *degree* of function h is defined as $\deg(h) = \max\{\mathrm{wt}(\boldsymbol{r}) : a_{\boldsymbol{r}} = 1, \boldsymbol{r} \in \mathbb{F}_2^n\}$, where $\mathrm{wt}(\boldsymbol{r})$ denotes the weight of vector $\boldsymbol{r}$. A special form of functions to be considered in the following sections are the *elementary symmetric polynomials* of degree s, defined as

$$\sigma_s(\boldsymbol{z}) = \sum_{\boldsymbol{r}\in\mathbb{F}_2^k,\, \mathrm{wt}(\boldsymbol{r})=s} z_1^{r_1} z_2^{r_2} \cdots z_k^{r_k}$$

where $\boldsymbol{z} \in \mathbb{F}_{2^n}^k$ [12]. Subsequently, we use the convention that $\sigma_s(\boldsymbol{z}) = 0$ if $s < 0$ or $s > k$.

For an integer $e \in \mathbb{Z}_N$, we define its *cyclotomic coset* as the distinct elements in $C_e = \{e, 2e, \ldots, 2^{n-1}e\}$ modulo N. The cardinality of C_e is always a divisor of n [11]. Hereinafter, we say that $\alpha^e \in \mathbb{F}_{2^n}$ has weight s if $\mathrm{wt}(e) = s$, that is $e = 2^{e_0} + \cdots + 2^{e_{s-1}}$. It is well-known that if the degree of the function h equals k, then the linear complexity of sequence y satisfies $L_y \leq \sum_{i=1}^{k} \binom{n}{i}$ [8]. When h is comprised of a single product of degree k, that is $y_j = x_{j-t_1} x_{j-t_2} \cdots x_{j-t_k}$, the root presence test for the elements $\alpha^e \in \mathbb{F}_{2^n}$ of weight k, first stated in [16], is shown below

$$T_e = \det\left(\alpha^{t_i\, 2^{e_{j-1}}}\right)_{i,j=1}^{k} \tag{3}$$

and asserts that α^e is a root of the minimal polynomial of sequence y if $T_e \neq 0$. It is well-known that (3) becomes a Vandermonde determinant in the case of equidistant filters. The root presence test has also been formed for elements $\alpha^e \in \mathbb{F}_{2^n}$ of weight $k-1$ [9], and is given by

$$T_e = \sum_{1\leq r<s\leq k} \det\left(\alpha^{l_i\, 2^{e_{j-1}+1}}\right)_{i,j=1}^{k-1} \tag{4}$$

where $(l_1, \ldots, l_{k-1}) = (t_1, \ldots, t_{r-1}, t_{r+1}, \ldots, t_{s-1}, t_{s+1}, \ldots, t_k, \frac{1}{2}(t_r + t_s))$. In [9] it was assumed $1 \leq t_1 < \cdots < t_k \leq n$; however, it is easily seen that (4) still holds in the general case where the phases t_i belong to $\mathbb{Z}_N$.

3 Generalised Vandermonde Determinants

In this section, we study generalised Vandermonde determinants over the finite field $\mathbb{F}_{2^n}$. Several generalisations of the Vandermonde determinant are found in the literature (*see* e.g. [18] and the references therein); our interest is on the generalisation treated in [17]. Let us consider the vector $\boldsymbol{x} = (x_1, \ldots, x_k)$ of nonzero elements in $\mathbb{F}_{2^n}$ and the increasing sequence of nonnegative integers $R = \{r_1, \ldots, r_k\}$. Then

$$\boldsymbol{V}(\boldsymbol{x}; R) = \det\,(x_j^{r_i})_{i,j=1}^k \tag{5}$$

is called *generalised Vandermonde determinant.* It is clear that the choice $R = \{0, \ldots, k-1\}$ leads to the ordinary Vandermonde determinant $\boldsymbol{V}(\boldsymbol{x})$ that is nonzero if and only if the elements of $\boldsymbol{x}$ are pairwise distinct, since

$$\boldsymbol{V}(\boldsymbol{x}) = \det\,(x_j^{i-1})_{i,j=1}^k = \prod_{1\le i<j\le k} (x_i + x_j) \ . \tag{6}$$

Let $\operatorname{diag}(\boldsymbol{x})$ be the diagonal matrix, with the elements $x_1, x_2, \ldots, x_k$ along its main diagonal. In order to analyse the properties of $\boldsymbol{V}(\boldsymbol{x}; R)$ we need only consider $\boldsymbol{V}(\boldsymbol{x}; R')$, with $R' = \{0, r_2 - r_1, \ldots, r_k - r_1\}$, since it holds

$$\boldsymbol{V}(\boldsymbol{x}; R) = \boldsymbol{V}(\boldsymbol{x}; R') \det\big(\operatorname{diag}(\boldsymbol{x})^{r_1}\big) = \boldsymbol{V}(\boldsymbol{x}; R')\,\big(x_1 x_2 \cdots x_k\big)^{r_1} \ .$$

Thus, we assume without loss of generality that $r_1 = 0$ in the rest of the section. Let us define the set $I = \{0, 1, \ldots, r_k\} \setminus R$ of distinct nonnegative integers; obviously, it contains the *discontinuities* that appear among the elements of R and we have $k - 1 \le r_k \le |I| + k - 1$. When the cardinality of I is less than k, we will find it convenient to write $\boldsymbol{V}_\perp(\boldsymbol{x}; I)$ instead of $\boldsymbol{V}(\boldsymbol{x}; R)$. Next, the elements of the set I are denoted by $l_1, l_2, \ldots, l_{|I|}$ and it is assumed that $0 \le l_1 < l_2 < \cdots < l_{|I|} \le |I| + k - 1$. Moreover, we use the notation $\boldsymbol{x}_s = (x_1, \ldots, x_{s-1}, x_{s+1}, \ldots, x_k)$, for $1 \le s \le k$.

Lemma 1. *With the above notation, let us assume $I = \{l\}$. Then we get the identity $\boldsymbol{V}_\perp(\boldsymbol{x}; I) = \boldsymbol{V}(\boldsymbol{x})\,\sigma_{k-l}(\boldsymbol{x})$.*

Proof. Let us define the polynomial $g(z) = \boldsymbol{V}(\boldsymbol{x}, z)$ over $\mathbb{F}_{2^n}$, in terms of the Vandermonde determinant $\boldsymbol{V}(\boldsymbol{x}, z)$ of order $k+1$. From (6) we have

$$g(z) = \boldsymbol{V}(\boldsymbol{x}) \prod_{i=1}^{k} (z + x_i) = \boldsymbol{V}(\boldsymbol{x}) \sum_{i=0}^{k} \sigma_{k-i}(\boldsymbol{x})\, z^i \ . \tag{7}$$

On the other hand, by expanding $\boldsymbol{V}(\boldsymbol{x}, z)$ along its $k+1$ column, i.e. the column corresponding to z, we get $g(z) = \sum_{i=0}^{k} \boldsymbol{V}_\perp(\boldsymbol{x}; \{i\})\, z^i$. Comparing the latter expression with (7) proves the claim. □

Lemma 2. *With the above notation, let us assume $I = \{l_1, l_2\}$. Then we get the identity $\boldsymbol{V}_\perp(\boldsymbol{x}; I) = \boldsymbol{V}(\boldsymbol{x}) \det\,(\sigma_{k-l_i+j-1}(\boldsymbol{x}))_{i,j=1}^2$.*

Proof. Let us similarly define the polynomial $g(z) = \boldsymbol{V}_{\perp}((\boldsymbol{x}, z); \{i\})$ over $\mathbb{F}_{2^n}$, for an integer $0 \leq i \leq k+1$. From Lemma 1, and by considering (7), the polynomial $g(z)$ is given by

$$g(z) = \boldsymbol{V}(\boldsymbol{x}, z)\, \sigma_{k-i+1}(\boldsymbol{x}, z) = \boldsymbol{V}(\boldsymbol{x}) \sum_{j=0}^{k} \sigma_{k-j}(\boldsymbol{x})\, \sigma_{k-i+1}(\boldsymbol{x}, z)\, z^j \ .$$

By using the identity $\sigma_{k-i+1}(\boldsymbol{x}, z) = \sigma_{k-i+1}(\boldsymbol{x}) + z\, \sigma_{k-i}(\boldsymbol{x})$ we have that

$$\begin{aligned} g(z) &= \boldsymbol{V}(\boldsymbol{x}) \sum_{j=0}^{k} \sigma_{k-j}(\boldsymbol{x}) \big(\sigma_{k-i+1}(\boldsymbol{x}) + z\, \sigma_{k-i}(\boldsymbol{x})\big)\, z^j \\ &= \boldsymbol{V}(\boldsymbol{x}) \left(\sum_{j=0}^{k} \sigma_{k-j}(\boldsymbol{x})\, \sigma_{k-i+1}(\boldsymbol{x})\, z^j + \sum_{j=1}^{k+1} \sigma_{k-j+1}(\boldsymbol{x})\, \sigma_{k-i}(\boldsymbol{x})\, z^j \right) \\ &= \boldsymbol{V}(\boldsymbol{x}) \sum_{j=0}^{k+1} \big(\sigma_{k-j}(\boldsymbol{x})\, \sigma_{k-i+1}(\boldsymbol{x}) + \sigma_{k-j+1}(\boldsymbol{x})\, \sigma_{k-i}(\boldsymbol{x})\big)\, z^j \end{aligned} \tag{8}$$

since by convention $\sigma_l(\boldsymbol{x}) = 0$ if $l < 0$ or $l > k$. Note that the coefficient of z^i in (8) vanishes, which agrees with the definition of $g(z)$. Expanding $\boldsymbol{V}_{\perp}((\boldsymbol{x}, z); \{i\})$ along its $k+1$ column we obtain

$$g(z) = \sum_{j=0}^{i-1} \boldsymbol{V}_{\perp}(\boldsymbol{x}; \{j, i\})\, z^j + \sum_{j=i}^{k+1} \boldsymbol{V}_{\perp}(\boldsymbol{x}; \{i, j\})\, z^j \ .$$

Since the coefficient of z^j in (8) is symmetric with respect to the integers i, j, comparison with the above expression yields the desired result. □

Expressing generalised Vandermonde determinants in terms of elementary symmetric polynomials and the discontinuities in the powers involved is proved in Theorem 3 for any number of elements in the set I. The proof uses only basic properties of the elementary symmetric polynomials.

Theorem 3. *With the above notation, let us assume* $I = \{l_1, l_2, \ldots, l_s\}$. *Then we get the identity* $\boldsymbol{V}_{\perp}(\boldsymbol{x}; I) = \boldsymbol{V}(\boldsymbol{x}) \det \left(\sigma_{k-l_i+j-1}(\boldsymbol{x})\right)_{i,j=1}^{s}$.

Proof. The proof is provided in the appendix. □

4 Improved Lower Bounds on Equidistant Filterings

In this section we present the new lower bounds on the linear complexity of nonlinearly filtered maximal length sequences. We focus on equidistant filters of degree k, that is we assume the resulting sequence y is given by

$$y_j = x_{j-t}\, x_{j-t-d} \cdots x_{j-t-(k-1)d}\,, \qquad j \geq 0 \tag{9}$$

where $t \geq 0$ and the distance $d \in \mathbb{Z}_N \setminus \{0\}$ satisfies $\gcd(d,N) = 1$. In the sequel, we consider the root presence test only for the elements $\alpha^e \in \mathbb{F}_{2^n}$ of weight $k-1$ since we know that $L_y \geq \binom{n}{k}$ [16]. Furthermore, we write $e = 2^{e_0} + \cdots + 2^{e_{k-2}}$. Prior to considering the root presence test, we first need to prove the following result.

Lemma 4. *With the notation of Section 3, let integer l satisfy $0 \leq l \leq k$ and let $m = \max\{0, 2l-k\}$. Then, it holds*

$$\sum_{i=m}^{l} \begin{vmatrix} \sigma_{k-i}(\boldsymbol{x}) & \sigma_{k-i+1}(\boldsymbol{x}) \\ \sigma_{k+i-2l-1}(\boldsymbol{x}) & \sigma_{k+i-2l}(\boldsymbol{x}) \end{vmatrix} = \sigma_{k-l}(\boldsymbol{x})^2 \ .$$

Proof. Expanding the determinants at the left-hand side we obtain that

$$\begin{aligned} &\sum_{i=m}^{l} \big(\sigma_{k-i}(\boldsymbol{x})\, \sigma_{k+i-2l}(\boldsymbol{x}) + \sigma_{k-i+1}(\boldsymbol{x})\, \sigma_{k+i-2l-1}(\boldsymbol{x})\big) \\ &\qquad = \sum_{i=m}^{l} \sigma_{k-i}(\boldsymbol{x})\, \sigma_{k+i-2l}(\boldsymbol{x}) + \sum_{i=m-1}^{l-1} \sigma_{k-i}(\boldsymbol{x})\, \sigma_{k+i-2l}(\boldsymbol{x}) \\ &\qquad = \sigma_{k-l}(\boldsymbol{x})^2 + \sigma_{k-m+1}(\boldsymbol{x})\, \sigma_{k+m-2l-1}(\boldsymbol{x}) \ . \end{aligned}$$

The last summand always vanishes, since for $m = 0$ and $m = 2l-k$ we get $\sigma_{k+1}(\boldsymbol{x})$ and $\sigma_{-1}(\boldsymbol{x})$ respectively, which by convention are zero. □

Theorem 5. *Let sequence y be given by (9), and consider the element $\alpha^e \in \mathbb{F}_{2^n}$ with $\mathrm{wt}(e) = k-1$. Then, α^e is a root of the minimal polynomial of sequence y if and only if $f_e(\alpha^d) \neq 0$, where*

$$f_e(z) = \sum_{i=0}^{k-2} \big(z\, g_{e,i}(z)^2\big)^{2^{e_i}}, \quad g_{e,i}(z) = \prod_{j \neq i} \frac{z + z^{2^{e_j - e_i + 1}}}{z + z^{2^{e_j - e_i}}} \ . \tag{10}$$

Proof. Substituting $t_i = t + i\,d$, for $0 \leq i \leq k-1$, in (4) and expanding the determinants along their last row, the root presence test becomes

$$T_e = \alpha^{2et} \sum_{0 \leq i < j \leq k-1} \sum_{s=0}^{k-2} \alpha^{(i+j)d2^{e_s}} \boldsymbol{V}_{\perp}(\boldsymbol{x}_s; \{i,j\}) \tag{11}$$

where the determinant $\boldsymbol{V}_{\perp}(\boldsymbol{x}_s; \{i,j\})$ has order $k-2$, and the vector $\boldsymbol{x}$ is equal to $\boldsymbol{x} = (\alpha^{d2^{e_0+1}}, \ldots, \alpha^{d2^{e_{k-2}+1}})$. When $i+j$ is even, that is $i+j = 2l$ for some $1 \leq l \leq k-2$, the rightmost sum of (11) vanishes, as this case corresponds to $\frac{1}{2}(t_i + t_j) = t_l$ and hence the last row of the determinant in (4) coincides with one of the rows above it. From Lemmas 2, 4 and the change of variables $(i,j) \mapsto (v, 2l+1-v)$ we get

$$T_e = \alpha^{2et} \sum_{s=0}^{k-2} \boldsymbol{V}(\boldsymbol{x}_s) \sum_{\substack{0 \le i < j \le k-1 \\ i+j \text{ odd}}} \alpha^{(i+j)d2^{e_s}} \begin{vmatrix} \sigma_{k-2-i}(\boldsymbol{x}_s) & \sigma_{k-1-i}(\boldsymbol{x}_s) \\ \sigma_{k-2-j}(\boldsymbol{x}_s) & \sigma_{k-1-j}(\boldsymbol{x}_s) \end{vmatrix}$$

$$= \alpha^{2et} \sum_{s=0}^{k-2} \boldsymbol{V}(\boldsymbol{x}_s) \sum_{l=0}^{k-2} \alpha^{(2l+1)d2^{e_s}} \sum_{v=m}^{l} \begin{vmatrix} \sigma_{k-2-v}(\boldsymbol{x}_s) & \sigma_{k-1-v}(\boldsymbol{x}_s) \\ \sigma_{k-3+v-2l}(\boldsymbol{x}_s) & \sigma_{k-2+v-2l}(\boldsymbol{x}_s) \end{vmatrix}$$

$$= \alpha^{2et} \sum_{s=0}^{k-2} \boldsymbol{V}(\boldsymbol{x}_s)\, \alpha^{d2^{e_s}} \left(\sum_{l=0}^{k-2} \alpha^{ld2^{e_s}} \sigma_{k-2-l}(\boldsymbol{x}_s) \right)^2$$

where $m = \max\{0, 2l - (k-2)\}$. The expression inside the parentheses is a polynomial on $\alpha^{d2^{e_s}}$, and is equal to $\prod_{l \ne s}(\alpha^{d2^{e_s}} + \alpha^{d2^{e_l+1}})$. Since from the definition of the Vandermonde determinant we also have the identity $\boldsymbol{V}(\boldsymbol{x}_s) = \boldsymbol{V}(\boldsymbol{x}) \prod_{l \ne s}(\alpha^{d2^{e_s+1}} + \alpha^{d2^{e_l+1}})^{-1}$, then T_e becomes

$$T_e = \alpha^{2et}\, \boldsymbol{V}(\boldsymbol{x}) \sum_{s=0}^{k-2} \alpha^{d2^{e_s}} \left(\prod_{l \ne s} \frac{\alpha^{d2^{e_s}} + \alpha^{d2^{e_l+1}}}{\alpha^{d2^{e_s}} + \alpha^{d2^{e_l}}} \right)^2 = \alpha^{2et}\, \boldsymbol{V}(\boldsymbol{x})\, f_e(\alpha^d) \ .$$

Obviously, $T_e \ne 0$ if and only if $f_e(\alpha^d) \ne 0$. □

Clearly, the simplification occuring in the root presence test given by (4) is important. Note that the poles of the functions $g_{e,i}(z)$, $0 \le i \le k-2$, do not include primitive elements $\alpha^d \in \mathbb{F}_{2^n}$ at which the function $f_e(z)$ is evaluated. Some interesting properties of $f_e(z)$ are illustrated next.

Theorem 6. *Let the function $f_e(z)$ be given by (10), and let us consider the element $\alpha^d \in \mathbb{F}_{2^n}$ with $\gcd(d, N) = 1$. The following hold for all $l \ge 0$*

1. *$f_e(\alpha^d) \ne 0$ if and only if $f_e(\alpha^{-d}) \ne 0$,*
2. *$f_e(\alpha^d) \ne 0$ if and only if $f_e(\alpha^{d2^l}) \ne 0$, and*
3. *$f_e(\alpha^d) \ne 0$ if and only if $f_{e2^l}(\alpha^d) \ne 0$.*

Proof. Let us set $\beta = \alpha^d$. We will prove only the first property; the last two properties are trivial since $f_e(\beta^{2^l}) = (f_e(\beta))^{2^l} = f_{e2^l}(\beta)$ for all $l \ge 0$. Multiplying the numerator and denominator of each fraction in $g_{e,i}(\beta^{-1})$ with $\beta^{2^{e_j-e_i+1}+1}$ we have

$$g_{e,i}(\beta^{-1}) = \prod_{j \ne i} \beta^{-2^{e_j-e_i}} \frac{\beta + \beta^{2^{e_j-e_i+1}}}{\beta + \beta^{2^{e_j-e_i}}} = \left(\beta^{2^{e_i}-e}\right)^{2^{-e_i}} g_{e,i}(\beta) \ .$$

Substituting the above result into the function $f_e(\beta^{-1})$ we finally get that

$$f_e(\beta^{-1}) = \sum_{i=0}^{k-2} \beta^{-2^{e_i}} \left(\left(\beta^{2^{e_i}-e}\right)^{2^{-e_i}} g_{e,i}(\beta) \right)^{2^{e_i+1}} = \sum_{i=0}^{k-2} \beta^{2^{e_i}-2e} g_{e,i}(\beta)^{2^{e_i+1}}$$

or equivalently $f_e(\beta^{-1}) = \beta^{-2e} f_e(\beta)$, which proves the claim. □

A direct result of Theorem 6 is that every root of $f_e(z)$ coexists with its inverse in $\mathbb{F}_{2^n}$, leading to $f_e(z) = f_e^*(z)$. This implies that we can identify degeneracies occuring at elements $\alpha^e \in \mathbb{F}_{2^n}$ of weight $k-1$ from filterings of the form $\tilde{y}_j = x_j x_{j+d} \cdots x_{j+(k-1)d}$ if we already know the corresponding degeneracies of (9). Next, $f_e(z)$ is further simplified by considering runs of 1s in the binary representation of the integer e.

Corollary 7. *Let the function $f_e(z)$ be given by (10), and let us consider the element $\alpha^e \in \mathbb{F}_{2^n}$ with $\mathrm{wt}(e) = k-1$. Further, let $w > 0$ be the number of runs of 1s in the binary representation of e, and assume that the i-th run has length $c_i > 0$ starting at position $b_i \geq 0$. Then*

$$f_e(z) = \sum_{i=0}^{w-1} \left(z\, g_{e,i}(z)^2\right)^{2^{b_i}}, \quad g_{e,i}(z) = \frac{z + z^{2^{c_i}}}{z + z^2} \prod_{j \neq i} \frac{z + z^{2^{b_j - b_i + c_j}}}{z + z^{2^{b_j - b_i}}} . \tag{12}$$

Proof. It is clear from (10) that when two consecutive 1s are encountered in the binary representation of e, say at positions $e_l = b$ and $e_{l+1} = b+1$ for some $0 \leq l \leq k-2$, then the term $g_{e,l+1}(z)$ vanishes (caused by $j = l$). This is readily generalised for runs of 1s of longer length $c > 1$, in which case all terms $g_{e,l+1}(z), \ldots, g_{e,l+c-1}(z)$ are zero. Hence, if the number of runs is w, starting at positions $e_{l_i} = b_i$, for $0 \leq i < w$, only the following terms $g_{e,l_0}(z), \ldots, g_{e,l_{w-1}}(z)$ survive. From (10), $g_{e,l_i}(z)$ becomes

$$\begin{aligned} g_{e,l_i}(z) &= \prod_{\substack{j=0 \\ j \neq l_i}}^{k-2} \frac{z + z^{2^{e_j - b_i + 1}}}{z + z^{2^{e_j - b_i}}} = \prod_{r=1}^{c_i - 1} \frac{z + z^{2^{e_{l_i} - b_i + r + 1}}}{z + z^{2^{e_{l_i} - b_i + r}}} \prod_{\substack{s=0 \\ s \neq i}}^{w-1} \prod_{r=0}^{c_s - 1} \frac{z + z^{2^{e_{l_s} - b_i + r + 1}}}{z + z^{2^{e_{l_s} - b_i + r}}} \\ &= \frac{\prod_{r=2}^{c_i} \left(z + z^{2^r}\right)}{\prod_{r=1}^{c_i - 1} \left(z + z^{2^r}\right)} \prod_{\substack{s=0 \\ s \neq i}}^{w-1} \frac{\prod_{r=1}^{c_s} \left(z + z^{2^{b_s - b_i + r}}\right)}{\prod_{r=0}^{c_s - 1} \left(z + z^{2^{b_s - b_i + r}}\right)} \end{aligned}$$

assuming the s-th run has length c_s. In all fractions only the first term of the denominator and the last term of the numerator remain. □

Remark 8. Let the number of runs w divide both n and the weight $k-1$ of the integer e. Let us set $b = n/w$ and $c = (k-1)/w$; clearly $0 < c < b$. If $b_i = ib$ and $c_i = c$ for $0 \leq i < w$, then it is easily seen that (12) becomes $f_e(z) = \mathrm{tr}_b^n\left(z\, g_e(z)^2\right)$, where

$$g_e(z) = \frac{z + z^{2^c}}{z + z^2} \prod_{j=1}^{n/b-1} \frac{z + z^{2^{jb+c}}}{z + z^{2^{jb}}} . \tag{13}$$

Obviously, such simplified versions of $f_e(z)$ occur if the cyclotomic coset corresponding to the element $\alpha^e \in \mathbb{F}_{2^n}$ belongs to the class of the so-called *regular cosets* [1]. These are cyclotomic cosets whose elements belong to subfields of the finite field $\mathbb{F}_{2^n}$. Since it is well-known that in this case the equation $f_e(z) = 0$

has 2^{n-b} solutions $\gamma \in \mathbb{F}_{2^n}$ [11], it is of great interest to find which of these are written as $\gamma = g_e(\alpha^d)$. Complete determination of the roots of $f_e(z)$ has resulted so far in the cases given below, leading to the improved lower bound $\binom{n}{k} + \binom{n}{k-1}$ on the linear complexity of y.

Theorem 9. *Let sequence y be given by (9). Then, for any distance d, with $\gcd(d, N) = 1$, and degree $k = 2, 3, n-1, n$ we have $L_y \geq \binom{n}{k} + \binom{n}{k-1}$.*

Proof. For $k = 2$ it has already been proved using different approaches in [8], [9] that sequence y attains maximum complexity $L_y = \binom{n}{2} + \binom{n}{1}$. In our notation, we simply have $f_e(z) = z^e$ with $\mathrm{wt}(e) = 1$, which is nonzero at all points $z = \alpha^d$.

For $k = 3$ the cyclotomic cosets with elements of weight 2 are those with coset leaders in $\{e = 1 + 2^s : \text{for } 1 \leq s \leq \lfloor n/2 \rfloor\}$. From (10) we get

$$f_e(z) = z\left(\frac{z + z^{2^{s+1}}}{z + z^{2^s}}\right)^2 + z^{2^s}\left(\frac{z^{2^s} + z^2}{z^{2^s} + z}\right)^2 = \frac{z^3 + z^{2^{s+2}+1} + z^{3\,2^s} + z^{2^s+4}}{z^2\left(1 + z^{2^s-1}\right)^2}$$
$$= \frac{z\left(1 + z^{2^s+1}\right)\left(1 + z^{3\,(2^s-1)}\right)}{\left(1 + z^{2^s-1}\right)^2}$$

leading to $d\,e \not\equiv 0 \pmod{N}$ and $3d\,(e-2) \not\equiv 0 \pmod{N}$ since we need to ensure that $f_e(\alpha^d)$ does not vanish. Due to $\gcd(d, N) = 1$ both conditions hold for all e and the linear complexity of y satisfies $L_y \geq \binom{n}{3} + \binom{n}{2}$.

For $k = n-1$ we proceed as in the above case. All cyclotomic cosets with elements of weight $n-2$ are exactly those with coset leaders in the set $\{e = 2^{n-1} - 1 - 2^s : \text{for } \lceil n/2 \rceil - 1 \leq s \leq n-2\}$. Alternatively, we can write $e = 2^{s+1}(2^{n-2-s} - 1) + (2^s - 1)$ and hence we have two runs in the binary representation of e (unless $s = n-2$, in which case we have only one run of 1s). With the notation of Corollary 7, we have $b_0 = 0$, $c_0 = s$, $b_1 = s+1$, and $c_1 = n-2-s$, and (12) gives

$$f_e(z) = z\left(\frac{z + z^{2^s}}{z + z^2} \cdot \frac{z + z^{2^{n-1}}}{z + z^{2^{s+1}}}\right)^2 + z^{2^{s+1}}\left(\frac{z^{2^{s+1}} + z^{2^{n-1}}}{z^{2^{s+1}} + z^{2^{s+2}}} \cdot \frac{z^{2^{s+1}} + z^{2^s}}{z^{2^{s+1}} + z}\right)^2$$
$$= \frac{z^2 + z^{2^{s+1}}}{(1+z)\left(z^2 + z^{2^{s+2}}\right)} + \frac{z^{2^{s+2}} + z}{\left(1 + z^{2^{s+1}}\right)\left(z^{2^{s+2}} + z^2\right)}$$
$$= \frac{z^{2^{s+1}+1} + z^{2^{s+1}-1} + z^{2^{s+2}} + 1}{z\,(1+z)^{2^{s+1}+1}\left(1 + z^{2^{s+1}-1}\right)^2} = \frac{1 + z^{2^{s+1}+1}}{z\,(1+z)^{2^{s+1}+1}\left(1 + z^{2^{s+1}-1}\right)}\,.$$

Note that if $s = n-2$, Corollary 7 would lead to the following expression $f_e(z) = z\,(1 + z^{2^{n-2}-1})^2(1+z)^{-2}$. In either case, no primitive element α^d would lead to degeneracy for any integer s, giving $L_y \geq \binom{n}{n-1} + \binom{n}{n-2}$.

Finally, for $k = n$ there is only one cyclotomic coset with elements of weight $n-1$, namely the one corresponding to $e = 2^{n-1} - 1$. Therefore, the binary representation of e presents only one run of 1s with $b_0 = 0$ and $c_0 = n-1$. In this case, (12) implies that $f_e(z) = z\,(z + z^{2^{n-1}})^2\,(z + z^2)^{-2} = (1+z)^{-1}$. Hence, $f_e(\alpha^d) \neq 0$ for all integers d. □

Theorem 10. *Let sequence y be given by* (9)*. Then, for any distance d, with $\gcd(d, N) = 1$, and degree $4 \le k \le n-2$ we have $L_y \ge \binom{n}{k} + n$.*

Proof. For each equidistant filter of degree k, with $4 \le k \le n-2$, let us consider the integer $e = 2^{k-1} - 1$. Clearly, the cardinality of the cyclotomic coset of e equals n. Corollary 7 gives $f_e(z) = z\,(1 + z^{2^{k-1}-1})^2(1+z)^{-2}$, and therefore $f_e(\alpha^d) \ne 0$ if α^d is a primitive element of $\mathbb{F}_{2^n}$. □

The above result ensures the improved lower bound $\binom{n}{k} + n$ for all cases not covered by Theorem 9. The methodology presented is easily extended to include the sum of equidistant filters, i.e. by adding shifted versions of the filter given in (9). From Theorem 5, the root presence test becomes

$$T_e = \left(\textstyle\sum_t v_t\, \alpha^{et}\right)^2 \boldsymbol{V}(\boldsymbol{x})\, f_e(\alpha^d), \quad v_t \in \mathbb{F}_2 \ . \tag{14}$$

Hence, we also need to ensure that for a particular choice of coefficients $v_t \in \mathbb{F}_2$, no element $\alpha^e \in \mathbb{F}_{2^n}$ with $\mathrm{wt}(e) = k-1$ is root of the polynomial $\sum_t v_t z^t$. Exhaustive search for $2 \le n \le 20$ verified the results obtained and revealed that most of the degeneracies occur when $\alpha^e \in \mathbb{F}_{2^n}$ belongs to regular cosets (*see* also Remark 8).

The methodologies developed facilitate the analysis of nonlinear filter classes more complex than the ones currently studied in the literature [9], [10], [16]. For $1 \le s \le N-k$, we generalise the definition of equidistant filters to the *s-th order semi-equidistant filters* of degree k as

$$y_j = x_{j-t} \cdots x_{j-t-(r-1)d}\, x_{j-t-(r+s)d} \cdots x_{j-t-(k+s-1)d}\,, \qquad j \ge 0 \tag{15}$$

where $t \ge 0$, $1 \le r \le k-1$, and the distance satisfies $\gcd(d, N) = 1$. From Theorem 3, the root presence test (3) for $\alpha^e \in \mathbb{F}_{2^n}$ of weight k becomes

$$T_e = \alpha^{et}\, \boldsymbol{V}(\boldsymbol{x}) \det\left(\sigma_{k-r+j-i}(\boldsymbol{x})\right)_{i,j=1}^{s} \tag{16}$$

with $\boldsymbol{x} = (\alpha^{d2^{e_0}}, \ldots, \alpha^{d2^{e_{k-1}}})$. Clearly, in the above case the determinant $\det\left(\sigma_{k-r+j-i}(\boldsymbol{x})\right)_{i,j=1}^{s}$ has the structure of a Toeplitz matrix. This allows to prove the following result about the complexity of sequence y.

Theorem 11. *With the above notation, let sequence y be given by* (15) *where $s=1$ and $r=k-1$. If $\{\alpha^d, \alpha^{d2}, \ldots, \alpha^{d2^{n-1}}\}$ is a normal basis of $\mathbb{F}_{2^n}$ over $\mathbb{F}_2$, then $L_y \ge \binom{n}{k}$.*

Proof. Let us consider the finite field element $\alpha^e \in \mathbb{F}_{2^n}$ with $\mathrm{wt}(e) = k$. From the hypothesis, the root presence test (16) leads to

$$T_e = \alpha^{et}\, \boldsymbol{V}(\alpha^{d2^{e_0}}, \ldots, \alpha^{d2^{e_{k-1}}})\, \sigma_1(\alpha^{d2^{e_0}}, \ldots, \alpha^{d2^{e_{k-1}}}) \ .$$

Obviously, the polynomial $\sigma_1(\alpha^{d2^{e_0}}, \ldots, \alpha^{d2^{e_{k-1}}}) = \sum_{i=0}^{k-1} \alpha^{d2^{e_i}}$ is nonzero if and only if $\{\alpha^{d2^{e_0}}, \ldots, \alpha^{d2^{e_{k-1}}}\}$ are linearly independent. In order to ensure the lower bound $\binom{n}{k}$ is attained, this has to hold for all possible choices of integers e with $\mathrm{wt}(e) = k$, which by hypothesis is satisfied. □

It is clear from the proof of Theorem 11 that the lower bound $L_y \geq \binom{n}{k}$ is also attained by sequence y if $\{\alpha^{-d}, \alpha^{-d2}, \ldots, \alpha^{-d2^{n-1}}\}$ is a normal basis of $\mathbb{F}_{2^n}$ over $\mathbb{F}_2$ and $s = r = 1$. Moreover, the preconditions of Theorem 11 impose no difficulty in choosing the distance d between the phases of x, since there always exists a normal basis of $\mathbb{F}_{2^n}$ over $\mathbb{F}_2$ [11].

5 Conclusions

Maximal length sequences nonlinearly filtered by equidistant filters were studied in this paper. It was shown that by using properties of generalised Vandermonde determinants simple conditions for testing the presence of roots in the minimal polynomial of filterings can be derived. As a result, the improved lower bound $\binom{n}{k} + \binom{n}{k-1}$ on the linear complexity of filterings was obtained and new class of filters was introduced that in some cases attain the bound $\binom{n}{k}$. Obviously, the results obtained can be considerably improved, by further exploiting Corollary 7 and Remark 8, or extended to nonlinear filters whose phases are chosen from a normal basis.

Acknowledgements

This work was partially supported by the Greek Ministry of Education and Religious Affairs under Pythagoras Grant.

References

1. Caballero-Gil, P.: Regular cosets and upper bounds on the linear complexity of certain sequences. In Ding, C., Helleseth, T., Niederreiter, H., ed.: Sequences and Their Applications. Discrete Mathematics and Theoretical Computer Science. Springer-Verlag, Berlin, Germany (1999) 242–256.
2. Caballero-Gil, P., Fúster-Sabater, A.: A wide family of nonlinear filter functions with large linear span. Inform. Sci. **164** (2004) 197–207.
3. García-Villalba, L. J., Fúster-Sabater, A.: On the linear complexity of the sequences generated by nonlinear filterings. Inform. Process. Lett. **76** (2000) 67–73.
4. Golomb, S. W.: Shift Register Sequences. Holden-Day Inc., San Francisco, CA (1967).
5. Göttfert, R., Niederreiter, H.: On the linear complexity of products of shift–register sequences. In Helleseth, T., ed.: Advances in Cryptology – Eurocrypt '93. Lecture Notes in Computer Science **765**. Springer-Verlag, Berlin, Germany (1994) 151–158.
6. Göttfert, R., Niederreiter, H.: On the minimal polynomial of the product of linear recurring sequences. Finite Fields Applic. **1** (1995) 204–218.
7. Groth, E. J.: Generation of binary sequences with controllable complexity. IEEE Trans. Inform. Theory **17** (1971) 288–296.
8. Key, E. L.: An analysis of the structure and complexity of nonlinear binary sequence generators. IEEE Trans. Inform. Theory **22** (1976) 732–736.
9. Kolokotronis, N., Kalouptsidis, N.: On the linear complexity of nonlinearly filtered PN-sequences. IEEE Trans. Inform. Theory **49** (2003) 3047–3059.

10. Lam, C., Gong, G.: A lower bound for the linear span of filtering sequences. In State of the Art of Stream Ciphers – SASC (2004) 220–233.
11. Lidl, R., Niederreiter, H.: Finite Fields. In Encyclop. Math. Its Applic. **20** 2nd ed. Cambridge Univ. Press, Cambridge, U.K. (1996).
12. Macdonald, I. G.: Symmetric Functions and Hall Polynomials. Oxford Univ. Press, 2nd ed., Oxford, U.K. (1995).
13. Massey, J. L.: Shift-register synthesis and BCH decoding. IEEE Trans. Inform. Theory **15** (1969) 122–127.
14. Massey, J. L., Serconek, S.: A Fourier transform approach to the linear complexity of nonlinearly filtered sequences. In Desmedt, Y. G., ed.: Advances in Cryptology – Crypto '94. Lecture Notes in Computer Science **839**. Springer-Verlag, Berlin, Germany (1994) 332–340.
15. Paterson, K. G.: Root counting, the DFT and the linear complexity of nonlinear filtering. Des. Codes Cryptogr. **14** (1998) 247–259.
16. Rueppel, R. A.: Analysis and Design of Stream Ciphers. Springer-Verlag, Berlin, Germany (1986).
17. Shparlinski, I. E.: On the singularity of generalised Vandermonde matrices over finite fields. Finite Fields Appl. **11** (2005) 193–199.
18. Tu, L. W.: A partial order on partitions and the generalised Vandermonde determinant. J. Algebra **278** (2004) 127–133.

A Proof of Theorem 3

We proceed by induction on the cardinality of the set $I = \{l_1, l_2, \ldots, l_m\}$. The validity of the identity has been proved in Lemmas 1 and 2 for $|I| = 1$ and $|I| = 2$ respectively. Let us assume it holds for $|I| = m$, that is

$$\boldsymbol{V}_{\perp}(\boldsymbol{x}; I) = \boldsymbol{V}(\boldsymbol{x}) \det\left(\sigma_{k-l_i+j-1}(\boldsymbol{x})\right)_{i,j=1}^{m} \tag{17}$$

where $\boldsymbol{x} = (x_1, x_2, \ldots, x_k)$. Subsequently, we prove that it also holds for $|I| = m+1$. We define the polynomial $g(z) = \boldsymbol{V}_{\perp}((\boldsymbol{x}, z); I)$ over $\mathbb{F}_{2^n}$, which from (6) and the induction hypothesis is written as

$$\begin{aligned} g(z) &= \boldsymbol{V}(\boldsymbol{x}, z) \det\left(\sigma_{k-l_i+j}(\boldsymbol{x}, z)\right)_{i,j=1}^{m} \\ &= \boldsymbol{V}(\boldsymbol{x}) \det\left(\sigma_{k-l_i+j}(\boldsymbol{x}) + z\,\sigma_{k-l_i+j-1}(\boldsymbol{x})\right)_{i,j=1}^{m} \sum_{r=0}^{k} \sigma_{k-r}(\boldsymbol{x})\, z^r \end{aligned} \tag{18}$$

as a result of the identity $\sigma_{k-l_i+j}(\boldsymbol{x}, z) = \sigma_{k-l_i+j}(\boldsymbol{x}) + z\,\sigma_{k-l_i+j-1}(\boldsymbol{x})$. It can be easily verified that the determinant appearing in (18) is equal to (*see* also remark at the end of the proof)

$$\begin{aligned} &\det\left(\sigma_{k-l_i+j}(\boldsymbol{x}) + z\,\sigma_{k-l_i+j-1}(\boldsymbol{x})\right)_{i,j=1}^{m} \\ &\qquad = \sum_{\boldsymbol{c} \in \mathbb{F}_2^m} \det\left(\sigma_{k-l_i+j-c_j}(\boldsymbol{x})\right)_{i,j=1}^{m} z^{\mathrm{wt}(\boldsymbol{c})} \end{aligned} \tag{19}$$

where $\boldsymbol{c} = (c_1, c_2, \ldots, c_m)$. From the above expression we conclude that if vector $\boldsymbol{c}$ is such that $c_s = 0$ and $c_{s+1} = 1$, for some $1 \leq s < m$, then we have

$\det(\sigma_{k-l_i+j-c_j}(\boldsymbol{x}))_{i,j=1}^m = 0$, since two of its columns are identical. Therefore, the nonzero determinants correspond to the integers in the set $\mathfrak{A}_m = \{u_s = (u_{s,1}, u_{s,2}, \ldots, u_{s,m}) : u_s = 2^s - 1 \text{ for } 0 \le s \le m\}$. From (19) and the above analysis, (18) leads to

$$\begin{aligned} g(z) &= \boldsymbol{V}(\boldsymbol{x}) \left(\sum_{s=0}^{m} \det(\sigma_{k-l_i+j-u_{s,j}}(\boldsymbol{x}))_{i,j=1}^m z^s \right) \left(\sum_{r=0}^{k} \sigma_{k-r}(\boldsymbol{x}) z^r \right) \\ &= \boldsymbol{V}(\boldsymbol{x}) \sum_{r=0}^{k+m} \left(\sum_{s=a_r}^{b_r} \sigma_{k-r+s}(\boldsymbol{x}) \det(\sigma_{k-l_i+j-u_{s,j}}(\boldsymbol{x}))_{i,j=1}^m \right) z^r \end{aligned} \tag{20}$$

where $a_r = \max\{0, r-k\}$ and $b_r = \min\{r, m\}$. Notice that when $r < m$ then $b_r = r$; however we may include $s = r+1, \ldots, m$ in (20) since then $k - r + s > k$ and by convention we have $\sigma_{k-r+s}(\boldsymbol{x}) = 0$. Moreover, when $r > k$ then $a_r = r - k$; but we can similarly include $s = 0, \ldots, r-k-1$ in (20) since then $k - r + s < 0$ and $\sigma_{k-r+s}(\boldsymbol{x}) = 0$. Therefore, if we also denote r by l_{m+1}, and recall that $u_{s,j} = 1$ for $1 \le j \le s$ and $u_{s,j} = 0$ otherwise, then (20) becomes

$$\begin{aligned} g(z) &= \boldsymbol{V}(\boldsymbol{x}) \sum_{l_{m+1}=0}^{k+m} \left(\sum_{s=0}^{m} \sigma_{k-l_{m+1}+s}(\boldsymbol{x}) \det(\sigma_{k-l_i+j-u_{s,j}}(\boldsymbol{x}))_{i,j=1}^m \right) z^{l_{m+1}} \\ &= \boldsymbol{V}(\boldsymbol{x}) \sum_{l_{m+1} \in \{0,1,\ldots,k+m\} \setminus I} \det(\sigma_{k-l_i+j-1}(\boldsymbol{x}))_{i,j=1}^{m+1} z^{l_{m+1}} \end{aligned} \tag{21}$$

where the sum inside the parentheses is identified with the expansion of the determinant $\det(\sigma_{k-l_i+j-1}(\boldsymbol{x}))_{i,j=1}^{m+1}$ along the row corresponding to l_{m+1}. Clearly, the determinant vanishes whenever $l_{m+1} \in I$. On the other hand, expanding $g(z) = \boldsymbol{V}_{\perp}((\boldsymbol{x}, z); I)$ along its $k+1$ column we obtain

$$g(z) = \sum_{l_{m+1} \in \{0,1,\ldots,k+m\} \setminus I} \boldsymbol{V}_{\perp}(\boldsymbol{x}; I \cup \{l_{m+1}\}) z^{l_{m+1}} \tag{22}$$

From (21), (22) we get $\boldsymbol{V}_{\perp}(\boldsymbol{x}; I \cup \{l_{m+1}\}) = \boldsymbol{V}(\boldsymbol{x}) \det(\sigma_{k-l_i+j-1}(\boldsymbol{x}))_{i,j=1}^{m+1}$ which concludes our proof. □

Remark. To simplify the notation, let us write $a_{i,j}$ in place of $\sigma_{k-l_i+j}(\boldsymbol{x})$, for $1 \le i, j \le m$. Further, let $\mathcal{P}_m$ be the set of all permutations $\pi \in \mathcal{P}_m$ of $\{1, 2, \ldots, m\}$. The determinant at the left-hand side of (19) becomes

$$\det(a_{i,j} + z\, a_{i,j-1})_{i,j=1}^m = \sum_{\pi \in \mathcal{P}_m} \prod_{j=1}^{m} (a_{\pi_j,j} + z\, a_{\pi_j,j-1}) \tag{23}$$

where $\pi(1, \ldots, m) = (\pi_1, \ldots, \pi_m)$. Let us define the integers $c_j \in \mathbb{F}_2$, for $1 \le j \le m$. The product at the right-hand side of (23) can be written as the sum of 2^m terms of the form $(z^{c_1} a_{\pi_1,1-c_1}) \cdots (z^{c_m} a_{\pi_m,m-c_m})$, where c_j indicates whether $a_{\pi_j,j}$ or $z\, a_{\pi_j,j-1}$ contributes for each $1 \le j \le m$. As a result, we have

$$\begin{aligned}
\det\left(a_{i,j} + z\,a_{i,j-1}\right)_{i,j=1}^{m} &= \sum_{\pi \in \mathcal{P}_m} \sum_{\boldsymbol{c} \in \mathbb{F}_2^m} a_{\pi_1,1-c_1} \cdots a_{\pi_m,m-c_m}\, z^{c_1+\cdots+c_m} \\
&= \sum_{\boldsymbol{c} \in \mathbb{F}_2^m} \left(\sum_{\pi \in \mathcal{P}_m} a_{\pi_1,1-c_1} \cdots a_{\pi_m,m-c_m} \right) z^{\mathrm{wt}(\boldsymbol{c})} \\
&= \sum_{\boldsymbol{c} \in \mathbb{F}_2^m} \det\left(a_{i,j-c_j}\right)_{i,j=1}^{m} z^{\mathrm{wt}(\boldsymbol{c})}
\end{aligned}$$

where $\boldsymbol{c} = (c_1, c_2, \ldots, c_m)$, thus establishing the validity of (19). □

Construction of Pseudo-random Binary Sequences from Elliptic Curves by Using Discrete Logarithm*

Zhixiong Chen[1,2], Shengqiang Li[1,3], and Guozhen Xiao[1]

[1] National Key Lab. of I.S.N, Xidian Univ., Xi'an 710071, China
[2] Depart. of Math., Putian Univ., Putian, Fujian 351100, China
ptczx@126.com
[3] University of Electronic Science and Technology of China
Chengdu, 610054, China
shqli@mail.xidian.edu.cn

Abstract. An upper bound is established for certain exponential sums with respect to multiplicative characters defined on the rational points of an elliptic curve over a prime field. The bound is applied to investigate the pseudo-randomness of a large family of binary sequences generated from elliptic curves by using discrete logarithm. That is, we use this estimate to show that the resulting sequences have the advantages of 'small' well-distribution measure and 'small' multiple correlation measure.

1 Introduction

In a series of papers Mauduit and Sárközy (partly with further coauthors) studied finite pseudo-random binary sequences

$$S_N = \{s_1, s_2, \cdots, s_N\} \in \{+1, -1\}^N.$$

They first introduced several measures to evaluate the pseudo-randomness of such sequences in [17]. Two main measures within these measures are the well-distribution measure and the correlation measure of order k. The *well-distribution measure* of S_N is defined as

$$W(S_N) = \max_{a,b,t} \left| \sum_{j=0}^{t-1} s_{a+jb} \right|,$$

where the maximum is taken over all a, b, t such that $a, b, t \in \mathbb{N}$ and $1 \leq a \leq a + (t-1)b \leq N$, while the *correlation measure of order k* of S_N is defined as

$$C_k(S_N) = \max_{M,D} \left| \sum_{n=1}^{M} s_{n+d_1} s_{n+d_2} \cdots s_{n+d_k} \right|,$$

* The work was supported in part by the National Natural Science Foundation of China (No. 60473028). Research of the first author was partially supported by the Natural Science Foundation of Fujian Province of China (No.A0540011), the Science and Technology Foundation of Fujian Educational Committee (No.JA04264) and the Science and Technology Foundation of Putian City (No.2005S04).

G. Gong et al. (Eds.): SETA 2006, LNCS 4086, pp. 285–294, 2006.

where the maximum is taken over all $D = (d_1, \cdots, d_k)$ with non-negative integers $0 \leq d_1 < \cdots < d_k$ and M such that $M + d_k \leq N$.

The sequence S_N is considered as a "good" pseudo-random sequence, if both these measures $W(S_N)$ and $C_k(S_N)$ (at least for small k) are "small" in terms of N (in particular, both are $o(N)$ as $N \to \infty$).

It was shown by Mauduit and Sárközy in [17] that the Legendre symbol forms a "good" pseudo-random sequence. That is, let p be an odd prime, and

$$N = p - 1, s_n = \left(\frac{n}{p}\right), S_N = \{s_1, s_2, \cdots, s_N\} \in \{+1, -1\}^N.$$

This sequence is also known as the *Legendre sequence.* According to Theorem 1 of [17], we have

$$W(S_N) = O(p^{1/2}\log(p))$$

and

$$C_k(S_N) = O(kp^{1/2}\log(p)).$$

Indeed, it was shown in [3] that for a "random" sequence $S_N \in \{+1, -1\}^N$ (i.e., choosing $S_N \in \{+1, -1\}^N$ with probability $1/2^N$), both $W(S_N)$ and $C_k(S_N)$ (for some fixed k) are around $N^{1/2}$ with "near 1" probability.

Later Goubin, Mauduit and Sárközy extended this construction in [7], they constructed binary sequences, which we call as *GMS-sequences* for short, by using a family of polynomials $f(x) \in \mathbb{F}_p[x]$ under some special conditions:

$$N = p - 1, s_n = \left(\frac{f(n)}{p}\right), S_N = \{s_1, s_2, \cdots, s_N\} \in \{+1, -1\}^N.$$

Gyarmati constructed a family of binary sequences by using the notion of discrete logarithm in [8], and we refer to the sequences as *G-sequences* for short. The GMS-sequences and the G-sequences also have very interesting pseudo-random behaviour. Many other binary sequences were designed in the literature, see for example [3,16,18] and references therein.

Goubin et al in [7] also constructed a large family of binary sequences by using elliptic curves. Indeed, the authors of [7] only listed some numerical data and didn't give any theoretical estimates. We apply exponential sums (with respect to additive characters) on elliptic curves to show that such sequences possess "good" pseudo-random properties in a separate paper. The essential tool is the character sums estimates of [11].

We note that recent developments point towards an interest in the elliptic curve analogues of pseudo-random number generators, such as the elliptic curve linear congruential generators [5,9,10,19], the elliptic curve power generators [14] and the elliptic curve Naor-Reingold generators [22,24]. For other number generators related to elliptic curves, the reader is referred to [1,6,13,15]. In elliptic curve cryptosystems, such number generators provide strong potential applications for generating pseudo-random numbers and session keys in encryption phases.

Motivated by [8], we construct a large family of binary sequences by using the notion of index (discrete logarithm) and show that the resulting sequences possess "good" pseudo-random properties. That is, both the well-distribution measure and the correlation measure of order k of such sequences are "small".

This article is organized as follows. In Section 2, the exponential sums with respect to multiplicative characters on elliptic curves are estimated. The construction of a large family of binary sequences is proposed and the pseudo-random properties are considered in Section 3. We compare the well-distribution measure and the correlation measure of order k of our sequence with some other sequences and draw a conclusion in Section 4.

We conclude this section with some notions and basic facts of elliptic curves over finite fields. Let $p > 3$ be a prime and $\mathbb{F}_p$ the finite field of p elements, which we identify with the set $\{0, 1, \cdots, p-1\}$. $\mathbb{F}_p^*$ is the set of non-zero elements of $\mathbb{F}_p$. Let $\mathcal{E}$ be an elliptic curve over $\mathbb{F}_p$, given by an affine Weierstrass equation of the standard form

$$y^2 = x^3 + Ax + B$$

with coefficients $A, B \in \mathbb{F}_p$ and nonzero discriminant, see [4] for details. It is known that the set $\mathcal{E}(\mathbb{F}_p)$ of $\mathbb{F}_p$-rational points of $\mathcal{E}$ forms an Abelian group under an appropriate composition rule denoted by $\oplus$ and with the point at infinity $\mathcal{O}$ as the neutral element. We recall that

$$|\#\mathcal{E}(\mathbb{F}_p) - p - 1| \leq 2p^{1/2},$$

where $\#\mathcal{E}(\mathbb{F}_p)$ is the number of $\mathbb{F}_p$-rational points, including the point at infinity $\mathcal{O}$. For any rational point R, a multiple of R is taken by $nR = \oplus_{i=1}^{n} R$.

It is known that, as a group, $\mathcal{E}(\mathbb{F}_p)$ is isomorphic to $\mathbb{Z}_M \times \mathbb{Z}_L$ for unique integers M and L with $L|M$ and $\#\mathcal{E}(\mathbb{F}_p) = ML$. Rational points P and Q in $\mathcal{E}(\mathbb{F}_p)$ are called *echelonized generators* (see [11]) if the order of P is M, the order of Q is L, and any point in $\mathcal{E}(\mathbb{F}_p)$ can be represented in the form $mP \oplus lQ$ with $1 \leq m \leq M$ and $1 \leq l \leq L$.

Let $\mathbb{F}_p(\mathcal{E})$ be the function field of $\mathcal{E}$ defined over $\mathbb{F}_p$. For any $f \in \mathbb{F}_p(\mathcal{E})$ and $R \in \mathcal{E}(\overline{\mathbb{F}}_p)$, R is called a *zero* (resp. *pole*) of f if $f(R) = 0$ (resp. $f(R) = \infty$). Any rational function has only a finite number of zeros and poles. The divisor of a rational function f is written as

$$\mathrm{Div}(f) = \sum_{R \in \mathcal{E}(\overline{\mathbb{F}}_p)} \mathrm{ord}_R(f)[R],$$

where each integer $\mathrm{ord}_R(f)$ is the order of f at R and $\mathrm{ord}_R(f) = 0$ for all but finitely many $R \in \mathcal{E}(\overline{\mathbb{F}}_p)$. Note that $\mathrm{ord}_R(f) > 0$ if R is a zero of f and $\mathrm{ord}_R(f) < 0$ if R is a pole of f.

We also write

$$\mathrm{Supp}(f_0) = \{R \in \mathcal{E}(\overline{\mathbb{F}}_p) | f(R) = 0\}$$

and

$$\mathrm{Supp}(f_\infty) = \{R \in \mathcal{E}(\overline{\mathbb{F}}_p) | f(R) = \infty\}.$$

Then $\mathrm{Supp}(f) = \mathrm{Supp}(f_0) \cup \mathrm{Supp}(f_\infty)$, which is called the *support* of $\mathrm{Div}(f)$. In particular, $\#\mathrm{Supp}(f)$, the cardinality of $\mathrm{Supp}(f)$, is 2 or 3 if $f = x$ and $\#\mathrm{Supp}(f) \leq 4$ if $f = y$.

The translation map by $W \in \mathcal{E}(\mathbb{F}_p)$ on $\mathcal{E}(\mathbb{F}_p)$ is defined as

$$\begin{array}{rl} \tau_W : \mathcal{E}(\mathbb{F}_p) \to & \mathcal{E}(\mathbb{F}_p) \\ P \mapsto & P \oplus W. \end{array}$$

It is obvious that $(f \circ \tau_W)(P) = f(\tau_W(P)) = f(P \oplus W)$. We denote by $\ominus$ the inverse operation of $\oplus$ in the rational points group of $\mathcal{E}$. From Lemma 3.16, Theorem 3.17 and Lemma 3.14 of [4], we have the following statement.

Lemma 1. *Let $f \in \mathbb{F}_p(\mathcal{E})$ be a nonconstant rational function. If $R \in \mathrm{Supp}(f)$ and the order of f at R is ρ, then $R \ominus W$ belongs to the support of $\mathrm{Div}(f \circ \tau_W)$ with the same order ρ.*

2 Exponential Sums on Elliptic Curves

For any positive n, an additive character of $\mathbb{Z}_n := \{0, 1, \cdots, n-1\}$, the residue ring modulo n, is defined as

$$e_n(z) = \exp(2\pi i z/n).$$

All additive characters of $\mathbb{F}_p$ can be described by the set:

$$\Phi = \{\phi_\alpha | \phi_\alpha(z) = e_p(\alpha z) \text{ for } \alpha \in \mathbb{F}_p\}.$$

Let $P, Q \in \mathcal{E}(\mathbb{F}_p)$ be the echelonized generators of order M and L, respectively. The group $\Omega = \mathrm{Hom}(\mathcal{E}(\mathbb{F}_p), \mathbb{C}^*)$ of the characters on $\mathcal{E}(\mathbb{F}_p)$ is defined as follows:

$$\Omega = \{\omega_{ab} | \omega_{ab}(mP \oplus lQ) = e_M(am)e_L(bl) \text{ for } 0 \leq a < M \text{ and } 0 \leq b < L\}.$$

The exponential sums

$$S(\omega, \phi, f) = {\sum}^*_{R \in \mathcal{E}(\mathbb{F}_p)} \omega(R)\phi(f(R))$$

have been investigated in [11], where $\omega \in \Omega$, $\phi \in \Phi$, $f \in \mathbb{F}_p(\mathcal{E})$ is a rational function and $\sum^*$ indicates that the poles of f are excluded from the summation. In fact, in [11], the exponential sums have been considered for an elliptic curve $\mathcal{E}$ defined over any extension field $\mathbb{F}_q$. And the exponential sums $S(\omega_{00}, \phi, f)$ with the trivial character $\omega_{00} \in \Omega$ have been estimated in [2,1,12,27].

Now we introduce the multiplicative characters of $\mathbb{F}_p^*$. In the sequel, let g be a fixed primitive root modulo p. For each $x \in \mathbb{F}_p^*$, let $\mathrm{ind}(x)$ denote the index (discrete logarithm) of x (to the base g) so that

$$g^{\mathrm{ind}(x)} \equiv x \pmod{p}.$$

We add the condition

$$1 \leq \mathrm{ind}(x) \leq p - 1$$

to make the value of index unique. A multiplicative character of F_p^* is defined by

$$\chi_a(x) := e_{p-1}(a \cdot \text{ind}(x)) = \exp\left(\frac{2\pi i a \cdot \text{ind}(x)}{p-1}\right)$$

where $a \in Z_{p-1}$. We denote by Ψ the set of all multiplicative characters of F_p^* with the trivial character χ_0:

$$\Psi := \{\chi_a | a \in Z_{p-1}\}.$$

Ψ forms a cyclic group under the multiplication of characters. For any $\chi \in \Psi$, we set $\overline{\chi}(x) = \chi(x^{-1})$, i.e., $\overline{\chi}$ is the inverse of χ. Since $\chi \in \Psi$ is defined over F_p^*, for convenience, we extend χ to $\mathbb{F}_p$ only by defining $\chi(0) = 0$. And the identity $\chi(xy) = \chi(x)\chi(y)$ will remain true for any $x, y \in \mathbb{F}_p$. We also denote by Ψ^* the set of all nontrivial multiplicative characters of F_p^*.

We define the exponential sums with respect to multiplicative characters as follows:

$$S(\omega, \chi, f) = \sum\nolimits^*_{R \in \mathcal{E}(\mathbb{F}_p)} \omega(R)\chi(f(R)),$$

where $\omega \in \Omega$, $\chi \in \Psi$, $f \in \mathbb{F}_p(\mathcal{E})$ is a rational function and $\sum^*$ indicates that the poles of f are excluded from the summation. In particular, a special case $S(\omega_{00}, \chi, f)$ with trivial character $\omega_{00} \in \Omega$ has been estimated in [1,20,21].

Theorem 1. *Let $f(x,y) \in \mathbb{F}_p(\mathcal{E})$ be a nonconstant rational function with $f(x,y) \neq z^l(x,y)$ for all $z(x,y) \in \overline{\mathbb{F}}_p(\mathcal{E})$ and all factors $l > 1$ of $p-1$. For any $\chi \in \Psi^*$ and any $\omega \in \Omega$, the following upper bound holds:*

$$|S(\omega, \chi, f)| < \#\text{Supp}(f)\sqrt{p}.$$

The proof is analogous to that of Theorem 1 of [11]. Instead of applying Artin-Schreier extensions, we obtain the desired result from Propositions 3.1 and 4.5 of [21] by using Kummer extensions.

Corollary 1. *With conditions as in Theorem 1. Then we have*

$$\left|\sum\nolimits^*_{R \in \mathcal{H}} \omega(R)\chi(f(R))\right| < \#\text{Supp}(f)\sqrt{p},$$

where $\mathcal{H}$ is an arbitrary subgroup of $\mathcal{E}(\mathbb{F}_p)$, and $\sum^*$ indicates that the poles of f are excluded from the summation.

3 Construction of Binary Sequences

Let $G \in \mathcal{E}(\mathbb{F}_p)$ be a point of order N, that is, N is the size of the cyclic group generated by G. We also suppose that $f(x,y) \in \mathbb{F}_p(\mathcal{E})$ is a rational function with $f(x,y) \neq z^l(x,y)$ for all $z(x,y) \in \overline{\mathbb{F}}_p(\mathcal{E})$ and all factors $l > 1$ of $p-1$. In particular, we are interested in the functions $f = x$ or $f = y$. If $R \in \mathcal{E}(\overline{\mathbb{F}}_p)$ is a pole of f, we select a fixed value (for example, zero) as its output.

Definition 1. *Let $G \in \mathcal{E}(\mathbb{F}_p)$ be a point of order N. We define the binary sequence $S_N = \{s_1, s_2, \cdots, s_N\}$ by*

$$s_n := \begin{cases} +1, \text{if } 1 \le \mathrm{ind}(f(nG)) \le (p-1)/2; \\ -1, \text{if } (p+1)/2 \le \mathrm{ind}(f(nG)) \le p-1 \text{ or } p|f(nG). \end{cases}$$

From Definition 1, it is easy to see that for any $n \ge 1$

$$\frac{1}{p-1} \sum_{i=1}^{(p-1)/2} \sum_{\chi \in \Psi} \overline{\chi}(f(nG))\chi(g^i) = \begin{cases} 1, \text{if } 1 \le \mathrm{ind}(f(nG)) \le (p-1)/2; \\ 0, \text{otherwise.} \end{cases}$$

And hence we have

$$s_n = \frac{2}{p-1} \sum_{i=1}^{(p-1)/2} \sum_{\chi \in \Psi^*} \overline{\chi}(f(nG))\chi(g^i). \tag{1}$$

Before we give some estimates for the parameters of the pseudorandom sequence S_N defined above, we present some necessary statements on character sums. The following lemma is a special case of Lemma 3 of [8].

Lemma 2. *Let g be a fixed primitive root modulo p. Ψ^* is the set of all nontrivial multiplicative characters (with respect to g) of F_p^*. The bound holds:*

$$\sum_{\chi \in \Psi^*} \left| \sum_{i=1}^{(p-1)/2} \chi(g^i) \right| < 2(p-1)\log(p).$$

Lemma 3. *Let N be a positive integer, and $a, b, t \in \mathbb{N}$ with $1 \le a \le a+(t-1)b \le N$. Then the following bound holds:*

$$\sum_{\lambda=0}^{N-1} \left| \sum_{x=0}^{t-1} e_N(\lambda(a+bx)) \right| < N(1+\log(N)).$$

Proof. Let $d = \gcd(b, N)$, $M = N/d$ and $b_1 = b/d$. Since $(t-1)b \le N-1$, we have $d(t-1) \le (t-1)b < N$, and hence $t-1 < M$. We derive

$$\begin{aligned} \sum_{\lambda=0}^{N-1} \left| \sum_{x=0}^{t-1} e_N(\lambda(a+bx)) \right| &= \sum_{\lambda=0}^{N-1} \left| e_N(\lambda a) \sum_{x=0}^{t-1} e_N(\lambda b x) \right| \\ &= d \sum_{\lambda=0}^{M-1} \left| \sum_{x=0}^{t-1} e_M(\lambda b_1 x) \right| < dM(1+\log(M)). \end{aligned}$$

Since $\gcd(M, b_1) = 1$, we complete the proof by Inequality (3.4) of [23]. □

Lemma 4. *With conditions as in Theorem 1. And let $G \in \mathcal{E}(\mathbb{F}_p)$ be a rational point of order N and $\chi \in \Psi^*$. Then for any fixed $a, b, t \in \mathbb{N}$ with $1 \le a \le a+(t-1)b \le N$, the following bound holds:*

$$\left| \sum_{x=0}^{t-1} \chi(f((a+bx)G)) \right| < \#\mathrm{Supp}(f) p^{1/2}(1+\log(N)),$$

where $\#\mathrm{Supp}(f)$ is the cardinality of the support of f.

Proof.

$$\begin{aligned}\left|\sum_{x=0}^{t-1}\chi(f((a+bx)G))\right| &= \left|\tfrac{1}{N}\sum_{n=1}^{N}\sum_{x=0}^{t-1}\chi(f(nG))\sum_{\lambda=0}^{N-1}e_N(\lambda(n-(a+bx)))\right|\\ &= \tfrac{1}{N}\left|\sum_{\lambda=0}^{N-1}\sum_{x=0}^{t-1}e_N(-\lambda(a+bx))\sum_{n=1}^{N}\chi(f(nG))e_N(\lambda n))\right|\\ &\le \tfrac{1}{N}\sum_{\lambda=0}^{N-1}\left|\sum_{x=0}^{t-1}e_N(-\lambda(a+bx))\right|\cdot\left|\sum_{n=1}^{N}\chi(f(nG))e_N(\lambda n))\right|.\end{aligned}$$

Now by Corollary 1 and Lemma 3, we derive the desired result. □

Theorem 2. *Let $f(x,y)\in\mathbb{F}_p(\mathcal{E})$ be a rational function with $f(x,y)\neq z^l(x,y)$ for all $z(x,y)\in\overline{\mathbb{F}}_p(\mathcal{E})$ and all factors $l>1$ of $p-1$. The sequence S_N is defined as in Definition 1. Then the upper bound of the well-distribution measure of S_N satisfies:*

$$W(S_N) < 4\#\mathrm{Supp}(f)p^{1/2}\log(p)(1+\log(N)),$$

where $\#\mathrm{Supp}(f)$ is the cardinality of the support of f.

Proof. According to Eq.(1), for any $a,b,t\in\mathbb{N}$ with $1\le a\le a+(t-1)b\le N$,

$$\begin{aligned}\left|\sum_{j=0}^{t-1}s_{a+jb}\right| &= \tfrac{2}{p-1}\left|\sum_{j=0}^{t-1}\sum_{i=1}^{(p-1)/2}\sum_{\chi\in\Psi^*}\overline{\chi}(f((a+jb)G))\chi(g^i)\right|\\ &\le \tfrac{2}{p-1}\sum_{\chi\in\Psi^*}\left|\sum_{i=1}^{(p-1)/2}\chi(g^i)\right|\cdot\left|\sum_{j=0}^{t-1}\overline{\chi}(f((a+jb)G))\right|.\end{aligned}$$

Now by Lemmas 2 and 4, we obtain the desired result. □

Theorem 3. *Let $f(x,y)\in\mathbb{F}_p(\mathcal{E})$ be a rational function with $f(x,y)\neq z^l(x,y)$ for all $z(x,y)\in\overline{\mathbb{F}}_p(\mathcal{E})$ and all factors $l>1$ of $p-1$. The sequence S_N is defined as in Definition 1. Then the bound of the correlation measure of order k holds:*

$$C_k(S_N) < k4^k\#\mathrm{Supp}(f)p^{1/2}\log^k(p)(1+\log(N)),$$

where $\#\mathrm{Supp}(f)$ is the cardinality of the support of f.

Proof. According to Eq.(1), for integers $D=(d_1,\cdots,d_k)$ and M with $0\le d_1<\cdots<d_k\le N-M$, we have

$$\begin{aligned}&\left|\sum_{n=1}^{M}s_{n+d_1}s_{n+d_2}\cdots s_{n+d_k}\right|\\ &= \tfrac{2^k}{(p-1)^k}\left|\sum_{n=1}^{M}\prod_{j=1}^{k}\left(\sum_{i=1}^{(p-1)/2}\sum_{\chi\in\Psi^*}\overline{\chi}(f((n+d_j)G))\chi(g^i)\right)\right|\\ &= \tfrac{2^k}{(p-1)^k}\left|\sum_{\chi_1,\cdots,\chi_k\in\Psi^*}\sum_{i_1=1}^{(p-1)/2}\chi_1(g^{i_1})\cdots\sum_{i_k=1}^{(p-1)/2}\chi_k(g^{i_k})\sum_{n=1}^{M}\prod_{i=1}^{k}\overline{\chi}_i(f((n+d_i)G))\right|\\ &= \tfrac{2^k}{(p-1)^k}\left|\sum_{\chi_1\in\Psi^*}\sum_{i_1=1}^{(p-1)/2}\chi_1(g^{i_1})\cdots\sum_{\chi_k\in\Psi^*}\sum_{i_k=1}^{(p-1)/2}\chi_k(g^{i_k})\sum_{n=1}^{M}\prod_{i=1}^{k}\overline{\chi}_i((f\circ\tau_{d_iG})(nG))\right|.(*)\end{aligned}$$

Let ψ be a generator of the cyclic group Ψ, i.e., the order of ψ is $p-1$. Then for each $\overline{\chi}_i$, $1 \le i \le k$, there exists an integer $\alpha_i \in [1, p-1]$ such that $\overline{\chi}_i = \psi^{\alpha_i}$. Now let $F = (f \circ \tau_{d_1 G})^{\alpha_1} \cdots (f \circ \tau_{d_k G})^{\alpha_k}$ be a rational function. From Lemma 1, it is easy to see that $\#\mathrm{Supp}(F) \le k\#\mathrm{Supp}(f)$. By Lemmas 2 and 4, we obtain

$$
\begin{aligned}
(*) &\le \frac{2^k}{(p-1)^k} \prod_{j=1}^{k} \left| \sum_{\chi_j \in \Psi^*} \sum_{i_j=1}^{(p-1)/2} \chi_j(g^{i_j}) \right| \cdot \left| \sum_{n=1}^{M} \prod_{j=1}^{k} \overline{\chi}_j((f \circ \tau_{d_j G})(nG)) \right| \\
&\le \frac{2^k}{(p-1)^k} \prod_{j=1}^{k} \sum_{\chi_j \in \Psi^*} \left| \sum_{i_j=1}^{(p-1)/2} \chi_j(g^{i_j}) \right| \cdot \left| \sum_{n=1}^{M} \prod_{j=1}^{k} \psi((f \circ \tau_{d_j G})^{\alpha_i}(nG)) \right| \\
&= \frac{2^k}{(p-1)^k} \prod_{j=1}^{k} \sum_{\chi_j \in \Psi^*} \left| \sum_{i_j=1}^{(p-1)/2} \chi_j(g^{i_j}) \right| \cdot \left| \sum_{n=1}^{M} \psi(F(nG)) \right| \\
&\le 4^k \#\mathrm{Supp}(F) p^{1/2} \log^k(p)(1+\log(N)) \\
&\le 4^k k \#\mathrm{Supp}(f) p^{1/2} \log^k(p)(1+\log(N)).
\end{aligned}
$$

We complete the proof of Theorem 3. □

As mentioned in the introduction, both $W(S_N)$ and $C_k(S_N)$ (for some fixed k) of a "random" sequence $S_N \in \{+1, -1\}^N$ are around $N^{1/2}$ with "near 1" probability. Theorems 2 and 3 indicate that the resulting binary sequence also forms a "good" pseudo-random sequence. The period N of the sequence is bounded by the size of $\mathcal{E}(\mathbb{F}_p)$. In particular, if $\mathcal{E}(\mathbb{F}_p)$ is a cyclic group, then $N \sim p$. We remark that the resulting sequences are very common enough. From Corollary 6.2 of [25], about 75% of the majority of (isomorphism classes of) elliptic curves have a cyclic point group. By Theorem 2.1 of [25], every cyclic group of order N satisfying $p - 1 - 2p^{1/2} \le N \le p - 1 + 2p^{1/2}$ can be realized as the point group of an elliptic curve over $\mathbb{F}_p$ ($p > 5$). The fact is also indicated in [10]. For more information on elliptic curves with cyclic groups, the reader is referred to [25,26].

We also note that in [1] and [6], a method for generating sequences was proposed by applying linear recurrence relations on elliptic curves, which may produce rational point sequences with long periods. But it seems that the exponential sums stated in Section 2 can not be extended to this case in a straight way.

4 Conclusion

By using the notion of discrete logarithm, We have constructed a large family of binary sequences from elliptic curves over finite fields. In Table 1, we compare our sequences with some other sequences, such as the Legendre sequence [17], the GMS-sequence [7] and the G-sequence [8] described in Section 1. We conclude that our sequences also have strong pseudo-random properties, they may be suitable for use in cryptography.

Table 1. Comparison of Our Sequence with Some Other Sequences

Sequences	Period	Well-distribution	Correlation of order k
Legendre sequence	$N = p - 1$	$O(p^{1/2}\log(p))$	$O(kp^{1/2}\log(p))$
GMS-sequence	$N = p - 1$	$O(p^{1/2}\log(p))$	$O(kp^{1/2}\log(p))$
G-sequence	$N = p - 1$	$O(p^{1/2}\log^2(p))$	$O(k4^k p^{1/2}\log^{k+1}(p))$
Our sequence	$N = O(p)$	$O(p^{1/2}\log(p)\log(N))$	$O(k4^k p^{1/2}\log^k(p)\log(N))$

Remark 1. The implied constant in the symbol "O" may sometimes depend on the degree $\deg(f)$ or $\#\mathrm{Supp}(f)$ of functions f adopted in the corresponding constructions and is absolute otherwise.

In elliptic curve cryptography, a rational point with large (prime) order will be chosen. It is natural to generate binary sequences by such point. We note that generating the sequences described in this paper relies on the implementation of group operations on elliptic curves over $\mathbb{F}_p$ and the computation of discrete logarithm in $\mathbb{F}_p^*$. Although one can borrow software/hardware from elliptic curve cryptosystems for computing the rational points efficiently, the generation is very slow since there is no fast algorithm for computing the discrete logarithm in $\mathbb{F}_p^*$, where subexponential-time index calculus methods are known.

Obtaining improvements of Theorems 2 and 3 is a challenging problem. It would also be interesting to study the linear complexity (profile) of the resulting sequences, which is an important cryptographic characteristic of pseudo-random sequences.

References

1. Beelen, P. H. T., Doumen, J. M.: Pseudorandom Sequences from Elliptic Curves. In: Finite Fields with Applications to Coding Theory, Cryptography and Related Areas. Springer-Verlag, Berlin Heidelberg New York (2002) 37–52
2. Bombieri, E.: On Exponential Sums in Finite Fields. Amer. J. Math. 88 (1966) 71–105
3. Cassaigne, J., Mauduit, C., Sárközy, A.: On Finite Pseudorandom Binary Sequences, VII: The Measures of Pseudorandomness. Acta Arithmetica 103 (2002) 97–118
4. Enge, A.: Elliptic Curves and Their Applications to Cryptography: an Introduction. Kluwer Academic Publishers, Dordrecht (1999)
5. Gong, G., Berson, T., Stinson, D.: Elliptic Curve Pseudorandom Sequence Generator. Available at http://www.cacr.math.uwaterloo.ca, Technical Reports, No. CORR1998-53 (1998)
6. Gong, G., Lam, C.Y.: Linear Recursive Sequences over Elliptic Curves. In: Proceedings of Sequences and Their Applications-SETA'01. DMTCS series. Springer-Verlag, Berlin Heidelberg New York (2001) 182–196
7. Goubin, L., Mauduit, C., Sárközy, A.: Construction of Large Families of Pseudorandom Binary Sequences. J. Number Theory 106(1) (2004) 56–69

8. Gyarmati, K.: On a Family of Pseudorandom Binary Sequences. Periodica Mathematica Hungarica 49(2) (2004) 45–63
9. Hallgren, S.: Linear Congruential Generators over Elliptic Curves. Technical Report, No. CS-94-143, Cornegie Mellon University (1994)
10. Hess, F., Shparlinski, I.E.: On the Linear Complexity and Multidimensional Distribution of Congruential Generators over Elliptic Curves. Designs, Codes and Cryptography 35(1) (2005) 111–117
11. Kohel, D., Shparlinski, I.E.: On Exponential Sums and Group Generators for Elliptic Curves over Finite Fields. In: Proc. Algorithmic Number Theory Symposium, Leiden, 2000. Lecture Notes in Computer Science, Vol. 1838. Springer-Verlag, Berlin Heidelberg New York (2000) 395–404
12. Lachaud, G.: Artin-Schreier Curves, Exponential Sums and the Carlitz-Uchiyama Bound for Geometric Codes. J. Number Theory 39(1) (1991) 18–40
13. Lam, C.Y., Gong, G.: Randomness of Elliptic Curve Sequences. Available at http://www.cacr.math.Uwaterloo.ca, Technical Reports, No. CORR 2002-18 (2002)
14. Lange, T. Shparlinski, I.E.: Certain Exponential Sums and Random Walks on Elliptic Curves. Canad. J. Math. 57(2) (2005) 338–350
15. Lee, L., Wong, K.: An Elliptic Curve Random Number Generator. In: Communications and Multimedia Security Issues of the New Century, Fifth Joint Working Conference on Communications and Multimedia Security-CMS'01. (2001) 127–133
16. Mauduit, C., Rivat, J., Sárközy, A.: Construction of Pseudorandom Binary Sequences Using Additive Characters. Mh. Math. 141(3) (2004) 197–208
17. Mauduit, C., Sárközy, A.: On Finite Pseudorandom Binary Sequences I: Measures of Pseudorandomness, the Legendre Symbol. Acta Arithmetica 82 (1997) 365–377
18. Mauduit, C., Sárközy, A.: On Finite Pseudorandom Binary Sequences II: The Champernowne, Rudin-Shapiro, and Thue-Morse Sequences, A Further Construction. J. Number Theory 73(2) (1998) 256–276
19. El Mahassni, E., Shparlinski, I.E.: On the Uniformity of Distribution of Congruential Generators over Elliptic Curves. In: Proc. Intern. Conf. on Sequences and Their Applications-SETA'01. Springer-Verlag, Berlin Heidelberg New York (2002) 257–264
20. Perret, M.: Multiplicative Character Sums and Nonlinear Geometric Codes. Eurocode '90. Springer-Verlag, Berlin Heidelberg New York (1991) 158–165
21. Perret, M.: Multiplicative Character Sums and Kummer Coverings. Acta Arithmetica 59 (1991) 279–290
22. Shparlinski, I.E.: On the Naor-Reingold Pseudo-random Number Function from Elliptic Curves. Appl. Algebra Engng. Comm. Comput. 11(1) (2000) 27–34
23. Shparlinski, I.E.: Cryptographic Applications of Analytic Number Theory: Complexity Lower Bounds and Pseudorandomness. Progress in Computer Science and Applied Logic, Vol. 22, Birkhauser Verlag, Basel (2003)
24. Shparlinski, I.E., Silverman, J.H.: On the Linear Complexity of the Naor-Reingold Pseudo-random Function from Elliptic Curves. Designs, Codes and Cryptography 24(3) (2001) 279–289
25. Vlăduţ, S.G.: Cyclicity Statistics for Elliptic Curves over Finite Fields. Finite Fields and Their Applications 5(1) (1999) 13–25
26. Vlăduţ, S.G.: On the Cyclicity of Elliptic Curves over Finite Field Extensions. Finite Fields and Their Applications 5(3) (1999) 354–363
27. Voloch, J.F., Walker, J.L.: Euclidean Weights of Codes from Elliptic Curves over Rings. Trans. Amer. Math. Soc. 352(11) (2000) 5063–5076

On the Discrepancy and Linear Complexity of Some Counter-Dependent Recurrence Sequences

Igor E. Shparlinski[1] and Arne Winterhof[2]

[1] Department of Computing
Macquarie University
North Ryde, NSW 2109, Australia
igor@ics.mq.edu.au

[2] Johann Radon Institute for Computational and Applied Mathematics
Austrian Academy of Sciences
Altenberger Straße 69, A-4040 Linz, Austria
arne.winterhof@oeaw.ac.at

In Memory of Hans Dobbertin

Abstract. We prove a discrepancy bound "on average" over all initial values $a_\alpha(0) = \alpha$ of congruential pseudorandom numbers obtained from the sequences $a_\alpha(n)$ over a finite field of prime order defined by $a_\alpha(n) = na_\alpha(n-1) + 1$, $n = 1, 2, \ldots$, using new bounds on certain exponential sums.

Moreover, we prove a lower bound on the linear complexity of this sequence showing that its structural properties are close to be best possible.

Keywords: Recurrence sequences, discrepancy, uniform distribution, linear complexity, nonlinear pseudorandom numbers, counter-dependent generator.

1 Introduction

Let p be a prime number and let $\mathbb{F}_p$ denote the finite field of p elements, which we always assume to be represented by the set $\{0, 1, \ldots, p-1\}$.

There is an extensive literature which studies pseudorandom properties of sequences $u_\vartheta(n)$ over $\mathbb{F}_p$ satisfying a recurrence relation $u_\vartheta(n) = f(u_\vartheta(n-1))$, $n = 1, 2, \ldots$, with the initial term $u_\vartheta(0) = \vartheta \in \mathbb{F}_p$, and some linear or nonlinear function f, see [7,12,13,14,15,18].

Recently, there has been a suggestion to consider more general, *counter-dependent* relations of the form $u_\vartheta(n) = f(u_\vartheta(n-1), n)$, see [16]. It is shown in [6] that, provided such a sequence is of large period, one can obtain nontrivial bounds of exponential sums with elements of such sequences and thus derive some conclusions about the uniformity of distribution of the corresponding *congruential pseudorandom numbers* $u_\vartheta(n)/p$, $n = 0, 1, \ldots$, in the unit interval.

Here we consider a very special case of this general construction and show that in this case much more can be deduced about the properties of the corresponding

G. Gong et al. (Eds.): SETA 2006, LNCS 4086, pp. 295–303, 2006.

sequence. Namely, here we consider the family of sequences $a_\alpha(n)$ satisfying the recurrence relation $a_\alpha(n) = na_\alpha(n-1) + 1$, $n = 1, 2, \ldots$, with the initial term $a_\alpha(0) = \alpha \in \mathbb{F}_p$.

It is easy to see that regardless of the initial value α, we always have $a_\alpha(p) = 1 = a_1(0)$ and thus more generally, $a_\alpha(n+p) = a_1(n)$, $n = 0, 1, \ldots$. Therefore, it is only interesting to study sequences $a_\alpha(n)$ on intervals $0 \le n \le N-1$ with an integer $N \le p$.

For such segments we obtain a nontrivial bound of exponential sums with $a_\alpha(n)$ "on average" over all initial values $\alpha \in \mathbb{F}_p$, that is, on the sums

$$W_s(N; \lambda_0, \ldots, \lambda_{s-1}) = \frac{1}{p} \sum_{\alpha=0}^{p-1} \left| \sum_{n=0}^{N-1} \mathbf{e}\left(\sum_{j=0}^{s-1} \lambda_j a_\alpha(n+j) \right) \right|,$$

where $\mathbf{e}(z) = \exp(2\pi i z/p)$ and $\lambda_0, \ldots, \lambda_{s-1} \in \mathbb{F}_p$. Using standard techniques one can then derive various results about the uniformity of distribution of the fractions $a_\alpha(n)/p$, $n = 0, 1, \ldots, N-1$, "on average" over $\alpha \in \mathbb{F}_p$.

We also obtain an "individual" (that is, for every $\alpha \in \mathbb{F}_p$) lower bound on the *linear complexity* $L_\alpha(N)$ of $a_\alpha(n)$, $n = 0, 1, \ldots, N-1$, which is the smallest positive integer L for which there are some $c_1, \ldots, c_L \in \mathbb{F}$ such that

$$a_\alpha(n+L) = c_{L-1} a_\alpha(n+L-1) + \ldots + c_0 a_\alpha(n), \qquad 0 \le n \le N-L-1.$$

Finally, we remark that for $\alpha = 0$ the sequence $a_0(n)$ (defined over the integers), has various combinatorial interpretations, see [17] and has also been studied in [11].

2 Exponential Sums

Theorem 1. *If* $\gcd(\lambda_0, \ldots, \lambda_{s-1}, p) = 1$ *then for any positive integer* $N \le p$*, the following bound holds:*

$$W_s(N; \lambda_0, \ldots, \lambda_{s-1}) < \left(s^{1/2} + 2\right) N^{3/4}.$$

Proof. Clearly, for every integer $k \ge 0$ we have

$$\left| \sum_{n=0}^{N-1} \mathbf{e}\left(\sum_{j=0}^{s-1} \lambda_j a_\alpha(n+j) \right) - \sum_{n=0}^{N-1} \mathbf{e}\left(\sum_{j=0}^{s-1} \lambda_j a_\alpha(n+k+j) \right) \right| \le 2k.$$

Hence for any integer $K \ge 1$ we have

$$\left| \sum_{n=0}^{N-1} \mathbf{e}\left(\sum_{j=0}^{s-1} \lambda_j a_\alpha(n+j) \right) \right| \le \frac{1}{K} \sum_{n=0}^{N-1} \left| \sum_{k=0}^{K-1} \mathbf{e}\left(\sum_{j=0}^{s-1} \lambda_j a_\alpha(n+k+j) \right) \right| + K - 1.$$

Accordingly

$$pW_s(N;\lambda_0,\dots,\lambda_{s-1}) \le \frac{1}{K}\sigma + (K-1)p, \tag{1}$$

where

$$\sigma = \sum_{\alpha=0}^{p-1}\sum_{n=0}^{N-1}\left|\sum_{k=0}^{K-1}\mathbf{e}\left(\sum_{j=0}^{s-1}\lambda_j a_\alpha(n+k+j)\right)\right|.$$

Applying the Cauchy-Schwarz inequality, we derive

$$\sigma^2 \le pN\sum_{\alpha=0}^{p-1}\sum_{n=0}^{N-1}\left|\sum_{k=0}^{K-1}\mathbf{e}\left(\sum_{j=0}^{s-1}\lambda_j a_\alpha(n+k+j)\right)\right|^2. \tag{2}$$

We now note that for $m \ge 0$ we have

$$a_\alpha(n+m) = f_m(n)a_\alpha(n) + g_m(n), \quad n = 0,1,\dots, \tag{3}$$

where

$$f_m(X) = \prod_{j=1}^{m}(X+j), \qquad g_m(X) = \sum_{\nu=1}^{m}\prod_{j=\nu+1}^{m}(X+j) \tag{4}$$

with the natural convention that $f_0(X) = 1$ and $g_0(X) = 0$. Substituting (3) into (2) we deduce

$$\sigma^2 \le pN\sum_{\alpha=0}^{p-1}\sum_{n=0}^{N-1}\left|\sum_{k=0}^{K-1}\mathbf{e}\left(\sum_{j=0}^{s-1}\lambda_j\left(f_{k+j}(n)a_\alpha(n) + g_{k+j}(n)\right)\right)\right|^2. \tag{5}$$

We now remark that the pairs $(n, a_\alpha(n))$, $0 \le n, \alpha \le p-1$, are pairwise distinct. Indeed, if

$$(n, a_\alpha(n)) = (m, a_\beta(m))$$

then certainly $n = m$ and then, if $n \ge 1$, $a_\alpha(n-1) = a_\beta(n-1)$. Thus, proceeding the same way, after n steps, we obtain $\alpha = a_\alpha(0) = a_\beta(0) = \beta$. Therefore, comparing the cardinalities we see that

$$\begin{aligned}\{(n, a_\alpha(n)) \mid 0 \le n \le N-1,\ 0 \le \alpha \le p-1\}\\ = \{(n,a) \mid 0 \le n \le N-1,\ 0 \le a \le p-1\}.\end{aligned}$$

We now derive from (5) that

$$\begin{aligned}\sigma^2 &\le pN\sum_{a=0}^{p-1}\sum_{n=0}^{N-1}\left|\sum_{k=0}^{K-1}\mathbf{e}\left(\sum_{j=0}^{s-1}\lambda_j\left(f_{k+j}(n)a + g_{k+j}(n)\right)\right)\right|^2\\ &= pN\sum_{k,l=0}^{K-1}\sum_{n=0}^{N-1}\mathbf{e}\left(\sum_{j=0}^{s-1}\lambda_j\left(g_{k+j}(n) - g_{l+j}(n)\right)\right)\\ &\qquad \sum_{a=0}^{p-1}\mathbf{e}\left(a\sum_{j=0}^{s-1}\lambda_j\left(f_{k+j}(n) - f_{l+j}(n)\right)\right).\end{aligned}$$

The inner sum is vanishing, unless

$$\sum_{j=0}^{s-1} \lambda_j \left(f_{k+j}(n) - f_{l+j}(n)\right) = 0,$$

in which case it is equal to p.

This certainly always happens if $k = l$. Otherwise, noticing that $\deg f_m = m$ we see that the above equation has at most $\max\{k,l\} \le K + s - 2$ solutions in n, $0 \le n \le p-1$. Therefore

$$\sigma^2 \le p^2 N(K(K-1)(K+s-2) + KN) < p^2 N(K^2(K+s-2) + KN).$$

Thus, substituting this bound in (1), we obtain

$$\begin{aligned} W_s(N;\lambda_0,\ldots,\lambda_{s-1}) &< \frac{1}{K}\sqrt{N(K^2(K+s-2)+KN)} + (K-1) \\ &= \sqrt{N(K+s-2+K^{-1}N)} + (K-1). \end{aligned}$$

We now choose $K = \lceil N^{1/2} \rceil$, getting

$$W_s(N;\lambda_0,\ldots,\lambda_{s-1}) < \sqrt{2N^{3/2} + N(s-1)} + N^{1/2},$$

and the result follows by simple calculations. □

Let us denote by $D_{\alpha,s}(N)$ the *discrepancy* of the s-tuples

$$\mathbf{a}_n = \left(\frac{a_\alpha(n)}{p}, \ldots, \frac{a_\alpha(n+s-1)}{p}\right), \qquad 0 \le n \le N-1,$$

that is,

$$D_{\alpha,s}(N) = \sup_{J \subseteq [0,1)^s} \left| \frac{A(J,N)}{N} - \lambda(J) \right|,$$

where the supremum is extended over all subintervals J of $[0,1)^s$, $A(J,N)$ is the number of points $\mathbf{a}_n$ in J for $0 \le n \le N-1$, and $\lambda(J)$ is the volume of J, see [5,10]. Using the celebrated *Erdös-Turan-Koksma inequality* (see also Theorem 1.21 of [5]), in a standard fashion we derive:

Corollary 1. *For any positive integer $N \le p$ and every fixed integer $s \ge 1$, the following bound holds:*

$$\frac{1}{p}\sum_{\alpha=0}^{p-1} D_{\alpha,s}(N) = O(N^{3/4}\log^s N),$$

where the implied constant depends only on s.

3 Linear Complexity

Theorem 2. *For any positive integer N, the following bound holds:*

$$L_\alpha(N) \geq \min\{(N-1)/2, p\}.$$

Proof. We see that for $l \geq k \geq 0$ the polynomials in (4) satisfy

$$f_l(X) = \prod_{j=l-k+1}^{l} (X+j) f_{l-k}(X), \tag{6}$$

and

$$g_l(X) = \prod_{j=l-k+1}^{l} (X+j) g_{l-k}(X) + \sum_{\nu=l-k+1}^{l} \prod_{j=\nu+1}^{l} (X+j). \tag{7}$$

Put $L = L_\alpha(N)$ and $c_L = 1$, and let

$$\sum_{l=0}^{L} c_l a_\alpha(n+l) = 0, \qquad 0 \leq n \leq N-L-1,$$

be a shortest recurrence relation satisfied by the first N elements of $a_\alpha(n)$.

Using (3) we derive

$$\sum_{l=0}^{L} c_l (f_l(n) a_\alpha(n) + g_l(n)) = 0, \qquad 0 \leq n \leq N-L-1,$$

Similarly, from

$$\sum_{l=0}^{L} c_l a_\alpha(n+l+1) = 0, \qquad 0 \leq n \leq N-L-2,$$

and (3) we also have

$$\sum_{l=0}^{L} c_l (f_l(n+1) a_\alpha(n) + g_l(n+1)) = 0, \qquad 0 \leq n \leq N-L-2.$$

Hence, we see that for $0 \leq n \leq N-L-2$,

$$\begin{aligned} 0 = &\left(\sum_{l=0}^{L} c_l f_{l+1}(n)\right)\left(\sum_{k=0}^{L} c_k (f_k(n) a_\alpha(n) + g_k(n))\right) \\ &- \left(\sum_{l=0}^{L} c_l f_l(n)\right)\left(\sum_{k=0}^{L} c_k (f_{k+1}(n) a_\alpha(n) + g_{k+1}(n))\right) \\ = &\left(\sum_{l=0}^{L} c_l f_{l+1}(n)\right)\left(\sum_{k=0}^{L} c_k g_k(n)\right) - \left(\sum_{l=0}^{L} c_l f_l(n)\right)\left(\sum_{k=0}^{L} c_k g_{k+1}(n)\right) \end{aligned}$$

$$= \sum_{l=0}^{2L} \sum_{m=\max\{0,l-L\}}^{\min\{l,L\}} c_{l-m}c_m(f_{l-m+1}(n)g_m(n) - f_{l-m}(n)g_{m+1}(n))$$

$$= \sum_{l=0}^{2L} \sum_{m=\max\{0,l-L\}}^{\min\{l,L\}} c_{l-m}c_m f_{l-m}(n)((l-2m)g_m(n) - 1),$$

where we have used (6) and (7) in the last step. Then the polynomial

$$F(X) = \sum_{l=0}^{2L} \sum_{m=\max\{0,l-L\}}^{\min\{l,L\}} c_{l-m}c_m f_{l-m}(X)\,((l-2m)g_m(X) - 1)$$

has at least $\min\{N-L-1, p\}$ distinct zeros.

We have to show that $F(X)$ is not identically zero.

For $0 \le l \le L-1$ each term $c_{l-m}c_m f_{l-m}(X)\,((l-2m)g_m(X)-1)$ is of degree at most $L-2$.

We now examine the terms corresponding to l in the range $L \le l \le 2L$.

If l is even then the term corresponding to $m = l/2$ is $-c_m^2 f_m(X)$ and thus either identically zero if $c_{l/2} = 0$ and of degree $l/2$ otherwise. In particular for $l = 2L$ we get the term

$$-c_L^2 f_L(X) = -f_L(X)$$

of degree L since $c_L = 1$.

For $L \le l \le 2L-1$ and $\mu = l-L, \ldots, \lceil l/2 \rceil - 1$ we add the terms for $m = \mu$ and $m = l - \mu$, getting

$$\begin{aligned} & c_{l-\mu}c_\mu\,(f_{l-\mu}(X)\,((l-2\mu)g_\mu(X)-1) + f_\mu(X)\,((2\mu-l)g_{l-\mu}(X)-1)) \\ &\quad = c_{l-\mu}c_\mu(l-2\mu)\,(f_{l-\mu}(X)g_\mu(X) - f_\mu(X)g_{l-\mu}(X)) \\ &\qquad - c_{l-\mu}c_\mu\,(f_{l-\mu}(X) + f_\mu(X)). \end{aligned}$$

Using (6) and (7), we also derive

$$\begin{aligned} & f_{l-\mu}(X)g_\mu(X) - f_\mu(X)g_{l-\mu}(X) \\ &\quad = f_{(l-2\mu)+\mu}(X)g_\mu(X) - f_\mu(X)g_{(l-2\mu)+\mu}(X) \\ &\quad = \prod_{j=\mu+1}^{l-\mu}(X+j)f_\mu(X)g_\mu(X) \\ &\qquad - f_\mu(X)\left(\prod_{j=\mu+1}^{l-\mu}(X+j)g_\mu(X) + \sum_{\nu=\mu+1}^{l-\mu}\prod_{j=\nu+1}^{l-\mu}(X+j)\right) \\ &\quad = f_\mu(X)\sum_{\nu=\mu+1}^{l-\mu}\prod_{j=\nu+1}^{l-\mu}(X+j). \end{aligned}$$

Thus for any $\mu = l-L, \ldots, \lceil l/2 \rceil - 1$, the above polynomial is of degree $l - \mu - 1 \le L - 1$. We also see that the term $-f_{l-\mu}(X) - f_\mu(X)$ is of degree

$l - \mu$ and thus is of degree at most $L-1$ except when $\mu = l - L$ when it is of degree L. Thus

$$F(X) = -X^L \sum_{l=L}^{2L} c_{l-L} + G(X),$$

where $\deg G(X) \le L-1$.

Therefore if

$$\sum_{l=0}^{L} c_l \neq 0 \qquad (8)$$

then we get

$$L = \deg F \ge \min\{N - L - 1, p\}.$$

Finally, assume that (8) fails and also that $L \le (N-1)/2$. Assuming that $L \le (N-1)/2$ we note that the shortest recurrence relation

$$\sum_{l=0}^{L} c_l a_\alpha(n+l) = 0, \qquad 0 \le n \le N - L - 1,$$

for the first N terms of $a_\alpha(n)$ is unique (if as before $c_L = 1$) and coincides with the unique recurrence relation for the first $N-1$ terms (for example, see [4, Proposition 2 and Lemma 3]). On the other hand, if (8) fails then for $0 \le n \le N - L - 2$, we have

$$\begin{aligned} 0 &= \sum_{l=0}^{L} c_l a_\alpha(n+l+1) = \sum_{l=0}^{L} c_l \left((n+l+1)a_\alpha(n+l) + 1\right) \\ &= (n+1)\sum_{l=0}^{L} c_l a_\alpha(n+l) + \sum_{l=0}^{L} l c_l a_\alpha(n+l) + \sum_{l=0}^{L} c_l = \sum_{l=0}^{L} l c_l a_\alpha(n+l). \end{aligned}$$

We may assume $L < p$. Hence, the uniqueness property implies that $c_l l / L = c_l$ for $l = 0, \dots, L$. Therefore, we immediately derive that $c_l = 0$ for $l = 0, \dots, L-1$, which implies $\sum_{l=0}^{L} c_l = 1$ in contradiction to our assumption that (8) is not valid. □

Certainly the proof of Theorem 2 applies to sequences satisfying $a_\alpha(n) = n a_\alpha(n-1) + 1$, $n = 1, 2, \dots$, over an arbitrary field $\mathbb{K}$ of characteristic p (including zero characteristic), leading to the bound

$$L_\alpha(N) \ge \begin{cases} \min\{(N-1)/2, p\}, & \text{if } p > 0, \\ (N-1)/2, & \text{if } p = 0. \end{cases}$$

4 Open Questions

We conclude with mentioning some possible directions for further research.

Certainly it would be interesting to get "individual" bounds for every $\alpha \in \mathbb{F}_p$ for exponential sums with $a_\alpha(n)$. This seems to be very hard since even for the presumably easier sequence $b(n) = n!$, satisfying $b(n) = nb(n-1)$, this is not known, see [8,9] for some recent progress in this direction.

It is easy to see that

$$a_\alpha(n) = na_\alpha(n-1) + 1 = \left(\frac{a_\alpha(n-1) - 1}{a_\alpha(n-2)} + 1\right) a_\alpha(n-1) + 1,$$

(at least when $a_\alpha(n-2) \neq 0$). It is natural to ask whether this relation can be used for breaking the "truncated" version of this generator in the same fashion as it is done in [1,2,3] for other nonlinear congruential generators.

Acknowledgements. This work was done during a pleasant visit by I. S. to the Johann Radon Institute for Computational and Applied Mathematics whose support and hospitality are gratefully acknowledged. During the preparation of this paper, I. S. was supported in part by ARC grant DP0556431.

References

1. S. R. Blackburn, D. Gomez-Perez, J. Gutierrez and I. E. Shparlinski, 'Predicting the inversive generator', *Lect. Notes in Comp. Sci.*, Springer-Verlag, Berlin, **2898** (2003), 264–275.
2. S. R. Blackburn, D. Gomez-Perez, J. Gutierrez and I. E. Shparlinski, 'Predicting nonlinear pseudorandom number generators', *Math. Comp.*, **74** (2005), 1471–1494.
3. S. R. Blackburn, D. Gomez-Perez, J. Gutierrez and I. E. Shparlinski, 'Reconstructing noisy polynomial evaluation in residue rings', *J. of Algorithms*, (to appear).
4. G. Dorfer and A. Winterhof, 'Lattice structure and linear complexity profile of nonlinear pseudorandom number generators', *Appl. Algebra Engrg. Comm. Comput.*, **13** (2003), 499–508.
5. M. Drmota and R. Tichy, *Sequences, discrepancies and applications*, Springer-Verlag, Berlin, 1997.
6. E. D. El Mahassni and A. Winterhof, 'On the distribution and linear complexity of counter-dependent nonlinear congruential pseudorandom number generators', *JP J. Algebra Number Theory Appl.*, (to appear).
7. G. Everest, A. J. van der Poorten, I. E. Shparlinski and T. Ward, *Recurrence sequences*, Amer. Math. Soc., 2003.
8. M. Z. Garaev, F. Luca and I. E. Shparlinski, 'Character sums and congruences with $n!$', *Trans. Amer. Math. Soc.*, **356** (2004), 5089–5102.
9. M. Z. Garaev, F. Luca and I. E. Shparlinski, 'Exponential sums and congruences with factorials', *J. Reine Angew. Math.*, **584**, (2005), 29–44.
10. L. Kuipers and H. Niederreiter, *Uniform distribution of sequences*, John Wiley, NY, 1974.
11. T. Müller, 'Prime and composite terms in Sloane's sequence A056542', *J. Integer Sequences*, **8** (2005), Article 05.3.3.
12. H. Niederreiter, *Random number generation and Quasi–Monte Carlo methods*, SIAM Press, 1992.
13. H. Niederreiter, 'Design and analysis of nonlinear pseudorandom number generators', *Monte Carlo Simulation*, A.A. Balkema Publishers, Rotterdam, 2001, 3–9.

14. H. Niederreiter and I. E. Shparlinski, 'Recent advances in the theory of nonlinear pseudorandom number generators', *Proc. Conf. on Monte Carlo and Quasi-Monte Carlo Methods, 2000*, Springer-Verlag, Berlin, 2002, 86–102.
15. H. Niederreiter and I. E. Shparlinski, 'Dynamical systems generated by rational functions', *Lect. Notes in Comp. Sci.*, Springer-Verlag, Berlin, **2643** (2003), 6–17.
16. A. Shamir and B. Tsaban, 'Guaranteeing the diversity of number generators', *Inform. and Comp.*, **171** (2001), 350–363.
17. N. J. A. Sloane, *On-line encyclopedia of integer sequences*, available from `http://www.research.att.com/~njas/sequences`.
18. A. Topuzoglu and A. Winterhof, 'Pseudorandom sequences', *Topics in Geometry, Cryptography and Coding Theory*, Springer-Verlag, Berlin, (to appear).

Nonexistence of a Kind of Generalized Perfect Binary Array

Zhang Xiyong[1,2], Guo Hua[3], and Han Wenbao[2]

[1] Department of Mathematics, Zhengzhou University,
450052 Zhengzhou, China
xyzhxy3711@sina.com

[2] Department of Applied Mathematics, Information Engineering University,
450002 Zhengzhou, China
xyzhxy3711@sina.com, wb.han@263.net

[3] School of Computer Science Engineering, Beihang University,
100083 Beijing, China
guohua80125@sina.com

Abstract. Generalized perfect binary array(GPBA) is a useful tool in the construction of perfect binary arrays. By investigating the character values of corresponding relative difference sets, we obtain some nonexistence results of GPBAs. In particular, we show that no $GPBA(2,2,p^n)$ of any type z exists for $n=1$ and any odd prime p, or for any n and any odd prime $p \not\equiv 1 \pmod 8$. For the case $p=2$, there exists a $GPBA(2,2,2^n)$ of type $z=(z_1,z_2,z_3)$ if and only if $z=(0,0,0)$ and $n=0,2,4$, or $z \neq (0,0,0)$ with $z_3=0$ and $0 \leq n \leq 5$, with $z_3=1$ and $0 \leq n \leq 3$.

1 Introduction

Perfect binary arrays(PBAs) have many applications in the theory of difference sets and communications, but this kind of array seems to be very rare. Thus as a generalization of perfect binary array, generalized perfect binary array was first introduced by J.Jedwab in [3] to obtain PBAs (Note that Yang [12] gave the equivalent definition: Quasi-perfect binary array). For example, J. Jedwab [4] found that $PBA(s,t)$ and $DQPBA(s,t)$ (DQPBA is some special case of GPBA) can be used to construct $PBA(2s,2t)$ and $PBA(4s,t)$ if $t/gcd(s,t)$ is odd. Moreover, Jedwab [3] gave the connection between $GPBA(2,2,t)$ with t odd and "binary supplementary quadruple" which is used in the construction of PBAs(or Hadamard difference sets [1]). In this paper, we will discuss the existence of such kind of generalized perfect binary array ($GPBA(2,2,t)$ with t odd). Let's begin with some basic definitions.

Definition 1. [3] *An* $r-$*dimensional array* $A=(a[j_1,\cdots,j_r])$ *with* $(0 \leq j_i < s_i)$ $(1 \leq i \leq r)$ *is called an* $s_1 \times \cdots \times s_r$ *binary array if* $a[j_1,\cdots,j_r]=\pm 1$.

Definition 2. [3] *Let* $A=(a[j_1,\cdots,j_r])$ *be an* $s_1 \times \cdots \times s_r$ *binary array, the periodic autocorrelation function of* A *is for all* u_i $(1 \leq i \leq r)$,

G. Gong et al. (Eds.): SETA 2006, LNCS 4086, pp. 304–312, 2006.

$$R_A(u_1,\cdots,u_r)=\sum_{j_1=0}^{s_1-1}\cdots\sum_{j_r=0}^{s_r-1}a[j_1,\cdots,j_r]a[j_1+u_1,\cdots,j_r+u_r].$$

Definition 3. [3] *A sequence* $z=(z_1,\cdots,z_r)$ *with* $z_i=0$ *or* 1 *for all* $1\leq i\leq r$ *is called a type vector.*

Definition 4. [3] *Let* $A=(a[j_1,\cdots,j_r])$ *be an* $s_1\times\cdots\times s_r$ *binary array and* $z=(z_1,\cdots,z_r)$ *be a type vector. The expansion of* A *with respect to* z *is the* $(z_1+1)s_1\times\cdots\times(z_r+1)s_r$ *binary array* $\varepsilon(A;z)=(a'[j_1,\cdots,j_r])$ *given by*

$$a'[j_1+y_1s_1,\cdots,j_r+y_rs_r]=(-1)^{\sum_i y_i}a[j_1,\cdots,j_r],\quad 0\leq j_i<s_i, 0\leq y_i\leq z_i.$$

Definition 5. [3] *Let* A *be an* $s_1\times\cdots\times s_r$ *binary array and* $z=(z_1,\cdots,z_r)$ *a type vector.* A *is a generalized perfect binary array of type* z*, abbreviated as a* $GPBA(m;s_1,\cdots,s_r)$ *type* z*, if*

$$R_{\varepsilon(A;z)}(u_1,\cdots,u_r)\neq 0,\ \ 0\leq u_i<(z_i+1)s_i\ \ \textit{only if}\ \ u_i\equiv 0\pmod{s_i}\ \forall i.$$

Remark 1. We call a $GPBA(s_1,\cdots,s_r)$ of type z non-splitting if there exists a $z_i=1$ with $2|s_i$, otherwise call it splitting. Note that our definition of "splitting" for GPBA is similar to the definition of splitting for RDS [9].

The concept of generalized perfect binary array has been introduced for many years. However, the existence problem of GPBAs is little studied, in particular, the existence of $GPBA(2,2,p^n)$ is not known. In [2], it was demonstrated that there exists no $GPBA(2,2,t)$, where $t\equiv 3\pmod 4$, and $t=5,9$. In this note, we will show that $GPBA(2,2,p^n)$ doesn't exist for $n=1$ and any odd prime p, or for any n and any odd prime $p\not\equiv 1\pmod 8$. Additionally, in the case $p=2$, there exists a $GPBA(2,2,2^n)$ of type $z=(z_1,z_2,z_3)$ if and only if $z=(0,0,0)$ and $n=0,2,4$, or $z\neq(0,0,0)$, $z_3=0$ and $0\leq n\leq 5$, or $z_3=1$ and $0\leq n\leq 3$.

2 Some Preliminaries

One of the tools in order to investigate the existence of GPBAs comes from difference set theory. We introduce some basic facts about relative difference sets in this section. Interested readers are referred to [1] for a detailed survey about the nonexistence theory of difference sets.

Let R be a k-element subset of a finite multiplicative group G of order mn containing a normal subgroup N of order n, R is called an (m,n,k,λ)-relative difference set(RDS) in G relative to N provided that the multiset $r_1r_2^{-1}(r_1\neq r_2\in R)$ replicates each element of $G\setminus N$ exactly λ times and replicates no element of N. If $G\cong G/N\oplus N$, then R is called splitting. If $k=n\lambda$, then R is called semi-regular.

Difference sets are usually studied in the context of the group ring $Z[G]$. In general, a subset D of a finite group G can be regarded as an element $\sum_{g\in D}g$ in $Z[G]$. $\forall\chi\in G^*$, where G^* denotes the character group of G, let $\chi(D)=$

$\sum_{g\in D}\chi(g)$, $D^{(-1)}=\sum_{g\in D}g^{-1}$. Throughout this paper, χ_0 denotes the principal character of G, ξ_q is the primitive q-th root of unity, and all groups will be implicitly assumed to be abelian and finite.

We mention two standard lemmas about RDSs. For more about RDSs see [9].

Lemma 1. *A k-element subset R of G is an (m,n,k,λ)-RDS in G relative to N, if and only if for every nonprincipal character χ of G ,*

$$|\chi(R)| = \begin{cases} \sqrt{k-\lambda n} & \text{if } \chi \text{ is principal on } N \\ \sqrt{k} & \text{if } \chi \text{ is nonprincipal on } N \end{cases}.$$

Lemma 2. *Let R be an (m,n,k,λ)-RDS in G relative to N and let $\rho: G\to G/U$ denote the canonical epimorphism, where U is a subgroup of G, then the image $\rho(R)$ satisfies*

$$\rho(R)\cdot\rho(R^{(-1)}) = k + |U|\cdot\lambda\cdot G/U - |U\cap N|\cdot\lambda\cdot N/U.$$

If U is a subgroup of N, then $\rho(R)$ is an $(m,n/u,k,\lambda\cdot\mu)$−RDS in G/U relative to N/U.

The following theorem establishes the equivalence between GPBAs and appropriate RDSs.

Theorem 1. [3] *Let $z=(z_1,\cdots,z_r)\neq(0,\cdots,0)$ be a type vector and $E_A = s_1\times\cdots\times s_r$. Let A be an $s_1\times\cdots\times s_r$ binary array and let $A'=\varepsilon(A;z)$. Define the following groups G, H and K, where H is a subgroup of G and K is a subgroup of H:*

$$\begin{aligned} G &= Z_{(z_1+1)s_1}\times\cdots\times Z_{(z_r+1)s_r} \\ H &= \{(h_1,\cdots,h_r) : h_i = y_i s_i \text{ and } 0\le y_i\le z_i\} \\ K &= \{(k_1,\cdots,k_r) : k_i = y_i s_i, 0\le y_i\le z_i \text{ and } \textstyle\sum_i y_i \text{ is even}\} \end{aligned}$$

Let D be the subset of the factor group G/K given by

$$D = \{K + (j_1,\cdots,\ j_r) : A'(j_1,\cdots,j_r) = -1\}.$$

Then A is a non-trivial $GPBA(s_1,\cdots,s_r)$ of type z if and only if D is an $(E_A, 2, E_A, E_A/2)$-RDS in G/K relative to H/K.

Remark 2. In the case $z=(0,\cdots,0)$, a $GPBA(s_1,\cdots,s_r)$ of type z is in fact a $PBA(s_1,\cdots,s_r)$.

In the end of this section, we introduce Turyn's *self-conjugacy* assumption and Ma's lemma, which are important in the theory of difference sets.

Definition 6. *Let p be a prime and m be a positive integer, where $m=p^a m'$ with $(m',p)=1$. p is called self-conjugate modulo m if there exists a positive integer i with $p^i\equiv -1 \pmod{m'}$.*

Lemma 3. [11] *Let p be a prime which is self-conjugate modulo m. If $X \in Z[\xi_m]$ satisfies $X \cdot \overline{X} \equiv 0 \pmod{p^{2a}}$, then we have*

$$X \equiv 0 \pmod{p^a}.$$

Lemma 4. [6] *Let A be an element in $Z[G]$ where G is an abelian group with a cyclic Sylow p-group P. Let P_1 denote the unique subgroup of order p. If $\chi(A) \equiv 0 \pmod{p^a}$ for all nonprincipal characters of G, then*

$$A = P_1 \cdot X + p^a \cdot Y$$

for suitable X and Y in $Z[G]$, where the coefficients of X and Y can be chosen to be nonnegative if the coefficients of A are nonnegative.

3 The Result

In this section we will discuss the nonexistence of $GPBA(2, 2, p^n)$. To this end, we give several lemmas.

Lemma 5. *Let G be a finite abelian group, N be a subgroup of order 2 contained in a cyclic subgroup of order 4 in G, and G_p be the Sylow $p-$subgroup of G(p is odd). Suppose $G/N \cong Z_{s_1} \times \cdots \times Z_{s_r}$, and p is self-conjugate modulo $expG$. If there exists a non-splitting $GPBA(s_1, \cdots, s_r)$ of type $z \neq 0$, then $|G_p| = p^{2b}$ and $expG_p \leq p^b$.*

Proof. If there exists a non-splitting $GPBA(s_1, \cdots, s_r)$ of type $z \neq 0$, then G contains a non-splitting $(2n, 2, 2n, n)$-RDS R, where $2n = s_1 \times \cdots \times s_r$. Let χ be a nonprincipal character of G, then $\chi(R) \cdot \overline{\chi(R)} = 2n$ or 0. Assume $p^c \| 2n$, we have $\chi(R) \cdot \overline{\chi(R)} \equiv 0 \pmod{p^c}$.

Let ρ be the canonical projection epimorphism $G \to G/G_p$, and $R_1 = \rho(R)$, then

$$\psi(R_1) \cdot \overline{\psi(R_1)} = 0 \pmod{p^c}. \quad (1)$$

Since p is *self-conjugate* modulo $expG$, $P = \overline{P}$ for every prime ideal factor P over p in $Z[\xi_{exp\rho(G)}]$, therefore $P^2 | p$. By (1), c must be even, say $c = 2b$, $b \geq 1$. Thus we get $|G_p| = p^{2b}$.

Let the maximal cyclic subgroup of G_p be $G'_p \cong Z'_{p^d}$. Let ρ' be the canonical projection epimorphism $G \to G/(G_p/G'_p)$, and $R_2 = \rho'(R)$, $N_2 = \rho'(N)$.

Suppose that $expG_p = p^d > p^b$, we can similarly get

$$\forall \psi \in \rho'(G)^*, \quad \psi(R_2) \equiv 0 \pmod{p^b}. \quad (2)$$

By $R \cdot R^{(-1)} = 2n + n(G - N)$, we have

$$R_2 \cdot R_2^{(-1)} = 2n + n(p^{2b-d}\rho'(G) - N_2). \quad (3)$$

By (2) and Lemma 4, $R_2 = p^b X + PY$, where $\rho'(G)$ is the unique subgroup of P of order p, $X, Y \in Z[\rho(G)]$, and the coefficients of X, Y are all nonnegative.

If $X = 0$, then $R_2 = PY$, so $R_2 \cdot R_2^{(-1)} = p \cdot P \cdot Y \cdot Y^{(-1)}$ which contradicts to (3).

So $|X| > 0$, hence some coefficients of R_2 are greater than p^b. On the other hand, coefficients of $R_2 = \rho'(R)$ are all smaller than $|G_p/G'_p| = p^{2b-d}$, so we get $p^b \leq p^{2b-d}$, which is impossible.

Remark 3. From the above result, there doesn't exist non-splitting $GPBA(2,2,t)$ ($t \equiv 3 \pmod 4$) of any type (see [2]). Especially, if p is an odd prime and $p \equiv 3 \pmod 4$, there doesn't exist non-splitting $GPBA(2,2,p^n)$ of any type.

Lemma 6. *Let R be an $(m,n,m,m/n)$-RDS in G relative to N, and ρ denote the canonical projection epimorphism $G \to G/U$, where U is a subgroup of G. Suppose $\rho(R) = \sum_{g\in\rho(G)} r_g \cdot g$ and $U \cap N = 1$, then $\forall x \in \rho(G)$, $\sum_{g \in x\rho(N)} r_g = |U|$.*

Proof. Assume $C = \{gN : g \in R\} \in Z[G/N]$. $\forall \chi \in G^* \setminus \{\chi_0\}$, and χ is principal on N, we have $\chi(R) = 0$ by the definition of RDS. So for every nonprincipal character ψ of G/N, $\psi(R) = 0$. By the Fourier inversion formula, $C = c \cdot G/N$.

Since $|C| = |R| = |G/N|$, we get $c = 1$, hence $C = G/N$. Thus there exists exactly one element of R in every coset of N.

If $U \cap N = 1$, every coset of $U \cdot N$ contains exactly $|U|$ cosets of N, i.e.,

$$\forall x \in \rho(G), \quad \sum_{g \in x\rho(N)} r_g = |U|.$$

Lemma 7. *Let R be a non-splitting $(2^{2e}h, 2, 2^{2e}h, 2^{2e-1}h)$-RDS in $G = G_2 \times H$ relative to a subgroup N of order 2, where G_2 is the Sylow 2-subgroup, and $|G_2| = 2^{2e+1}$, $|H| = h$ is odd. If $expG_2 = 2^{e+1}$ and 2 is self-conjugate modulo $expG$, then there exists a subset R_1 of order h in $G_0 = Z_{2^{e+1}} \times H$, satisfying $R_1 \cdot R_1^{(-1)} = h + N_1 \cdot Y$, where N_1 is a subgroup of G_0 of order 2. Moreover, elements of R_1 are distributed in different cosets of N_1 .*

Proof. Denote ρ the canonical projection epimorphism $G \to G_0$, and let $R_0 = \rho(R), N_1 = \rho(N)$. Because R is a $(2^{2e}h, 2, 2^{2e}h, 2^{2e-1}h)$-RDS, $\forall \psi \in G_0^* \setminus \{\psi_0\}$, $\psi(R_0) \cdot \overline{\psi(R_0)} = 2^{2e}h$ or 0, i.e.,

$$\psi(R_0) \cdot \overline{\psi(R_0)} \equiv 0 \pmod{2^{2e}}. \quad (4)$$

Since 2 is *self-conjugate* modulo $expG$, 2 is *self-conjugate* modulo $expG_0$. By Lemma 3, we get

$$\forall \psi \in G_0^*, \quad \psi(R_0) \equiv 0 \pmod{2^e}. \quad (5)$$

Then by Lemma 4, there exist R_1, X in $Z[G_0]$, where the coefficients of R_1, X are nonnegative, satisfying

$$R_0 = 2^e R_1 + N_1 X. \quad (6)$$

Obviously the coefficients of R_0 are all smaller than $|G/G_0| = 2^e$, so the coefficients of R_1 are 0 or 1, and $R_1 \cap N_1 X = \emptyset$.

By Lemma 6, the sum of all elements of R_0 in every coset of N_1 is 2^e, hence every coefficient of X is 2^{e-1}. Let $X = 2^{e-1}X_1$, where X_1 is a subset of G_0. So $R_1 \cup X_1$ is the complete coset representation of N_1.

Thus $N_1X = 2^{e-1}(G_0 - N_1R_1)$. Together with (6), we get

$$R_0 = 2^{e-1}G_0 + 2^{e-1}R_1(2 - N_1). \tag{7}$$

Because R is a $(2^{2e}h, 2, 2^{2e}h, 2^{2e-1}h)$-RDS, it can be easily deduced that

$$R_0R_0^{(-1)} = 2^{2e}h + 2^{2e-1}h(2^eG_0 - N_1). \tag{8}$$

Furthermore,

$$R_1R_1^{(-1)}(2 - N_1) = h \cdot (2 - N_1). \tag{9}$$

Therefore

$$R_1R_1^{(-1)} = h + N_1 \cdot Y, \quad Y \in Z[G_0]. \tag{10}$$

Assume $|R_1| = r_1$, then $|X_1| = 2^eh - r_1$. Noting that there are no units in $R_1(N_1X)^{(-1)}$ and $R_1^{(-1)}N_1X$ and calculating the number of units of equation (8), we have

$$(2^e)^2r_1 + (2^{(e-1)})^2 \cdot 2 \cdot (2^eh - r_1) = 2^{2e}h + 2^{2e-1}h(2^e - 1). \tag{11}$$

Hence

$$r_1 = h.$$

Let $p = e_0f + 1$ be an odd prime, Γ be the set of all primitive e_0−th roots of unity, g be a fixed primitive root modulo p. $G(p, e_0) = \{\sum_{i=0}^{p-2} \gamma^i \xi_p^{g^i} | \gamma \in \Gamma\}$ is the set of all Gauss sums, and $G(2, 2) = \{1 + \xi_4\}$.

Lemma 8. [10] *Let $m = p^am'$, where p is a prime, $(p, m') = 1$ and $m \neq 2 \pmod 4$. If $x \in Z[\xi_m]$ is a solution of $x \cdot \overline{x} = p^b, b \geq 1$, then there is an integer j such that*

$$x\xi_m^j \in Z[\xi_{m'}] \quad \text{or} \quad x = \xi_m^j y x_0,$$

where $x_0 \in Z[\xi_{m'}]$, $x_0 \cdot \overline{x_0} = p^{b-1}$ and $y \in G(p, e_0)$ for some divisor $e_0 \neq 1$ of w_0 with $w_0 = 2$ if $p = 2$, $w_0 = (p - 1, m')$ if m' is even and $w_0 = (p - 1, 2m')$ if both p and m' are odd .

Now we are ready to give the main result.

Theorem 2. *For any n and any odd prime $p \equiv 5 \pmod 8$, there is no $GPBA(2, 2, p^n)$ of any type z.*

Proof. Let $z = (z_1, z_2, z_3)$, we consider two cases.

1) If $z_1 = 0$ and $z_2 = 0$, $GPBA(2, 2, p^n)$ of type z is splitting, so its existence yields the existence of $PBA(2, 2, p^n)$. However, it is well known that $PBA(2, 2, p^n)$ doesn't exist (see [1]).

2) If $z_1 = 1$ or $z_2 = 1$, $GPBA(2,2,p^n)$ of type z is equivalent to the existence of a non-splitting $(4p^n, 2, 4p^n, 2p^n)$-RDS in $G = Z_4 \times Z_2 \times Z_{p^n}$ relative to a subgroup N of order 2.

When $p \equiv 5 \pmod 8$, 2 is *self-conjugate* modulo $expG$. Let R be the RDS. By Lemma 7, $G_0 = Z_4 \times Z_{p^n}$ contains a subset R_1 of order p^n, satisfying $R_1 \cdot R_1^{(-1)} = p^n + N_1 \cdot Y$, where N_1 is a subgroup of G_0 of order 2, and elements of R_1 are distributed in different cosets of N_1.

Let χ be a character of G and χ is nonprincipal on N_1. Denote $x = \chi(R_1)$. Since $R_1 \cdot R_1^{(-1)} = p^n + N_1 \cdot Y$, we get $x \cdot \overline{x} = p^n$, where x is the sum of p^n different $4p^n$-th roots of unity.

By Lemma 8, $x\xi_{4p^n}^j \in Z[\xi_4]$; or $x = \xi_{4p^n}^j y x_0$, where $x_0 \in Z[\xi_4], x_0 \cdot \overline{x_0} = p^{n-1}$.

2.1) In the case $x\xi_{4p^n}^j \in Z[\xi_4]$, w.l.o.g., suppose $j = 0$, thus $x = a + b\xi_4, a, b \in Z$. Obviously, x cannot be the sum of p^n different $4p^n$-th roots of unity.

2.2) In the case $x = \xi_{4p^n}^j y x_0$, where $x_0 \in Z[i]$, $x_0 \cdot \overline{x_0} = p^{n-1}$, suppose $x_0 = a + b\xi_4$, so $p^{n-1} = a^2 + b^2$, $a, b \in Z$. Hence the solutions of $x \cdot \overline{x} = p^n$ are in the form of $x = \xi_{4p^n}^j y(a + b\xi_4)$, where $y \in G(p, e_0)$. Assume $j = 0$ for convenience. Let $y = \sum_{j=0}^{p-2} \gamma^j \xi_{p^n}^{p^{n-1}g^j}$, $\gamma = -1, \xi_4$.

If $n = 1$, $x_0 \cdot \overline{x_0} = 1$. x_0 is a 4-th root of unity by Kronecker's theorem. So we can assume $x = \sum_{i=1}^{p-1} (\frac{i}{p}) \cdot \xi_p^i$. Obviously x is the sum of $p-1$ different $4p$-th roots of unity, and it cannot be the sum of p different $4p$-th roots of unity under the cyclotomic polynomial $x^{p-1} + \cdots + x + 1$.

If $n \geq 2$, the coefficients of $x = y(a+b\xi_4)$ are $\pm a$ or $\pm b$. Because $p^{n-1} = a^2+b^2$, $|a|$ or $|b|$ is greater than 1. So, x cannot be the sum of p^n different $4p^n$-th roots of unity.

When $p \equiv 1 \pmod 8$, we need the technique of Ma [7] to prove the nonexistence of $GPBA(2,2,p)$.

Lemma 9. [7] *Let $G =< \alpha > \times < \beta >$ be a cyclic group of order $v = p^{t_1} q^{t_2}$, where $o(\alpha) = p^{t_1}$, $o(\beta) = q^{t_2}$, $t_1, t_2 \geq 1$, and p, q are distinct primes. Suppose there exists $y \in Z[G]$ such that*

(i) $\chi(y)\overline{\chi(y)} = p^e a$ for a character χ of G such that $\chi(\alpha) = \xi_{p^{t_1}}$ and $\chi(\beta) = \xi_{q^{t_2}}$, where a is an integer, and

(ii) $\chi_1(y)\overline{\chi_1(y)} = pq^c$ for a character χ_1 of G such that $\chi_1(\alpha) = 1$ and $\chi_1(\beta) = \xi_{q^{t_2}}$.

Then

$$y = f(\beta)^e x_0 + < \alpha^{p^{t_1-1}} > x_1 + < \beta^{q^{t_2-1}} > x_2,$$

where $x_0, x_1, x_2 \in Z[G]$ and $f(X) \in Z[X]$ such that $f(\xi_{q^{t_2}}) f(\xi_{q^{t_2}}^{-1}) = p$.

Theorem 3. *There is no $GPBA(2,2,p)$ of any type z for any odd prime $p \equiv 1 \pmod 4$.*

Proof. By Theorem 1, it suffices to prove the nonexistence of non-splitting $(4p, 2, 4p, 2p) - RDS$ in group of exponent $4p$. Suppose the multiplicative group

$G = <\alpha> \times <\beta> \times <\gamma>$ contains such a RDS R relative to $<\alpha^2>$ where $o(\alpha) = 4$, $o(\beta) = p$ and $o(\gamma) = 2$.

Let $\rho : G \longrightarrow K = <\alpha> \times <\beta>$ denote the homomorphism. By Lemma 9 and the fact that R is a $(4p, 2, 4p, 2p)$-RDS, we can deduce that

$$\rho(R) = (a + b\alpha)x_0 + <\alpha> x_1 + <\beta> x_2, \tag{12}$$

where $p = a^2 + b^2$, a, b are integers, and $x_0, x_1, x_2 \in Z[K]$.

So there exists $x \in Z[K]$ such that

$$(1 - \alpha)(1 - \beta)\rho(R) = (a + b\alpha)(1 - \alpha)x. \tag{13}$$

Since $|R \bigcap <\alpha^2> \beta^i\gamma^j| = 1$ for all i, j, it can be deduced that the coefficients of $(\beta^j, \alpha\beta^j, \alpha^2\beta^j, \alpha^3\beta^j)$ in $(1 - \alpha)(1 - \beta)\rho(R)$ are $(1, 1, -1, -1), (2, 2, -2, -2)$, $(2, 0, -2, 0), (0, 0, 0, 0)$ or their shifts. However the corresponding coefficients in $(a + b\alpha)(1 - \alpha)x$ are of the form $(au + bv, av + bw, aw + bs, as + bu)$ where u, v, w, s are integers satisfying $u + v + w + s = 0$, which implies that $|a|, |b| \leq 2$, contradicting the assumption that p is an odd prime, $p \equiv 1 \pmod 4$ and $p = a^2 + b^2$.

Combining Remark 3, Theorem 2 and Theorem 3, we have:

Theorem 4. *There is no $GPBA(2, 2, p^n)$ of any type z for $n = 1$ and any odd prime p, or for any n and any odd prime $p \not\equiv 1 \pmod 8$.*

In the end of this paper, we deal with the existence of $GPBA(2, 2, 2^n)$. Let's state a result of RDSs, which is due to Ma and Schmidt [8].

Lemma 10. [8] *Let G be an abelian group of order p^{2c+1} and N be its subgroup of order p, then a $(p^{2c}, p, p^{2c}, p^{2c-1})$-RDS exists in G relative to N if and only if $exp(G) \leq c + 1$. For an abelian group of order 2^{2c+2} and a subgroup N of G of order 2, a $(2^{2c+1}, 2, 2^{2c+1}, 2^{2c})$-RDS in G relative to N if and only if $exp(G) \leq 2^{c+2}$ and N is contained in a cyclic subgroup of G of order 4 .*

By Theorem 1 and Lemma 10, we can easily get

Theorem 5. *There exists $GPBA(2, 2, 2^n)$ of type $z = (z_1, z_2, z_3)$ if and only if $z = (0, 0, 0)$ and $n = 0, 2, 4$, or $z \neq (0, 0, 0)$, $z_3 = 0$ and $0 \leq n \leq 5$, or $z_3 = 1$ and $0 \leq n \leq 3$.*

Proof. When $z = (0, 0, 0)$, a $GPBA(2, 2, 2^n)$ of type z is in fact a $PBA(2, 2, 2^n)$ whose existence is equivalent to the existence of Hadamard difference sets with parameters $(2^{n+2}, 2^{n+1} \pm \sqrt{2^n}, 2^n \pm \sqrt{2^n})$. It is well known that there exists a Hadamard difference set in an abelian group G of order 2^{2d+2} if and only if $expG \leq 2^{d+2}$ [1,5]. So, there exists $PBA(2, 2, 2^n)$, i.e. $GPBA(2, 2, 2^n)$ of type $(0, 0, 0)$ if and only if $n = 0, 2, 4$.

For $z \neq (0, 0, 0)$, the existence of $GPBA(2, 2, 2^n)$ of type z is equivalent to the existence of a non-splitting $(2^{n+2}, 2, 2^{n+2}, 2^{n+1})$-RDS in group $G \cong Z_4 \times Z_{2^n} \times Z_2$ or $G \cong Z_{2^{n+1}} \times Z_2 \times Z_2$ relative to a subgroup of order 2. By Lemma 10, it is a simple matter to verify that $n = 0, 1, 2, 3, 4, 5$ if $z_3 = 0$ and $n = 0, 1, 2, 3$ if $z_3 = 1$.

Acknowledgments

The authors are indebted to the referees for their beneficial comments.

References

1. Beth T., Jungnickel D., and Lenz H., Design Theory, 2nd edition. Cambridge University Press, Cambridge (1999).
2. Hughes, G.: Cocyclic Theory of Generalized Perfect Binary Arrays. Royal Melbourne Institute of Technology, Department of Mathematics, Research Report No. 6 (1998).
3. Jedwab, J.: Generalized Perfect Arrays and Menon Difference Sets. *Designs Codes and Cryptography*, 2 (1992) 19-68.
4. Jedwab J. and Mitchell C.: Constructing new perfect binary arrays. *Electronics Letters*, 24 (1988) 650-652.
5. Kraemer R.G.: Proof of a conjecture on Hadamard 2-groups. *Journal of Combinatorial Theory(A)*, 63 (1993) 1-10.
6. Ma, S. L.: Polynomial addition sets. Ph.D. thesis. University of Hong Kong (1985).
7. Ma, S.L.: Planar Functions, Relative Difference Sets, and Character Theory. *Journal of Algebra*, 185(1996) 342-356.
8. Ma, S. L. and Schmidt, B.: On (p^a, p, p^a, p^{a-1})-relative difference sets. *Designs, Codes and Cryptography*, 6 (1995) 57-71.
9. Pott, A.: A survey on relative difference sets. *In "Groups, Difference sets and the Monster"(eds K.T.Arasu et al.)*, deGruyter Verlag, Berlag-New York (1996) 195-232.
10. Schmidt, B.: Cyclotomic Integers of Prescribed Absolute Value and the Class Group. *J.Number Theory*, 72 (1998) 269-281.
11. Turyn, R. J.: Character sums and difference sets. *Pacific J.Math*, 15 (1965) 319-346.
12. Yang, Y. X.: Quasi-perfect binary arrays. *Acta Electronica Sinica*, 20(4) (1992) 37-44 (in Chinese).

On the Distinctness of Decimations of Generalized l-Sequences*

Hong Xu and Wen-Feng Qi

Department of Applied Mathematics, Zhengzhou Information Engineering University, Zhengzhou, 450002, China
xuhong0504@163.com, wenfeng.qi@263.net

Abstract. For an odd prime number p and positive integer e, let $\underline{a}$ be an l-sequence with connection integer p^e. Goresky and Klapper conjectured that when $p^e \notin \{5, 9, 11, 13\}$, all decimations of $\underline{a}$ are cyclically distinct. For any primitive sequence $\underline{u}$ of order n over $\mathbb{Z}/(p^e)$, call $\underline{u}(\text{mod } 2)$ a generalized l-sequence. In this article, we show that almost all decimations of any generalized l-sequence are also cyclically distinct.

Keywords: Feedback-with-carry shift registers (FCSRs), l-sequences, generalized l-sequences, 2-adic numbers, integer residue ring, primitive sequences.

1 Introduction

Recently, lots of research has been done on feedback-with-carry shift register (FCSR) sequences [1], [2], [5]-[7], [9]-[12], [14]- [15] and linear recurring sequences over integer residue ring [3], [8], [13], [16]-[20].

An FCSR is a feedback shift register together with a small amount of auxiliary memory. The contents (0 or 1) of the tapped cells of the shift register are added as integers to the current contents of the memory to form a sum σ. The parity ($\sigma(\text{mod } 2)$) of σ is fed back into the first cell, and the higher order bits ($\lfloor \sigma/2 \rfloor$) are retained for the new value of the memory. The r taps $q_1, q_2, ..., q_r$ on the cells of an r-stage FCSR define a connection integer $q = -1 + q_1 2 + q_2 2^2 + ... + q_r 2^r$. The period of the FCSR sequence is at most $\varphi(q)$, where $\varphi(\cdot)$ is Euler's phi function. For a detailed descriptions of FCSR sequences, please see [10].

An l-sequence is a periodic sequence (of period $T = \varphi(q)$) which is obtained from an FCSR with connection integer q for which 2 is a primitive root. Thus q is of the form $q = p^e$, where p is an odd prime, and $e \geq 1$. Such a sequence $\underline{a} = (a(t))_{t \geq 0}$ has the following exponential representation [10]

$$a(t) = (A \cdot 2^{-t}(\text{mod } q))(\text{mod } 2),\ t \geq 0, \gcd(A, q) = 1.$$

The l-sequences are known to have several remarkable statistical properties similar to m-sequences. They are 0-1 balanced, have fine run properties [9], [10], [11]

* This work was supported by National Nature Science Foundation of China under Grant number 60373092.

G. Gong et al. (Eds.): SETA 2006, LNCS 4086, pp. 313–322, 2006.

and the arithmetic correlations between any two cyclically distinct decimations of l-sequences are precisely zero [5].

If $\underline{a} = (a(t))_{t\geq 0}$ is a binary periodic sequence with period T, let $\underline{a}^{(d)} = (a(dt))_{t\geq 0}$ denote its d-fold decimation and $x^\tau \underline{a} = (a(t+\tau))_{t\geq 0}$ denote the τ-shifted sequence. If $\underline{a}, \underline{b}$ are binary periodic sequences with the same period T, we say they are cyclically distinct if $x^\tau \underline{a} \neq \underline{b}$, for every shift τ with $0 < \tau < T$. In this article, whenever a d-decimation of a sequence $\underline{a}$ is referred, it's required that d is always relatively prime to T.

On the basis of extensive experimental evidence, Goresky and Klapper made the following conjecture.

Conjecture 1. [5] Let $\underline{a}$ be an l-sequence with connection number p^e and period T. Suppose $p^e \notin \{5, 9, 11, 13\}$, let c and d be relatively prime to T and incongruent modulo T. If $\underline{c}$ is the c-fold decimation of $\underline{a}$ and $\underline{d}$ is the d-fold decimation of $\underline{a}$, then $\underline{c}$ and $\underline{d}$ are cyclically distinct.

Note that the c-fold decimation $\underline{c}$ and d-fold decimation $\underline{d}$ of $\underline{a}$ can be represented as $c(t) = (A \cdot 2^{-ct} (\text{mod } p^e))(\text{mod } 2)$ and $d(t) = (A \cdot 2^{-dt} (\text{mod } p^e))(\text{mod } 2)$ respectively, where $2^{-c}(\text{mod } p^e)$ and $2^{-d}(\text{mod } p^e)$ are both primitive roots modulo p^e. More generally, let ξ be a primitive root modulo p^e, and set $u(t) = A \cdot \xi^t (\text{mod } p^e)$. Then the sequence $\underline{u} = (u(t))_{t\geq 0}$ is a primitive sequence of order 1 over $\mathbb{Z}/(p^e)$ generated by $x - \xi$, and $\underline{u}(\text{mod } 2)$ is an l-sequence or its decimation. For the definition of primitive sequences, please see Section 2.

Similarly, let $\underline{u}$ be a primitive sequence over $\mathbb{Z}/(p^e)$ of order n, and call the modulo 2 derivative sequence $\underline{u}(\text{mod } 2)$ a *generalized l-sequence.* As a natural generalization of l-sequences, such sequences may also share many fine pseudo-random properties similar to l-sequences.

For any monic polynomial $f(x)$ over $\mathbb{Z}/(p^e)$, denote $G(f(x), p^e)$ for the set of all sequences over $\mathbb{Z}/(p^e)$ generated by $f(x)$ and set $G'(f(x), p^e) = \{\underline{u} \in G(f(x), p^e) \mid \underline{u} \not\equiv \underline{0}(\text{mod } p)\}$. Detailed description of these two notations, see also Section 2. Note that if $\underline{u} \in G(f(x), p^e)$, then $x^k \underline{u} \in G(f(x), p^e)$. Thus Conjecture 1 can be restated as follows.

Conjecture 2. Let $p^e \notin \{5, 9, 11, 13\}$ with p an odd prime and $e \geq 1$ such that 2 is a primitive root modulo p^e. Suppose ξ, ζ are two different primitive roots modulo p^e, and set $f(x) = x - \xi$, $g(x) = x - \zeta$. Then for any $\underline{u} \in G'(f(x), p^e)$, $\underline{v} \in G'(g(x), p^e)$, we have

$$\underline{u} \not\equiv \underline{v}(\text{mod } 2).$$

When $e = 1$, it's shown in [7] that almost all decimations of l-sequences are cyclically distinct, and when $e \geq 2$, $p^e \neq 9$, it's shown in [15] that all decimations of l-sequences are cyclically distinct. In this article, we further show that almost all decimations of any generalized l-sequence are also cyclically distinct.

Let $f(x), g(x)$ be two different primitive polynomials of degree n over $\mathbb{Z}/(p^e)$ satisfying $f(x) \not\equiv g(x)(\text{mod } p)$, and suppose $\underline{u} \in G'(f(x), p^e)$, $\underline{v} \in G'(g(x), p^e)$. We show that when $e \geq 2$, $\underline{u} \not\equiv \underline{v}(\text{mod } 2)$ holds for all primes p, and when $e = 1$, $\underline{u} \not\equiv \underline{v}(\text{mod } 2)$ holds for almost all primes p.

The rest of this article is organized as follows. Firstly, an introduction to primitive sequences over integer residue ring and some of their important properties are given in Section 2. Next, the main result on the distinctness of decimations of generalized l-sequences is shown in Section 3.

Throughout the article, for any positive integers a and n, the sign "$a(\bmod\ n)$"refers to the minimal nonnegative residue of a modulo n, that is, reducing the number a modulo n to obtain a number between 0 and $n-1$. The notation "$x \equiv a(\bmod\ n)$"is the usual congruent equation, and the notation "$x = a(\bmod\ n)$"means x is equal to the minimal nonnegative residue of $a(\bmod\ n)$.

2 Preliminaries

For any odd prime number p and positive integer e, let $\mathbb{Z}/(p^e) = \{0, 1, \ldots, p^e - 1\}$ be the integer residue ring modulo p^e, and $(\mathbb{Z}/(p^e))^*$ its multiplicative group. Particularly, $\mathbb{Z}/(p) = \mathrm{GF}(p)$ is the Galois field with p elements.

Let $f(x) = x^n + c_{n-1}x^{n-1} + \cdots + c_0$ be a monic polynomial of degree $n \geq 1$ over $\mathbb{Z}/(p^e)$. If $f(0) \not\equiv 0(\bmod\ p)$, then there exists a positive integer P such that $f(x)$ divides $x^P - 1$ over $\mathbb{Z}/(p^e)$. The least such P is called the period of $f(x)$ over $\mathbb{Z}/(p^e)$ and denoted by $\mathrm{per}(f(x), p^e)$, which is upper bounded by $p^{e-1}(p^n - 1)$ [13]. Moreover, if $\mathrm{per}(f(x), p^e) = p^{e-1}(p^n - 1)$, then say $f(x)$ is a *primitive polynomial* of degree n over $\mathbb{Z}/(p^e)$. In this case, $f(x)(\bmod\ p^i)$ is also a primitive polynomial over $\mathbb{Z}/(p^i)$, whose period is $\mathrm{per}(f(x), p^i) = p^{i-1}(p^n - 1)$, $i = 1, 2, \ldots, e-1$. Especially, $f(x)(\bmod\ p)$ is a primitive polynomial over the prime field $\mathrm{GF}(p)$.

The sequence $\underline{u} = (u(t))_{t \geq 0}$ over $\mathbb{Z}/(p^e)$ satisfying the recursion

$$u(t+n) = -[c_0 u(t) + c_1 u(t+1) + \cdots + c_{n-1} u(t+n-1)](\bmod\ p^e), t \geq 0,$$

is called a *linear recurring sequence* of order n over $\mathbb{Z}/(p^e)$, generated by $f(x)$. Such a sequence is called a *primitive sequence* if $f(x)$ is a primitive polynomial and $\underline{u} \not\equiv \underline{0}(\bmod\ p)$. Particularly, the primitive sequences over $\mathbb{Z}/(p)$ are called m-sequences.

Denote $G(f(x), p^e)$ for the set of all sequences over $\mathbb{Z}/(p^e)$ generated by $f(x)$, and $G'(f(x), p^e) = \{\underline{u} \in G(f(x), p^e) \mid \underline{u} \not\equiv \underline{0}(\bmod\ p)\}$ for the set of all primitive sequences over $\mathbb{Z}/(p^e)$ generated by $f(x)$.

Any element v in $\mathbb{Z}/(p^e)$ has a unique p-adic decomposition as $v = v_0 + v_1 \cdot p + \cdots + v_{e-1} \cdot p^{e-1}$, where $v_i \in \mathbb{Z}/(p)$. Similarly, a sequence $\underline{u}$ over $\mathbb{Z}/(p^e)$ has a unique p-adic decomposition as

$$\underline{u} = \underline{u}_0 + \underline{u}_1 \cdot p + \cdots + \underline{u}_{e-1} \cdot p^{e-1},$$

where $\underline{u}_i$ is a sequence over $\mathbb{Z}/(p)$. The sequence $\underline{u}_i$ is called the i-th level sequence of $\underline{u}$, and $\underline{u}_{e-1}$ the highest-level sequence of $\underline{u}$. They can be naturally considered as sequences over the prime field $\mathrm{GF}(p)$. Particularly, $\underline{u}_0$ is an m-sequence over $\mathbb{Z}/(p)$ generated by $f(x)(\bmod\ p)$ with period $\mathrm{per}(\underline{u}_0) = p^n - 1$.

The following are two important results on primitive polynomials and primitive sequences over $\mathbb{Z}/(p^e)$.

Proposition 1. *[8] Let $f(x)$ be a primitive polynomial of degree n over $\mathbb{Z}/(p^e)$ with p an odd prime and $e \geq 1$. Then there exists a unique nonzero polynomial $h_f(x)$ over $\mathbb{Z}/(p)$ with $\deg(h_f(x)) < n$, such that*

$$x^{p^{i-1}T_0} \equiv 1 + p^i \cdot h_f(x)(\text{mod } f(x), p^{i+1}), i = 1, 2, \ldots, e-1, \quad (1)$$

where $T_0 = p^n - 1$, the notation "(mod $f(x), p^{i+1}$)" means this congruence equation holds modulo $f(x)$ and p^{i+1} simultaneously. In other words, we can say $x^{p^{i-1}T_0} \equiv 1 + p^i \cdot h_f(x)(\text{mod } f(x))$ holds over $\mathbb{Z}/(p^{i+1})$ for all $i = 1, 2, \ldots, e-1$.

Proposition 2 ([16]). *Let $f(x)$ be a primitive polynomial of degree n over $\mathbb{Z}/(p^e)$ with p an odd prime and $e \geq 2$. Let $\underline{u} \in G'(f(x), p^e)$, and denote $\underline{\alpha} = h_f(x)\underline{u}_0(\text{ mod } p)$, where $h_f(x)$ is defined as in equation (1). Then*

$$u_{e-1}(t + j \cdot p^{e-2}T_0) \equiv u_{e-1}(t) + j \cdot \alpha(t)(\text{mod } p), t \geq 0, \quad (2)$$

holds for all $j = 0, 1, \ldots, p-1$, where $T_0 = p^n - 1$. Furthermore, if $\alpha(t) \neq 0$ for some $t \geq 0$, then

$$\{u_{e-1}(t + j \cdot p^{e-2}T_0) | j = 0, 1, ..., p-1\} = \{0, 1, ..., p-1\}. \quad (3)$$

Else if $\alpha(t) \neq 0$ for some $t \geq 0$, then

$$u_{e-1}(t + j \cdot p^{e-2}T_0) = u_{e-1}(t) \text{ for all } j = 0, 1, ..., p-1. \quad (4)$$

Remark 1. Since $\underline{u}_0$ is an m-sequence over $\mathbb{Z}/(p)$ generated by $f(x)(\text{mod } p)$, and $\deg(h_f(x)) < \deg(f(x))$, then $\underline{\alpha}$ is also an m-sequence over $\mathbb{Z}/(p)$ generated by $f(x)(\text{mod } p)$.

3 Distinctness of Decimations

Before giving the main results, we first show some necessary lemmas. As reference [20] has not yet been published, their proofs are given in Appendix for completeness.

Lemma 1. *[20] Let $f(x)$ be a primitive polynomial over $\mathbb{Z}/(p)$ with p an odd prime. Then for any $\underline{u}, \underline{v} \in G'(f(x), p)$, $\underline{u} = \underline{v}$ if and only if $\underline{u} \equiv \underline{v}(\text{mod } 2)$.*

Remark 2. In other words, if $\underline{u}$ and $\underline{v}$ are two different primitive sequences generated by the same polynomial, then $\underline{u} \not\equiv \underline{v}(\text{mod } 2)$.

Lemma 2. *[20] Let p be an odd prime, $\lambda, \alpha, \beta \in (\mathbb{Z}/(p))^*$ with $\alpha \equiv \lambda\beta(\text{ mod } p)$, and $\delta \in \mathbb{Z}/(p)$ with $\delta \equiv 0(\text{mod } 2)$. If $1 \leq \lambda \leq p-2$, then there exists a positive integer j, $1 \leq j \leq p-1$, such that*

$$(j\alpha(\text{mod } p))(\text{mod } 2) \neq ((j\beta + \delta)(\text{mod } p))(\text{mod } 2).$$

The first main result is as follows.

Theorem 1. *Let p be an odd prime and $e \geq 2$. Suppose $f(x)$ and $g(x)$ are two different primitive polynomials of degree n over $\mathbb{Z}/(p^e)$ satisfying $f(x) \not\equiv g(x)(\bmod\ p)$. Then for any $\underline{u} \in G'(f(x), p^e), \underline{v} \in G'(g(x), p^e)$, we have*

$$\underline{u} \not\equiv \underline{v}(\bmod\ 2).$$

Proof. Set $T_0 = p^n - 1$. Let $\underline{\alpha} = h_f(x)\underline{u}_0(\bmod\ p)$ and $\underline{\beta} = h_g(x)\underline{v}_0(\bmod\ p)$, where $h_f(x)$ and $h_g(x)$ are defined as (1). Since $\underline{\alpha}, \underline{\beta}$ are m-sequences of order n over $\mathbb{Z}/(p)$ generated by $f(x)(\bmod\ p)$ and $g(x)(\bmod\ p)$ respectively, and $f(x) \not\equiv g(x)(\bmod\ p)$, then $\underline{\alpha}, \underline{\beta}$ are linearly independent over $\mathbb{Z}/(p)$. Thus $\underline{\alpha} \not\equiv (p-1)\underline{\beta}(\bmod\ p)$. That is, there exists an integer $t_0, t_0 \geq 0$, such that $\alpha(t_0) \not\equiv (p-1)\beta(t_0)(\bmod\ p)$. It's obvious that $\alpha(t_0)$ and $\beta(t_0)$ can not be equal to 0 simultaneously.

Case 1. If exactly one of $\alpha(t_0)$ and $\beta(t_0)$ is equal to 0, without loss of generality, let $\alpha(t_0) \neq 0$ and $\beta(t_0) = 0$. Then by Proposition 2 and (3), (4), we have

$$\{u_{e-1}(t_0 + j \cdot p^{e-2}T_0) | j = 0, 1, ..., p-1\} = \{0, 1, ..., p-1\},$$

and

$$v_{e-1}(t_0 + j \cdot p^{e-2}T_0) = v_{e-1}(t_0) \text{ for all } j = 0, 1, ..., p-1.$$

Thus there exist integers j_1, j_2, $0 \leq j_1, j_2 \leq p-1$, such that

$$u_{e-1}(t_0 + j_1 \cdot p^{e-2}T_0) \equiv v_{e-1}(t_0 + j_1 \cdot p^{e-2}T_0)(\bmod\ 2),$$

and

$$u_{e-1}(t_0 + j_2 \cdot p^{e-2}T_0) \not\equiv v_{e-1}(t_0 + j_2 \cdot p^{e-2}T_0)(\bmod\ 2).$$

On the other hand, for $e \geq 2$ we have

$$\underline{u} = \underline{u}(\bmod\ p^{e-1}) + \underline{u}_{e-1} \cdot p^{e-1}, \text{ where } \mathrm{per}(\underline{u}(\bmod\ p^{e-1})) = p^{e-2}(p-1),$$

and

$$\underline{v} = \underline{v}(\bmod\ p^{e-1}) + \underline{v}_{e-1} \cdot p^{e-1}, \text{ where } \mathrm{per}(\underline{v}(\bmod\ p^{e-1})) = p^{e-2}(p-1).$$

Thus the two congruence equations $u(t_0+j_1 \cdot p^{e-2}T_0) \equiv v(t_0+j_1 \cdot p^{e-2}T_0)(\bmod\ 2)$ and $u(t_0+j_2 \cdot p^{e-2}T_0) \equiv v(t_0+j_2 \cdot p^{e-2}T_0)(\bmod\ 2)$ can not hold simultaneously. That is, there exists an integer t, $t \geq 0$, such that

$$u(t) \not\equiv v(t)(\bmod\ 2).$$

So we get

$$\underline{u} \not\equiv \underline{v}(\bmod\ 2).$$

Case 2. If $\alpha(t_0) \neq 0$ and $\beta(t_0) \neq 0$, then by Proposition 2 and (2) we know that

$$u_{e-1}(t_0 + j \cdot p^{e-2}T_0) \equiv u_{e-1}(t_0) + j \cdot \alpha(t_0)(\bmod\ p),$$

and

$$v_{e-1}(t_0 + j \cdot p^{e-2}T_0) \equiv v_{e-1}(t_0) + j \cdot \beta(t_0)(\bmod\ p),$$

hold for all $j = 0, 1, \ldots, p-1$.

On the other hand, by Proposition 2 and (3), we have

$$\{u_{e-1}(t_0 + j \cdot p^{e-2}T_0) | j = 0, 1, ..., p-1\} = \{0, 1, ..., p-1\},$$
$$\{v_{e-1}(t_0 + j \cdot p^{e-2}T_0) | j = 0, 1, ..., p-1\} = \{0, 1, ..., p-1\}.$$

Without loss of generality, let $u_{e-1}(t_0) = 0$, and set $v_{e-1}(t_0) = \delta$.

If $\delta \not\equiv 0 (\mathrm{mod}\ 2)$, then $u_{e-1}(t_0) \not\equiv v_{e-1}(t_0) (\mathrm{mod}\ 2)$. When $j = 0, 1, ..., p-1$, as $u_{e-1}(t_0 + j \cdot p^{e-2}T_0)$ and $v_{e-1}(t_0 + j \cdot p^{e-2}T_0)$ belong to the same set $\{0, 1, ..., p-1\}$ with odd cardinality p, thus there exists an integer j_0, $1 \le j_0 \le p-1$, such that

$$u_{e-1}(t_0 + j_0 \cdot p^{e-2}T_0) \equiv v_{e-1}(t_0 + j_0 \cdot p^{e-2}T_0) (\mathrm{mod}\ 2).$$

Similar as Case 1, we know that the two congruence equations $u(t_0) \equiv v(t_0) (\mathrm{mod}\ 2)$ and $u(t_0 + j_0 \cdot p^{e-2}T_0) \equiv v(t_0 + j_0 \cdot p^{e-2}T_0) (\mathrm{mod}\ 2)$ can not hold simultaneously. That is, there exists an integer t, such that

$$u(t) \not\equiv v(t) (\mathrm{mod}\ 2).$$

So we get

$$\underline{u} \not\equiv \underline{v} (\mathrm{mod}\ 2).$$

Otherwise, $\delta \equiv 0 (\mathrm{mod}\ 2)$, and $u_{e-1}(t_0) \equiv v_{e-1}(t_0) (\mathrm{mod}\ 2)$. Let $\alpha = \alpha(t_0)$ and $\beta = \beta(t_0)$ for simplicity, then $\alpha \neq 0$ and $\beta \neq 0$. Set $\lambda = \alpha\beta^{-1} (\mathrm{mod}\ p)$, *i.e.*, $\alpha \equiv \lambda\beta (\mathrm{mod}\ p)$, then $1 \le \lambda \le p-2$. Thus from Lemma 2 we know that there exists a positive integer j_1, $1 \le j_1 \le p-1$, such that

$$(j_1\alpha (\mathrm{mod}\ p)) (\mathrm{mod}\ 2) \neq ((j_1\beta + \delta) (\mathrm{mod}\ p)) (\mathrm{mod}\ 2).$$

That is,

$$u_{e-1}(t_0 + j_1 \cdot p^{e-2}T_0) \not\equiv v_{e-1}(t_0 + j_1 \cdot p^{e-2}T_0) (\mathrm{mod}\ 2).$$

Similar as above, we know that the two congruence equations $u(t_0) \equiv v(t_0)$ (mod 2) and $u(t_0 + j_1 \cdot p^{e-2}T_0) \equiv v(t_0 + j_1 \cdot p^{e-2}T_0)$ (mod 2) can not hold simultaneously. That is, there exists an integer t, $t \ge 0$, such that

$$u(t) \not\equiv v(t) (\mathrm{mod}\ 2).$$

So we get

$$\underline{u} \not\equiv \underline{v} (\mathrm{mod}\ 2).$$

Remark 3. Note that it's shown in [15] that when $n = 1$, this result also holds if the condition that $f(x) \not\equiv g(x) (\mathrm{mod}\ p)$ is omitted.

Let $p > 13$ be a prime such that for any sequences $\underline{u} \in G'(x - \xi, p), \underline{v} \in G'(x - \zeta, p)$, $\underline{u} \not\equiv \underline{v} (\mathrm{mod}\ 2)$ holds, where ξ and ζ are two different primitive roots modulo p. The following is the second main result of this article.

Theorem 2. *Suppose $f(x), g(x)$ are two different primitive polynomials of degree n over $\mathbb{Z}/(p)$ with p an odd prime as above. Then for any $\underline{u} \in G'(f(x), p), \underline{v} \in G'(g(x), p)$, we have*

$$\underline{u} \not\equiv \underline{v} (\mathrm{mod}\ 2).$$

Proof. If $n = 1$, then the result holds from the assumption on p.

If $n \geq 2$, set $s = (p^n - 1)/(p-1) > 1$. Then there exists uniquely one constant $\delta_f \in \mathbb{Z}/(p)$, relatively to $f(x)$ only, such that

$$x^s \equiv \delta_f (\bmod f(x)). \tag{5}$$

Actually, the constant δ_f is a primitive element in $\mathbb{Z}/(p)$. Since $\underline{u} \in G'(f(x), p)$, that is, $f(x)\underline{u} = \underline{0}$, thus by (5) we have $(x^s - \delta_f)\underline{u} = \underline{0}$. So we get $u(t+s) = \delta_f \cdot u(t)$ for all $t \geq 0$. Generally, we have

$$u(t + j \cdot s) = \delta_f^j \cdot u(t), \text{ for all } t \geq 0 \text{ and } j \geq 0.$$

Similarly, there exists uniquely one constant $\delta_g \in \mathbb{Z}/(p)$, also a primitive element in $\mathbb{Z}/(p)$ and relatively to $g(x)$ only, such that $x^s \equiv \delta_g(\bmod\ g(x))$. Thus

$$v(t + j \cdot s) = \delta_g^j \cdot v(t), \text{ for all } t \geq 0 \text{ and } j \geq 0.$$

For any fixed $t \geq 0$, the sequences $\underline{u}' = (u(t+j \cdot s))_{j \geq 0}$ and $\underline{v}' = (v(t+j \cdot s))_{j \geq 0}$ can be considered as the t-shifted sequences of the s-fold decimations of $\underline{u}$ and $\underline{v}$, respectively. They are m-sequences with period $\mathrm{per}(\underline{u}') = \mathrm{per}(\underline{v}') = p - 1$ generated by $x - \delta_f$ and $x - \delta_g$, respectively.

As $\underline{u} \neq \underline{v}$, there exists an integer $t_0, t_0 \geq 0$, such that $u(t_0) \neq v(t_0)$. We will show the result holds according to the following two cases, respectively.

Case 1. If exactly one of $u(t_0)$ and $v(t_0)$ is equal to 0, without loss of generality, let $u(t_0) = 0$ and $v(t_0) \neq 0$, then

$$u(t_0 + j \cdot s) = 0 \text{ for all } j = 0, 1, ..., p-2,$$

and

$$\{v(t_0 + j \cdot s) | j = 0, 1, ..., p-2\} = \{1, 2, ..., p-1\}.$$

Thus there exists an integer $j_0, 0 \leq j_0 \leq p-2$, such that $u(t_0 + j_0 \cdot s) \not\equiv v(t_0 + j_0 \cdot s)(\bmod 2)$, so we get $\underline{u} \not\equiv \underline{v}(\bmod 2)$.

Case 2. If neither $u(t_0)$ nor $v(t_0)$ are equal to 0, then the sequences $\underline{u}' = (u(t_0 + j \cdot s))_{j \geq 0}$ and $\underline{v}' = (v(t_0 + j \cdot s))_{j \geq 0}$ are m-sequences generated by $x - \delta_f$ and $x - \delta_g$, respectively. As $u(t_0) \neq v(t_0)$, then $\underline{u}' \neq \underline{v}'$. If $\delta_f \neq \delta_g$, then from the assumption on p we know that $\underline{u}' \not\equiv \underline{v}'(\bmod 2)$, thus $\underline{u} \not\equiv \underline{v}(\bmod 2)$. If $\delta_f = \delta_g$, then from Lemma 1 we know that $\underline{u}' \not\equiv \underline{v}'(\bmod 2)$, thus $\underline{u} \not\equiv \underline{v}(\bmod 2)$.

Remark 4. Note that when 2 is a primitive root modulo p, the assumption on p corresponds to the case of Conjecture 1 when $e = 1$. It's shown in [7] that Conjecture 1 was verified by experiments for all primes $p < 2,000,000$, and asymptotically for large prime p, the collection of counterexamples to Conjecture 1 is a vanishingly small fraction of the set of all decimations. In fact, this result also holds if 2 is not a primitive root modulo p. Thus there exists large numbers of such primes p.

Combining Theorem 1 and Theorem 2, we actually show that almost all decimations of any generalized l-sequence are cyclically distinct.

4 Conclusions

In this article, a new kind of sequences called generalized l-sequences is introduced and the distinctness of their decimations is shown. It's well-known that l-sequences can be easily generated by the FCSRs, thus how to effectively generate these generalized l-sequences is an open problem. Moreover, research on the pseudorandom properties of these generalized l-sequences is also an interesting thing.

References

1. Arnault, F., Berger, T.P.: Design and Properties of a New Pseudorandom Generator Based on a Filtered FCSR Automaton. IEEE Transactions on computer, Vol. 54, (2005) 1374–1383
2. Arnault, F., Berger, T.P.: F-FCSR, Design of a new class of stream ciphers, Fast Software Encryption, Lecture Notes in Computer Science, Vol. 3557, Springer Verlag, (2005) 83–97
3. Dai, Z.-D.: Binary Sequences Derived from ML-Sequences over rings I: Periods and minimal polynomials, J. Cryptology, Vol. 5, **4** (1992) 193–207
4. Golomb, S.: Shift Register Sequences. Laguna Hills, CA: Aegean Park, 1982.
5. Goresky, M., Klapper, A.: Arithmetic crosscorrelations of feedback with carry shift register sequences. IEEE Trans. Inform. Theory, Vol. 43, (1997) 1342–1345
6. Goresky, M., Klapper, A., Murty, R.: On the distinctness of decimations of l-sequences. Sequences and their Applications – SETA 01, T. Helleseth, P.V. Kumar, K. Yang, ed. Springer Verlag, N.Y., (2001) 197–208
7. Goresky, M., Klapper, A., Murty, R., Shparlinski, I.: On decimations of l-sequences. SIAM J. Discrete Math., Vol. 18, (2004) 130–140
8. Huang, M.-Q., Dai, Z.-D.: Projective maps of linear recurring sequences with maximal p-adic periods. Fibonacci Quart, Vol. 30, **2** (1992) 139–143
9. Klapper, A., Goresky, M.: Large period nearly deBruijn FCSR sequences. Advances in Cryptology – Eurocrypt 1995, Lecture Notes in Computer Science, Vol. 921, Springer Verlag, N.Y., (1995) 263–273
10. Klapper, A., Goresky, M.: Feedback shift registers, 2-adic span, and combiners with memory. J. Cryptology 10 (1997), pp. 111–147.
11. Qi, Q.-W., Xu, H.: Partial period distribution of FCSR sequences. IEEE Trans. Inform. Theory, Vol. 49, (2003) 761–765
12. Seo, C., Lee, S., Sung, Y., Han, K., Kim, S.: A lower bound on the linear span of an FCSR. IEEE Trans. Inform. Theory, Vol. 46, (2000) 691-693
13. Ward, M.: The Arithmetical Theory of Linear Recurring Series. Trans. Amer. Math. Soc., Vol. 35, (1933) 600–628
14. Xu, H., Qi, W.-F.: Autocorrelations of maximum period FCSR sequences. SIAM Journal on Discrete Mathematics, to appear.
15. Xu, H., Qi, W.-F.: Further results on the distinctness of decimations of l-sequences. IEEE Trans. Inform. Theory, to appear.
16. Zhu, X.-Y., Qi, W.-F.: Compression mappings on primitive sequences over $Z/(p^e)$. IEEE Trans. Inform. Theory, Vol. 50, (2004) 2442–2448
17. Zhu, X.-Y.: Some Results on Injective Mappings of Primitive Sequences Modulo Prime Powers. Doctoral dissertation of ZhengZhou Information Engineering University, (2004), (in Chinese)

18. Zhu, X.-Y., Qi, W.-F.: Uniqueness of the distribution of zeroes of primitive level sequences over $Z/(p^e)$. Finite Fields and Their Applications, Vol. 11, **1** (2005) 30–44
19. Zhu, X.-Y., Qi, W.-F.: Uniqueness of the distribution of zeroes of primitive level sequences over $Z/(p^e)$ (II). Finite Fields and Their Applications, to appear.
20. Zhu, X.-Y., Qi, W.-F.: Compression Mappings of Modulo 2 on Primitive Sequences over $Z/(p^e)$. submitted.

Appendix

In this section, we give the proofs of Lemma 1 and Lemma 2 that appeared in Section 3. Such lemmas also appeared in the doctoral dissertation of Zhu, see [17].

Lemma 1. [20] *Let $f(x)$ be a primitive polynomial over $Z/(p)$ with p an odd prime. Then for any $\underline{u}, \underline{v} \in G'(f(x), p), \underline{u} = \underline{v}$ if and only if $\underline{u} \equiv \underline{v} (\text{mod } 2)$.*

Proof. The necessary condition is obvious. We need only to show if $\underline{u} \equiv \underline{v}(\text{mod } 2)$, then $\underline{u} = \underline{v}$.

If $\underline{u}$ and $\underline{v}$ are linear dependent over $\mathbb{Z}/(p)$, that is, there exists an integer $\lambda \in (\mathbb{Z}/(p))^*$, such that $\underline{v} \equiv \lambda \cdot \underline{u}(\text{mod } p)$. If λ is even, let t be an integer such that $u(t) = 1$, then $u(t) \not\equiv v(t)(\text{mod } 2)$, which is in contradiction with $\underline{u} \equiv \underline{v}(\text{mod } 2)$. If λ is odd and $\lambda \neq 1$, let k be the least positive integer such that $(k-1)\lambda < p < k\lambda$, and let t be an integer such that $u(t) = k$. Since $(k\lambda(\text{mod } p))(\text{mod } 2) = (k\lambda - p)(\text{mod } 2) \neq k(\text{mod } 2)$, then $u(t) \not\equiv v(t)(\text{mod } 2)$, which is in contradiction with $\underline{u} \equiv \underline{v}(\text{mod } 2)$. Thus $\lambda = 1$ and $\underline{u} = \underline{v}$.

If $\underline{u}$ and $\underline{v}$ are linear independent over $\mathbb{Z}/(p)$, since $\underline{u}$ and $\underline{v}$ are m-sequences generated by the same polynomial $f(x)$, then there exists an integer t such that $u(t) = 0$ and $v(t) = 1$. So we have $u(t) \not\equiv v(t)(\text{mod } 2)$, which is also in contradiction with $\underline{u} \equiv \underline{v}(\text{mod } 2)$. Thus $\underline{u} = \underline{v}$.

Lemma 2. [20] *Let p be an odd prime, $\lambda, \alpha, \beta \in (Z/(p))^*$ with $\alpha \equiv \lambda\beta(\text{mod } p)$, and $\delta \in Z/(p)$ with $\delta \equiv 0(\text{mod } 2)$. If $1 \leq \lambda \leq p-2$, then there exists a positive integer j, $1 \leq j \leq p-1$, such that*

$$(j\alpha(\text{mod } p))(\text{mod } 2) \neq ((j\beta + \delta)(\text{mod } p))(\text{mod } 2).$$

Proof. Since $\alpha \equiv \lambda\beta(\text{mod } p)$, we have

$$\begin{aligned}&\{(\ j\ \cdot\alpha(\text{mod } p), (j \cdot \beta + \delta)(\text{mod } p)) \mid j = 0, 1, \ldots, p-1\}\\ &\quad = \{(j \cdot \lambda(\text{mod } p), (j + \delta)(\text{mod } p)) \mid j = 0, 1, \ldots, p-1\}.\end{aligned}$$

Thus we need only to show there exists a positive integer j, $1 \leq j \leq p-1$, such that

$$(j\lambda(\text{mod } p))(\text{mod } 2) \neq ((j + \delta)(\text{mod } p))(\text{mod } 2). \qquad (6)$$

1. $\lambda = 1$.

As δ is even, set $j = p - \delta$, then $j\lambda(\text{mod } p) = p - \delta$ is odd, but $(j + \delta)(\text{mod } p) = 0$ is even, thus (6) holds.

2. $2 \le \lambda \le p-2$, and $\delta < p-1$.

If λ is even, set $j=1$, then $j\lambda(\bmod\ p) = \lambda$ is even, but $(j+\delta)(\bmod\ p) = 1+\delta$ is odd, thus (6) holds.

If λ is odd, let k_1 be the least positive integer such that $(k_1-1)\lambda < p < k_1\lambda < 2p$ and k_2 be the least positive integer such that $(k_2-1)\lambda < 2p < k_2\lambda < 3p$. It's clear that $2 \le k_1 < k_2 < p$.

(2.1) If $k_1 < p-\delta$, then $(k_1\lambda(\bmod p))(\bmod 2) = (k_1\lambda - p)(\bmod 2) \ne k_1(\bmod 2)$, but $(k_1+\delta)(\bmod 2) = k_1(\bmod 2)$. Set $j = k_1$, then (6) holds.

(2.2) If $k_1 = p-\delta$, from $k_2 > k_1 = p-\delta$ and $2p < k_2\lambda < 3p$ we have $(k_2\lambda(\bmod p))(\bmod 2) = k_2\lambda - 2p(\bmod 2) = k_2(\bmod 2)$, and $((k_2+\delta)(\bmod p))(\bmod 2) = (k_2+\delta-p)(\bmod 2) \ne k_2(\bmod 2)$. Set $j = k_2$, then (6) holds.

(2.3) If $k_1 > p-\delta$, then from the definition of k_1 we know that $0 < (p-\delta)\lambda < p$. Set $j = p-\delta$, then $j\lambda(\bmod p) = (p-\delta)\lambda$ is odd, but $(j+\delta)(\bmod p) = 0$ is even, thus (6) holds.

3. $2 \le \lambda \le p-2$, and $\delta = p-1$.

In this case, we need only to show there exists a positive integer j, $1 \le j \le p-1$, such that

$$(j\lambda(\bmod p))(\bmod 2) \ne ((j-1)(\bmod p))(\bmod 2). \tag{7}$$

If λ is odd, set $j=1$, then $j\lambda(\bmod p) = \lambda$ is odd, but $(j-1)(\bmod p) = 0$ is even, thus (7) holds.

If λ is even, then $2 \le \lambda \le p-3$. Let k be the least positive integer such that $k\lambda(\bmod p) < p-\lambda$. As $2 \le \lambda \le p-3$, then $p-\lambda \ge 3$, and $1 \le k \le p-2$. Thus $(k\lambda(\bmod p))(\bmod 2) = ((k+1)\lambda(\bmod p))(\bmod 2)$. Since $(k-1)(\bmod 2) \ne k(\bmod 2)$, set $j_1 = k$, $j_2 = k+1$, then either $(j_1\lambda(\bmod p))(\bmod 2) \ne ((j_1-1)(\bmod p))(\bmod 2)$ or $(j_2\lambda(\bmod p))(\bmod 2) \ne ((j_2-1)(\bmod p))(\bmod 2)$, thus (7) holds.

Remark 5. The case when $\delta = 0$ is not included in the original result of [20], but the proof is the same, so we include it here.

On FCSR Memory Sequences*

Tian Tian and Wen-Feng Qi

Department of Applied Mathematics,
Zhengzhou Information Engineering University, Zhengzhou, P.R. China
wenfeng.qi@263.net

Abstract. In this paper we investigate FCSR memory sequences in two aspects, period and complementarity property. We show that an FCSR memory sequence shares the same period with its associated binary sequence for a special kind of connection integers. Especially, binary sequences generated by an FCSR with such connection integers contain most of the l-sequences. Furthermore, for an l-sequence $\underline{a}$ with the minimum connection integer q and $\underline{m} = (m_0, m_1, ...)$ its memory sequence, we prove $m_i + m_{i+T/2} = w - 1$ for $i \geq 0$, where $T = \text{per}(\underline{a})$ and w is the Hamming weight of $q + 1$.

1 Introduction

Feedback with Carry Shift Registers (FCSRs) were introduced by M. Goresky and A. Klapper in [1]. The main characteristic of an FCSR is the fact that the elementary additions are not additions modulo 2 but with propagation of carries.

Assume q is a positive odd integer, $r = \lfloor \log_2(q+1) \rfloor$(where $\lfloor\ \rfloor$ denotes the integer part), and $q + 1 = q_1 2 + q_2 2^2 + ... + q_r 2^r$ is the binary representation of $q + 1$. Let $wt(q + 1)$ be the number of nonzero q_i for $1 \leq i \leq r$, the Hamming weight of $q + 1$. Figure 1 depicts an r-stage FCSR with connection integer q, where $\sum$ denotes integer addition:

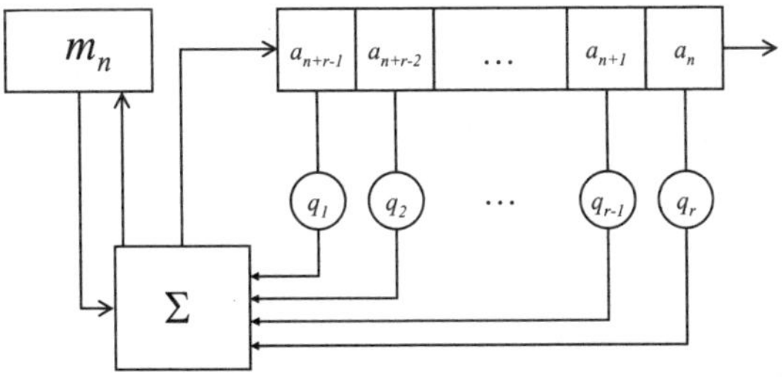

Fig. 1. Feedback with carry shift register

The shift registers and the memory register at any given clock time consist of r bits and a memory integer respectively, which is denoted by $(m_n; a_{n+r-1}, a_{n+r-2},..., a_n)$ and called the *state* of the FCSR at the nth clock time or just

* This work was supported by NSF of China under Grant number 60373092.

G. Gong et al. (Eds.): SETA 2006, LNCS 4086, pp. 323–333, 2006.

state for short. Especially, $(m_0; a_{r-1}, a_{r-2}, ..., a_0)$ is called *the initial state.* The operation of the FCSR at the nth clock time is defined as follows:

(1) Compute integer addition: $\sigma_n = \sum_{k=1}^{r} q_k a_{n+r-k} + m_n$;
(2) Shift the contents of r shift registers one step to the right, outputting the rightmost bit a_n;
(3) Place $a_{n+r} = \sigma_n \pmod 2$ into the leftmost shift register;
(4) Replace the memory integer m_n with $m_{n+1} = (\sigma_n - a_{n+r})/2 = \lfloor \sigma_n/2 \rfloor$.

We shall say that a state is *periodic* if, left to run, the FCSR will eventually return to that same state.

Recently, many research papers have been done on pseudorandom properties of FCSR sequences, such as [2]-[5]. On the other hand, new pseudorandom generators based on FCSRs have also been designed and studied, such as F-FCSR [6]. Although so much progress has been achieved on FCSRs and their output sequences, in point of developing the design of stream ciphers, there are still some most basic properties inside FCSR architecture worth our discovering.

The memory register ingeniously introduces the nonlinear architecture into FCSR, which greatly separates the FCSR sequences from traditional LFSR sequences. In spite of this, we should also see that its effects are limited, for FCSR sequences share many of the important properties with LFSR sequences and they are comparable in many aspects. In this paper, we shift our attention on FCSR sequences to their memory sequences and mainly concerns the periodicity and the complementarity property of them. For the periodicity, in Section 2.2 we show that an FCSR memory sequence shares the same period with its associated binary sequence for a special kind of connection integers. l-sequences, the counterpart of m-sequences in the domain of LFSR sequences, are an important kind of FCSR sequences. Especially, binary sequences generated by FCSRs with such connection integers contain most of the l-sequences. Then, for an l-sequence $\underline{a}$ with the minimum connection integer q and $\underline{m} = (m_0, m_1, ...)$ its memory sequence, in Section 2.3 we prove $m_i + m_{i+T/2} = w - 1$ for $i \geq 0$, where $T = \text{per}(\underline{a})$ is the period of $\underline{a}$ and w is the Hamming weight of $q + 1$. This property is quite similar with the fact that the second half of one period of an l-sequence is the bitwise complement of the first half.

2 Main Results

2.1 Preminilaries

The following two lemmas are from [7].

Lemma 1 ([7]). *Let $\underline{a}$ be an ultimately periodic binary sequence. Then $\alpha = \sum_{i=0}^{\infty} a_i 2^i$ is a quotient of two integers, p/q, and the denominator q is the connection integer of an FCSR which generates the sequence $\underline{a}$. Furthermore, $\underline{a}$ is periodic if and only if $\alpha \leq 0$ and $\mid \alpha \mid \leq 1$.*

Lemma 2. [7] *Let integer* $q = q_0 + q_1 2 + q_2 2^2 + \ldots + q_r 2^r$*, where* $q_0 = -1$*,* $q_r = 1$ *and* $q_i \in \{0,1\}$ *for* $1 \le i \le r-1$*. Let* $\underline{a} = (a_0, a_1, a_2, \ldots)$ *be an output sequence of an FCSR with connection integer* q *and initial state* $(m_0; a_{r-1}, \ldots, a_0)$*. Then* $\sum_{i=0}^{\infty} a_i 2^i = p/q$*, where* $p = \sum_{k=0}^{r-1} \sum_{i=0}^{k} q_i a_{k-i} 2^k - m_0 2^r$*.*

From Lemma 1 and Lemma 2, it is clear that the initial memory m_0 of an FCSR with connection integer q which generates the sequence $\underline{a}$ is uniquely determined by $\underline{a}$ and q, and so the memory sequence $\underline{m}$ is also uniquely determined by them. Therefore, from now on, $\underline{m} = (m_0, m_1, \ldots)$ is called *the memory sequence of* $(\underline{a}, q)$ and always referred as an integer sequence, while other sequences are referred as binary sequences.

If a sequence $\underline{a}$ can be generated by an FCSR with connection integer q, then we also call q a *connection integer* of $\underline{a}$. The least connection integer of $\underline{a}$ is called *the minimum connection integer* of $\underline{a}$. Consequently, we have following well-known corollary and the proof is omitted.

Corollary 1. *If* $\underline{a}$ *is an FCSR sequence with the minimum connection integer* q*, then any connection integer* q' *of* $\underline{a}$ *is divisible by* q*. Suppose* $k \ge 0$ *and* q'' *is the minimum connection integer of* $L^k\underline{a}(L^k\underline{a} = (a_k, a_{k+1}, \ldots))$*, then* $q'' \mid q$*. If* $\underline{a}$ *is periodic, then* $q'' = q$*.*

Theorem 1. *Let* $\underline{a}$ *be an FCSR sequence with connection integer* q *and* $\underline{m}$ *the memory sequence of* (a, q)*. Then* $L^i\underline{m} = (m_i, m_{i+1}, \ldots)$ *is the memory sequence of* $(L^i\underline{a}, q)$ *for* $i \ge 0$*. Furthermore, the period of the state sequence*

$$(m_0; a_{r-1}, \ldots, a_0), (m_1; a_r, \ldots, a_1), \ldots,$$

where $r = \lfloor \log_2(q+1) \rfloor$*, is equal to that of the sequence* $\underline{a}$*.*

Proof. Since $L^i\underline{a}$ can be generated by an FCSR with connection integer q and initial state $(m_i; a_{r+i-1}, \ldots, a_i)$, $L^i\underline{m} = (m_i, m_{i+1}, \ldots)$ must be the unique memory sequence of $(L^i\underline{a}, q)$. Suppose the period of the state sequence is T_1 and that of the sequence $\underline{a}$ is T_2. It is clear that $T_2 \mid T_1$. Since $\underline{a}$ is ultimately periodic, there exists i_0 such that $L^{i_0}\underline{a}$ is periodic, and so the memory sequence of $(L^i\underline{a}, q)$ is the same with that of $(L^{i+T_2}\underline{a}, q)$ for any $i \ge i_0$. Thus, $m_i = m_{i+T_2}$ for $i \ge i_0$, which implies $T_1 \mid T_2$, and so we have $T_1 = T_2$.

By Theorem 1, we can immediately get

Corollary 2. *Let* $\underline{a}$ *be an FCSR sequence with connection integer* q *and* $\underline{m}$ *the memory sequence of* $(\underline{a}, q)$*. Then per*$(\underline{m})$ $\mid$ *per*$(\underline{a})$*.*

It is natural to ask whether per$(\underline{m})$ is equal to per$(\underline{a})$. Experimental results show that the answer is not always affirmative, but it is most likely to be valid for l-sequences.

2.2 Periods of Memory Sequences

In this subsection, we try to theoretically solve the question proposed at the end of the last subsection for sequences with connection integers of which 2 is primitive modulo every prime factor p but 2^{p-1} is not divisible by p^2.

Definition 1. *Let ξ be a primitive nth root of unity over* $\mathbf{Q}$. *Then the polynomial*

$$Q_n(x) = \sum_{\substack{s=1 \\ \gcd(s,n)=1}}^{n} (x - \xi^s)$$

is called the nth cyclotomic polynomial over $\mathbf{Q}$.

Remark 1. $Q_n(x)$ is an irreducible polynomial over $\mathbf{Z}$ with degree $\phi(n)$.

Lemma 3. *Let $\underline{a}$ be an FCSR sequence with connection integer $q = -1 + q_1 2 + q_2 2^2 + ... + q_r 2^r$ and $\underline{m}$ the memory sequence of $(\underline{a}, q)$. If $per(\underline{m}) \neq per(\underline{a})$, then there exists a positive factor $t > 1$ of $per(\underline{a})$ such that $Q_t(2)$ divides q.*

Proof. Suppose $\underline{a}$ is periodic, and so is $\underline{m}$ by Theorem 1. Setting $\mathrm{per}(\underline{a}) = T$ and $\mathrm{per}(\underline{m}) = S$, we obtain $S \mid T$ by Corollary 2. Put

$$\delta_i = q_1 a_{i+r-1} + ... + q_r a_i,$$

for $i \geq 0$. Thus

$$a_{i+r} \equiv \delta_i + m_i \pmod 2, m_{i+1} = \lfloor (\delta_i + m_i)/2 \rfloor,$$

for $i \geq 0$. Assume there exists a pair of integers i and j such that

$$m_i = m_j, m_{i+1} = m_{j+1}. \tag{1}$$

(1) If $\delta_i \equiv \delta_j \pmod 2$, then $\delta_j = \delta_i + 2k$ for some integer k. Using (1), we have

$$m_{i+1} = \lfloor (\delta_i + m_i)/2 \rfloor = k + \lfloor (\delta_i + m_i)/2 \rfloor = m_{j+1},$$

so that, $k = 0$, $\delta_j = \delta_i$ and $a_{i+r} = a_{j+r}$.

(2) If $\delta_i \equiv \delta_j + 1 \pmod 2$, then $\delta_j = \delta_i + 1 + 2k$ for some integer k. Again using (1), we have

$$m_{i+1} = \lfloor (\delta_i + m_i)/2 \rfloor = k + \lfloor (\delta_i + 1 + m_i)/2 \rfloor = m_{j+1}.$$

Hence, if $\delta_i + m_i \equiv 0 \pmod 2$, then $k = 0$, $\delta_j = \delta_i + 1$, $a_{r+i} = 0$ and $a_{r+j} = 1$. Otherwise, $\delta_i + m_i \equiv 1 \pmod 2$, and $k = -1$, $\delta_j = \delta_i - 1$, $a_{r+i} = 1$, $a_{r+j} = 0$.

Combining above two aspects, we can conclude that

$$\text{if } m_i = m_j \text{ and } m_{i+1} = m_{j+1}, \text{ then } \delta_j - \delta_i = a_{r+j} - a_{r+i}. \tag{2}$$

For any $i \geq 0$, we have

$$q_r a_i + q_{r-1} a_{i+1} + ... + q_1 a_{i+r-1} = \delta_i, \tag{3}$$

$$q_r a_{i+S} + q_{r-1} a_{i+S+1} + ... + q_1 a_{i+S+r-1} = \delta_{i+S}. \tag{4}$$

Subtracting (4) from (3) leads to

$$q_r(a_i - a_{i+S}) + q_{r-1}(a_{i+1} - a_{i+S+1}) + ... + q_1(a_{i+r-1} - a_{i+S+r-1}) = \delta_i - \delta_{i+S}.$$

Since $\underline{m}$ is periodic and $\mathrm{per}(\underline{m}) = S$, it follows from the conclusion (2) that

$$\delta_i - \delta_{i+S} = a_{i+r} - a_{i+r+S},$$

and so

$$q_r(a_i - a_{i+S}) + q_{r-1}(a_{i+1} - a_{i+S+1}) + \dots + q_1(a_{i+r-1} - a_{i+S+r-1}) = a_{i+r} - a_{i+r+S} \quad (5)$$

for any $i \geq 0$.

Let integer sequence $\underline{c} = \underline{a} - L^S\underline{a}$(componentwise integer subtraction). Since $S \neq T$, $\underline{c} \neq \underline{0}$. By (5) we get that $\underline{c}$ satisfies following linear recurrence relation

$$q_r c_i + q_{r-1}c_{i+1} + \dots + q_1 c_{i+r+1} = c_{i+r}$$

for $i \geq 0$, over $\mathbf{Z}$, implying $f(x) = x^r - (q_1 x^{r-1} + q_2 x^{r-2} + \dots + q_r) \in \mathbf{Z}[x]$ is a characteristic polynomial of $\underline{c}$.

Consider $\underline{a}$ and $\underline{c}$ as sequences over $\mathbf{Q}$. Let $m_{\underline{a}}(x) \in \mathbf{Q}[x]$ be the minimum polynomial of $\underline{a}$ and $m_{\underline{c}}(x) \in \mathbf{Q}[x]$ the minimum polynomial of $\underline{c}$. Then $m_{\underline{c}}(x) = m_{\underline{a}}(x)/\gcd(m_{\underline{a}}(x), 1 - x^S)$. $\mathrm{per}(\underline{a}) = T$ implies $m_{\underline{a}}(x) \mid (x^T - 1)$, and so we may assume

$$m_{\underline{a}}(x) = Q_{t_1}(x)Q_{t_2}(x)\cdots Q_{t_h}(x), \ \mathrm{lcm}(t_1, ..., t_h) = T \quad (6)$$

and

$$m_{\underline{c}}(x) = Q_{t_1}(x)Q_{t_2}(x)\cdots Q_{t_g}(x), \ 1 \leq g \leq h. \quad (7)$$

Since $x^T - 1$ has no multiple factor and $(x-1) \mid (x^S - 1)$, all $t_i > 1$ for $1 \leq i \leq g$. We have shown that $f(x)$ is a characteristic polynomial of $\underline{c}$, so $Q_{t_1}(x)Q_{t_2}(x)\cdots Q_{t_g}(x) \mid f(x)$. Besides all cyclotomic polynomials are self-reciprocal except $Q_1(x)$, thus

$$Q_{t_1}(x)Q_{t_2}(x)\cdots Q_{t_g}(x) \mid f^*(x) = x^n f(1/x). \quad (8)$$

Then it follows that

$$Q_{t_1}(2)Q_{t_2}(2)\cdots Q_{t_g}(2) \mid f^*(2) = -q. \quad (9)$$

Note that $g \geq 1$ and $t_i > 1$ for $1 \leq i \leq g$, therefore the proof is complete for periodic sequences.

If $\underline{a}$ is not periodic but ultimately periodic, then $L^{i_0}\underline{a}$ is periodic for some i_0. After replacing $\underline{a}$ with $L^{i_0}\underline{a}$ and $\underline{m}$ with $L^{i_0}\underline{m}$, the above proof is also valid.

In the sequel, we will see that Lemma 3 plays a role in converting the problem of periods of memory sequences to some solved classical number theory problems. For any integers a and b, $\mathrm{ord}_b(a)$ denotes the multiplicative order of a modulo b, that is, $\mathrm{ord}_b(a)$ is the least positive integer m such that $a^m \equiv 1 \pmod{b}$.

Definition 2. [8] *Let a and n be integers greater than 1, then a prime p is called a Zsigmondy prime for $< a, n >$ if p does not divide a and $\mathrm{ord}_p(a) = n$. A Zsigmondy prime p for $< a, n >$ is called a large Zsigmondy prime for $< a, n >$ if $p > n + 1$ or $p^2 \mid (a^n - 1)$.*

Lemma 4. [8] *For any integer $n > 1$, there exists a Zsigmondy prime for $< 2, n >$ except $n = 6$. If p is an odd prime with $ord_p(2) = p - 1$ and e a positive integer, then p^2 does not divide $Q_{p^e(p-1)}(2)$, and p is the unique prime that divides $Q_{p^e(p-1)}(2)$ but non Zsigmondy prime for $< 2, p^e(p-1) >$.*

Lemma 5. [8] *Let p be an odd prime with $ord_p(2) = p - 1$ and e a positive integer. There exists a large Zsigmondy prime for $< 2, p^{e-1}(p-1) >$ except $p^e \in \{3, 5, 7, 9, 11, 13, 19\}$.*

Remark 2. $Q_6(2) = 3$, $Q_{18}(2) = 3 \times 19$, $Q_{p-1}(2) = p$ for $p \in \{3, 5, 11, 13\}$.

Lemma 6. *Let $q = p_1^{e_1} p_2^{e_2} \cdots p_b^{e_b}$, where e_i is non negative integer and p_i is prime greater than 13 such that $ord_{p_i}(2) = p_i - 1$ and $(2^{p_i - 1} - 1)$ is not divisible by p_i^2, $1 \le i \le b$. Then for any positive integer $Y > 1$, q is not divisible by $Q_Y(2)$.*

Proof. Assume $Q_Y(2) \mid q$. By Lemma 4, $Q_Y(2)$ has a Zsigmondy prime of $< 2, Y >$ except $Y = 6$. Since $Q_2(2) = 3$, $Q_6(2) = 3$, $Q_{18}(2) = 3 \times 19$ and $p_i \neq 3$ for $1 \le i \le b$, $Y \notin \{2, 6, 18\}$. Suppose for some $1 \le j \le b$, p_j is a Zsigmondy prime of $< 2, Y >$. Then $Y = p_j - 1$, since $\mathrm{ord}_{p_j}(2) = p_j - 1$, which implies p_j is the unique prime factor of q that is a Zsigmondy prime of $< 2, Y >$. But $Q_Y(2)$ has a large Zsigmondy prime $r \neq p_j$ of $< 2, Y >$ by Lemma 5. Thus, $r \neq p_i$ for any $1 \le i \le b$, which contradicts with the assumption. Therefore, $Q_Y(2)$ does not divide q and the proof is complete.

Combing Lemma 3 and Lemma 6, we can easily get

Theorem 2. *Let $\underline{a}$ be a sequence with connection integer q in Lemma 6 and $\underline{m}$ the memory sequence of $(\underline{a}, q)$. Then $per(\underline{a}) = per(\underline{m})$.*

Recall that a periodic sequence $\underline{a}$ with the minimum connection integer q is called an l-sequence if $\mathrm{per}(\underline{a}) = \phi(q)$, that is, 2 is primitive modulo q. Such q must be a prime power p^e with $\mathrm{ord}_p(2) = p - 1$ and $(2^{p-1} - 1)$ is not divisible by p^2. Then especially for l-sequences, we have

Corollary 3. *Let $\underline{a}$ be an l-sequence with the minimum connection integer p^e and $\underline{m}$ the memory sequence of $(\underline{a}, p^e)$. If $p \notin \{3, 5, 11, 13\}$, then $per(\underline{m}) = per(\underline{a})$.*

As for $p \in \{3, 5, 11, 13\}$ and integer $e > 1$, from the proof of Lemma 3, we can deduce that

Corollary 4. *Let $\underline{a}$ be an l-sequence with the minimum connection integer p^e and $\underline{m}$ the memory sequence of $(\underline{a}, p^e)$, where $e > 1$, then $per(\underline{m})$ is divisible by p^{e-1} except $p^e = 9$.*

Proof. Since $\underline{a}$ is an l-sequence, $\mathrm{per}(\underline{a}) = p^{e-1}(p-1)$. Then at least one t_i in (6) is divisible by p^{e-1}, $1 \le i \le h$. So it suffices to prove any t_i in (7) is not divisible by p^{e-1} for $1 \le i \le g$. On one hand, since $\mathrm{ord}_p(2) = p - 1$, t_i is divisible by $p - 1$ for all $1 \le i \le g$ by (9). On the other hand, from Lemma 4, for $1 \le i \le e - 1$, $Q_{p^i(p-1)}(2)$ must has at least one Zsigmondy prime of $< 2, p^i(p-1) >$ that is not equal to p except $p = 3$ and $i = 1$. Thus, (9) is valid only for $g = 1$, $t_1 = p - 1$ or $g = 2$, $t_1 = p - 1$, $t_2 = p(p-1)$ if $p = 3$. Therefore, except $p^e = 9$, $p^{e-1} \mid \mathrm{per}(\underline{m})$.

Remark 3. We completely solved the case of $p = 3$ and 5 at the end of next subsection.

Other than Theorem 2 and Corollary 3, we have some sufficient conditions that the period of the memory sequence is equal to that of the FCSR sequence for more general connection integers, but they are quite incomplete and need further improvements. Besides, the cases where they are not equal indeed exist, for example, a sequence with the minimum connection integer 135 has period 12, while the period of its memory sequence is equal to 6.

2.3 Complementarity Property of Memory Sequences

Following is the well known complementarity property of l-sequences.

Lemma 7. [7] *If $\underline{a}$ is an l-sequence, then $a_i + a_{i+T/2} = 1$ for $i \geq 0$, where $T = per(\underline{a})$.*

We will prove that the memory sequence of an l-sequence possesses similar complementarity property as what is showed in Lemma 7.

Lemma 8. [7] *Let $\underline{a}$ be a periodic FCSR sequence with connection integer q and $\underline{m}$ the memory sequence of $(\underline{a}, q)$. Then $0 \leq m_i < wt(q+1)$ for $i \geq 0$.*

Theorem 3. *Let $\underline{a}$ be an l-sequence with the minimum connection integer $q = -1 + q_1 2 + q_2 2^2 + ... + q_r 2^r$ and $\underline{m}$ the memory sequence of $(\underline{a}, q)$. Then $m_i + m_{i+T/2} = w - 1$ for $i \geq 0$, where $T = per(\underline{a})$ and $w = wt(q+1)$.*

Proof. Recall the definition of δ_i in Theorem 2, then it follows from Lemma 7 that $\delta_{T/2+i} = w - \delta_i$ for $i \geq 0$ and

$$(m_i + \delta_i) + (m_{i+T/2} + w - \delta_i) = 1 \pmod 2.$$

Thus, we get

$$m_i + m_{i+T/2} = w - 1 \pmod 2. \tag{10}$$

If $w \leq 2$, then $m_i = 1$ or 0 by Lemma 8, so that the result follows from (10). Next, we consider the case $w \geq 3$.

By Lemma 2, $\sum_{i=0}^{\infty} a_i 2^i = p/q$ such that p is given by

$$\begin{aligned} p = a_0 q_1 2 + (a_0 q_2 + a_1 q_1) 2^2 + (a_0 q_3 + a_1 q_2 + a_2 q_1) 2^3 + ... + \\ (a_0 q_{r-1} + ... + a_{r-2} q_1) 2^{r-1} - (a_0 + a_1 2 + a_2 2^2 + ... + a_{r-1} 2^{r-1}) - m_0 2^r. \end{aligned}$$

Set

$$\begin{aligned} A = a_0 q_1 2 + (a_0 q_2 + a_1 q_1) 2^2 + (a_0 q_3 + a_1 q_2 + a_2 q_1) 2^3 + ... + \\ (a_0 q_{r-1} + ... + a_{r-2} q_1) 2^{r-1} - (a_0 + a_1 2 + a_2 2^2 + ... + a_{r-1} 2^{r-1}). \end{aligned}$$

Since $\underline{a}$ is periodic, it follows from Lemma 1 that

$$-q < A - m_0 2^r < 0.$$

Thus we obtain

$$-m_0 2^r < -A < q - m_0 2^r. \tag{11}$$

By Corollary 1, q is also the minimum connection integer of $L^{T/2}\underline{a}$ and $L^{T/2}\underline{m}$ is the memory sequence of $(L^{T/2}\underline{a}, q)$. Then by Lemma 1 and Lemma 7, $\sum_{i=T/2}^{\infty} a_i 2^i = p'/q$, where p' is given by

$$\begin{aligned} p' &= (1-a_0)q_1 2 + ((1-a_0)q_2 + (1-a_1)q_1)2^2 + ... + \\ &\quad ((1-a_0)q_{r-1} + ... + (1-a_{r-2})q_1)2^{r-1} - ((1-a_0) + \\ &\quad (1-a_1)2 + (1-a_2)2^2 + ... + (1-a_{r-1})2^{r-1}) - m_{T/2}2^r \\ &= q_1 2 + (q_1+q_2)2^2 + ... + (q_1+q_2+...+q_{r-1})2^{r-1} \\ &\quad -(1+2+...+2^{r-1}) - A - m_{T/2}2^r. \end{aligned}$$

Set

$$B = q_1 2 + (q_1+q_2)2^2 + ... + (q_1+q_2+...+q_{r-1})2^{r-1} - (2^r - 1). \tag{12}$$

Then we have

$$p' = B - A - m_{T/2}2^r.$$

Since $L^{T/2}\underline{a}$ is periodic, again it follows from Lemma 1 that

$$-q < B - A - m_{T/2}2^r < 0. \tag{13}$$

For the upper bound of B, we obtain from (12) that
(1) If $w = r$, then $B = (w-3)2^w + 3$.
(2) If $w \leq r-1$, then

$$\begin{aligned} B &\leq 2 + 2\cdot 2^2 + ... + (w-1)2^{w-1} + (w-1)2^w + ... + (w-1)2^{r-1} - (2^r - 1) \\ &= (w-2)2^w + 2 + (w-1)(2^r - 2^w) - (2^r - 1) \\ &= (w-2)2^r + 3 - 2^w \end{aligned}$$

Above combined with $-A < q - m_0 2^r$ by (11), we have
(1) If $w = r$, then

$$\begin{aligned} &B - A - m_{T/2}2^r \\ &< (w-3)2^w + 3 + q - (m_0 + m_{T/2})2^w \end{aligned}$$

Thus, it follows from (13) that

$$(w-3)2^w + 3 + q - (m_0 + m_{T/2})2^w \geq -q + 2,$$

that is

$$(w-3)2^w + 1 + 2q \geq (m_0 + m_{T/2})2^w.$$

Since $q + 1 = 2^{w+1} - 2$, we obtain

$$(w-3)2^w + 1 + 2(2^{w+1} - 3) \geq (m_0 + m_{T/2})2^w,$$

that's also

$$(w+1) - 5/2^w \geq m_0 + m_{T/2}. \tag{14}$$

Since $w \geq 3$ implies $5/2^w < 1$, it follows from (14) that $m_0 + m_{T/2} \leq w$. But (10) implies $m_0 + m_{T/2} \neq w$, and so $m_0 + m_{T/2} \leq w - 1$.

(2) If $w \leq r - 1$, then

$$\begin{aligned} &B - A - m_{T/2}2^r \\ &< (w-2)2^r + 3 - 2^w + q - (m_0 + m_{T/2})2^w. \end{aligned}$$

It follows from (13) that

$$(w-2)2^r + 3 - 2^w + q - (m_0 + m_{T/2})2^r \geq -q + 2,$$

that is

$$(w-2)2^r + 1 - 2^w + 2q \geq (m_0 + m_{T/2})2^r,$$

Since $q + 1 \leq 2^{r+1} - 4$, we obtain

$$(w-2)2^r + 1 - 2^w + 2(2^{r+1} - 5) \geq (m_0 + m_{T/2})2^r,$$

that is

$$(w+2) - (9 + 2^w)/2^r \geq m_0 + m_{T/2}.$$

Therefore, we get $w + 1 \geq m_0 + m_{T/2}$.

For any $i \geq 0$, it is easy seen that the above proof is also valid for $L^i\underline{a}$. Therefore, we can conclude that

$$m_i + m_{i+T/2} \leq w + 1$$

for $i \geq 0$.

If $m_i + m_{T/2+i} = w + 1$ for any $i \geq 0$, then considering the case of $i = 1$, we have

$$\begin{aligned} m_1 &= \lfloor (m_0 + \delta_0)/2 \rfloor \\ m_{T/2+1} &= \lfloor (m_{T/2} + \delta_{T/2})/2 \rfloor \\ &= w - \lceil (m_0 + \delta_0 - 1)/2 \rceil \end{aligned}$$

and so

$$m_1 + m_{T/2+1} = w - \lceil (m_0 + \delta_0 - 1)/2 \rceil + \lfloor (m_0 + \delta_0)/2 \rfloor = w,$$

a contradiction. Therefore, we must have some integer $i \geq 0$ such that $m_i + m_{i+T/2} \leq w - 1$.

Let us take $i = 0$, and clearly the discussion for any other case is the same. Suppose $m_0 + m_{T/2} = w - k$, $w \geq k \geq 1$. Thus

$$\begin{aligned} m_1 &= \lfloor (m_0 + \delta_0)/2 \rfloor \\ m_{T/2+1} &= \lfloor (m_{T/2} + \delta_{T/2})/2 \rfloor \\ &= \lfloor (w - k - m_0 + w - \delta_0)/2 \rfloor \\ &= w - \lceil k + m_0 + \delta_0)/2 \rceil \end{aligned}$$

where $\lceil \, \rceil$ denotes the next largest integer. Therefore,

$$m_1 + m_{T/2+1} = w - \lceil k + m_0 + \delta_0)/2 \rceil + \lfloor (m_0 + \delta_0)/2 \rfloor.$$

(a) If $m_0 + \delta_0 \equiv 0 \pmod 2$, then $m_1 + m_{T/2+1} = w - \lceil k/2 \rceil$;
(b) If $m_0 + \delta_0 \equiv 1 \pmod 2$, then $m_1 + m_{T/2+1} = w - \lceil (k+1)/2 \rceil$.
Then similarly, we can get

$$m_2 + m_{T/2+2} \in \{w - \lceil \frac{\lceil \frac{k}{2} \rceil}{2} \rceil, w - \lceil \frac{\lceil \frac{k}{2} \rceil + 1}{2} \rceil, w - \lceil \frac{\lceil \frac{k+1}{2} \rceil}{2} \rceil, w - \lceil \frac{\lceil \frac{k+1}{2} \rceil + 1}{2} \rceil\},$$

and so on for $i \geq 3$.

According to above deduced recurrence relation, we can obtain

$$m_i + m_{T/2+i} \leq w - k/2^i$$

for any $i \geq 0$, so that

$$m_i + m_{T/2+i} \leq w - 1 \tag{15}$$

for any $i \geq 0$.

On the other hand, we can also obtain

$$\begin{aligned} m_i + m_{T/2+i} &\geq w - k/2^i - (1 + 1/2 + 1/2^2 + ... + 1/2^{i-1}) \\ &= w - 2 - (k-2)/2^i \end{aligned}$$

for any $i \geq 1$. Since i is arbitrary and $\underline{m}$ is periodic, we have $m_i + m_{T/2+i} \geq w-2$ for $i \geq 0$. Note that (10) implies $m_i + m_{T/2+i} \neq w - 2$, and so

$$m_i + m_{T/2+i} \geq w - 1 \tag{16}$$

for $i \geq 0$.

Hence, it follows from (15) and (16) that $m_i + m_{T/2+i} = w - 1$ for $i \geq 0$.

By Theorem 3, for an l-sequence $\underline{a}$ and its memory sequence $\underline{m}$, if $\text{per}(\underline{m}) \mid \text{per}(\underline{a})/2$, then $\text{per}(\underline{m}) = 1$ and $w-1$ is even. From this fact, we can get

Corollary 5. *Let $\underline{a}$ be an l-sequence with the minimum connection integer p^e and $\underline{m}$ the memory sequence of $(\underline{a}, p^e)$. If $p = 5$ or $p = 3$ and $e > 2$, then $per(\underline{m}) = per(\underline{a})$.*

Proof. We only need to prove the case of $p = 5$, since the proof for the other case is completely the same. On one hand, we have $\text{per}(\underline{a}) = 5^{e-1} \cdot 4$ and $\text{per}(\underline{m}) \mid \text{per}(\underline{a})$. On the other hand, from Corollary 4, we have $5^{e-1} \mid \text{per}(\underline{m})$. So, it suffices to prove $\text{per}(\underline{a})/2$ is not divisible by $\text{per}(\underline{m})$. If $wt(5^e + 1) - 1$ is odd, then it is evident that $\text{per}(\underline{a})/2$ is not divisible by $\text{per}(\underline{m})$. If $wt(5^e + 1) - 1$ is even and $\text{per}(\underline{m}) \mid \text{per}(\underline{a})/2$, then $\text{per}(\underline{m}) = 1$, which contradicts with Corollary 4 except $e = 1$. So, we again come to the conclusion that $\text{per}(\underline{a})/2$ is not divisible by $\text{per}(\underline{m})$ except $e = 1$. For $e = 1$ and $\text{per}(\underline{m}) = 1$, by the proof of Lemma 3 and (8), $x^2 - x - 1$(corresponds to $5 + 1 = 2 + 2^2$) is divisible by $Q_4(x) = x^2 + 1$ over $\mathbf{Q}$, which is clearly impossible. Therefore, when $e = 1$, $\text{per}(\underline{m}) = \text{per}(\underline{a})$.

Remark 4. Similarly, for $p^e = 11^e$ and 13^e, where $e > 1$, we have $\text{per}(\underline{a})/5 \mid \text{per}(\underline{m})$ and $\text{per}(\underline{a})/3 \mid \text{per}(\underline{m})$ respectively. For $p^e = 3$ and 9, experimental results show that $\text{per}(\underline{m}) = 1$ and 6 respectively.

Acknowledgments

The authors would like to thank the anonymous referees for their helpful comments and suggestions.

References

1. Klapper, A., Goresky, M.: 2-Adic Shift Registers. In Proc. of 1993 Cambridge Security Workshop, Fast Software Encryption, Lecture Notes in Computer Science, Vol. 809. Springer-Verlag, Cambridge, UK (1994) 174–178
2. Goresky, M., Klapper, A.: Arithmetic Crosscorrelations of Feedback with Carry Shift Register Sequences. IEEE Transactions on Information Theory **43** (1997) 1342–1345
3. Seo, C., Lee, S., Sung, Y., Han, K., Kim, S.: A Lower Bound on The Linear Span of an FCSR. IEEE Transactions on Information Theory **46** (2000) 691–693
4. Qi, W.F., Xu, H.: Partial Period Distribution of FCSR Sequences. IEEE Transactions on Information Theory **49** (2003) 761–765
5. Xu, H., Qi, W.F.: Autocorrelations of Maximum Period FCSR Sequences. SIAM Journal on Discrete Mathematics, to appear.
6. Arnault, F., Berger, T.P.: Design and Properties of a New Pseudorandom Generator Based on a Filtered FCSR Automaton. IEEE Transactions on Computers **54** (2005) 1374–1383
7. Klapper, A., Goresky, M.: Feedback Shift Registers, 2-Adic Span, and Combiners with Memory. J. Cryptology **10** (1997) 111–147
8. Roitman, M.: On Zsigmondy Primes. Proceedings of The American Mathematical Society **125** (1997) 1913–1919

Periodicity and Distribution Properties of Combined FCSR Sequences

Mark Goresky[1,*] and Andrew Klapper[2,**]

[1] Institute for Advanced Study, Princeton NJ
www.math.ias.edu/~goresky
[2] Dept. of Computer Science, University of Kentucky, Lexington KY
www.cs.uky.edu/~klapper

Abstract. This is a study of some of the elementary statistical properties of the bitwise exclusive or of two maximum period feedback with carry shift register sequences. We obtain conditions under which the resulting sequences has the maximum possible period, and we obtain bounds on the variation in the distribution of blocks of a fixed length. This may lead to improved design of stream ciphers using FCSRs.

Keywords: Feedback with carry shift register, pseudorandom sequence, stream cipher.

1 Introduction

The *summation combiner* [8] is a stream cipher in which two binary m-sequences are combined using addition-with-carry. This cipher attracted considerable attention during the 1980's because it was fast, simple to construct in hardware, and the linear span of the resulting combined sequence was known to approach its period, which is approximately the product of the periods of the constituent sequences.

The security of the summation combiner was called into question following the introduction of feedback-with-carry shift registers, or FCSRs [4], [5], and the associated rational approximation algorithm [5]. This is because the *2-adic complexity* of the output of the summation combiner is no more than the sum of the 2-adic complexities of the constituent sequences. Nevertheless, the summation combiner remains an interesting and difficult to analyze procedure for generating pseudorandom sequences and many basic questions concerning this combiner have never been satisfactorily addressed.

One might just as well consider the reverse procedure, and combine two binary FCSR sequences using binary addition ("XOR"). Sequences of this type are just as difficult to analyze, which perhaps explains why they have been largely ignored despite having been suggested ten years ago [5], [9].

* Partially supported by DARPA grant no. HR0011-04-1-0031.
** Partially supported by N.S.F. grant no. CCF-0514660.

G. Gong et al. (Eds.): SETA 2006, LNCS 4086, pp. 334–341, 2006.

Recall that a binary ℓ-sequence is a maximal length FCSR sequence [4] of 0's and 1's. Such a sequence is obtained whenever the *connection integer* $q \geq 3$ is a prime number such that 2 is a primitive root modulo q. The period of such an ℓ-sequence is $q-1$ and it is known to have a number of desirable statistical properties, one of which is that the number of occurrences of any given block $f = (f_0, f_1, \cdots, f_{k-1})$ of size k differs at most by one, as f ranges over all 2^k possibilities [4].

In this paper we consider "combining" two distinct ℓ-sequences $\mathbf{a} = (a_0, a_1, \cdots)$ and $\mathbf{b} = (b_0, b_1, \cdots)$ using addition modulo 2 (or "XOR", denoted $\oplus$) to obtain a sequence $\mathbf{c} = (c_0, c_1, \ldots)$ with $c_j = a_j \oplus b_j$. Suppose $\mathbf{a}$ is the ℓ-sequence that is generated by an FCSR with connection polynomial q and that $\mathbf{b}$ is the ℓ-sequence that is generated by an FCSR with connection polynomial r. We are interested in the resulting sequence $\mathbf{c}$, perhaps as a possible constituent in a stream cipher — there is experimental evidence (not reported on in this paper) that the 2-adic complexity is close to half its period.

We first show that the combined sequence $\mathbf{c}$ will have maximal period if one of the periods, say, $q-1$ is divisible by 4, if the other period, $r-1$ is not divisible by 4, and if no odd prime divides both.

We also consider the distribution properties of these sequences. That is, we bound the number of occurrences of each block of size k within such a sequence. We are able to show that by careful choice of the constituent sequences it is possible to guarantee good distribution properties for the resulting combined sequence. The precise statement is given in Theorem 3.

2 Recollections on Binary FCSR Sequences

Let $q > 2$ be a prime number, the *connection integer*. Let $s = \text{ord}_q(2)$ be the smallest integer such that $2^s \equiv 1 \pmod{q}$ or equivalently, such that q divides $2^s - 1$.

For any integer h, with $0 \leq h < q$, the base-2 expansion of the fraction h/q will be periodic with (minimal) period s. It is a *binary sequence*, meaning that its symbols are taken from the alphabet $\Sigma = \mathbf{Z}/(2)$. These sequences have been studied since the time of Gauss [3], [2] (p. 163). The reverse of this sequence is known as an *FCSR sequence* [4], [5] since it is the output sequence of a *feedback with carry shift register* with connection integer q, with cell contents taken from $\mathbf{Z}/(2)$, and with initial loading that depends on h, cf. [5]. This FCSR sequence can also be described as the 2-adic expansion of the fraction $-h/q$. To be explicit, let $0 \leq h \leq q$ and suppose the 2-adic expansion

$$-\frac{h}{q} = a_0 + a_1 2 + a_2 2^2 + \cdots \tag{1}$$

(with $a_i \in \{0, 1\}$) is periodic with period s. Then the sequence $\mathbf{a} = a_0, a_1, \cdots$ is an FCSR sequence. Its reverse is the base 2 expansion of the fraction h/q :

$$\frac{h}{q} = \frac{a_{s-1}}{2} + \frac{a_{s-2}}{2^2} + \cdots + \frac{a_0}{2^{s-1}} + \frac{a_{s-1}}{2^s} + \cdots \tag{2}$$

as may easily be seen by summing the geometric series in (1) and (2).

The period s of such a sequence satisfies $0 \leq s \leq q-1$. The period is maximal ($s = q-1$) if and only if 2 is a *primitive root* modulo q, meaning that the distinct powers 2^j modulo q, account for all the nonzero elements in $\mathbf{Z}/(q)$. In this case the base 2 expansion of h/q is known as a $1/q$ *sequence* [1] or as a Barrows-Mandelbaum codeword [7]. Its reverse, the corresponding FCSR sequence, is known as a (binary) *ℓ-sequence.*

It is also known [5] that there exists $B \in \mathbf{Z}/(q)$ (the choice of which depends on the value of h) such that

$$a_j = B2^{-j} \pmod{q} \pmod{2} \tag{3}$$

for all j, meaning that first $B2^{-j} \in \mathbf{Z}/(q)$ is computed; this number is represented as an integer between 0 and $q-1$, and it is then reduced modulo 2. The $q-1$ possible different non-zero choices of $B \in \mathbf{Z}/(q)$ give cyclic shifts of the resulting sequence $\mathbf{a}$, and this accounts for all the binary ℓ-sequences with connection integer q. The following fact was observed over a hundred years ago [2, p. 163].

Lemma 1. *Let* $\mathbf{a} = a_0, a_1, a_2, \cdots$ *be the binary ℓ-sequence corresponding to the fraction* $-h/q$ *where 2 is primitive modulo the (odd) prime* q*, and where* $0 < h < q$*. Then*

$$a_{j+\frac{q-1}{2}} \equiv q - a_j \equiv q_0 - a_j \pmod{2},$$

where $q_0 = q \pmod{2}$*. In other words, within any period of the ℓ-sequence* $\mathbf{a}$*, the second half is the complement of the first half, [6].*

Proof. Since 2 is primitive mod q, we have: $2^{q-1} \equiv 1 \pmod{q}$ hence $2^{\frac{q-1}{2}} \equiv -1 \pmod{q}$ so $2^{-\frac{q-1}{2}} \equiv -1 \pmod{q}$. It suffices to prove the lemma for any single shift of the sequence $\mathbf{a}$. Accordingly, we may take $B = 1$ in equation (3), then calculate

$$\begin{aligned} a_{j+\frac{q-1}{2}} &\equiv -2^{-j} \pmod{q} \pmod{2} \\ &\equiv (q - 2^{-j}) \pmod{q} \pmod{2} \end{aligned}$$

If $A_j \in \{1, 2, \cdots, q-1\}$ is the positive integer representation of the number $2^{-j} \pmod{q} \in \mathbf{Z}/(q)$ then $0 < q - A_j < q$ so $q - A_j$ is the positive integer representation of the number $q - 2^{-j} \pmod{q} \in \mathbf{Z}/(q)$. Therefore, reducing this equation modulo 2 gives

$$a_{j+\frac{q-1}{2}} \equiv q_0 - a_j \pmod{2}$$

where $q_0 = q \pmod{2} \in \mathbf{Z}/(2)$.

3 Period

In this section we describe a very general criterion which guarantees that the period of a sequence $\mathbf{c}$ obtained by "combining" two periodic sequences $\mathbf{a}, \mathbf{b}$ is the least common multiple of the periods of $\mathbf{a}$ and $\mathbf{b}$. It would surprise us to find that this theorem is unknown, but we are not aware of its having appeared in print.

Let Σ be an alphabet (i.e., a finite set). Let $\odot$ be a binary operation on Σ. That is, $\odot : \Sigma \times \Sigma \to \Sigma$. We write $a \odot b$ for the value of $\odot$ at (a, b).

Definition 1. *The operation $\odot$ is* cancellative *if for all $a, b, c \in \Sigma$, if $a \odot b = a \odot c$, then $b = c$.*

Theorem 1. *Let $\mathbf{a} = (a_0, a_1, \cdots)$ be a periodic sequence of (minimal) period n with each $a_i \in \Sigma$, and let $\mathbf{b} = (b_0, b_1, \cdots)$ be a periodic sequence of (minimal) period m with each $b_i \in \Sigma$. Let $\mathbf{c} = (c_0, c_1, \cdots)$ be the sequence with $c_i = a_i \odot b_i$ for each i. Suppose that for every prime r, the largest power of r that divides n is not equal to the largest power of r that divides m. Then $\mathbf{c}$ is periodic and the period of $\mathbf{c}$ is the least common multiple of n and m.*

Proof. It is straightforward to see that $\mathbf{c}$ is periodic and its (least) period divides the least common multiple of n and m. Let t denote the (least) period of $\mathbf{c}$. Suppose that $t < \mathrm{lcm}(n, m)$. Then there is some prime r so that t divides $\mathrm{lcm}(n, m)/r$. In particular, $\mathbf{c}$ has $\mathrm{lcm}(n, m)/r$ as a period.

Suppose that the largest power of r dividing n is r^e and the largest power of r dividing m is r^f. By symmetry we may assume that $e < f$. Thus the largest power of r dividing $\mathrm{lcm}(n, m)/r$ is r^{f-1}, so n divides $\mathrm{lcm}(n, m)/r$ and m does not divide $\mathrm{lcm}(n, m)/r$. For every i we have

$$\begin{aligned} a_i \odot b_i &= c_i \\ &= c_{i+\mathrm{lcm}(n,m)/r} \\ &= a_{i+\mathrm{lcm}(n,m)/r} \odot b_{i+\mathrm{lcm}(n,m)/r} \\ &= a_i \odot b_{i+\mathrm{lcm}(n,m)/r}. \end{aligned}$$

By the cancellative property of $\odot$, it follows that for every i,

$$b_i = b_{i+\mathrm{lcm}(n,m)/r}.$$

But this contradicts the fact that $\mathrm{lcm}(n, m)/r$ is not a multiple of the minimal period of $\mathbf{b}$, and thus proves the theorem. □

Corollary 1. *Let $\mathbf{a} = (a_0, a_1, \cdots)$, $\mathbf{b} = (b_0, b_1, \cdots)$ be binary ℓ-sequences with connection integers q and r respectively. Suppose that 4 divides $q - 1$ but does not divide $r - 1$ and that no odd prime divides both $q - 1$ and $r - 1$ (so that $\gcd(q - 1, r - 1) = 2$). Then the sequence $\mathbf{c} = \mathbf{a} \oplus \mathbf{b} \pmod{p}$ obtained by taking the termwise sum, modulo 2, of $\mathbf{a}$ and $\mathbf{b}$ has period $(q - 1)(r - 1)/2$.*

4 Distributions

By an *occurrence* of a block $e = (e_0, \cdots, e_{k-1})$ in a sequence $\mathbf{a}$ of period n we mean an index i, $0 \le i < n$ so that $a_i = e_0, a_{i+1} = e_1, \cdots, a_{i+k-1} = e_{k-1}$. Recall the following result of [1] (Theorem 1). See also [5].

Theorem 2. *Let $\mathbf{a} = (a_0, a_1, \cdots)$ be a binary ℓ-sequence with connection integer q. Then the number of occurrences of any block $e = (e_0, e_2, \cdots, e_{k-1})$ of size k in $\mathbf{a}$ varies at most by 1 as the block e varies over all 2^k possibilities. That is, there is an integer w so that every block of length k occurs either w times or $w+1$ times in $\mathbf{a}$. The number of blocks of length k that occur $w+1$ times is $q-1 \pmod{2^k}$, and the number of blocks of length k that occur w times is $2^k - (q - 1 \pmod{2^k})$.*

Proof. The first statement is explicitly given in [1] Theorem 1 (for the corresponding $1/q$ sequence). The second statement follows immediately: let Q be the number of blocks of length k that occur $w+1$ times in $\mathbf{a}$. Then

$$\begin{aligned} q - 1 &= Q(w+1) + (2^k - Q)w \\ &= 2^k w + Q. \end{aligned}$$

It follows that $Q = q - 1 \pmod{2^k}$, as claimed. □

Throughout the remainder of this section we fix prime numbers q and r such that 2 is a primitive root modulo q and also modulo r. Let $\mathbf{a} = (a_0, a_1, \cdots)$ and $\mathbf{b} = (b_0, b_1, \cdots)$ be binary ℓ-sequences with connection integers q and r respectively, (and thus periods $q-1$ and $r-1$ respectively). We will further assume that 4 divides $q-1$, and that 4 does not divide $r-1$, so that $\gcd(q-1, r-1) = 2$. Let $\mathbf{c} = \mathbf{a} \oplus \mathbf{b}$ be the sequence obtained as sum, modulo 2 (or the exclusive or) of these two sequences: $c_i = a_i \oplus b_i \pmod 2$. According to Corollary 1, the period of the sequence $\mathbf{c}$ is maximal, and is equal to $(q-1)(r-1)/2$.

Lemma 2. *Let $0 \le i < q-1$ and $0 \le j < r-1$. Then in a full period of $\mathbf{c}$, a_i is combined with b_j if and only if j and i have the same parity. That is, there are integers k and l with $i + k(q-1) = j + l(r-1)$ if and only if $i \equiv j \pmod 2$.*

Proof. This is an application of the Euclidean theorem. The integer 2 is the greatest common divisor $q-1$ and $r-1$. The integers i and j have the same parity if and only if $i - j$ is a multiple of 2, which by the Euclidean theorem is equivalent to the existence of k and l. □

Lemma 3. *Within any single period, the second half of the sequence $\mathbf{c} = \mathbf{a} \oplus \mathbf{b}$ is the complement of the first half.*

Proof. The second half of a period of the sequence $\mathbf{a}$ is the complement of the first half and the same is true for the sequence $\mathbf{b}$. Let $T = (q-1)(r-1)/2$ be the period of $\mathbf{c}$. Then

$$\frac{T}{2} = \frac{q-1}{2} \cdot \frac{r-1}{2} = \frac{q-1}{2} \cdot \text{ odd } = \frac{r-1}{2} \cdot \text{ even.}$$

Therefore $a_{j+T/2} = \overline{a}_j$ and $b_{j+T/2} = b_j$ whenever $0 \le j < T/2$. Here, $\overline{a}_j$ denotes the complement of $a_j \in \mathbb{Z}/(2)$. Hence, for these values of j,

$$c_{j+T/2} = \overline{a}_j \oplus b_j = \overline{c}_j$$

which proves the lemma. □

Theorem 3. *Fix $k \ge 0$. Let $Q = q - 1 \pmod{2^k}$ and let $R = r - 1 \pmod{2^k}$. Define*

$$s = \frac{\min(Q, R) - \max(0, Q + R - 2^k)}{2}.$$

Then the number of occurrences of a block $e = (e_0, e_2, \cdots, e_{k-1})$ of size k in the sequence $\mathbf{c} = \mathbf{a} \oplus \mathbf{b}$ varies at most by s as the block e varies over all 2^k possibilities.

Proof. Let $\mathbf{b}^{(1)} = (b_1, b_2, \cdots)$ be the shift of the sequence $\mathbf{b}$ by one. Then we claim that the sequence

$$\mathbf{d} = \mathbf{a} \oplus \mathbf{b}^{(1)}$$

is a shift of the sequence $\mathbf{c} = \mathbf{a} \oplus \mathbf{b}$.

To prove this claim, note that because $(r-3)/2$ is even and $\gcd(r-1, q-1) = 2$, there exist integers ℓ and m such that

$$\frac{r-3}{2} = m(q-1) - \ell(r-1).$$

That is,

$$m(q-1) = \frac{r-3}{2} + \ell(r-1).$$

Therefore, for all j,

$$\begin{aligned} d_{j+m(q-1)} &= a_{j+m(q-1)} \oplus b^{(1)}_{j+\frac{r-3}{2}+\ell(r-1)} \\ &= a_{j+m(q-1)} \oplus b_{j+\frac{r-1}{2}+\ell(r-1)} \\ &= a_j \oplus b_{j+\frac{r-1}{2}} \\ &= a_j \oplus \overline{b}_j \\ &= \overline{c}_j. \end{aligned}$$

since $\mathbf{d}$ is obtained by shifting $\mathbf{b}$ by one before adding it to $\mathbf{a}$. By Lemma 3 the sequence $\mathbf{c}$ is a shift of its complement, so $\mathbf{d}$ is also a shift of $\mathbf{c}$.

Therefore, if we count the occurrences of each block of a fixed length k in both $\mathbf{c}$ and $\mathbf{d}$, then for each block we will have exactly twice the number of occurrences of that block in $\mathbf{c}$. However, in the construction of these two sequences, each occurrence of each block of length k in $\mathbf{a}$ is matched with each occurrence of each block of length k in $\mathbf{b}$. Thus to count the occurrences of a block e of length k in $\mathbf{c}$, we want to count the number of pairs (i, j) where i is an occurrence of a block f in $\mathbf{a}$, j is an occurrence of a block g in $\mathbf{b}$, and $f \oplus g = e$. That is, $g = f \oplus e$. Thus we sum over all blocks f of length k the number of occurrences of f in $\mathbf{a}$ times the number of occurrences of $f \oplus e$ in $\mathbf{b}$.

Let w denote the minimum number of occurrences of a block of length k in $\mathbf{a}$, so that by Theorem 2 every possible block of length k occurs either w or $w+1$ times. Similarly, let z denote the minimum number of occurrences of a block of length k in $\mathbf{b}$, so that every possible block of length k occurs either z or $z+1$ times. For a fixed block e of length k, as we have seen, the occurrences of a block f of length k in $\mathbf{a}$ are matched with the occurrences of block $e \oplus f$ in $\mathbf{b}$. There are four possibilities:

1. f occurs w times in $\mathbf{a}$ and $e \oplus f$ occurs z times in $\mathbf{b}$;
2. f occurs $w+1$ times in $\mathbf{a}$ and $e \oplus f$ occurs z times in $\mathbf{b}$;
3. f occurs w times in $\mathbf{a}$ and $e \oplus f$ occurs $z+1$ times in $\mathbf{b}$;
4. f occurs $w+1$ times in $\mathbf{a}$ and $e \oplus f$ occurs $z+1$ times in $\mathbf{b}$.

Let Y_i denote the number of fs in case i above, $i = 1, 2, 3, 4$. Then the number of occurrences of e in $\mathbf{c}$ is

$$N_e = \frac{wzY_1 + (w+1)zY_2 + w(z+1)Y_3 + (w+1)(z+1)Y_4}{2}. \tag{4}$$

We have $Y_2 + Y_4 = Q$ since cases (2) and (4) together account for all the blocks f that occur $w+1$ times in $\mathbf{a}$. Similarly, $Y_3 + Y_4 = R$, and $Y_1 + Y_2 + Y_3 + Y_4 = 2^k$. Thus $Y_1 = 2^k - Q - R + Y_4$, $Y_2 = Q - Y_4$, and $Y_3 = R - Y_4$. Therefore, substituting these values into (4) gives

$$N_e = \frac{wz2^k + zQ + wR + Y_4}{2}.$$

It follows that the possible variation in N_e is one half the possible variation in Y_4. By the definition of Y_4 we have $Y_4 \le \min(Q, R)$ and $Y_4 \ge 0$. Also, $Y_2 \le 2^k - R$, so that $Y_4 = Q - Y_2 \ge Q + R - 2^k$. It follows that the possible variation in Y_4 for various e is at most

$$\min(Q, R) - \max(0, Q + R - 2^k).$$

The theorem follows immediately from this. □

Corollary 2. *The sequence* $\mathbf{c}$ *is balanced and the distribution of consecutive pairs in* $\mathbf{c}$ *is uniform.*

Proof. Balance follows from the case of Theorem 3 when $k = 1$. The uniform distribution of pairs follows from Theorem 3 with $k = 2$. In both cases the bound s in the theorem equals zero. □

It follows from Theorem 3 that the sequence $\mathbf{c} = \mathbf{a} \oplus \mathbf{b}$ is highly uniform if $\min(Q, R) - \max(0, Q + R - 2^k)$ is small for all small k.

A small amount of experimental evidence indicates that this bound is very close to optimal, in the sense that there are blocks of length k whose numbers of occurrences differ by almost $\min(Q, R) - \max(0, Q + R - 2^k)/2$. Further experimentation is planned.

5 Conclusions

It is apparent from these results how to look for pairs of ℓ-sequences whose exclusive ors have large period and for small k have near uniform distribution of blocks of length k. This situation is an improvement over the situations for many sequence generators that have been proposed previously as components of stream ciphers – in many cases the period has not even been computed. On the basis of experimentation we believe that our exclusive or sequences have other good properties such as large 2-adic complexity. Before they are used as components in stream cipher construction, however, we need to test them with the NIST test suite and examine their resistance to other attacks such as correlation attacks and algebraic attacks.

References

1. L. Blum, M. Blum, and M. Shub, A simple unpredictable pseudorandom number generator, SIAM J. Comput. **15** (1986), 364-383.
2. L. E. Dickson, *History of the Theory of Numbers*, vol. 1, Chelsea, New York, 1950.
3. C. F. Gauss, *Disquisitiones Arithmeticae*, Leipzig, 1801, English translation, Yale, New haven, 1966.
4. A. Klapper and M. Goresky, 2-adic shift registers, in *Fast Software Encryption: Proceedings of 1993 Cambridge Algorithms Workshop*, Lecture Notes in Computer Science **809**, Springer Verlag, 1994, 174-178.
5. A. Klapper and M. Goresky, Feedback Shift Registers, Combiners with Memory, and 2-Adic Span, *Journal of Cryptology* **10** (1997), 111-147.
6. A. Klapper and M. Goresky, Arithmetic crosscorrelation of feedback with carry shift registers, *IEEE Trans. Info. Theory* **43** (1997), 1342-1345.
7. D. Mandelbaum, Arithmetic codes with large distance, *IEEE Trans. Info. Theory* **IT-13** (1967), 237-242.
8. R. Rueppel, *Analysis and Design of Stream Ciphers*. Springer Verlag, New York, 1986.
9. B. Schneier, *Applied Cryptography*. John Wiley & Sons, New York, 1996.

Generalized Bounds on Partial Aperiodic Correlation of Complex Roots of Unity Sequences*

Lifang Feng and Pingzhi Fan

Institute of Mobile Communications, Southwest Jiaotong University, Chengdu, 610031, China
lf_feng03@yahoo.com.cn, p.fan@ieee.org

Abstract. Partial correlation properties of sets of sequences are important in CDMA system as well as in ranging, channel estimation and synchronization applications. In general, it is desirable to have sequence sets with small absolute values of partial correlations. In this paper, generalized lower bounds on partial aperiodic correlation of complex roots of unity sequence sets with respect to family size, sequence length, subsequence length, maximum partial aperiodic autocorrelation sidelobe, maximum partial aperiodic crosscorrelation value and the zero or low correlation zone are derived. It is shown that the previous aperiodic sequence bounds such as Sarwate bounds, Welch bounds, Levenshtein bounds, Tang-Fan bounds and Peng-Fan bounds can be considered as special cases of the new partial aperiodic bounds derived.

1 Introduction

Sets of sequences with good correlation properties are important in code-division multiple access (CDMA), spread-spectrum communications, as well as in ranging and synchronization applications. Traditionally it is the periodic or aperiodic auto- and crosscorrelation functions that have received most attention [1]-[8]. Partial correlations of sequences (where correlations are computed over only subsequences of sequences) are much less well understood, but sequence sets having low absolute values of partial correlation are important in certain types of communications systems. In CDMA systems where many data bits are spread by each copy of a user's spreading sequences, it was shown in [9][10] how the multiple-access capability of CDMA systems in which the period of the signature sequences was much larger than the number of chips per data and multiple data bits are spread by each sequence can be related to the mean square value of partial correlation for sequence sets. In [11][12], a long sequence is used for

* This work was supported by the National Science Foundation of China (NSFC) (No.90604035 and No. 60472089) and the Foundation for the Author of National Excellent Doctoral Dissertation of PR China (FANEDD) No.200341. The authors would also like to thank Prof. Daiyuan Peng and Prof. Xiaohu Tang for their useful discussions and comments.

G. Gong et al. (Eds.): SETA 2006, LNCS 4086, pp. 342–350, 2006.

synchronization, but the correlations are computed over only a short subsequence of that sequence. It was shown that the performance of acquisition such as mean acquisition time can be improved.

In 1998, Kenneth G. Paterson and Paul J. G. Lothian [13] derived a lower partial periodic bound based on Welch's technique [1]. But Paterson-Lothian bounds does not apply to low correlation zone (LCZ) sequences or generalized orthogonal (GO) sequences [14]-[19], which can be employed in quasi-synchronous CDMA (QS-CDMA) to eliminate the multiple access interference and multipath interference. In 2000, Tang and Fan established bounds on the periodic and aperiodic correlations of GO sequences based on Welch's technique [1][2]. Peng and Fan derived generalized Sarwate bounds on the periodic and aperiodic correlation for binary and complex roots of unity sequences [3][4][6]-[8] including GO sequences and pseudonoise sequences based on Levenshtein's technique [20]. It was shown that Peng-Fan bounds can include all the previous periodic and aperiodic bounds as special cases, it is because that the pseudonoise sequences are special cases of GO sequences, and the binary sequences are special cases of complex roots of unity sequences. However, these bounds apply only to periodic and aperiodic correlation over sequence length, and cannot cover partial aperiodic correlation bounds of GO sequences. As far as the authors are aware, there is no research work on the theoretical limits for GO sequences among the sequence length n, subsequence length l, sequence set size M, maximum partial aperiodic autocorrelation sidelobe value APl_A, maximum partial aperiodic crosscorrelation value APl_C, and low correlation zone L_{CZ}.

In this paper, our attention will be paid only to the partial aperiodic correlation bounds and complex roots of unity sequences, not to the partial periodic correlation bounds which have been discussed by the authors elsewhere. It will be shown in the following sections that all the previous aperiodic sequence bounds such as Sarwate bounds, Welch bounds, Levenshtein bounds, Tang-Fan bounds and Peng-Fan bounds can be considered as special cases of the new partial aperiodic bounds derived.

2 Preliminaries

Let q be an arbitrary, positive, integer greater than 1, Z_q={0,1,...,q-1}, $i = \sqrt{-1}$, ω=exp[$i2\pi/q$], $E = \{1, \omega^1, \ldots, \omega^{q-1}\}$. Then $x=\{x_0,x_1,\ldots,x_{n-1}\} \in E^n$ is called a complex roots of unity sequence of length n, l denotes its subsequence length. When, q=2, then the complex roots of unity sequence becomes the binary sequence. For any two such sequences $x=\{x_0,x_1,\ldots,x_{n-1}\}$ and $y = (y_0, y_1, \cdots y_{n-1})$, the partial aperiodic correlation functions $APl(x, y; d)$ of x and y are defined as follows:

$$APl(x,y;d) = \begin{cases} \sum\limits_{i=1}^{l} x_i y_{i+d}^*, d = 0, 1, \cdots, n-l \\ \sum\limits_{i=1}^{n-d} x_i y_{i+d}^*, d = n-l+1, \cdots, n-1 \end{cases} \quad (1)$$

where y^* denotes the complex conjugate of y.

For $C \subseteq E^n$, $M = |C|$, the aperiodic low correlation zone L_{CZ}, the aperiodic low autocorrelation zone L_{ACZ} and the aperiodic low crosscorrelation zone L_{CCZ} of $C$are defined, respectively, as follows:

$$L_{CZ} = \min\{L_{ACZ}, L_{CCZ}\}$$

$$L_{ACZ} = \max\{T|\,|APl(x,x;d)| \leq APl_A, \forall x \in C, 0 < |d| \leq T\}$$

and

$$L_{CCZ} = \max\{T|\,|APl(x,y;d)| \leq APl_C, \forall x,y \in C, x \neq y, |d| \leq T\}$$

A sequence set C with L_{CZ} >0 is called aperiodic low correlation zone (LCZ) set if $APl_M \geq 0$ where $APl_M = max\{APl_A, APl_C\}$, and APl_A denotes maximum partial aperiodic autocorrelation sidelobe, APl_C denotes maximum partial aperiodic crosscorrelation value respectively.

For any sequence $x = \{x_0, x_1, \ldots, x_{n-1}\} \in E^n$, let T denote the operator which shifts sequence cyclically to the right by one place, that is $Tx = (x_{n-1}, x_0, \ldots x_{n-2})$, and let $T_0 x = x$, $T_{i+1}x = T(T_i x)$ for positive integer $i \geq 1$. Given any positive integer k, a sequence $x0^k = (x_0, x_1, \ldots, x_{n-1}, 0, 0, \ldots, 0)$ is obtained by appending k zeros to the right-hand of x.

Throughout this paper, it is assumed that the partial inner product of $x0^{L_{CZ}}$ and $y0^{L_{CZ}}$ is given by

$$\langle T_s(x0^{L_{CZ}}), T_t(y0^{L_{CZ}})\rangle_l = \begin{cases} \sum_{i=0}^{s+l} x0_i^{L_{CZ}}(y0_{i+s-t}^{L_{CZ}})^*, s \geq t \\ \sum_{i=0}^{t+l} y0_i^{L_{CZ}}(x0_{i+t-s}^{L_{CZ}})^*, s < t \end{cases} \quad (2)$$

$w_i \geq 0$, $i = 0, 1, \cdots, L_{CZ}$, $\sum_{i=0}^{L_{CZ}} w_i = 1$ and $w = (w_0, w_1, \cdots, w_{L_{CZ}})$.

For $x \in E^n, A, B \subseteq E^n, |A||B| > 0$, let $W(x) = \{T_i x | i = 0, 1, \cdots, L_{CZ}\}$, $W(A) = \cup_{x \in A} W(x)$ and

$$F(A,B) := \frac{1}{|A|\,|B|} \sum_{x \in A} \sum_{y \in B} \sum_{s=0}^{L_{CZ}} \sum_{t=0}^{L_{CZ}} \left|\langle T_s(x0^{L_{CZ}}), T_t(y0^{L_{CZ}})\rangle\right|^2 w_s w_t \quad (3)$$

Lemma 1. For any sequence $x \in E^n$, and any integer $d = 0, 1, \ldots, n-1$, we have

$$\sum_{y \in E^n} |APl(x,y;d)|^2 = \begin{cases} lq^n, & 0 \leq d \leq n-l \\ (n-d)q^n, & n-l < d \leq n-1 \end{cases} \quad (4)$$

Lemma 2. For any $x \in E^n$, $A \subseteq E^n$

$$\begin{aligned} F(\{x\}, E^n) &= F(A, E^n) = F(E^n, E^n) \\ &= \sum_{\substack{|s-t| \leq n-l \\ 0 \leq s,t \leq L_{CZ}}} l w_s w_t + \sum_{\substack{|s-t| > n-l \\ 0 \leq s,t \leq L_{CZ}}} (n - |s-t|) w_s w_t \end{aligned} \quad (5)$$

Lemma 3. For $C \subseteq E^n$

$$F(C,C) \geq F(E^n,E^n) = \sum_{\substack{|s-t|\leq n-l \\ 0\leq s,t\leq L_{CZ}}} l w_s w_t + \sum_{\substack{|s-t|>n-l \\ 0\leq s,t\leq L_{CZ}}} (n-|s-t|) w_s w_t \quad (6)$$

3 Lower Bounds on Partial Aperiodic Correlation of LCZ Complex Roots-of-Unity Sequences

Let C be a set of M complex roots-of unity sequences of length n, APl_A denotes maximum partial aperiodic autocorrelation sidelobe, APl_C, denotes maximum partial aperiodic crosscorrelation value, $APl_M = max\{APl_A, APl_C\}$, L_{CZ} denotes low correlation zone, l denotes subsequence length. Then we can derive the partial aperiodic correlation bounds of LCZ complex roots-of-unity sequences in this section.

Theorem 1. *For any $C \subseteq E^n$, $M = |C| > 0$, we have*

$$\frac{1}{M}(1-\sum_{s=0}^{L_{CZ}} w_s^2) APl_A^2 + (1-\frac{1}{M}) APl_C^2 \geq R(n,l,L_{CZ}) - \frac{l^2}{M}\sum_{s=0}^{L_{CZ}} w_s^2 \quad (7)$$

$$(1-\frac{1}{M}\sum_{s=0}^{L_{CZ}} w_s^2) APl_M^2 \geq R(n,l,L_{CZ}) - \frac{l^2}{M}\sum_{s=0}^{L_{CZ}} w_s^2 \quad (8)$$

where

$$R(n,l,L_{CZ}) = \sum_{\substack{|s-t|\leq n-l \\ 0\leq s,t\leq L_{CZ}}} l w_s w_t + \sum_{\substack{|s-t|>n-l \\ 0\leq s,t\leq L_{CZ}}} (n-|s-t|) w_s w_t \quad (9)$$

Proof. By lemma 2 and 3, we have

$$M^2 F(C,C) = \sum_{x\in C}\sum_{s=0}^{L_{CZ}} \left|\langle T_s(x0^{L_{CZ}}), T_s(x0^{L_{CZ}})\rangle_l\right|^2 w_s w_s$$

$$+ \sum_{x\in C}\sum_{\substack{s,t=0 \\ s\neq t}}^{L_{CZ}} \left|\langle T_s(x0^{L_{CZ}}), T_t(x0^{L_{CZ}})\rangle_l\right|^2 w_s w_t$$

$$+ \sum_{\substack{x,y\in C \\ x\neq y}}\sum_{s,t=0}^{L_{CZ}} \left|\langle T_s(x0^{L_{CZ}}), T_t(y0^{L_{CZ}})\rangle_l\right|^2 w_s w_t$$

$$\leq Ml^2 \sum_{s=0}^{L_{CZ}} w_s^2 + M APl_A^2 (1-\sum_{s=0}^{L_{CZ}} w_s^2) + M(M-1) APl_C^2$$

where

$$1 = \sum_{s,t=0}^{L_{CZ}} w_s w_t = \sum_{s=0}^{L_{CZ}} w_s^2 + \sum_{s,t=0;s\neq t}^{L_{CZ}} w_s w_t$$

Noting $APl_M = max\{APl_A, APl_C\}$, the inequality (8) follows immediately from inequality (7). Q.E.D.

Based on Theorem 1, we can derive some useful results as follows:

Corollary 1. For $C \subseteq E^n$, any integer $0 \le L \le L_{CZ}$, we have

$$(1 - \frac{1}{L+1})APl_A^2 + (M-1)APl_C^2$$
$$\ge \frac{M}{L+1}\left[2nl - l^2 - n^2 + n + nL - \frac{1}{3}(L^2+2L)\right] - \frac{l^2}{L+1} \tag{10}$$

$$APl_M^2 \ge \frac{M\left[3(2nl - l^2 - n^2 + n + nL) - (L^2+2L)\right] - 3l^2}{3(ML+M-1)} \tag{11}$$

Proof. Put the weight vector $w = (w_0, w_1, \cdots, w_{L_{CZ}})$, where

$$w_s = \begin{cases} \frac{1}{L+1}, 0 \le s \le L \\ 0, L < s \le L_{CZ} \end{cases}$$

then $\sum_{s=0}^{L_{CZ}} w_s^2 = \frac{1}{L+1}$, and $R(n, l, L_{CZ})$can be partitioned into 4 parts:

1. $s \ge t, s - t \le n - l$, then $t \le s \le n - l + t$

$$R(n, l, L_{CZ})_1 = \frac{1}{(L+1)^2}\sum_{t=0}^{L} l(n-l+1) \tag{12}$$

2. $s \ge t, s - t > n - l$, then $n - l + t < s \le L$

$$R(n, l, L_{CZ})_2 = \frac{1}{(L+1)^2}\sum_{t=0}^{L}\left[\frac{1}{2}(L-n+l-t)(n+t-L+l-1)\right] \tag{13}$$

3. $s < t, t - s \le n - l$, then $t - n + l \le s < t$

$$R(n, l, L_{CZ})_3 = \frac{1}{(L+1)^2}\sum_{t=0}^{L} l(n-l) \tag{14}$$

4. $s < t, t - s > n - l$, then $0 \le s < t - n + l$

$$R(n, l, L_{CZ})_4 = \frac{1}{(L+1)^2}\sum_{t=0}^{L}\left[\frac{1}{2}(t-n+l)(n-t+l-1)\right] \tag{15}$$

By Eqn.(12), (13), (14) and (15), we have

$$R(n, l, L_{CZ}) = \sum_{\substack{|s-t| \le n-l \\ 0 \le s,t \le L_{CZ}}} l w_s w_t + \sum_{\substack{|s-t| > n-l \\ 0 \le s,t \le L_{CZ}}} (n - |s-t|) w_s w_t$$
$$= \frac{1}{(L+1)^2}\sum_{t=0}^{L}\left[\sum_{s=0}^{t-n+l-1}[n-(t-s)] + \sum_{t-n+l}^{t-1} l + \sum_{t}^{n-l+t} l + \sum_{n-l+t+1}^{L}[n-(s-t)]\right]$$
$$= \frac{1}{L+1}\left[2nl - l^2 - n^2 + n + nL - \frac{1}{3}(L^2+2L)\right]$$

Noting $APl_M = max\{APl_A, APl_C\}$, the inequality (11) follows immediately from (10). Q.E.D.

Corollary 2. For $C \subseteq E^n$, any integer $0 \leq L \leq L_{CZ}$, we have

$$2(1-4^{-L})APl_A^2 + 3(M-1)APl_C^2 \geq 3MR(n,l,L_{CZ}) - (1+2\times 4^{-L})l^2 \quad (16)$$

$$APl_M^2 \geq \frac{3MR(n,l,L_{CZ}) - (1+2\times 4^{-L})l^2}{2(1-4^{-L}) + 3(M-1)} \quad (17)$$

$$R(n,l,L_{CZ}) = n + 2^{-L}(2L-2n+2l) - 4\times 2^{l-n-L} + 4^{-L}(l-n+2L+4) - \frac{1}{3}(1-4^{-L})(2^{1+l-n} + 2^{1+n-l}) \quad (18)$$

Proof. Let the weight vector $w = (w_0, w_1, \cdots, w_{L_{CZ}})$, where

$$w_s = \begin{cases} 2^{-L}, s=0 \\ 2^{-s}, 1 \leq s \leq L \\ 0, L < s \leq L_{CZ} \end{cases}$$

Thus, we have,

$$\sum_{s=0}^{L_{CZ}} w_s^2 = \frac{1}{3} + \frac{2}{3}4^{-L}$$

where $R(n,l,L_{CZ})$ can be partitioned into 4 parts:

1. $s=0, t=0$:
$$R(n,l,L_{CZ})_1 = l4^{-L} \quad (19)$$
2. $s=0, t\neq 0$:
$$R(n,l,L_{CZ})_2 = l2^{-L} - 2^{-L-n+l+1} + (L+2-n)4^{-L} \quad (20)$$
3. $s\neq 0, t=0$:
$$R(n,l,L_{CZ})_3 = l2^{-L} - 2^{-L-n+l+1} + (L+2-n)4^{-L} \quad (21)$$
4. $s\neq 0, t\neq 0$:
$$R(n,l,L_{CZ})_4 = n + 2^{-L}(2L-2n+2l) - 4\times 2^{l-n-L} + 4^{-L}(l-n+2L+4) - \frac{1}{3}(1-4^{-L})(2^{l-n+1} + 2^{n-l+1}) \quad (22)$$

By Eqn.(19), (20), (21) and (22), the Eqn. (18) can be derived. By Theorem 1 and (18), the inequality (16) and (17) can be proved. Q.E.D.

4 Lower Bounds on Partial Aperiodic Correlation of Normal Complex Roots-of-Unity Sequences

Because the normal correlation operation can be considered as a special case of the correlation operation for LCZ sequences for $L_{CZ} = n - 1$, based on the general results presented in Section 3, lower bounds for normal complex roots-of-unity sequences can also be established.

Let $l = n$ in corollary 1 and 2, we have the following lower bounds on aperiodic correlation of LCZ complex roots-of-unity sequences obtained by Peng and Fan [6].

$$3LAPl_A^2 + 3(L+1)(M-1)APl_C^2 \geq 3Mn + 3MnL - ML^2 - 2ML - 3n^2 \quad (23)$$

$$APl_M^2 \geq \frac{3Mn + 3MnL - ML^2 - 2ML - 3n^2}{3(ML + M - 1)} \quad (24)$$

$$2(4^L - 1)APl_A^2 + 3(M-1)4^L APl_C^2$$
$$\geq (3Mn - n^2 - 4M)4^L + 6(L-2)2^L M + 6ML + 16M - 2n^2 \quad (25)$$

The bounds (23) and (24) follow from corollary 1, and bound (25) follows from corollary 2, respectively. Because the complex roots-of-unity sequences include the binary sequences as special cases, and generalized orthogonality (GO) sequences include pseudonoise sequences as special cases, all the previous aperiodic sequence bounds such as Sarwate bounds, Welch bounds, Levenshtein bounds, Tang-Fan bounds and Peng-Fan bounds can be considered as special cases of the above results.

Besides, we have the following lower bounds on partial aperiodic correlation of normal complex roots-of-unity sequences:

Corollary 3. For any $C \subseteq E^n$, $M = |C| > 0$, we have

$$(n-1)APl_A^2 + n(M-1)APl_C^2 \geq 2Mnl - Ml^2 - \frac{1}{3}(Mn^2 - M) - l^2 \quad (26)$$

$$APl_M^2 \geq \frac{6Mnl - 3Ml^2 - Mn^2 + M - 3l^2}{3(Mn-1)} \quad (27)$$

Corollary 4. For any $C \subseteq E^n$, $M = |C| > 0$, we have

$$2(1-4^{-n+1})APl_A^2 + 3(M-1)APl_C^2 \geq 3MR(n, n-1) - (1 + 2 \times 4^{-n+1})n^2 \quad (28)$$

$$APl_M^2 \geq \frac{3MR(n, n-1) - (1 + 2 \times 4^{-n+1})n^2}{2(1 - 4^{-n+1}) + 3(M-1)} \quad (29)$$

$$R(n, n-1) = n + 2^{-n+2}(n-1) - 4 \times 2^{-n+1} + 4^{-n+1}(2n+2) - \frac{4}{3}(1 - 4^{-n+1}) \quad (30)$$

Based on the lower bounds on partial aperiodic correlation of normal complex roots-of-unity sequences, let $l = n$ in corollary 3 and 4 then we have the lower bounds on aperiodic correlation of normal complex roots-of-unity sequences obtained by Peng and Fan[6].

$$\frac{3(n-1)}{2Mn^2+M-3n^2}APl_A^2+\frac{3n(M-1)}{2Mn^2+M-3n^2}APl_C^2\geq 1 \tag{31}$$

$$APl_M^2\geq\frac{2Mn^2+M-3n^2}{3(Mn-1)} \tag{32}$$

$$\begin{aligned}&2(4^{n-1}-1)APl_A^2+3(M-1)4^{n-1}APl_C^2\\&\geq(3Mn-n^2-4M)4^{n-1}+3(n-3)M2^n+(6Mn-2n^2+10M)\end{aligned} \tag{33}$$

The bounds (31) and (32) follow from corollary 3, and bound (33) follows from corollary 4 respectively. Because the complex roots-of-unity sequences include the binary sequences as special cases, the lower bounds derived here are appropriate to the normal pseudonoise sequences such as m-sequence, Kassami sequence, Gold sequence and so on.

5 Conclusion

Sequence sets having low absolute values of nontrivial partial correlations are important in CDMA system as well as in ranging, channel estimation and synchronization applications. The lower bounds on the partial correlation are important criteria on the selection and design of good sequence sets in CDMA systems and other applications. In this paper, generalized lower bounds on the partial aperiodic correlations of both GO sequences and normal complex roots-of-unity sequences are established, which discloses theoretical relationship among the sequence length, subsequence length, sequence set size, maximum partial aperiodic autocorrelation sidelobe value, maximum partial aperiodic crosscorrelation value and low correlation zone. Because the complex roots-of-unity sequences include binary sequences as special cases, and GO sequences include pseudonoise sequences as special cases, all the previous aperiodic correlation bounds such as Peng-Fan bounds, Sarwate bounds, Welch bounds and Levenshtein bounds can be considered as special cases of the bounds derived in this paper.

References

1. L.R.Welch: Lower bounds on the maximum cross-correlation of signals. IEEE Trans. Information Theory, Vol.20 (1974) 397–399
2. X.H. Tang, P.Z. Fan, S. Matsufuji: Lower bounds on maximum correlation of sequence set with low or zero correlation zone, ELECTRONICS LETTER, Vol.36, No.6 (2000) 551–552
3. D.Y. Peng, P.Z. Fan: Generalized Sarwate bounds on the periodic correlation of complex roots of unity sequences. Proc. of IEEE Int. Symp. on Personal, Indoor and Mobile Radio Comms (PIMRC'2003), Beijing, China, IEEE Press (2003) 449–452

4. D.Y. Peng, P.Z. Fan: Generalized Sarwate bounds on periodic autocorrelations and cross-correlations of binary sequences. IEE Electronics Letters, Vol.38, No.24 (2002) 1521–1523
5. Sarwate, D.V.: Bounds on crosscorrelation and auto- correlation of sequences. IEEE Trans. Inform. Theory, vol.25 (1979) 720–724
6. Daiyuan Peng, Pingzhi Fan: Generalised Sarwate bounds on the aperiodic correlation of sequences over complex roots of unity. IEE Proc.-Commun., Vol.151, No.4 (2004) 375–382
7. Daiyuan Peng, Pingzhi Fan, Naoki Suehiro: Bounds on aperiodic autocorrelation and crosscorrelation of binary LCZ/ZCZ sequences. IEICE Trans. on Fundamentals (2005) 1–9
8. Daiyuan Peng, Pingzhi Fan: Bounds on aperiodic auto- and cross-correlation of binary sequences with low or zero correlation zone. IEEE 0-7803-7840-7/03 (2003) 882–886
9. M. B. Pursley: On the mean-square partial correlation of periodic sequences, in Proc. Conf. Information Science and Systems (Johns Hopkins Univ., Baltimore MD, 1979), 377–379.
10. M. B. Pursley, D.V. Sarwate, T. U. Basar: Partial correlation effects in direct-sequence spread-spectrum multiple-access communications systems, IEEE Trans. Commun., vol. COM-32 (1984) 567–573.
11. D. E. Cartier: Partial correlation properties of pseudonoise (PN) codes in non-coherent synchronization/detection schemes, IEEE Trans. On Commun. (1976) 898–903.
12. Yong-Hwan Lee, Sawasd Tantaratana: Sequential acquisition of PN sequences for DS/SS communications: design and performance, IEEE Journal on selected areas in communications, vol. 10, No.4 (1992) 750–759.
13. Kenneth G. Paterson, Paul J. G. Lothian: Bounds on partial correlations of sequences. IEEE Trans. Information Theory, Vol 44. No. 3 (1998) 1164–1175.
14. P.Z. Fan, N. Suehiro, N. Kuroyanagi, X.M. Deng: A class of binary sequences with zero correlation zone. IEE Electron. Lett., Vol.35 (1999) 777–779.
15. Pingzhi Fan, Li Hao: Generalized orthogonal sequences and their applications in synchronous CDMA. IEICE TRANS. FUNDAMENTAL, VolE83-A, No.11 (2000) 2054–2069.
16. Xinmin Deng, Pingzhi Fan: Spreading sequence sets with zero and low correlation zone for quasi-synchronous CDMA communication systems. IEEE 0-7803-6507-0/00 (2000) 1698–1703.
17. Hideyuki Torii, Makoto Nakamura, Naoki Suehiro: A new class of zero-correlation zone sequences, IEEE TRANSACTIONS ON INFORMATION THEORY, Vol. 50, No.3 (2004) 559–565.
18. X.H. Tang, P.Z. Fan, D.B. Li, N. Suehiro: Binary array set with zero correlation zone, IEE ELECTRONICS LETTERS Vol. 37, No.13 (2001) 841–842.
19. P.Z. Fan: Spreading sequence design and theoretical limits for quasis-ynchronous CDMA systems. EURASIP Journal on Wireless Communications and Networking (JWCN, USA), Vol.1, No.1 (2004) 19–31.
20. Vladimir I. Levenshtein: New lower bounds on aperiodic crosscorrelation of binary codes, IEEE Trans. on Information Theory, Vol. 45, NO. 1 (1999) 284–288.

Chip-Asynchronous Version of Welch Bound: Gaussian Pulse Improves BER Performance

Yutaka Jitsumatsu and Tohru Kohda*

Dept. of Computer Science and Communication Engineering, Kyushu University, 6-10-1 Hakozaki, Higashi-ku, Fukuoka, 812-8581, Japan
{jitumatu, kohda}@csce.kyushu-u.ac.jp

Abstract. We give a quadratic form expression of the mean squared multiple-access interference (MAI) averaged over relative time delays for chip-asynchronous DS/CDMA systems. A lower bound on the mean squared MAI is referred to as *chip-asynchronous version of Welch bound*, which depends on chip pulse shapes. Real analysis tells us that a pair of rectangular and sinc functions is one of Fourier transform and its inverse Fourier transform and vice versa. On the other hand, Gaussian pulses have the self-duality property. Gaussian chip pulses sacrifice inter-symbol interference, however, they give smaller mean squared MAI, as well as lower bit error rate, than the conventional Nyquist pulses.

Keywords: Welch bound equality (WBE) sequence, total squared correlation (TSC), pulse shaping filter, asynchronous DS/CDMA system.

1 Introduction

Constructing a spread spectrum (SS) code set minimizing cross-correlation values between any two codes in the set is a goal of SS code design for direct sequence/code division multiple access (DS/CDMA) communications. It is also an important and interesting problem to give a tight lower bound of cross-correlations. The Welch bound is a lower bound of such cross-correlation values [1].

For symbol-synchronous DS/CDMA systems, cross-correlations reduce to inner products among the code set. The optimal SS codes are the set of orthogonal sequences when the number of users K is less than or equal to the code length N. Thus, the overloaded system ($K > N$) attracts many researchers' attention. An SS code set achieving the Welch's lower bound is called Welch bound equality (WBE) sequences [2,3,4,5]. For real- and complex-valued SS codes, the Welch bound is always achievable but this is not the case for binary antipodal code sets [6]. Recently, Karystinos and Pados gave a new lower bound for binary codes [6,7].

* This work was supported in part by Grants-in-Aid for Scientific Research of Japan Society for the Promotion of Science, no. 15360206, 17650065, and no. 17760312.

G. Gong et al. (Eds.): SETA 2006, LNCS 4086, pp. 351–363, 2006.

For asynchronous systems, we have three correlation functions, i.e., periodic (or, even), odd, and aperiodic correlation functions. (Odd correlation is defined only if data symbol is antipodal binary. If data symbol is polyphase, odd correlation is generalized to polyphase correlations [3].) Welch gave lower bounds on periodic and aperiodic correlations as well as a bound on inner products. A lower bound on odd correlation functions was given in [8]. Pursley [9] exploited an aperiodic cross-correlation function and gave a useful expression of signal to interference ratio (SIR) parameter averaged over all possible time delays. Note that cross-correlation values depend on a relative time delay between the two sequences. Recently, Ulukus and Yates [10] defined a quantity to measure the user capacity of a CDMA system. The quantity is called the total squared asynchronous correlation (TSAC) and depends on the users' delay profile. A lower bound on TSAC was given in [10] based on the assumption that the system is symbol-asynchronous but chip-synchronous. It must be stressed that there are two types asynchronisms: chip-synchronous and chip-asynchronous and that characteristics of cross-correlation functions are different between the two systems.

For chip-asynchronous (i.e., completely asynchronous) DS/CDMA systems, the bit error rate (BER) performance depends on pulse shaping filters as well as signature sequences. It was found in [11] and proven in [12, 13, 14] that SS codes generated by a Markov chain have less BER than linear feedback shift register (LFSR) sequences, such as Gold and Kasami codes. One of criticisms to this surprising fact was that rectangular chip waveforms are wrongly assumed, because they have infinite band-widths and are not used in practical systems. However, it was shown that Markov codes also reduce the BER when pulse shapes are the band-limited root raised cosine pulses [15, 16].

Pulse shape optimization was thoroughly discussed in [17], where spreading sequences are modeled as independent and identically distributed (i.i.d.) random variables. The pulses in [17] were assumed to have zero inter-symbol interference (ISI). However, reduction of multiple-access interference (MAI) is more important than ISI-free property because BER is mainly increased by MAI.

In this paper, we evaluated the mean squared MAI averaged over uniformly distributed relative time delays. The mean squared MAI is given in a positive definite quadratic form, which defines a lower bound. Such a lower bound is referred to as "chip-asynchronous version of Welch bound", which depends on the chip pulse shape. As a related work, the quantity called continuous-time equivalent of total squared correlation (CTE-TSC) was studied by Cho and Gao [18]. They gave a lower bound on CTE-TSC by assuming that signature waveforms are completely band-limited signals without excess bandwidth, while we assume signature waveforms are produced by code sequences and pulse shaping filters having unavoidable excess bandwidth. This paper examines Gaussian pulse shaping filters [19]. Gaussian pulse has the self-duality property, that is, it has the same expression in both time and frequency domains [20]. Gaussian chip pulse sacrifices ISI but it has lower BER than root raised cosine pulses.

2 The Welch Bound

Denote a set of SS codes by

$$\mathcal{X} = \{\boldsymbol{x}^{(1)}, \ldots, \boldsymbol{x}^{(K)}\}, \quad \boldsymbol{x}^{(k)} = (x_1^{(k)}, \cdots, x_N^{(k)})^T, \tag{1}$$

where N is the code length, also referred to as the spreading factor, and K the family size. It is assumed that SS codes are complex-valued with $\sum_{n=1}^{N} \left|x_n^{(k)}\right|^2 = N$. Let data and chip durations be T_d and $T_c = T_d/N$. SS code and data signals and for i-th user are, respectively,

$$x^{(i)}(t) = \sum_{n=0}^{N-1} x_n^{(i)} \delta(t - nT_c), \tag{2}$$

$$d^{(i)}(t) = \sum_{p=-\infty}^{\infty} d_p^{(i)} \delta(t - pT_d). \tag{3}$$

Aperiodic cross-correlation is defined as

$$x^{(i)}(t) * \overline{x^{(j)}(-t)} = \sum_{k=1-N}^{N-1} c_k^{(i,j)} \delta(t - kT_c), \tag{4}$$

where $c_k^{(i,j)} = \sum_{n=0}^{N-1-k} x_n^{(i)} \overline{x_{n+k}^{(j)}}$ and $\bar{z}$ denotes the complex conjugate of z. The SS code set $\mathcal{X}$ is required to be designed so that $|c_k^{(i,j)}|$ is as small as possible for every pair of users and delays. The Welch bound on aperiodic correlations is known as [3]

Theorem 1 (Welch). *Let $S_a = \{(i,i,k) | 1 \leq i \leq K, k = \pm 1, \pm 2, \ldots, \pm N-1\}$ denote the set of variables representing out-of-phase auto-correlations and $S_c = \{(i,j,k) | i \neq j, 1-N \leq k \leq N-1\}$ the one representing cross-correlations. Then*

$$\max_{(i,j,k) \in S_a \cup S_c} |c_k^{(i,j)}| \geq \sqrt{\frac{N^2(K-1)}{K(2N-1)-1}}. \tag{5}$$

The right hand side of (5) gives *chip-synchronous version of Welch bound.* Roughly speaking, this bound is $\sqrt{N/2}$ asymptotically.

The Welch bound gives a lower bound of maximum correlation values among all cross-correlations and out-of-phase auto-correlations. Welch bound equality (WBE) signal set is optimum in the sense that (5) holds with equality. This is true when the system is chip-synchronous. However, in chip-asynchronous systems, time delay takes continuous values and correlation function of non-integer time delay, say $\ell + \varepsilon$, is an intermediate value between $c_\ell^{(i,j)}$ and $c_{\ell+1}^{(i,j)}$. Therefore multiple access interference (MAI) depends on pulse shapes. The maximum value of correlation functions is bounded by (5). However, such maximum value infrequently occurs since it is rare that the relative time delay takes an integer value in the chip time. Hence, it seems more reasonable to consider a bound on the time-averaged MAI with a weight function given by the chip pulse shape.

3 Chip-Asynchronous DS/CDMA Systems

Let us consider a chip-asynchronous DS/CDMA system with pulse shaping filters. Denote the transmitter's pulse shape by $g(t)$ with $\int_{-\infty}^{\infty} |g(t)|^2 dt = 1$. The k-th user's propagation delay is $0 \leq t_k < T_d$. The received signal is given by

$$r(t) = \sum_{j=1}^{K} d^{(j)}(t) * x^{(j)}(t) * g(t) * \delta(t - t_j) + n_0(t), \tag{6}$$

where an asterisk sign denotes a convolution and $n_0(t)$ is an additive white Gaussian noise with two-sided spectral density σ_N^2. A single-user receiver consists of a chip-matched filter and a code-matched filter. Note that it is often supposed that chip-matched filter is given by $g(-t)$ but it is reported that using other pulse shapes than $g(-t)$ can suppress the MAI [21,22]. Hence the chip-matched filter is denoted by $h(t)$, normalized as $\int_{-\infty}^{\infty} |h(t)|^2 dt = 1$, while code-matched filter is denoted by $\overline{x^{(i)}(-t)}$.

Output of the i-th user's receiver is $z^{(i)}(t) = r(t) * h(t) * \overline{x^{(i)}(-t)} * \delta(t + t_i)$. The multiple access interference is defined as

$$\mathrm{MAI}^{(i,j)}(t) = x^{(j)}(t) * g(t) * \overline{x^{(i)}(-t)} * h(t) \quad \text{for } j \neq i. \tag{7}$$

Let us consider, without loss of generality, the correlator output of the 0-th bit:

$$\begin{aligned} z^{(i)}(0) =& d_0^{(i)} \cdot \mathrm{MAI}^{(i,i)}(0) + \sum_{p \neq 0} d_p^{(i)} \cdot \mathrm{MAI}^{(i,i)}(-pT_d) \\ &+ \sum_{\substack{j=1 \\ j \neq i}}^{K} \sum_{p=-\infty}^{\infty} d_p^{(j)} \cdot \mathrm{MAI}^{(i,j)}(t_{ij} - pT_d) + \eta^{(i)}(0), \end{aligned} \tag{8}$$

where $t_{ij} = t_i - t_j$ denotes the relative time delay between i-th and j-th users' signal. The first, second, third, and fourth term of the right hand side of (8) are the signal component, inter-symbol interference (ISI), MAI, and noise component, respectively. The transmitted data symbol is estimated as $\widehat{d_0^{(i)}} = \mathrm{sgn}(z^{(i)}(0))$.

Calculation of the bit error rate (BER) of a CDMA receiver is a difficult task in general because it includes a multiple integration of Q-function with all unknown parameters [23,24]. The standard Gaussian approximation (SGA) is a widely accepted way to approximate the BER, where the MAI terms in (8) is regarded as a Gaussian random vector. Such an approximation is based on the central limit theorem and therefore it is necessary that the number of users K is sufficiently large. The bit error rate based on the SGA is

$$\widetilde{P}_e^{(i)} = Q\left(\sqrt{\frac{|\mathrm{MAI}^{(i,i)}(0)|^2/N}{\sigma_N^2 + \{\sigma_{\mathrm{ISI}}^{(i,i)}\}^2 + \sum_{j\neq i}\{\sigma_{\mathrm{MAI}}^{(i,j)}\}^2}}\right), \tag{9}$$

where $Q(x) = \frac{1}{2\pi}\int_x^\infty \exp(-u^2/2)du$ and

$$\sigma_N^2 = \frac{N_0}{2N}\sum_{k=1-N}^{N-1}\left[\boldsymbol{x}^{(i)} \star \boldsymbol{x}^{(i)}\right]_k \cdot h_R * h_R(kT_c), \tag{10}$$

$$\{\sigma_{\text{ISI}}^{(i,i)}\}^2 = \frac{1}{N}\sum_{p\neq 0}\left|\text{MAI}^{(i,i)}(-pT_d)\right|^2, \tag{11}$$

$$\begin{aligned}\{\sigma_{\text{MAI}}^{(i,j)}\}^2 &= \mathbf{Var}_D\left[\frac{1}{\sqrt{N}}\sum_{p=-\infty}^{\infty} d_p^{(j)}\text{MAI}^{(i,j)}(t_{ij}-pT_d)\right] \\ &= \frac{1}{N}\sum_{p=-\infty}^{\infty}\left|\text{MAI}^{(i,j)}(t_{ij}-pT_d)\right|^2, \end{aligned}\tag{12}$$

where $\mathbf{Var}_D[\cdot]$ shows the variance of a random variable with respect to data symbols assumed to be independent and identically distributed (i.i.d.) random variables.

Let us minimize the total amount of (12) over $1 \le i, j \le K$ because the bit error rate $\widetilde{P}_e^{(i)}$ should be suppressed for every users. This quantity is used to measure the system performance of multiuser detection (MUD) receivers [10]. For a single user receiver, the relative time delay t_{ij} ($j \neq i$) is supposed to be unknown and uniformly distributed in $[0, T_d]$. Hence we consider

$$\sum_{i=1}^{K}\sum_{j=1}^{K}\frac{1}{T_d}\int_0^{T_d}\{\sigma_{\text{MAI}}^{(i,j)}\}^2 dt_{ij}. \tag{13}$$

Note that $\sigma_{\text{MAI}}^{(i,i)}$ gives auto-correlation function of i-th user's spreading signal, where timing error is denoted by t_{ii}. We get

Lemma 1. *Let $u_0 = 2NK$, $\boldsymbol{u} = (u_1, \ldots, u_{N-1})^T$ and $\boldsymbol{v} = (v_1, \ldots, v_{N-1})^T$ are vectors comprised of real and imaginary parts of auto-correlation functions of SS codes such that $u_k = \sum_{i=1}^{K}[c_k^{(i,i)} + c_{-k}^{(i,i)}]$ and $v_k = \sqrt{-1}\sum_{i=1}^{K}[c_k^{(i,i)} - c_{-k}^{(i,i)}]$, respectively. Then, the mean squared MAI averaged over time delay is expressed in a positive definite quadratic form as*

$$\begin{aligned}&\sum_{i=1}^{K}\sum_{j=1}^{K}\frac{1}{T_d}\int_0^{T_d}\{\sigma_{\text{MAI}}^{(i,j)}\}^2 dt_{ij} \\ &= \frac{1}{NT_d}\left\{(\tfrac{u_0}{2})^2 a_0 + u_0\boldsymbol{b}^T\boldsymbol{u} + \tfrac{1}{2}\boldsymbol{u}^T\boldsymbol{A}^{(c)}\boldsymbol{u} + \tfrac{1}{2}\boldsymbol{v}^T\boldsymbol{A}^{(s)}\boldsymbol{v}\right\}, \end{aligned}\tag{14}$$

where $\boldsymbol{A}^{(c)}$ and $\boldsymbol{A}^{(s)}$ are $(N-1)$ by $(N-1)$ matrices and $\boldsymbol{b} - (b_1, b_2, \ldots b_{N-1})$ with

$$\begin{aligned} a_0 =& g*g*h*h(0), & (15)\\ b_n =& g*g*h*h(nT_c), & (16)\\ A^{(c)}_{m,n} =& g*g*h*h((n+m)T_c) \\ &+ g*g*h*h((n-m)T_c). & (17)\\ A^{(s)}_{m,n} =& g*g*h*h((n+m)T_c) \\ &- g*g*h*h((n-m)T_c). & (18) \end{aligned}$$

Proof: From (4), (7) and (12), we have

$$\begin{aligned} &\frac{1}{T_d}\int_0^{T_d} \{\sigma^{(i,j)}_{\mathrm{MAI}}\}\, dt_{ij} \\ =&\frac{1}{NT_d}\sum_{p=-\infty}^{\infty}\int_{-pT_d}^{-pT_d+T_d} |\mathrm{MAI}^{(i,j)}(t_{ij})|^2\, dt_{ij} \\ =&\frac{1}{NT_d}\int_{-\infty}^{\infty} |\mathrm{MAI}^{(i,j)}(\tau)|^2\, d\tau \\ =&\frac{1}{NT_d} x^{(i)}(t) * \overline{x^{(i)}(-t)} * x^{(j)}(t) * \overline{x^{(j)}(-t)} \\ &\qquad * g(t) * \overline{g(-t)} * h(t) * \overline{h(-t)}\Big|_{t=0} \\ =&\frac{1}{NT_d}\sum_{k=1-N}^{N-1} c_k^{(i,i)} \sum_{\ell=1-N}^{N-1} \overline{c_\ell^{(j,j)}}\, g*g*h*h((k-\ell)T_c). \qquad (19) \end{aligned}$$

Substituting $\sum_{i=1}^K c_k^{(i,i)} = \frac{u_k - \sqrt{-1}v_k}{2}$ into the summation of (19) from $i=1$ to K and $j=1$ to K together with the relation $c_{-k}^{(i,i)} = \overline{c_k^{(i,i)}}$ gives the quadratic form (14). □

The coefficient matrix $\boldsymbol{A}^{(c)}$ and $\boldsymbol{A}^{(s)}$ are positive definite. Therefore, Lemma 1 immediately gives a lower bound on (13).

Theorem 2. *The mean squared MAI averaged over time delay is bounded from below as*

$$\sum_{i=1}^{K}\sum_{j=1}^{K}\frac{1}{T_d}\int_0^{T_d}\{\sigma^{(i,j)}_{\mathrm{MAI}}\}^2 dt_{ij} \ge \frac{K^2}{T_c}(a_0 - 2\boldsymbol{b}^T\{\boldsymbol{A}^{(c)}\}^{-1}\boldsymbol{b}). \qquad (20)$$

Equality holds if and only if $\boldsymbol{u} = -u_0\{\boldsymbol{A}^{(c)}\}^{-1}\boldsymbol{b}$ *and* $\boldsymbol{v} = \boldsymbol{0}$. *We refer the right hand side of* (20) *to as "asynchronous version of Welch bound."*

It is observed in [3] that any WBE signal set should satisfy

$$\boldsymbol{u} = \boldsymbol{v} = \boldsymbol{0}. \qquad (21)$$

In this case, the mean squared MAI, Eq. (14) is equal to $a_0 K^2/T_c$. Theorem 2 implies that we can reduce the mean squared MAI by $2\boldsymbol{b}^T\{\boldsymbol{A}^{(c)}\}^{-1}\boldsymbol{b}$. Let us define the reduction ratio as

$$\frac{2\boldsymbol{b}^T\{\boldsymbol{A}^{(c)}\}^{-1}\boldsymbol{b}}{a_0}. \tag{22}$$

4 Discussions

4.1 Rectangular Pulse Case

A rectangular pulse is given by

$$g_{\mathrm{rec}}(t) = \begin{cases} 1/\sqrt{T_c} & \text{for } 0 \le t \le T_c. \\ 0 & \text{otherwise,} \end{cases} \tag{23}$$

Pursley defined the SIR parameter for a rectangular pulse by [9], [3, 25]

$$\frac{2}{3}\mu^{(i)}(0) + \frac{1}{3}\mu^{(i)}(1), \tag{24}$$

where

$$\mu^{(i)}(n) \stackrel{\mathrm{def}}{=} \sum_{\substack{j=1 \\ j\neq i}}^{K} \sum_{k=1-N}^{N-1-n} c_k^{(i,j)} \overline{c_{k+n}^{(i,j)}}. \tag{25}$$

The second term of (24) is very close to zero for Gold and Kasami codes. Using SS codes satisfying (5) with equality gives

$$\mu^{(i)}(0) = (K-1)N^2\left(1 - \tfrac{2(N-1)}{K(2N-1)-1}\right). \tag{26}$$

On the other hand, using i.i.d. codes gives

$$\mathbb{E}_{\mathcal{X}}\left[\mu^{(i)}(0)\right] = (K-1)N^2, \tag{27}$$

which means SIR parameter of WBE signal set does not have significant difference from that of random i.i.d. codes and is slightly less than $\frac{2}{3}(K-1)N^2$ [3].

However, it follows from (24) that minimization of $\mu^{(i)}(0)$ does not mean minimization of the SIR parameter. Optimization of (24) is actually done by making $\mu^{(i)}(0)$ more than (26) and making $\mu^{(i)}(1)$ negative. This approach is considered as *variance reduction method* and SS code set provide us *antithetic variates* [26]. For a rectangular pulse, we have $a_0 = 2/3$, $\boldsymbol{b}_{\mathrm{rec}} = (1/6, 0, \ldots, 0)^T$ and

$$\boldsymbol{A}_{\mathrm{rec}}^{(c)} = \boldsymbol{A}_{\mathrm{rec}}^{(s)} = \begin{pmatrix} \frac{2}{3} & \frac{1}{6} & 0 & \cdots & 0 \\ \frac{1}{6} & \frac{2}{3} & \frac{1}{6} & \ddots & \vdots \\ 0 & \frac{1}{6} & \ddots & \ddots & 0 \\ \vdots & \ddots & \ddots & \frac{2}{3} & \frac{1}{6} \\ 0 & \cdots & 0 & \frac{1}{6} & \frac{2}{3} \end{pmatrix}. \tag{28}$$

Hence, Theorem 2 gives the mean squared MAI for a rectangular pulse as

$$\sum_{i=1}^{K}\sum_{j=1}^{K}\frac{1}{T_d}\int_0^{T_d}\{\sigma_{\mathrm{MAI}}^{(i,j)}\}^2 dt_{ij} > \frac{K^2}{T_c}\left(\frac{2}{3}-\frac{2-\sqrt{3}}{3}\right)=\frac{K^2}{T_c}\frac{1}{\sqrt{3}}, \tag{29}$$

which suggests that we can reduce SIR parameter to $\frac{1}{\sqrt{3}}(K-1)N^2$. The reduction ratio for rectangular pulse is $(2-\sqrt{3})/2 = 0.134$. Variables $\mu^{(i)}(0)$ and $\mu^{(i)}(1)$ for Kasami and Gold sequences are shown in Table 1. In case of i.i.d. codes, the expectation value is listed, whereas optimum codes are given by $\boldsymbol{u} = -u_0\{\boldsymbol{A}^{(c)}\}^{-1}\boldsymbol{b}$ and $\boldsymbol{v} = \boldsymbol{0}$.

Table 1. Mean squared of MAI for LFSR codes, where code length is $N = 31$ for Gold codes and $N = 63$ for Kasami codes

	Gold	Kasami (small set)	i.i.d.	optimum
$\mu^{(i)}(0)$	0.970674	1.017601	1	$2/\sqrt{3}$
$\mu^{(i)}(1)$	-0.003153	0.002176	0	$-1/\sqrt{3}$

Such an optimization in terms of the SIR parameter, however, has a criticism that the assumption of rectangular chip waveforms is improper: rectangular waveform has infinite band-widths and are not used in practical systems. In the next subsection, a completely band-limited case is considered.

4.2 Sinc Pulse Case

If the chip waveforms $G(\omega)$ and $H(\omega)$ have flat spectra over the frequency band $\omega \in [-\pi/T_c : \pi/T_c]$, their expressions in time domain are sinc (or, sampling) functions. In this case, the coefficient vector $\boldsymbol{b}$ becomes zero vector and matrices $\boldsymbol{A}^{(c)}$ and $\boldsymbol{A}^{(s)}$ turn into identity matrices. Hence, the mean squared MAI for sinc pulse is

$$\begin{aligned}&\frac{1}{NT_d}((\tfrac{u_0}{2})^2+\tfrac{1}{2}\|\boldsymbol{u}\|^2+\tfrac{1}{2}\|\boldsymbol{v}\|^2)\\ =&\frac{1}{NT_d}((\tfrac{u_0}{2})^2+2\textstyle\sum_{k=1}^{K}\left|\sum_{i=1}^{N}c_k^{(i,i)}\right|^2)\geq\frac{K^2}{T_c}.\end{aligned} \tag{30}$$

Now we can give a brief proof of Theorem 1. Using the Pursley-Sarwate's identity [27], $\sum_{k=1-N}^{N-1} c_k^{(i,j)}\overline{c_{k+\ell}^{(i,j)}} = \sum_{k=1-N}^{N-1} c_k^{(i,i)}\overline{c_{k+\ell}^{(j,j)}}$, gives

$$\sum_{i=1}^{K}\sum_{j=1}^{K}\sum_{k=1-N}^{N-1}|c_k^{(i,j)}|^2 = K^2N^2+\sum_{k\neq 0}\left|\sum_{i=1}^{K}c_k^{(i,i)}\right|^2. \tag{31}$$

It is easy to see $\sum_{i=1}^{K}\sum_{j=1}^{K}\sum_{k=1-N}^{N-1}|c_k^{(i,j)}|^2 = KN^2 + \sum_{(i,j,k)\in S_a\cup S_c}|c_k^{(i,j)}|^2$. Thus,

$$\max_{(i,j,k)\in S_a\cup S_c}\left|c_k^{(i,j)}\right| \geq \sqrt{\frac{N^2(K-1)}{K(2N-1)-1}} \tag{32}$$

Eq. (31) is identical to the left hand side of (30) because we have $\|\boldsymbol{u}\|^2+\|\boldsymbol{v}\|^2 = \|\boldsymbol{u}+\sqrt{-1}\boldsymbol{v}\|^2 = \sum_{k=1}^{N-1}|2\sum_{i=1}^{K}c_k^{(i,i)}|^2$.

4.3 Root Raised Cosine Pulses

Completely band-limited pulses have infinite impulse responses and do not satisfy the causality of the filter. Hence a time truncated version of the band-limited pulse is used in actual situations. For practical communication systems, root raised-cosine pulse with excess bandwidth $\beta > 0$ is frequently employed for band-limited DS/CDMA systems. In practical systems, excess bandwidth is typically $\beta = 0.22$. The convolution of $g(t)$ and $h(t)$ becomes a raised-cosine waveform, i.e.

$$g * h(t) = \frac{\sin(\frac{\pi t}{T_c})}{\pi t}\frac{\cos(\frac{\pi\beta t}{T_c})}{1-4(\frac{\beta t}{T_c})^2}, \tag{33}$$

where β is also called the roll-off factor. Then

$$g * g * h * h(kT_c) = \begin{cases} 1-\beta/4 & \text{if } k=0 \\ (-1)^{k+1}\cdot\beta/8 & \text{if } k=\pm 1/\beta \\ \frac{\sin(k\beta\pi)}{4\pi kT_c}\cdot\frac{(-1)^{k+1}}{1-(k\beta)^2} & \text{otherwise.} \end{cases} \tag{34}$$

Optimum Markovian SS codes for RRC pulse shaping with different roll-off factors are discussed in [16] and [15].

Recently, Beaulieu et. al proposed a "better than" Nyquist pulse, which has a good eye-diagram and is expected to reduce timing jitter [28]. These pulses are of practical importance but we analyze the following pulse shaping filter, which is optimum in terms of BER [24].

$$|G_{\text{opt}}(\omega)|^2 = \begin{cases} T_c & \text{for } |\omega| < (1-\beta)\pi/T_c \\ 0 & \text{for } |\omega| > (1+\beta)\pi/T_c \\ T_c/2 & \text{otherwise} \end{cases} \tag{35}$$

Note that such a sharp power spectral density function is difficult to realize. We use this pulse because it gives a lower bound of correlation values for a given excess bandwidth. The impulse response of $|G_{\text{opt}}(\omega)|^2$ is $g(t) * g(-t) = \text{sinc}(\pi t/T_c)\cos(\beta\pi t/T_c)$. We have

$$g * g * h * h(nT_c) = \begin{cases} T_c(1-\frac{\beta}{2}) & \text{for } n=0 \\ \frac{(1-\beta)T_c}{2}\text{sinc}((1-\beta)n\pi) & \text{for } n=\pm1,\pm2,\ldots. \end{cases} \tag{36}$$

The amount of reduction $2\boldsymbol{b}^T\{\boldsymbol{A}^{(c)}\}^{-1}\boldsymbol{b}$ and the reduction ratio $2\boldsymbol{b}^T\{\boldsymbol{A}^{(c)}\}^{-1}\boldsymbol{b}/a_0$ for $G_{\text{opt}}(\omega)$ are illustrated in Fig. 1, which shows the mean squared MAI can be reduced 5% when $\beta = 0.5$ and 3 % when $\beta = 0.25$.

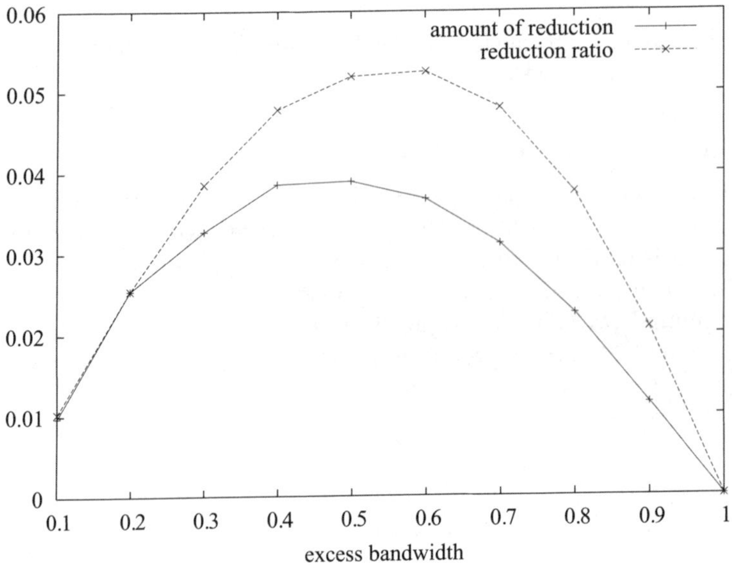

Fig. 1. Reduction of the mean squared MAI by optimization of autocorrelation of SS codes: A pulse shaping filters with excess bandwidth β is used

4.4 Gaussian Pulse Case

If pulse shaping filter is completely and ideally band-limited with flat spectrum, WBE sequence is the best in terms of TSC. However, as stated above, pulse shaping filters in practical systems employs nonzero excess bandwidth: typically it is $\beta = 0.22$. For RRC pulses, reduction ratio is only approximately 3%. In order to enhance the variance reduction effect, function of $g * g * h * h(k \cdot T_c)$ for $k \neq 0$ should take large value.

In this paper, we introduce a Gaussian pulse, which has nonzero inter-symbol interference (ISI). Gaussian pulses have the self-duality property: they have the same expression in both time and frequency domains. Gaussian chip pulses sacrifice ISI, however, they give lower bit error rate than the conventional RRC pulses. Let us define a Gaussian pulse with a parameter $\alpha > 0$ as

$$g_\alpha(t) = \sqrt{\frac{\alpha}{T_c}} \text{gauss}\left(\frac{\alpha t}{T_c}\right), \tag{37}$$

where $\text{gauss}(x) = \sqrt[4]{2}\exp(-\pi x^2)$. The energy of this pulse is concentrated with 92.4% both in time domain $t \in [-T_c/2 : T_c/2]$ and in frequency domain $\omega \in [-\pi/T_c : \pi/T_c]$ when $\alpha = 1$. We obtain

$$g_\alpha * g_\alpha(t) = \frac{1}{\sqrt[4]{2}} \text{gauss}(\alpha t/\sqrt{2}T_c), \tag{38}$$

$$g_\alpha * g_\alpha * g_\alpha * g_\alpha(t) = \frac{T_c}{\sqrt[4]{2}\alpha} \text{gauss}(\alpha t/2T_c). \tag{39}$$

Bit error probabilities for Gaussian and root raised cosine pulse are illustrated in Fig. 2. Simulation results shows the superiority of Gaussian pulses.

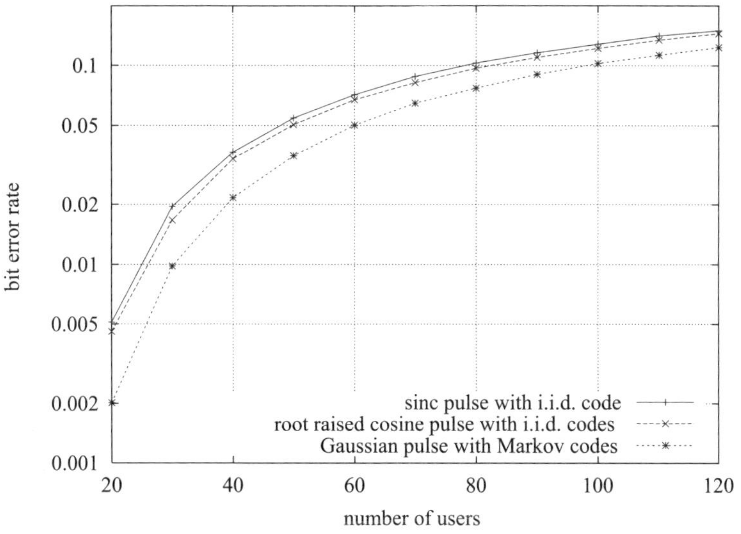

Fig. 2. Bit error rates of a chip-asynchronous CDMA system with spreading factor $N = 127$, where ideal sinc pulse and Gaussian pulse with $\alpha = 1$ are examined

Finally, we would like to emphasize that although chip waveform is often assumed to be band-limited, time-truncation is necessary for pulse shaping filter to be realized. Hence, the waveform after truncation is not band-limited in a strict sense. One can find a better solution to a problem on time-limitedness and band-limitedness in [29], i.e. prolate spheroidal wave function (PSWF). Applications of PSWF to CDMA and ultra wideband communications were given [30, 31, 32].

5 Concluding Remarks

The mean squared MAI for chip-asynchronous (completely asynchronous) DS/CDMA systems was expressed in a positive definite quadratic form, which gives a lower bound. This lower bound is referred to as "chip-asynchronous version of Welch bound" and is identical to the chip-synchronous one if the chip pulse is the ideal sinc function. However, using a chip pulse having excess bandwidth $\beta > 0$ makes the WBE sequences not optimum, which implies that we can reduce mean squared MAI. Moreover, we examined a Gaussian pulse as an example of nonzero-ISI pulses. Simulation results show Gaussian pulse sacrifices the ISI but it improves BER performance.

References

1. Welch, L.R.: Lower bounds on the maximum cross correlation of signals. IEEE Trans. Inform. Theory **20** (1974) 397–399
2. Rupf, M., Massey, J.L.: Optimum sequence multisets for synchronous code-divition multiple-access channels. IEEE Trans. Inform. Theory **40**(4) (1994) 1261–1266
3. Sarwate, D.: Meeting the Welch bound with equality. In C. Ding, T.H., Niederreiter, H., eds.: Sequences and Their Applications: Proceedings of SETA '98. Springer-Verlag, London, U.k. (1999)
4. Viswanath, P., Anatharam, V., Tse, D.N.C. IEEE Trans. Inform. Theory **45**(6) (1999) 1968–1983
5. Viswanath, P., Anatharam, V. IEEE Trans. Inform. Theory **45**(6) (1999) 1984–1991
6. Karystinos, G.N., Pados, D.A.: New bounds on the total squared correlation and optimum design of DS-CDMA binary signature sets. IEEE Trans. Commun. (2003) 48–51
7. Karystinos, G.N., Pados, D.A.: The maximum squared correlation, sum capacity, and total asymptotic efficiency of minimum total-squered-correlation binary signature sets. IEEE Trans. Inform. Theory (2005) 348–355
8. Mow, W.H.: On the bounds on odd correlation of sequences. IEEE Trans. Inform. Theory **40**(3) (1994) 954–955
9. Pursley, M.B.: Performance evaluation for phase-coded SS multiple-access communication-part-I: system analysis. IEEE Trans. Commun. **25**(8) (1977) 795–799
10. Ulukus, S., Yates, R.D.: User capacity of asynchronous CDMA systems with matched filter receivers and optimum signature sequences. IEEE Trans. Inform. Theory (2004) 903–909
11. Mazzini, G., Setti, G., Rovatti, R.: Chaotic complex spreading sequences for asynchronous DS-CDMA -part I: System modeling and results. IEEE Trans. Circuits Syst. I **44**(10) (1997) 937–947
12. Mazzini, G., R.Rovatti, Setti, G.: Interference minimisation by autocorrelation shaping in asynchronous DS/CDMA systems: Chaos based spreading is nearly optimal. Electronics Letters, IEE **35** (1999) 1054–55
13. Chen, C.C., Yao, K., Umeno, K., Biglieri, E.: Design of spread-spectrum sequences using chaotic dynamical systems and ergodic theory. IEEE Trans. Circuits Syst. I **48**(9) (2001) 1110–1114
14. Kohda, T., Fujisaki, H.: Variances of multiple access interference: Code average against data average. Electronics Letters, IEE **36** (2000) 1717–1719
15. Rovatti, R., Setti, G., Mazzini, G.: Pulse shaping and SIR-energy trade-off in chaos-based asynchronous DS-CDMA. In: Proc. of the 2004 Int. Symp. on Circuits and Systems. (2004) IV 613–616
16. Setti, G., Rovatti, R., Mazzini, G.: Performance of chaos-based asynchronous DS-CDMA with different pulse shapes. IEEE Commun. Lett. **8**(7) (2004) 416–418
17. Cho, J.H., Lehnert, J.S.: An optimal signal design for band-limited asynchronous DS-CDMA communications. IEEE Trans. Inform. Theory **48**(5) (2002) 1172–1185

18. Cho, J.H., Gao, W.: Continuous-time equivalents of Welch bound equality sequences. IEEE Trans. Inform. Theory **51** (2005) 3176–3185
19. Jitsumatsu, Y., Kohda, T.: Gaussian chip shaping enhances the superiority of markovian codes in DS/CDMA systems. In: Proc. of 2006 IEEE Int. Symp. on Circuits and Systems, (Island of Kos, Greece)
20. Blahut, R.E.: Digital Transmission of Information. Addison-Wesley (1990)
21. Monk, A.M., Davis, M., Milstein, L.B., Helstrom, C.W.: A noise-whitening approach to multiple access noise rejection–part I: Theory and background. IEEE Journal on Selected Area in Communications **12**(5) (1994) 817–827
22. Huang, Y., Ng, T.S.: A DS-CDMA system using despreading sequences weighted by adjustable chip waveforms. IEEE Trans. Commun. **47**(12) (1999) 1884–1896
23. Pursley, M.B., Sarwate, D.V., Stark, W.E.: Error probability for direct-sequence spread-spectrum multiple-access communications–part i: Upper and lower bounds (1982)
24. Morrow, R.K., Lehnert, J.S.: Bit-to-bit error dependence in slotted DS/SSMA packet systems with random signature sequences. IEEE Trans. Commun. **37** (1989) 1052–1061
25. Schotten, H.D.: Sequence families with optimum aperiodic mean-square correlation parameters. In C. Ding, T.H., Niederreiter, H., eds.: Sequences and Their Applications: Proceedings of SETA '98. Springer-Verlag, London, U.k. (1999)
26. Ripley, B.D.: Stochastic Simulation. John Wiley & Sons Inc, New York (1987)
27. Pursley, M.B., Sarwate, D.V.: Performance evaluation for phase-coded SS multiple-access communication-part–II: code sequence analysis. IEEE Trans. Commun. **25**(8) (1977) 800–803
28. Beaulieu, N.C., Tan, C.C., Damen, M.O.: A "better than" Nyquist pulse. IEEE Commun. Lett. **5**(9) (2001) 367–368
29. Slepian, D., Pollak, H.O.: Prolate spheroidal wave functions, fourier analysis, and uncertainty-i. Bell Syst. Tech. J. **40**(1) (1961) 43–46
30. Landolsi, M.A., Stark, W.E.: DS-CDMA chip waveform design for minimal interference under bandwidth, phase, and envelope constraints. IEEE Trans. Commun. **47**(11) (1999) 1737–1746
31. Nguyen, H.H., Shwedyk, E.: A new construction of signature waveforms for synchronous CDMA systems. IEEE Trans. Broadcast. **51** (2005) 520–529
32. Zhang, H., Zhou, X., Yazdandoost, K.Y., Chlamtac, I.: Multiple signal waveforms adaptation in cognitive ultra-wideband radio evolution. IEEE J. Select. Areas Commun. **24** (2006) 878–884

On Immunity Profile of Boolean Functions

Claude Carlet[1,2], Philippe Guillot[1], and Sihem Mesnager[1]

[1] MAATICAH, Université de Paris 8, France
[2] INRIA, Projet Codes, Rocquencourt, France
Claude.Carlet@inria.fr, ph.guillot@wanadoo.fr, hachai@math.jussieu.fr

Abstract. The notion of resilient function has been recently weakened to match more properly the features required for Boolean functions used in stream ciphers. We introduce and we study an alternate notion of almost resilient function. We show that it corresponds more closely to the requirements that make the cipher more resistant to precise attacks.

Introduction

The Boolean functions defined on the vector space $\mathbb{F}_2^n$ of binary vectors of a given length n are used in the pseudo-random generators of stream ciphers. They play a central role in their security. The generation of the keystream generally consists of a linear part, producing a sequence with a large period, usually composed of one or several LFSR's, and of a nonlinear combining or filtering function f which produces the output, given the state of the linear part. In the combiner generator model, the outputs to several Linear Feedback Shift Registers are combined by a Boolean function giving, at each clock cycle, one bit of the pseudo-random sequence. The combining function must be balanced for the good statistical properties of the generated stream sequence. Moreover, to avoid a divide and conquer attack (see *e.g.*[1,6,13,16]), the combining function must avoid low order correlation . This is the reason why such a combining function is often chosen with a rather high correlation immunity order.

There are two equivalent ways for characterising the correlation immunity: either by means of the Walsh transform or by means of the sub-functions. Originally, an n-variable Boolean function f is said to be *correlation immune* of order t (or t-th order correlation immune) if any sub-function deduced from f by fixing at most t inputs has the same output distribution as f. On the other hand, correlation immunity can be characterised by means of the Walsh transform of f. Recall that, given an integer-valued (or real-valued, or complex-valued) function f over $\mathbb{F}_2^n$, the *Fourier transform* of f is the function defined over $\mathbb{F}_2^n$ by $\widehat{f}(\omega) = \sum_{x\in\mathbb{F}_2^n} f(x)(-1)^{\omega\cdot x}$. The *Walsh transform* of a Boolean function f is by definition the Fourier transform of the sign function $\chi_f(x) = (-1)^{f(x)}$: $\widehat{\chi_f}(\omega) = \sum_{x\in\mathbb{F}_2^n}(-1)^{f(x)+\omega\cdot x}$. A Boolean function f is said correlation immune of order t if and only if the Walsh transform of f vanishes at all non zero vector of Hamming weight at most t, *cf.* [17]. If f is moreover balanced, then f is said to be t-resilient.

G. Gong et al. (Eds.): SETA 2006, LNCS 4086, pp. 364–375, 2006.

Siegenthaler's bound [16] states that the algebraic degree of an n-variable t-th order correlation immune Boolean function is necessarily less than or equal to $n-t$ [16]. On the other hand, the nonlinearity of a t-th order correlation immune Boolean function is necessarily less than or equal to $2^{n-1} - 2^t$ if $t > \frac{n}{2} - 1$ and $2^{n-1} - 2^{\frac{n}{2}-1} - 2^t$ otherwise [5]. When the Boolean function is moreover balanced, the upper bounds on its algebraic degree and its nonlinearity are lower. Indeed, the algebraic degree is less than or equal to $n - t - 1$ and the nonlinearity is upper bounded by $2^{n-1} - 2^{t+1}$ if $\frac{n}{2} - 1 < t < n - 1$ and $2^{n-1} - 2^{\frac{n}{2}-1} - 2^{t+1}$ if $t \leq \frac{n}{2} - 1$. Therefore, the correlation immunity criterion is not compatible with an high algebraic degree (necessary to withstand Berlekamp-Massey attack) and a high nonlinearity (necessary for avoiding attacks using linear approximation of the function). Moreover, the recent algebraic attacks, *e.g* [7,8], highlighted the need for having an high algebraic degree as well as an high algebraic immunity so that stream ciphers can resist to these attacks. Now, there seems to be some kind of contradiction for Boolean functions between having high correlation immunity and optimum or nearly optimum algebraic immunity; also, much attention having been given to algebraic immunity recently, several examples of functions having optimum algebraic immunity could be found but no example of correlation immune Boolean function with optimum algebraic immunity.

Fortunately, As observed in [11], strict correlation immunity is not absolutely required. The work factor to reconstitute the sequences coming from several registers increases with the number of registers, and a strict correlation immunity is necessary for small orders only. For higher orders, low non-zero correlations are sufficient (the lower the order, the lower the allowed correlations). In [11], the authors allow the restrictions to have output distributions slightly differing from the distribution of the global function. We propose here an alternate way of relaxing the constraint of correlation immunity. We allow the Walsh transform to take low values for low orders instead of being null. We introduce the concept of *immunity profile* of a Boolean function (Section 2, Definition 1). As we shall see, both definitions go in the same direction, but with non-negligible differences. In [11], the resiliency constraints have been relaxed by introducing the notion of *almost-resiliency*, that amounts to saying that the values of the Walsh transform is upper bounded on all vectors of weight lower than some positive integer. The notion introduced here is slightly more general as we are interested in the whole profile and in a way which sticks more precisely to the effective difficulty of the correlation attack. In Section 4, we study the relationship between the immunity profile and the approach of *almost resiliency* introduced by Kurosawa [11]. Moreover we wonder which immunity profile should have the combining Boolean function to have a good resistance to fast correlation attacks [1,6] (Section 3). Fast correlation attacks model the combining function as a noise on a communication channel and the cryptanalysis as a decoding problem of the keystream. There are then two ways of making hard the task of the cryptanalyst: either to oblige him to have a very large amount of the keystream or make the decoding step having a too high complexity. By considering these two points of view separately, we find two possible types of immunity profile: with arguments taken from the information theory, we explain that the immunity profile could increase in proportion to the square root of the order; next, considering the complexities of the decoding procedures used in fast correlation attacks, we see that the combining Boolean function could have an exponential immunity profile. In subsection 3.2, we consider another class of ciphers that are iterated ciphers (for example, self synchronizing stream ciphers). We explain that in this kind of cipher, a round ciphering function with an exponential immunity profile may provide a better resistance to linear cryptanalysis.We present in Sections 5 and 6 primary and secondary constructions of Boolean functions with such immunity profiles.

1 Basic Notation

Let n be any positive integer. An n-variable Boolean function is a map from $\mathbb{F}_2^n$ to $\mathbb{F}_2$. We denote by $\mathcal{B}_n$ the set of all n variable Boolean functions. An n-variable Boolean function f can be represented as a multivariate polynomial over $\mathbb{F}_2^n$, called the algebraic normal form of f:

$$f(x_1,\dots,x_n) = a_0 \oplus \sum_{i=1}^{n} a_i x_i \oplus \sum_{1\le i<j\le n} a_{ij}x_ix_j \oplus \cdots \oplus a_{12\dots n}x_1\cdots x_n$$

where the coefficients a_0, a_i, a_{ij}, …, $a_{1\dots n}$ are in $\mathbb{F}_2^n$. For cryptographic applications, there are several characteristics of Boolean functions that are interesting to investigate. The *support* of an n-variable Boolean function f, denoted by $\mathrm{supp}(f)$, is the set $f^{-1}(1) = \{x \in \mathbb{F}_2^n \mid f(x) = 1\}$. The *Hamming weight* of a Boolean function f, denoted by $\mathrm{wt}(f)$, is the cardinality of its support. An n-variable Boolean function is said *balanced* if its truth table contains an equal number of 0's and 1's. The *algebraic degree*, or simply *degree*, of a Boolean function f is by definition the degree of its algebraic normal form; it is denoted by $\deg(f)$. Functions of degree at most one are called affine Boolean functions. The *Hamming distance* between two functions f and $g \in \mathcal{B}_n$, denoted by $\mathrm{dist}(f,g)$, is defined as the cardinality of the set $\{x \in \mathbb{F}_2^n \mid f(x) \neq g(x)\}$. The *nonlinearity* of a Boolean function, denoted by $\mathrm{nl}(f)$, is the Hamming distance to the nearest affine function. It can be expressed by means of its Walsh transform: $\mathrm{nl}(f) = 2^{n-1} - \frac{1}{2}\max_{\omega\in\mathbb{F}_2^n} |\widehat{\chi_f}(\omega)|$.

2 φ-Correlation Immune Boolean Functions

Definition 1. *Let n be any integer, $n \geq 2$. Let φ be any integer valued mapping over the set $\{0,\dots,n\}$. A Boolean function f over $\mathbb{F}_2^n$ is said to be φ-correlation immune if, for any vector $\omega \in \mathbb{F}_2^n$,*

$$|\widehat{\chi_f}(\omega)| \le \varphi(\mathrm{wt}(\omega))$$

where $\mathrm{wt}(\omega)$ denotes the Hamming weight of vector ω which is by definition the number of non zero components. If f is moreover balanced then f is said to be φ-resilient. The integer mapping φ is called the immunity profile *of f.*

This definition generalizes correlation immunity as t-th order correlation immune Boolean functions are φ-correlation immune with $\varphi(i) = 0$ for $1 \le i \le t$ and $\varphi(i) = 2^n$ for $i \ge t+1$ or $i = 0$. Every Boolean function is clearly φ-correlation immune for some φ. It is advisable to carefully choose the integer mapping φ. It seems natural to consider increasing mappings φ, which take low values for low orders.

Remark 1. Let f be a φ-correlation immune Boolean function for some integer valued mapping φ over $\{0,\ldots,n\}$. Because of Parseval's identity that states that $\sum_{\omega\in\mathbb{F}_2^n} \widehat{\chi_f}^2(\omega) = 2^{2n}$, the immunity profile φ of f must satisfy

$$\sum_{l=0}^{n} \binom{n}{l} \varphi^2(l) \ge 2^{2n}\,. \tag{1}$$

Remark 2. The constraint on the algebraic degree stated by Siegenthaler's bound can be avoided for φ-correlation immune Boolean function if φ is carefully chosen. More precise statements on algebraic degree of φ-correlation immune Boolean functions shall be given in a full paper.

3 Which Immunity Profile?

3.1 Fast Correlation Attacks

Consider a stream generator constituted of n LFSR's. Each of them is of dimension about k and they are combined by an n-variable Boolean function f.

The adversary observes a sample of N bits of the keystream and must recover the initial state of each register. He may have several strategies. He can try to get initial state of a single LFSR, of two at once, or more.

Fast correlation attacks model the nonlinear function (in all models) as a noise on a communication channel with error probability $p = \frac{1}{2} - \varepsilon$, and the cryptanalysis as a decoding problem of length N, the amount of available keystream, and of dimension at most $k\ell$, where ℓ denotes the number of registers the adversary decides to recover by this decoding process [1,6,13,14,15]. More precisely, the nonlinear function is modelled by a Binary Symmetric Channel with transition probability p given by

$$p = \frac{1}{2} - \varepsilon \text{ with } \varepsilon = \frac{\varphi(\ell)}{2^{n+1}} \text{ and } \varphi(\ell) = \max_{u,\mathrm{wt}(u)=\ell} |\widehat{\chi_f}(u)|\,, \tag{2}$$

that corresponds to the maximum correlation between the output to f and the combination of ℓ LFSR states, by means of some ℓ-variable Boolean function, that the adversary decides to recover. When f is ℓ-th order correlation immune (this corresponds to the case where $p = \frac{1}{2}$), the adversary has no chance to recover the internal state of ℓ registers while if f is not ℓ-resilient the cryptanalyst can theoretically recover the state of the registers. Obviously, from a practical viewpoint, the success of the cryptanalyst will not be guaranteed either if he does not know enough keystream bits or if the complexity of the decoding procedure is too high. We will consider separately each of the two situations. This will lead us to different immunity profiles for the Boolean function combining the LFSR's (indeed, the limitation factor can come from the data or from the computation power).

From an information theory point of view, the data at the disposal of the adversary must be sufficient to recover the initial state of the registers and the error vector, thus, the size N of the sample must satisfy the following inequality (Shannon's channel coding theorem, see *e.g.* [12]):

$$N \geq \frac{k\ell}{1 - h_2(p)}, \tag{3}$$

where $h_2 : p \mapsto -p\log_2(p) - (1-p)\log_2(1-p)$ is the binary entropy function. Whenever ε is small, one has $1 - h_2(p) \approx \frac{2}{\ln(2)}\left(p - \frac{1}{2}\right)^2$. Thus, the condition (3) of success for the adversary becomes $k\ell \leq \frac{2N}{\ln(2)}\frac{\varphi(\ell)^2}{2^{2n+2}}$. Consequently, if the resiliency profile φ satisfies at ℓ

$$\varphi(\ell) \leq 2^n \sqrt{\frac{2k\ln(2)}{N}} \cdot \sqrt{\ell} \tag{4}$$

then, the adversary has no chance to decode if he has only N bits of the keystream. In conclusion, get an ℓ-resilient nonlinear function may not be the best choice as this implies, among other drawbacks, that this function has higher correlations of order $\geq \ell + 1$. Such a function may allow the adversary to apply with greater success a decoding strategy to $\ell+1$ LFSR at once. Choosing a function with a resiliency profile that increases in proportion to the square root of the order ℓ makes the resistance to correlation attack more homogeneous. We stress that this immunity profile is defined from the point of view of information theory independently from the complexity of the decoding procedure. Because of (1), there could exist Boolean functions whose immunity profile satisfies inequality (4) only if, for fixed N and k, n satisfies $n2^n \geq \frac{N}{k\ln(2)}$.

An alternative approach is to define the immunity profile according to the complexity of the decoding procedures. Mainly two different approaches have been proposed in the literature. The first approach [6] consists in associating a smaller linear code of dimension $\alpha k\ell$ (with $\alpha < 1$) to the keystream on which a maximum-likelihood procedure is performed. The resulting complexity of the decoding step is about $\mathcal{O}(\varepsilon^{-2t} \cdot 2^{\alpha k\ell} \cdot k\alpha\ell)$ (where t is some positive integer). The second approach [1] uses the existence of low-density parity-check equations to perform an efficient iterative decoding algorithm. When parity-check equations with weight w are used, the complexity of their decoding procedure is about $\mathcal{O}\Big(\left(\frac{1}{\varepsilon}\right)^{\frac{2w(w-2)}{w-1}} 2^{\frac{k\ell}{w-1}}\Big)$. Consequently, if the immunity profile φ is such that $\varphi(\ell) \leq 2^{\beta\ell}$ for $\ell > 0$ then the complexity of the decoding procedure of the first approach [6] would be greater than $\mathcal{O}\Big(2^{2t(n+1)} \cdot 2^{(\alpha k - 2t\beta)\ell} \cdot \alpha k\ell\Big)$ while the complexity of the second approach [1] would be greater than $\mathcal{O}\Big(2^{\frac{2w(w-2)}{w-1}(n+1)}$.

$2^{\frac{k-2w(w-2)\beta}{w-1}\ell}\Big)$. Basically, decoding the keystream of a combiner generator could be a very hard problem even if the LFSR registers are combined through a Boolean function f with an exponential immunity profile, that is, of the form $\varphi(\ell) = \lambda 2^{\beta\ell}$, $\ell \neq 0$ (the lower the values of β and λ, the more secure the stream cipher).

3.2 Composition of Boolean Functions

Let k be an integer greater than or equal to 2 and consider a Boolean function f over $\mathbb{F}_2^k$. For each $i \in \{1, \ldots, k\}$, let n_i be an integer greater than or equal to 2 and f_i be a Boolean function over $\mathbb{F}_2^{n_i}$. The composition of f by the f_i's is by definition the Boolean function F over $\mathbb{F}_2^{n_1} \times \cdots \times \mathbb{F}_2^{n_k}$ defined by $(x_1, \ldots, x_k) \mapsto f\big(f_1(x_1), \ldots, f_k(x_k)\big)$. Such a construction appears in iterated ciphers where a high complexity ciphering function is required, for example in a self synchronizing stream cipher. In a block cipher, vector valued functions are used, but the analysis principle is quite similar. In order to apply a linear cryptanalysis, a linear approximation of the ciphering function is required and the best approximation is given by the analysis of the above construction. On the other hand, the designer must take care at constructing highly nonlinear ciphering function. By definition, the Walsh transform of the iterated function F is, for any vector $(u_1, \ldots, u_k) \in \mathbb{F}_2^{n_1} \times \cdots \times \mathbb{F}_2^{n_k}$,

$$\widehat{\chi_F}(u_1, \ldots, u_k) = \sum_{(x_1,\ldots,x_k)\in\mathbb{F}_2^{n_1}\times\cdots\times F_2^{n_k}} (-1)^{f\big(f_1(x_1),\ldots,f_k(x_k)\big)+u_1\cdot x_1+\cdots+u_k\cdot x_k}$$

This expression can be expressed by means of the Walsh transform of f and of the f_i's. For this purpose, the inverse Walsh transform formula is used.

$$\begin{aligned}\widehat{\chi_F}(u_1, \ldots, u_k) &= \sum_{(x_1,\ldots,x_k)} \frac{1}{2^k} \sum_{v\in\mathbb{F}_2^k} \widehat{\chi_f}(v)(-1)^{v_1 f_1(x_1)+u_1\cdot x_1+\cdots+v_k f_k(x_k)+u_k\cdot x_k} \\ &= \frac{1}{2^k} \sum_{v\in\mathbb{F}_2^k} \widehat{\chi_f}(v) \prod_{i|v_i=0} (2^{n_i}\delta_0(u_i)) \prod_{i|v_i=1} \widehat{\chi_{f_i}}(u_i)\end{aligned}$$

where δ_0 denotes the Boolean function that takes value 1 at the zero vector and 0 elsewhere. Relation (5) shows that the major contribution to the Walsh transform of F is the 2^{n_i} factor that appears if $u_i = 0$, and this contribution grows exponentially with the number of zero components u_i. This implies that the best linear approximations of F are heuristically those of low weights. In order to counterbalance this effect, the idea is to choose a function f with a Walsh transform that grows exponentially with the weight of the variable. In this case, a very approximate bound on the nonlinearity of F can even be obtained. Suppose that all n_i's equal n, that the Walsh transform of each f_i is bounded by M, that is, $|\widehat{\chi_{f_i}}(u_i)| \leq M$ and that there exists a constant a such that, for any vector $v \in \mathbb{F}_2^k$, one has $|\widehat{\chi_f}(v)| \leq a^{\mathrm{wt}(v)}$. For $(u_1, \ldots, u_k) \in \mathbb{F}_2^n \times \cdots \times \mathbb{F}_2^n$, let

S denote the set $\{i \in \{1,\dots,k\} \mid u_i \neq 0\}$ and s denote the cardinality of S. As the nonzero terms of sum (5) are those for which $v_i = 0$ implies $u_i = 0$, the summation can be limited to vectors v whose support supp(v) includes S. Thus,

$$|\widehat{\chi_F}(u_1,\dots,u_k)| \leq \frac{1}{2^k} \sum_{v \mid S \subset \mathrm{supp}(v)} a^{\mathrm{wt}(v)} M^{\mathrm{wt}(v)} (2^n)^{k-\mathrm{wt}(v)}$$
$$\leq \frac{1}{2^k} \sum_{t=s}^{k} \binom{k-s}{t-s} (2^n)^{k-t} M^t a^t = \frac{M^s a^s}{2^k} (2^n + aM)^{k-s}$$

In consequence, in some iterated cipher, a round ciphering function with an immunity profile that grows exponentially may provide a better resistance to linear cryptanalysis.

4 Almost Resilient Boolean Functions and φ-Correlation Immune Boolean Functions

An alternative approach was proposed by Kurosawa [11] that relaxes the constraints of balancedness of the sub-functions and introduces the concept of *almost resiliency*. Each of the two approaches relies on one of the characterizations of correlation immunity that are equivalent. Consequently, it is advisable to wonder the possible connections between these two approaches. We clarify these connections in this section. We first recall the definition of almost resilient Boolean functions.

Definition 2 ([10]). *Let $f \in \mathcal{B}_n$. Let t be any positive integer less than n. Let ε be any positive real less than 1. Then f is said to be ε-almost $(n,1,t)$-resilient if $\left|\Pr\left(f(X) = y \mid X^I = \sigma\right) - \frac{1}{2}\right| \leq \varepsilon$ for any subset $I = \{i_1,\dots,i_t\}$ of $\{1,\dots,n\}$ whose cardinality equals t, $\sigma \in \mathbb{F}_2^t$ and $y \in \mathbb{F}_2$. Here $\{X^I = \sigma\}$ denotes the event $\{X_{i_1} = \sigma_1,\dots,X_{i_t} = \sigma_t\}$.*

The restrictions of an ε-almost $(n,1,t)$-resilient Boolean function f, obtained by fixing t input bits, lie at distance at most $\varepsilon 2^{n-t}$ from balanced functions. A sufficient condition for almost resiliency involving the Walsh transform has been proposed in [9].

Proposition 1 ([9, Corollary 4.1]). *Let $f \in \mathcal{B}_n$, ε be a positive real and t be a positive integer less than n. Suppose that f is balanced and that, for all $\omega \in \mathbb{F}_2^n$ such that $1 \leq \mathrm{wt}(\omega) \leq t$, $\left|\widehat{\chi_f}(\omega)\right| \leq 2^{n+1}\varepsilon$. Then f is $((2^t-1)\varepsilon)$-almost $(n,1,t)$ resilient.*

This result can be stated in a much more precise way for φ-correlation immune Boolean functions. First, some notation is introduced. Let f be an n-variable Boolean function and $\sigma = (\sigma_1,\dots,\sigma_r) \in \mathbb{F}_2^r$. For any subset $I = \{i_1,\dots,i_r\}$ of

$\{1,\ldots,n\}$, we denote by f_I^σ the sub-function on $\mathbb{F}_2^{n-r}$ obtained by setting the i_jth input to σ_j for every $j \in \{1,\ldots,r\}$. We finally recall the *Poisson summation formula* [2, Corollary 1]. Let f be a Boolean function on $\mathbb{F}_2^n$. Then, for any vector space E of $\mathbb{F}_2^n$, and any a, $b \in \mathbb{F}_2^n$, we have

$$\sum_{u\in a+E} (-1)^{b\cdot u}\widehat{\chi_f}(u) = |E|\,(-1)^{a\cdot b} \sum_{x\in b+E^\perp} (-1)^{a\cdot x+f(x)} \tag{5}$$

where $E^\perp = \{x \in \mathbb{F}_2^n \,/\, \forall y \in E,\ x \cdot y = 0\}$ is the dual of E. We then prove

Proposition 2. *Let $f \in \mathcal{B}_n$ and let φ be any integer-valued mapping over $\{0,\ldots,n\}$. Assume that f is φ-correlation immune. Let $r \in \{1,\ldots,n-1\}$, $\sigma \in \mathbb{F}_2^r$ and $\{i_1,\ldots,i_r\} \subset \{1,\ldots,n\}$. Then f_I^σ is φ_r-correlation immune with $\varphi_r(k) = \frac{1}{2^r}\sum_{j=0}^{r}\binom{r}{j}\varphi(k+j)$, $k \in \{0,\ldots,n-r\}$.*

Proof. In this proof, $I = \{i_1,\ldots,i_r\}$ is an arbitrary subset of $\{1,\ldots,n\}$ $(r<n)$, σ is an element of $\mathbb{F}_2^r$ and ω is an element of $\mathbb{F}_2^{n-r}$. Let E be the vector space whose dual equals $E^\perp = \{x \in \mathbb{F}_2^n \mid x_{i_1} = \cdots = x_{i_r} = 0\}$. Let $b \in \mathbb{F}_2^n$ be such that $b_{i_j} = \sigma_j$ for every $j \in \{1,\ldots,r\}$ and 0 otherwise. Set $\{k_1,\ldots,k_{n-r}\} = \{1,\ldots,n\} \setminus \{i_1,\ldots,i_r\}$. Assume that $k_1 < \cdots < k_{n-r}$. Let $a \in \mathbb{F}_2^n$ be such that $a_{k_j} = \omega_j$ for $j \in \{1,\ldots,n-r\}$ and 0 otherwise. With such notation, we have $\sum_{x\in b+E^\perp}(-1)^{a\cdot x+f(x)} = \widehat{\chi_{f_I^\sigma}}(\omega)$. Then, we deduce from the Poisson summation formula (5) that

$$|\widehat{\chi_{f_I^\sigma}}(\omega)| = \frac{1}{|E|}\left|\sum_{u\in a+E}(-1)^{b\cdot u}\widehat{\chi_f}(u)\right|.$$

The Hamming weights of the elements of $a+E$ range from $\mathrm{wt}(\omega)$ to $\mathrm{wt}(\omega)+r$. Therefore

$$|\widehat{\chi_{f_I^\sigma}}(\omega)| \le \frac{1}{|E|} \times \sum_{j=0}^{r}\binom{r}{j}\varphi(\mathrm{wt}(\omega)+j).$$

This proposition is a generalization of the well-known result : if a n-variable Boolean function is correlation immune of order t then any sub-function obtained by fixing r inputs with $r<t$ is $(t-r)$-th order correlation immune. We then prove thanks to this Proposition the following statement.

Proposition 3. *Let n be any integer, $n \ge 2$. Let φ be any integer-valued mapping over the set $\{1,\ldots,n\}$. Let f be a Boolean function over $\mathbb{F}_2^n$. Assume that f is φ-correlation immune. Then f is $\varepsilon_{\varphi,t}$-almost $(n,1,t)$ resilient for any positive integer t less than n where $\varepsilon_{\varphi,t} = \frac{1}{2^{n+1}}\sum_{j=1}^{t}\binom{t}{j}\varphi(j)$.*

Proof. Let $I = \{i_1,\ldots,i_t\}$ be an arbitrary subset of $\{1,\ldots,n\}$ $(t<n)$ and σ is an element of $\mathbb{F}_2^t$. Note that $\Pr\left(f(X)=y \mid X^I=\sigma\right) = \frac{1}{2} \pm \frac{\widehat{\chi_{f_I^\sigma}}(0)}{2^{n-t+1}}$. According to Proposition 2, f_I^σ is φ_t-correlation immune with $\varphi_t(k) = \frac{1}{2^t}\sum_{j=0}^{t}\binom{t}{j}\varphi(k+j)$, $k \in \{0,\ldots,n-t\}$. Therefore,

$$|\Pr\left(f(X)=y \mid X^I=\sigma\right)-\frac{1}{2}|=\frac{1}{2^{n-t+1}}|\widehat{\chi_{f_I^\sigma}}(0)|\leq\frac{1}{2^{n+1}}\sum_{j=0}^{t}\binom{t}{j}\varphi(j)\,.$$

Remark 3. We obtain a better result than the one directly deduced from Proposition 1 for φ-correlation immune Boolean functions. Indeed, Proposition 1 only allows to conclude that f is $\varepsilon'_{\varphi,t}$-almost $(n,1,t)$ resilient with $\varepsilon'_{\varphi,t}=\frac{2^t-1}{2^{n+1}}\cdot\max_{j\in\{1,\ldots,t\}}\left(\varphi(j)\right)$.

5 Primary Constructions of φ-Correlation Immune Boolean Functions

5.1 Maiorana-McFarland's Construction

The Maiorana-McFarland's class is the set of all n-variable Boolean functions which can be written as follows (n being a positive integer) :

$$\forall(x,y)\in\mathbb{F}_2^r\times\mathbb{F}_2^s,\quad f(x,y)\;=\;\pi(y)\cdot x\oplus g(y)\,,\tag{6}$$

where r and s are two positive integers such that $r+s=n$, where π is a Boolean map from $\mathbb{F}_2^s$ to $\mathbb{F}_2^r$ and g is a s-variable Boolean function. The Walsh transform of such a Boolean function is

$$\forall(a,b)\in\mathbb{F}_2^r\times\mathbb{F}_2^s,\quad \widehat{\chi_f}(a,b)\;=\;2^r\sum_{y\in\pi^{-1}(a)}(-1)^{b\cdot y+g(y)}\,.$$

Resilient Boolean functions whose immunity profile increases in proportion to the square root of the order can be designed from Maiorana-McFarland's class. Indeed, suppose that we can find π such that, for every $a\in\mathbb{F}_2^r$,

$$\begin{array}{ll}\#\pi^{-1}(a)=0 & \text{if } \mathrm{wt}(a)\leq t, \text{ and}\\ \#\pi^{-1}(a)\leq\lambda\lfloor\sqrt{\mathrm{wt}(a)}\rfloor & \text{otherwise}\end{array}\tag{7}$$

for some positive integer t less than r and some positive integer λ. Then any Boolean function f of the form (6) is φ-correlation immune with $\varphi(\ell)=0$ if $\ell\in\{0,\ldots,t\}$, $\varphi(\ell)=2^r\lambda\lfloor\sqrt{\ell}\rfloor$ for $\ell\in\{t+1,\ldots,r\}$ and $\varphi(\ell)=2^r\lambda\sqrt{r}$ otherwise. The existence of such an application π requires that r, s and t fulfil the following inequality deduced from $\bigcup_{a\in\mathbb{F}_2^r,\mathrm{wt}(a)\geq t+1}\pi^{-1}(a)=\mathbb{F}_2^s$:

$$\lambda\sum_{l=t+1}^{r}\binom{r}{l}\lfloor\sqrt{\ell}\rfloor\geq 2^s\,.\tag{8}$$

The nonlinearity of f is greater than or equal to $2^{n-1}-2^{r-1}\lfloor\sqrt{r}\rfloor$ and its algebraic degree equals $\max(\deg(\pi_1)+1,\ldots,\deg(\pi_r)+1,s)$. More generally, one can design ψ-correlation immune Boolean function from the class of Maiorana-McFarland with $\psi(i)=\lambda 2^r\varphi(\min(i,r))$ for $i\in\{0,\ldots,n\}$ provided that $\lambda\sum_{l=0}^{r}\binom{r}{l}\varphi(\ell)\geq 2^s$ under the assumption $\#\pi^{-1}(a)\leq\lambda\varphi(\mathrm{wt}(a))$ for every $a\in\mathbb{F}_2^r$.

Note that it is possible to design φ-correlation immune Boolean functions from the effective partial spreads class [4] with the same nice properties. Because of length limits, we do not develop this further in the present paper.

5.2 Symmetric Boolean Functions with Exponential Correlation Immunity Profile

The condition of φ-correlation immunity only deals with the weight of the argument of the Walsh transform. It is then natural to consider symmetric functions, that is, Boolean functions whose output only depends on the weight of the input vector. If f is an n-variable symmetric Boolean function (n is a positive integer), then there exists a function $\nu_f : \{0, \ldots, n\} \to \mathbb{F}_2$ such that $f(x) = \nu_f(\mathrm{wt}(x))$ for every $x \in \mathbb{F}_2^n$. In the sequel, the function ν_f is called the *simplified value vector* of the symmetric function f.

The Fourier transform of an n-variable symmetric Boolean function f is symmetric too and can be expressed by means of Krawtchouk polynomials for all $\omega \in \mathbb{F}_2^n$ by $\widehat{f}(\omega) = \sum_{k=0}^{n} \nu_f(k) K_k(\mathrm{wt}(\omega), n)$ where ν_f denotes the simplified value vector associated to f and where $K_k(X, n) = \sum_{j=0}^{n} (-1)^j \binom{X}{j}\binom{n-X}{k-j}$, $k = 0, 1, \ldots n$, are the so-called Krawtchouk polynomials. For every $k \in \{0, 1, 2\}$, we denote by $s_{k,3}$ the n-variable symmetric Boolean function whose simplified vector value $\nu_{s_{k,3}}$ is defined by $\nu_{s_{k,3}}(i) = 1$ if $i \equiv k \pmod 3$ and 0 otherwise. The values of the Fourier transform of such functions can easily be calculated. For every $u \in \mathbb{F}_2^n$, denoting $\ell = wt(u)$ and every $k \in \{0, 1, 2\}$, we have $\widehat{s_{k,3}}(u) = \sum_{\substack{0 \le j \le n \\ j \equiv k \ (mod\ 3)}} K_j(\ell, n)$. Let us denote by ω the primitive third root of unity $\omega = e^{2i\pi/3}$. Since we have $\omega^3 = 1$, we deduce that $\sum_{k=0}^{2} \omega^{ke} \widehat{s_{k,2}}(u) = \sum_{0 \le j \le n} \omega^{je} K_j(\ell, n)$, for every $e \in \{0, 1, 2\}$. The generating function of the Krawtchouk polynomials is $\sum_{k=0}^{n} K_k(w, n) z^k = (1 - z)^w (1 + z)^{n-w}$, for $w \in \{0, \ldots, n\}$, and $z \in \mathbb{C}$. This implies that $\sum_{k=0}^{2} \omega^{ke} \widehat{s_{k,3}}(u) = (1 - \omega^e)^\ell (1 + \omega^e)^{n-\ell}$. It is well-known that the inverse of the 3×3 matrix whose term at row k and column e equals ω^{ke} is the matrix whose term at row k and column e equals $\frac{1}{3}\omega^{-ke}$. Thus, For every $k \in \{0, 1, 2\}$ and every $u \in \mathbb{F}_2^n$, denoting $\ell = wt(u)$, the value at u of the Fourier transform $\widehat{s_{k,3}}(u)$ of the function $s_{k,3}$ equals $\frac{1}{3}\sum_{e=0}^{2} (1 - \omega^e)^\ell (1 + \omega^e)^{n-\ell} \omega^{-ke}$. Hence $\widehat{s_{k,3}}(u) = \frac{2}{3}\Re\left((1 - \omega)^\ell (1 + \omega)^{n-\ell} \omega^{-k}\right)$ (where $\Re(z)$ is the real part of $z \in \mathbb{C}$) because ω^2 is the complex conjugate of ω.

We deduce finally from $1 + \omega + \omega^2 = 0$ and $(1 - \omega)\omega^{-2} = \mathrm{i} \cdot \sqrt{3}$ (where i is the primitive square root of unity in $\mathbb{C}$) that $\widehat{\chi_{s_{k,3}}}(u) = -2\widehat{s_{k,3}}(u) = (-1)^{n+1-\ell} \cdot \frac{4}{3} \cdot 3^{\frac{\ell}{2}} \cdot \Re\left(\mathrm{i}^\ell \omega^{2n-k}\right)$. Now, $\Re\left(\mathrm{i}^\ell \omega^{2n-k}\right)$ equals ± 1 if ℓ is even and $2n - k \equiv 0 \pmod 3$, $\pm\frac{1}{2}$ if ℓ is even and $2n - k \not\equiv 0 \pmod 3$, 0 if ℓ is odd and $2n - k \equiv 0 \pmod 3$, $\pm\frac{\sqrt{3}}{2}$ if ℓ is odd and $2n - k \not\equiv 0 \pmod 3$. Then the n-variable symmetric Boolean functions $s_{k,3}$, $k \in \{0, 1, 2\}$, are φ-correlation immune where φ is the integer valued mapping over $\{0, \ldots, n\}$ defined by $\varphi(i) = 4 \cdot 3^{\lfloor \frac{i-1}{2} \rfloor}$ for every $i \in \{1, \ldots, n\}$ and $\varphi(0) = 2^n$.

6 Secondary Constructions of φ-Correlation Immune Boolean Functions

6.1 The Generalized Tarannikov et *al.* Construction

A series of secondary constructions of highly nonlinear resilient functions has been proposed in the literature. This series has led to the very general following construction [3] : Let r, s, t and m be positive integers such that $t < r$ and $m < s$. Let f_1 and f_2 be two r-variable t-resilient functions. Let g_1 and g_2 be two s-variable m-resilient functions. Then the function $h(x,y) = f_1(x) \oplus g_1(y) \oplus (f_1 \oplus f_2)(x)\,(g_1 \oplus g_2)(y)$, $x \in F_2^r, y \in F_2^s$ is an $(r+s)$-variable $(t+m+1)$-resilient function. The Walsh transform of h takes value $\widehat{\chi_h}(a,b) = \frac{1}{2}\widehat{\chi_{f_1}}(a)\,[\widehat{\chi_{g_1}}(b) + \widehat{\chi_{g_2}}(b)] + \frac{1}{2}\widehat{\chi_{f_2}}(a)\,[\widehat{\chi_{g_1}}(b) - \widehat{\chi_{g_2}}(b)]$. Assume then that f_1 and f_2 (resp. g_1 and g_2) are φ-correlation immune (resp. φ'-correlation immune), where φ and φ' are exponential, say $\varphi(\ell) = \lambda 2^{\beta\ell}$ and $\varphi'(\ell) = \lambda' 2^{\beta'\ell}$. Then, since $wt(a,b) = wt(a) + wt(b)$, h is φ''-correlation immune with $\varphi''(\ell) = 2\lambda\lambda' 2^{(\beta+\beta')\ell}$. Note that if $f_1 = f_2$ or $g_1 = g_2$, that is, in the case of a *direct sum*, we have $\varphi''(\ell) = \lambda\lambda' 2^{(\beta+\beta')\ell}$.

6.2 A Recent Secondary Construction Without Extension of the Number of Variables

Given three Boolean functions f_1, f_2 and f_3, there is a nice relationship between their Walsh transforms and the Walsh transforms of two of their elementary symmetric related functions [4]: let us denote by σ_1 the Boolean function equal to $f_1 \oplus f_2 \oplus f_3$ and by σ_2 the Boolean function equal to $f_1 f_2 \oplus f_1 f_3 \oplus f_2 f_3$; then we have $f_1 + f_2 + f_3 = \sigma_1 + 2\sigma_2$ (where these additions are calculated in the ring of integers, that is, not mod 2). This implies $\widehat{\chi_{f_1}} + \widehat{\chi_{f_2}} + \widehat{\chi_{f_3}} = \widehat{\chi_{\sigma_1}} + 2\widehat{\chi_{\sigma_2}}$. If f_1, f_2 and f_3 are k-th order correlation immune (resp. k-resilient), then σ_1 is k-th order correlation immune (resp. k-resilient) if and only if σ_2 is k-th order correlation immune (resp. k-resilient). Moreover, if f_1, f_2 and f_3 are φ-correlation immune as well as σ_1, then, σ_2 is 2φ-correlation immune, whatever is φ. This construction of σ_2 from f_1, f_2, f_3 and σ_1 has the interest of increasing the algebraic complexity of the functions (e.g. their algebraic immunity) without decreasing their nonlinearity (see [4]).

References

1. A. Canteaut and M. Trabbia. Improved fast correlation attacks using parity-check equations of weight 4 and 5. In B. Preneel, editor, *Advances in Cryptology - EUROCRYPT 2000*, volume 1807 of *Lecture Notes in Computer Science*, pages 573+, 2000. Proceedings.
2. C. Carlet. Boolean functions for cryptography and error correcting codes. chapter of the monograph "Boolean methods and models", to be published by Cambridge University Press (Peter Hammer and Yves Crama editors), `http://www-rocq.inria.fr/codes/Claude.Carlet/chap-fcts-Bool.pdf`.

3. C. Carlet. On the secondary constructions of resilient and bent functions. In *Workshop on Coding, Cryptography and Combinatorics*, pages 3–28. Birkhäuser Verlag, 2004.
4. C. Carlet. On bent and highly nonlinear balanced/resilient functions and their algebraic immunities. In *AAECC 16, Las Vegas, february 2006*, volume 3857 of *Lecture Notes in Computer Science*, pages 1–28. Springer, 2006.
5. C. Carlet and P. Sarkar. Spectral domain analysis of correlation immune and resilient boolean functions. *Finite Fields and their Applications*, 8:120–130, 2002.
6. V. V. Chepyzhov, T. Johansson, and B. Smeets. A simple algorithm for fast correlation attacks on stream ciphers. In B. Schneier, editor, *FSE 2000*, volume 1978 of *Lecture Notes in Computer Science*, pages 181–195. Springer-Verlag, April 10-12 2001.
7. N. Courtois. General principles of algebraic attacks and new design criteria for components of symmetric ciphers. In *AES 4 Conference, Bonn, May 10-12*, volume 3373 of *Lecture Notes in Computer Science*, page 67–83. Springer, 2004.
8. N. Courtois and W. Meier. Algebraic attacks on stream ciphers with linear feedback. In *Eurocrypt 03*, volume 2656 of *Lecture Notes in Computer Science*, page 345–359. Springer, 2003.
9. Pin-Hui Ke, Jie Zhang, and Qiao-Yan Wen. Further constructions of almost resilient functions. Cryptology ePrint Archive, Report 2005/453, 2005.
10. K. Kurosawa, T. Johansson, and D. Stinson. Almost k-wise Independent Sample Spaces and Their Cryptologic Applications. *Journal of Cryptology*, 14(4):301–324, 2001.
11. K. Kurosawa and R. Matsumoto. Almost Security of Cryptographic Boolean Functions. *IEEE Transactions on Information Theory*, 50(11):2752–2761, 2004.
12. D. J. C. MacKay. *Information Theory, Inference, and Learning Algorithms.* Cambridge University Press, 2003.
13. W. Meier and O. Staffelbach. Fast correlation attacks on certain stream ciphers. *Journal of Cryptology*, 1(3):159–176, 1989.
14. M. J. Mihaljevic, M. P. C. Fossorier, and H. Imai. A Low-Complexity and High-Performance Algorithm for the Fast Correlation Attack. In B. Schneier, editor, *FSE 2000*, volume 1978 of *Lecture Notes in Computer Science*, pages 196–210. Springer-Verlag, 2001.
15. H. Molland, J. E. Mathiassen, and T. Helleseth. Improved fast correlation attack using low rate codes. In *Cryptography and Coding*, volume 2898 of *Lecture Notes in Computer Science*, pages 67–81. Springer-Verlag GmbH, 2003.
16. T. Siegenthaler. Correlation-immunity of nonlinear combining boolean fucntions for cryptographic applications. *IEEE Transactions on Information Theory*, 30(5):776–779, 1984.
17. Guo-Zhen Xiao and J. L. Massey. A spectral characterization of correlation-immune combining functions. *IEEE Transactions on Information Theory*, 34(3) : 569+, 1988.

Reducing the Number of Homogeneous Linear Equations in Finding Annihilators

Deepak Kumar Dalai and Subhamoy Maitra

Applied Statistics Unit, Indian Statistical Institute,
203, B T Road, Calcutta 700 108, India
{deepak_r, subho}@isical.ac.in

Abstract. Given a Boolean function f on n-variables, we find a reduced set of homogeneous linear equations by solving which one can decide whether there exist annihilators at degree d or not. Using our method the size of the associated matrix becomes $\nu_f \times (\sum_{i=0}^{d} \binom{n}{i} - \mu_f)$, where, $\nu_f = |\{x|wt(x) > d, f(x) = 1\}|$ and $\mu_f = |\{x|wt(x) \leq d, f(x) = 1\}|$ and the time required to construct the matrix is same as the size of the matrix. This is a preprocessing step before the exact solution strategy (to decide on the existence of the annihilators) that requires to solve the set of homogeneous linear equations (basically to calculate the rank) and this can be improved when the number of variables and the number of equations are minimized. As the linear transformation on the input variables of the Boolean function keeps the degree of the annihilators invariant, our preprocessing step can be more efficiently applied if one can find an affine transformation over $f(x)$ to get $h(x) = f(Bx + b)$ such that $\mu_h = |\{x|h(x) = 1, wt(x) \leq d\}|$ is maximized (and in turn ν_h is minimized too). We present an efficient heuristic towards this. Our study also shows for what kind of Boolean functions the asymptotic reduction in the size of the matrix is possible and when the reduction is not asymptotic but constant.

Keywords: Algebraic Attacks, Algebraic Normal Form, Annihilators, Boolean Functions, Homogeneous Linear Equations.

1 Introduction

Results on algebraic attacks have received a lot of attention recently in studying the security of crypto systems [2,4,6,9,11,12,13,14,15,21,1,20,16]. Boolean functions are important primitives to be used in the crypto systems and in view of the algebraic attacks, the annihilators of a Boolean function play considerably serious role [5,7,10,17,18,19,22,23].

Denote the set of all n-variable Boolean functions by B_n. One may refer to [17] for the detailed definitions related to Boolean functions, e.g., truth table, algebraic normal form (ANF), algebraic degree (deg), weight (wt), nonlinearity (nl) and Walsh spectrum of a Boolean function. Any Boolean function can be uniquely represented as a multivariate polynomial over $GF(2)$, called the algebraic normal form (ANF), as

G. Gong et al. (Eds.): SETA 2006, LNCS 4086, pp. 376–390, 2006.

$$f(x_1, \ldots, x_n) = a_0 + \sum_{1 \le i \le n} a_i x_i + \sum_{1 \le i < j \le n} a_{i,j} x_i x_j + \ldots + a_{1,2,\ldots,n} x_1 x_2 \ldots x_n,$$

where the coefficients $a_0, a_i, a_{i,j}, \ldots, a_{1,2,\ldots,n} \in \{0,1\}$. The algebraic degree, $\deg(f)$, is the number of variables in the highest order term with non zero coefficient. Given $f \in B_n$, a nonzero function $g \in B_n$ is called an annihilator of f if $f * g = 0$. A function f should not be used if f or $1 + f$ has a low degree annihilator. It is also known [14,22] that for any function f or $1 + f$ must have an annihilator at the degree $\lceil \frac{n}{2} \rceil$. Thus the target of a good design is to use a function f such that neither f nor $1 + f$ has an annihilator at a degree less than $\lceil \frac{n}{2} \rceil$. Thus there is a need to construct such functions and the first one in this direction appeared in [18]. Later symmetric functions with this property has been presented in [19] followed by [7]. However, all these constructions are not good in terms of other cryptographic properties.

Thus there is a need to study the Boolean functions, which are rich in terms of other cryptographic properties, in terms of their annihilators. One has to find out the annihilators of a given Boolean function for this. Initially a basic algorithm in finding the annihilators has been proposed in [22, Algorithm 2]. A minor modification of [22, Algorithm 2] has been presented very recently in [8] to find out relationships for algebraic and fast algebraic attacks. In [7], there is an efficient algorithm to find the annihilators of symmetric Boolean functions, but symmetric Boolean functions are not cryptographically promising. Algorithms using Gröbner bases are also interesting in this area [3], but still they are not considerably efficient. Recently more efficient algorithms have been designed in this direction [1,20]. The algorithm presented in [1] can be used efficiently to find out relationships for algebraic and fast algebraic attacks. In [1], matrix triangularization has been exploited nicely to solve the annihilator finding problem (of degree d for an n-variable function) in $O(\binom{n}{d}^2)$ time complexity. In [20] a probabilistic algorithm having time complexity $O(n^d)$ has been proposed where the function is divided to its sub functions recursively and the annihilators of the sub functions are checked to study the annihilators of the original function.

The main idea in our effort is to reduce the size of the matrix (used to solve the system of homogeneous linear equations) as far as possible, which has not yet been studied in a disciplined manner to the best of our knowledge. We could successfully improve the handling of equations associated with small weight inputs of the Boolean function. This uses certain structure of the matrix that we discover here. We start with a matrix $M_{n,d}(g)$ (see Theorem 1) which is self inverse and its discovered structure allows to compute the new equations efficiently by considering the matrix UA^r (see Theorem 3 in Section 3). Moreover, each equation associated with a low weight input point directly provides the value of an unknown coefficient of the annihilator, which is the key point that allows to lower the number of unknowns. Further reduction in the size of the matrix is dependent on getting a proper linear transformation on the input variables of the Boolean function, which is discussed in Section 4.

One may wonder whether the very recently available strategies in [1,20] can be applied after the initial reduction proposed in this paper to get further improvements in finding the lowest degree annihilators. The standard Gaussian reduction technique ([20, Algorithm 1]) is used in the main algorithm [20, Algorithm 2], and in that case our idea of reduction of the matrix size will surely provide improvement. However, the ideas presented in [1, Algorithm 1, 2] and [20, Algorithm 3] already exploit the structure of the linear system in an efficient way. In particular, the algorithms in [1] by themselves deal with the equations of small weight efficiently. Thus it is not clear whether the reduction of matrix size proposed by us can be applied to exploit further efficiency from these algorithms.

2 Preliminaries

Consider all the n-variable Boolean functions of degree at most d, i.e., $\mathcal{R}(n,d)$, the Reed-Muller code of order d and length 2^n. Any Boolean function can be seen as a multivariate polynomial over $GF(2)$. Note that $\mathcal{R}(n,d)$ is a vector subspace of the vector space $\mathcal{B}_n$, the set of all n-variable Boolean functions. Now if we consider the elements of $\mathcal{R}(n,d)$ as the multivariate polynomials over $GF(2)$, then the standard basis is the set of all nonzero monomials of degree $\leq d$. That is, the standard basis is

$$S_{n,d} = \{x_{i_1} \dots x_{i_l} : 1 \leq l \leq d \text{ and } 1 \leq i_1 < i_2 < \dots < i_l \leq n\} \cup \{1\},$$

where the input variables of the Boolean functions are $x_1, \dots, x_n$.

The ordering among the monomials is considered in lexicographic ordering ($<_l$) as usual, i.e., $x_{i_1}x_{i_2} \dots x_{i_k} <_l x_{j_1}x_{j_2} \dots x_{j_l}$ if either $k < l$ or $k = l$ and there is $1 \leq p \leq k$ such that $i_k = j_k, i_{k-1} = j_{k-1}, \dots, i_{p+1} = j_{p+1}$ and $i_p < j_p$. So, the set $S_{n,d}$ is a totally ordered set with respect to this lexicographical ordering ($<_l$). Using this ordering we refer the monomials according their order, i.e., the k-th monomial as m_k, $1 \leq k \leq \sum_{i=0}^{d} \binom{n}{i}$ following the convention $m_l <_l m_k$ if $l < k$.

Definition 1. *Given $n > 0$, $0 \leq d \leq n$, we define a mapping*

$$v_{n,d} : \{0,1\}^n \mapsto \{0,1\}^{\sum_{i=0}^{d} \binom{n}{i}},$$

such that $v_{n,d}(x) = (m_1(x), m_2(x), \dots, m_{\sum_{i=0}^{d} \binom{n}{i}}(x))$. Here $m_i(x)$ is the ith monomial as in the lexicographical ordering ($<_l$) evaluated at the point $x = (x_1, x_2, \dots, x_n)$.

To evaluate the value of the t-th coordinate of $v_{n,d}(x_1, x_2, \dots, x_n)$ for $1 \leq t \leq \sum_{i=0}^{d} \binom{n}{i}$, i.e., $[v_{n,d}(x_1, \dots, x_n)]_t$, one requires to calculate the value of the monomial m_t (either 0 or 1) at $(x_1, x_2, \dots, x_n)$. Now we define a matrix $M_{n,d}$ with respect to a n-variable function f. To define this we need another similar ordering ($<^l$) over the elements of vector space $\{0,1\}^n$. We say for $u, v \in \{0,1\}^n$, $u <^l v$ if either $wt(u) < wt(v)$ or $wt(u) = wt(v)$ and there is some $1 \leq p \leq n$ such that $u_n = v_n, u_{n-1} = v_{n-1}, \dots, u_{p+1} = v_{p+1}$ and $u_p = 0, v_p = 1$.

Definition 2. *Given $n > 0$, $0 \le d \le n$ and an n-variable Boolean function f, we define a $wt(f) \times \sum_{i=0}^{d} \binom{n}{i}$ matrix*

$$M_{n,d}(f) = \begin{bmatrix} v_{n,d}(X_1) \\ v_{n,d}(X_2) \\ \vdots \\ v_{n,d}(X_{wt(f)}) \end{bmatrix}$$

where any X_i is an n-bit vector and $supp(f) = \{X_1, X_2, \ldots, X_{wt(f)}\}$ and $X_1 <^l X_2 <^l \cdots <^l X_{wt(f)}$; $supp(f)$ is the set of inputs for which f outputs 1.

Note that the matrix $M_{n,d}(f)$ is the transpose of the restricted generator matrix for Reed-Muller code of length 2^n and order d, $\mathcal{R}(d,n)$, to the support of f (see also [9, Page 7]). Any row of the matrix $M_{n,d}(f)$ corresponding to an input vector $(x_1, \ldots, x_n)$ is

$$\overbrace{1}^{0\,\text{deg}} \ \overbrace{x_1,\ \ldots,\ x_i,\ \ldots,\ x_n}^{1\,\text{deg}} \ \ldots \ \overbrace{x_1 \ldots x_d,\ \ldots,\ x_{i_1} \ldots x_{i_d},\ \ldots,\ x_{n-d+1} \ldots x_n}^{d\,\text{deg}}.$$

Each column of the matrix is represented by a specific monomial and each entry of the column tells whether that monomial is satisfied by the input vector which identifies the row, i.e., the rows of this matrix correspond to the evaluations of the monomials having degree at most d on support of f. As already discussed, here we have one-to-one correspondence from the input vectors $x = (x_1, \ldots, x_n)$ to the row vectors $v_{n,d}(x)$ of length $\sum_{i=0}^{d} \binom{n}{i}$. So, each row is fixed by an input vector.

2.1 Annihilator of f and Rank of the Matrix $M_{n,d}(f)$

Let f be an n-variable Boolean function. We are interested to find out the lowest degree annihilators of f. Let $g \in B_n$ be an annihilator of f, i.e., $f(x_1, \ldots, x_n) * g(x_1, \ldots, x_n) = 0$. In terms of truth table, this means that the function f *AND* g will be a constant zero function, i.e., for each vector $(x_1, \ldots, x_n) \in \{0,1\}^n$, the output of f *AND* g will be zero. That means,

$$g(x_1, \ldots, x_n) = 0 \text{ if } f(x_1, \ldots, x_n) = 1. \tag{1}$$

Suppose degree of the function g is $\le d$, then the ANF of g is of the form $g(x_1, \ldots, x_n) = a_0 + \sum_{i=0}^{n} a_i x_i + \cdots + \sum_{1 \le i_1 < i_2 \cdots < i_d \le n} a_{i_1,\ldots,i_d} x_{i_1} \cdots x_{i_d}$ where the subscripted a's are from $\{0,1\}$ and not all of them are zero. Following Equation 1, we get the following $wt(f)$ many homogeneous linear equations

$$a_0 + \sum_{i=0}^{n} a_i x_i + \cdots + \sum_{1 \le i_1 < i_2 \cdots < i_d \le n} a_{i_1,\ldots,i_d} x_{i_1} \cdots x_{i_d} = 0, \tag{2}$$

considering the input vectors $(x_1, \ldots, x_n) \in supp(f)$. This is a system of homogeneous linear equations on a's with $\sum_{i=0}^{d} \binom{n}{i}$ many a's as variables. The matrix form of this system of equations is $M_{n,d}(f)\ A^{tr} = O$, where $A = (a_0, a_1, a_2, \ldots, a_{n-d+1,\ldots,n})$, the row vector of coefficients of the monomials which are ordered according to the lexicographical order $<_l$. Each nonzero solution of the system of equations formed by Equation 2 gives an annihilator g of degree $\leq d$. This is basically the Algorithm 1 presented in [22]. Since the number of solutions of this system of equations are connected to the rank of the matrix $M_{n,d}(f)$, it is worth to study the rank and the set of linear independent rows/columns of matrix $M_{n,d}(f)$. If the rank of matrix $M_{n,d}(f)$ is equal to $\sum_{i=0}^{d} \binom{n}{i}$ (i.e., number of columns) then the only solution is the zero solution. So, for this case f has no annihilator of degree $\leq d$. This implies that the number of rows $\geq$ number of columns, i.e., $wt(f) \geq \sum_{i=0}^{d} \binom{n}{i}$ which is the Theorem 1 in [17]. If the rank of matrix is equal to $\sum_{i=0}^{d} \binom{n}{i} - k$ for $k > 0$ then the number of linearly independent solutions of the system of equations is k which gives k many linearly independent annihilators of degree $\leq d$ and $2^k - 1$ many number of annihilators of degree $\leq d$. However, to implement algebraic attack one needs only linearly independent annihilators. Hence, finding the degree of lowest degree annihilator of either f or $1 + f$, one can use the following algorithm.

Algorithm 1
for($i = 1$ *to* $\lceil \frac{n}{2} \rceil - 1$*) {*
  find the rank r_1 *of the matrix* $M_{n,i}(f)$*;*
  find the rank r_2 *of the matrix* $M_{n,i}(1 + f)$*;*
  if $\min\{r_1, r_2\} < \sum_{j=0}^{i} \binom{n}{j}$ *then output* i*;*
}
output $\lceil \frac{n}{2} \rceil$*;*

Since either f or $1 + f$ has an annihilator of degree $\leq \lceil \frac{n}{2} \rceil$, we are interested only to check till $i = \lceil \frac{n}{2} \rceil$. This algorithm is equivalent to Algorithm 1 in [22].

The simplest and immediate way to solve the system of these equations or find out the rank of $M_{n,d}(f), M_{n,d}(1 + f)$ is the Gaussian elimination process. To check the existence or to enumerate the annihilators of degree $\leq \lceil \frac{n}{2} \rceil$ for a balanced function, the complexity is approximately $(2^{n-2})^3$. Considering this time complexity, it is not encouraging to check annihilators of a function of 20 variables or more using the presently available computing power. However, given n and d, the matrix $M_{n,d}(f)$ has pretty good structure, which we explore in this paper towards a better algorithm (that is solving the set of homogeneous linear equations in an efficient way by decreasing the size of the matrix involved).

3 Faster Strategy to Construct the Set of Homogeneous Linear Equations

In this section we present an efficient strategy to reduce the set of homogeneous linear equations. First we present a technical result.

Theorem 1. *Let g be an n-variable Boolean function defined as $g(x) = 1$ iff $wt(x) \leq d$ for $0 \leq d \leq n$. Then $M_{n,d}(g)^{-1} = M_{n,d}(g)$, i.e., $M_{n,d}(g)$ is a self inverse matrix.*

Proof. Suppose $\mathcal{F} = M_{n,d}(g)M_{n,d}(g)$. Then the i-th row and j-th column entry of $\mathcal{F}$ (denoted by $\mathcal{F}_{i,j}$) is the scalar product of i-th row and j-th column of $M_{n,d}(g)$. Suppose the i-th row is $v_{n,d}(x)$ for $x \in \{0,1\}^n$ having $x_{q_1}, \ldots, x_{q_l}$ as 1 and others are 0. Further consider that the j-th column is the evaluation of the monomial $x_{r_1} \ldots x_{r_k}$ at the input vectors belonging to the support of g. If $\{r_1, \ldots, r_k\} \not\subseteq \{q_1, \ldots, q_l\}$ then $\mathcal{F}_{ij} = 0$. Otherwise, $\mathcal{F}_{i,j} = \binom{l-k}{0} + \binom{l-k}{1} + \cdots + \binom{l-k}{l-k} \bmod 2 = 2^{l-k} \bmod 2$. So, $\mathcal{F}_{i,j} = 1$ iff $\{x_{r_1}, \ldots, x_{r_k}\} = \{x_{q_1}, \ldots, x_{q_l}\}$. That implies, $\mathcal{F}_{i,j} = 1$ iff $i = j$ i.e., $\mathcal{F}$ is identity matrix. Hence, $M_{n,d}(g)$ is its own inverse. □

See the following example for the structure of $M_{n,d}(g)$ when $n = 4$ and $d = 2$.

Example 1. Let us present an example of $M_{n,d}(g)$ for $n = 4$ and $d = 2$. We have $\{1, x_1, x_2, x_3, x_4, x_1x_2, x_1x_3, x_2x_3, x_1x_4, x_2x_4, x_3x_4\}$, the list of 4-variable monomials of degree ≤ 2 in ascending order ($<_l$).

Similarly, $\{(0,0,0,0), (1,0,0,0), (0,1,0,0), (0,0,1,0), (0,0,0,1), (1,1,0,0), (1,0,1,0), (0,1,1,0), (1,0,0,1), (0,1,0,1), (0,0,1,1)\}$ present the 4 dimensional vectors of weight ≤ 2 in ascending order ($<^l$). So the matrix

$$M_{4,2}(g) = \begin{bmatrix} 1&0&0&0&0&0&0&0&0&0&0 \\ 1&1&0&0&0&0&0&0&0&0&0 \\ 1&0&1&0&0&0&0&0&0&0&0 \\ 1&0&0&1&0&0&0&0&0&0&0 \\ 1&0&0&0&1&0&0&0&0&0&0 \\ 1&1&1&0&0&1&0&0&0&0&0 \\ 1&1&0&1&0&0&1&0&0&0&0 \\ 1&0&1&1&0&0&0&1&0&0&0 \\ 1&1&0&0&1&0&0&0&1&0&0 \\ 1&0&1&0&1&0&0&0&0&1&0 \\ 1&0&0&1&1&0&0&0&0&0&1 \end{bmatrix}$$

One may check that $M_{4,2}(g)$ is self inverse.

Lemma 1. *Let A be a nonsingular $m \times m$ binary matrix where the row vectors are denoted as $v_1, v_2, \ldots, v_m$. Let U be a $k \times m$ binary matrix, $k \leq m$, where the rows are denoted as $u_1, u_2, \ldots, u_k$. Let $W = UA^{-1}$, a $k \times m$ binary matrix. Consider that a matrix A' is formed from A by replacing the rows $v_{i_1}, v_{i_2}, \ldots, v_{i_k}$ of A by the vectors $u_1, u_2, \ldots, u_k$. Further consider that a $k \times k$ matrix W' is formed by taking the i_1-th, i_2-th, $\ldots, i_k$-th columns of W (out of m columns). Then A' is nonsingular iff W' is nonsingular.*

Proof. Without loss of generality, we can take $i_1 = 1, i_2 = 2, \ldots, i_k = k$. So, the row vectors of A' are $u_1, \ldots, u_k, v_{k+1}, \ldots, v_m$.

We first prove that if the row vectors of A' are not linearly independent then the row vectors of W' are also not linearly independent. As the row vectors of A' are not linearly independent, we have $\alpha_1, \alpha_2, \ldots, \alpha_m \in \{0,1\}$ (not all zero) such that $\sum_{i=1}^{k} \alpha_i u_i + \sum_{i=k+1}^{m} \alpha_i v_i = 0$. If $\alpha_i = 0$ for all i, $1 \leq i \leq k$ then $\sum_{i=k+1}^{m} \alpha_i v_i = 0$ which implies $\alpha_i = 0$ for all i, $k+1 \leq i \leq m$ as $v_{k+1}, v_{k+2}, \ldots, v_m$ are linearly independent. So, all α_i's for $1 \leq i \leq k$ can not be zero.

Further, we have $UA^{-1} = W$, i.e., $U = WA$, i.e.,

$$\begin{pmatrix} u_1 \\ u_2 \\ \vdots \\ u_k \end{pmatrix} = \begin{pmatrix} w_1 \\ w_2 \\ \vdots \\ w_k \end{pmatrix} \begin{pmatrix} v_1 \\ v_2 \\ \vdots \\ v_m \end{pmatrix}, \text{ i.e., } u_i = w_i \begin{pmatrix} v_1 \\ v_2 \\ \vdots \\ v_m \end{pmatrix}.$$

Hence, $\sum_{i=1}^{k} \alpha_i u_i = \sum_{i=1}^{k} \alpha_i w_i \begin{pmatrix} v_1 \\ v_2 \\ \vdots \\ v_m \end{pmatrix} = r \begin{pmatrix} v_1 \\ v_2 \\ \vdots \\ v_m \end{pmatrix}$

where $r = (r_1, r_2, \ldots, r_m) = \sum_{i=1}^{k} \alpha_i w_i$.

If the restricted matrix W' were nonsingular, the vector $r' = (r_1, r_2, \ldots, r_k)$ is non zero as $(\alpha_1, \alpha_2, \ldots, \alpha_k)$ is not all zero. Hence, $\sum_{i=1}^{k} \alpha_i u_i + \sum_{i=k+1}^{m} \alpha_i v_i = 0$, *i.e.*, $\sum_{i=1}^{k} r_i v_i + \sum_{i=k+1}^{m} (r_i + \alpha_i) v_i = 0$. This contradicts that $v_1, v_2, \ldots, v_m$ are linearly independent as $r' = (r_1, r_2, \ldots, r_k)$ is nonzero. Hence W' must be singular. This proves one direction.

On the other direction if the restricted matrix W' is singular then there are $\beta_1, \beta_2, \ldots, \beta_k$ not all zero such that $\sum_{i=0}^{k} \beta_i w_i = (0, \ldots, 0, s_{k+1}, \ldots, s_m)$. Hence,

$\sum_{i=0}^{k} \beta_i u_i = \sum_{i=1}^{k} \beta_i w_i \begin{pmatrix} v_1 \\ v_2 \\ \vdots \\ v_m \end{pmatrix} = s_{k+1} v_{k+1} + \cdots + s_m v_m$, i.e., $\sum_{i=0}^{k} \beta_i u_i + \sum_{i=k+1}^{m} s_i v_i = 0$ which says matrix A' is singular. □

Following Lemma 1, one can check the nonsingularity of the larger matrix A' by checking the nonsingularity of the reduced matrix W'. Thus checking the nonsingularity of the larger matrix A' will be more efficient if the computation of matrix product $W = UA^{-1}$ can be done efficiently. The self inverse nature of the matrix $M_{n,d}(g)$ presented in Theorem 1 helps to achieve this efficiency. In the rest of this section we will study this in detail. In the following result we present the Lemma 1 in more general form.

Theorem 2. *Let A be a nonsingular $m \times m$ binary matrix with m-dimensional row vectors $v_1, v_2, \ldots, v_m$ and U be a $k \times m$ binary matrix with m-dimensional row vectors $u_1, u_2, \ldots, u_k$. Consider $W = UA^{-1}$, a $k \times m$ matrix. The matrix A', formed from A by removing the rows $v_{i_1}, v_{i_2}, \ldots, v_{i_l}$ $(l \leq m)$ from A and adding the rows $u_1, u_2, \ldots, u_k$ $(k \geq l)$, is of rank m iff the rank of restricted $k \times l$ matrix W' including only the i_1-th, i_2-th, $\ldots, i_l$-th columns of W is l.*

Proof. Here, the rank of matrix W' is l. So, there are l many rows of W', say $w'_{p_1}, \ldots, w'_{p_l}$ which are linearly independent. So, following the Lemma 1 we have the matrix A'' formed by replacing the rows $v_{i_1}, \ldots, v_{i_l}$ of A by $u_{p_1}, \ldots, u_{p_l}$ is nonsingular, i.e., rank is m. Hence the matrix A' where some more rows are added to A'' has rank m. The other direction can also be shown similar to the proof of the other direction in Lemma 1. □

Now using Theorem 1 and Theorem 2, we describe a faster algorithm to check the existence of annihilators of certain degree d of a Boolean function f. Suppose g be the Boolean function described in Theorem 1, i.e., $supp(g) = \{x | 0 \leq wt(x) \leq d\}$. In Theorem 1, we have already shown that $M_{n,d}(g)$ is nonsingular matrix (in fact it is self inverse). Let $\{x | wt(x) \leq d$ and $f(x) = 0\} = \{x_1, x_2, \ldots, x_l\}$ and $\{x | wt(x) > d$ and $f(x) = 1\} = \{y_1, y_2, \ldots, y_k\}$. Then we consider $M_{n,d}(f)$ as

A, $v_{n,d}(x_1), \ldots, v_{n,d}(x_l)$ as $v_{i_1}, \ldots, v_{i_l}$ and $v_{n,d}(y_1), \ldots, v_{n,d}(y_k)$ as $u_1, \ldots, u_k$. Then following Theorem 2 we can ensure whether $M_{n,d}(f)$ is nonsingular. If it is nonsingular, then there is no annihilator of degree $\leq d$, else there are annihilator(s). We may write this in a more concrete form as the following corollary to Theorem 2.

Corollary 1. *Let f be an n-variable Boolean function. Let A^r be the restricted matrix of $A = M_{n,d}(g)$, by taking the columns corresponding to the monomials $x_{i_1}x_{i_2}\ldots x_{i_l}$ such that $l \leq d$ and $f(x) = 0$ when $x_{i_1} = 1, x_{i_2} = 1, \ldots, x_{i_l} = 1$ and rest of the input variables are 0. Further* $U = \begin{pmatrix} v_{n,d}(y_1) \\ v_{n,d}(y_2) \\ \vdots \\ v_{n,d}(y_k) \end{pmatrix}$, *where $\{y_1, \ldots, y_k\} = \{x | wt(x) > d \text{ and } f(x) = 1\}$. If rank of UA^r is l then there is no annihilator of degree $\leq d$, else there are annihilator(s) of degree $\leq d$.*

Proof. As per Theorem 2, here $W = UA^{-1} = UA$, since A is self inverse following Theorem 1 and hence W' is basically UA^r. Thus the proof follows. □

Now we can use the following technique for fast computation of the matrix multiplication UA^r. For this we first present a technical result and its proof is similar in the line of the proof of Theorem 1.

Proposition 1. *Consider g as in Theorem 1. Let $y \in \{0,1\}^n$ such that i_1, i_2, $\ldots$, i_p-th places are 1 and other places are 0. Consider the j-th monomial $m_j = x_{j_1}x_{j_2}\ldots x_{j_q}$ according the ordering $<_l$. Then the j-th entry of $v_{n,d}(y)M_{n,d}(g)$ is 0 if $\{j_1, \ldots, j_q\} \not\subseteq \{i_1, \ldots, i_p\}$ else the value is $\sum_{i=0}^{d-q} \binom{p-q}{i} \bmod 2$.*

Following Proposition 1, we can get each row of U as some $v_{n,d}(y)$ and each column of A^r as m_j and construct the matrix UA^r. One can precompute the sums $\sum_{i=0}^{d-q} \binom{p-q}{i} \bmod 2$ for $d+1 \leq p \leq n$ and $0 \leq q \leq d$, and store them and the total complexity for calculating them is $O(d^2(n-d))$. These sums will be used to fill up the matrix UA^r which is an $l \times k$ matrix according to Corollary 1. Let us denote $\mu_f = |\{x | wt(x) \leq d, f(x) = 1\}|$ and $\nu_f = |\{x | wt(x) > d, f(x) = 1\}|$. Then $wt(f) = \mu_f + \nu_f$ and the matrix UA^r is of dimension $\nu_f \times (\sum_{i=0}^{d} \binom{n}{i} - \mu_f)$. Clearly $O(d^2(n-d))$ can be neglected with respect to $\nu_f \times (\sum_{i=0}^{d} \binom{n}{i} - \mu_f)$. Thus we have the following result.

Theorem 3. *Consider U and A^r as in Corollary 1. The time (and also space) complexity to construct the matrix UA^r is of the order of $\nu_f \times (\sum_{i=0}^{d} \binom{n}{i} - \mu_f)$. Further checking the rank of UA^r (as given in Corollary 1) one can decide whether f has an annihilator at degree d or not.*

In fact, to check the rank of the matrix UA^r using Gaussian elimination process, we need not store the ν_f many rows at the same time. One can add one row (following the calculation to compute a row of the matrix given in Proposition 1) at a time incrementally to the previously stored linearly independent rows by checking whether the present row is linearly independent with respect to the already stored rows. If the current row is linearly independent with the existing ones, then we do row operations and add the new row to the previously stored matrix. Otherwise we reject the new row. Hence, our matrix size never crosses the size $(\sum_{i=0}^{d}\binom{n}{i} - \mu_f) \times (\sum_{i=0}^{d}\binom{n}{i} - \mu_f)$.

If ν_f (the number of rows) is less than $(\sum_{i=0}^{d}\binom{n}{i} - \mu_f)$ (the number of variables), then there will be nontrivial solutions and we can directly say that the annihilators exist. Thus we always need to concentrate on the case $\nu_f \geq (\sum_{i=0}^{d}\binom{n}{i} - \mu_f)$, where the matrix size $(\sum_{i=0}^{d}\binom{n}{i} - \mu_f) \times (\sum_{i=0}^{d}\binom{n}{i} - \mu_f)$ provides a further reduction than the matrix size $\nu_f \times (\sum_{i=0}^{d}\binom{n}{i} - \mu_f)$ and one can save more space. This will be very helpful when one tries to check the annihilators of small degree d.

One may refer to Subsection 3.1 of the extended version of this paper at IACR eprint server (eprint.iacr.org, number 2006/032) to get detailed description why our strategy provides asymptotic improvement than [22] in terms of constructing this reduced set of homogeneous linear equations. In terms of the overall algorithm to find the annihilators, our algorithm works around eight times faster than [22] in general. Using our strategy to find the reduced matrix first and then using the standard Gaussian elimination technique, we could find the annihilators of any random balanced Boolean functions on 16 variables in around 2 hours in a Pentium 4 personal computer with 1 GB RAM. Note that, the very recently known efficient algorithms [1,20] can work till 20 variables.

4 Further Reduction in Matrix Size Applying Linear Transformation over the Input Variables of the Function

To check for the annihilators, we need to compute the rank of the matrix UA^r. Following Theorem 3, it is clear that the size of the matrix UA^r will decrease if μ_f increases and ν_f decreases. Let B be an $n \times n$ nonsingular binary matrix and b be an n-bit vector. The function $f(x)$ has an annihilator at degree d iff $f(Bx+b)$ has an annihilator at degree d. Thus one will try to get the affine transformation on the input variables of $f(x)$ to get $h(x) = f(Bx+b)$ such that $|\{x|h(x) = 1, wt(x) \leq d\}|$ is maximized. This is because in this case μ_h will be maximized and ν_h will be minimized and hence the dimension of the matrix UA^r, i.e., $\nu_f \times (\sum_{i=0}^{d}\binom{n}{i} - \mu_f)$ will be minimized. This will indeed decrease the complexity at the construction step (discussed in the previous section). More importantly, it will decrease the complexity to solve the system of homogeneous linear equations.

See the following example that explains the efficiency for a 5-variable function.

Example 2. We present an example for this purpose. Consider the 5-variable Boolean function f constructed using the method presented in [18] such that neither f nor $1+f$ has an annihilator at a degree < 3. The standard truth table representation of the function is 01010110010101100101011001101001, i.e., the outputs are corresponding to the inputs which are of increasing value. One can check that $|\{x \in \{0,1\}^5 \mid f(x) = 1 \ \& \ wt(x) < 3\}| = 6$. Now if we consider the function $h(x) = f(Bx+b)$ such that $B = \begin{bmatrix} 1&1&1&0&1 \\ 1&1&1&1&0 \\ 1&0&1&0&0 \\ 1&1&0&0&0 \\ 1&0&0&0&0 \end{bmatrix}$, and $b = \{1,1,0,0,1\}$, then $|\{x \in \{0,1\}^5 \mid h(x) = 1 \ \& \ wt(x) < 3\}| = 16$ and one can immediately conclude (from the results in [19]) that neither h nor $1+h$ has an annihilator of degree < 3. This is an example where after finding the affine transformation there is even no need for the solution step at all. For the function f, here $h(x) = f(Bx+b)$ such that $|\{x|h(x) = 1, wt(x) \le d\}|$ is maximized.

We also present an example for a sub optimal case. In this case we consider $B = \begin{bmatrix} 1&0&1&0&0 \\ 1&1&0&0&0 \\ 1&1&1&0&1 \\ 0&0&0&1&1 \\ 0&1&1&1&0 \end{bmatrix}$, and b an all zero vector, then $|\{x \in \{0,1\}^5 \mid h(x) = 1 \ \& \ wt(x) < 3\}| = 14$. Thus the dimension of the matrix UA^r becomes 2×2 as $\nu_f = 2$ and $\sum_{i=0}^{d} \binom{n}{i} - \mu_f = 2$. Thus one needs to check the rank of a 2×2 matrix only.

Now the question is how to find such an affine transformation (for the optimal or even for sub optimal cases) efficiently.

For exhaustive search to get the optimal affine transform one needs to check $f(Bx+b)$ for all $n \times n$ nonsingular binary matrices B and n bit vectors b. Since there are $\prod_{i=0}^{n-1}(2^n - 2^i)$ many nonsingular binary matrices and 2^n many n bit vectors, one needs to check $2^n \prod_{i=0}^{n-1}(2^n - 2^i)$ many cases for an exhaustive search. As weight of the input vectors are invariant under permutation of the arguments, checking for only one nonsingular matrix from the set of all nonsingular matrices whose rows are equivalent under certain permutation will suffice. Hence the exact number of search options is $\frac{1}{n!} 2^n \prod_{i=0}^{n-1}(2^n - 2^i)$. One can check for $n \times n$ nonsingular binary matrices B where $row_i < row_j$ for $i < j$ (row_i is the decimal value of binary pattern of ith row). It is clear that the search is infeasible for $n \ge 8$.

Now we present a heuristic towards this. Our aim is to find out an affine transformation $h(x)$ of $f(x)$, i.e., $h(x) = f(Bx+b)$, which maximizes the value of μ_h. This means the weight of the most of the input vectors having weight $\le d$ should be in *supp*(h). So we attempt to get an affine transformation for a Boolean function f such that the transformation increases the probability that an input vector, having output 1, will be translated to a low weight input vector.

Consider $h(Vx+v) = f(x)$, where V is an $n \times n$ binary matrix and $v = (v_1, v_2, \ldots, v_n) \in \{0,1\}^n$. Suppose $r_1, r_2, \ldots, r_n \in \{0,1\}^n$ are the row vectors of the transformation V. By $Vx + v = y$ we mean $Vx^{tr} + v = y^{tr}$, where $x = (x_1, x_2, \ldots, x_n), y = (y_1, y_2, \ldots, y_n) \in \{0,1\}^n$. Given an x, we find a y by this transformation and then $h(y)$ is assigned to the value of $f(x)$. If $f(x) = 1$, we like that the corresponding $y = Vx + v$ should be of low weight. The chance of $(y_1, y_2, \ldots, y_n)$ getting low weight increases if the probability of $y_i = 0, 1 \leq i \leq n$ is increased. That means the probability of $r_i \cdot (x_1, x_2, \ldots, x_n) + v_i = 0$ for $1 \leq i \leq n$ needs to be increased. Hence we will like to choose a linearly independent set $r_i \in \{0,1\}^n$, $1 \leq i \leq n$ and $v \in \{0,1\}^n$ such that the probability $r_i \cdot (x_1, x_2, \ldots, x_n) + b_i = 0$, $1 \leq i \leq n$ is high when $(x_1, x_2, \ldots, x_n) \in supp(f)$. Since we use the relations $h(Vx+v) = f(x)$, and $h(x) = f(Bx+b)$, that means $B = V^{-1}$ and $b = V^{-1}v$.

The heuristic is presented below. By $bin[i]$ we denote the n-bit binary representation of the integer i.

Heuristic 1

1. *loop* $= 0$; *max* $= |\{x|f(x) = 1, wt(x) \leq d\}|$;
2. *For* $(i = 1; i < 2^n; i++)$ {
   (a) $t = |\{x = (x_1, x_2, \ldots, x_n) \in supp(f) | bin[i] \cdot x = 0\}|$
   (b) *if* $t \geq \frac{wt(f)}{2}$, $val[i] = t$ *and* $a_i = 0$ *else* $val[i] = wt(f) - t$ *and* $a_i = 1$.
   }
3. *Arrange the triplets* $(bin[i], a_i, val[i])$ *in descending order of* $val[i]$.
4. *Choose suitable* n *many triplets* (r_j, v_j, k_j) *for* $1 \leq j \leq n$ *such that* r_j*s are linearly independent and* k_j*'s are high.*
5. *Construct the nonsingular matrix* V *taking* r_j, $1 \leq j \leq n$ *as* j*-th row and* $v = (v_1, v_2, \ldots, v_n)$.
6. *Increment loop by* 1; *while* (*loop* $<$ *maxval*)
   (a) $B = V^{-1}$, $b = V^{-1}v$.
   (b) *if max* $< |\{x|f(Bx+b) = 1, wt(x) \leq d\}|$ *replace* $f(x)$ *by* $f(Bx+b)$ *and update max by* $|\{x|f(Bx+b) = 1, wt(x) \leq d\}|$.
   (c) *Go to step 2.*

The time complexity of this heuristic is $(maxval \times n2^{2n})$. See the following example, where we trace Heuristic 1 for the 5-variable function f given in Example 2.

Example 3. We have $f =$ 01010110010101100101011001101001 and check that $|\{x \in \{0,1\}^5 \mid f(x) = 1 \;\&\; wt(x) \leq 2\}| = 6$. In step 2, we get $(val[i], a_i)$ for $1 \leq i \leq 31$ as $1 : (11,1)$, $2 : (8,1)$, $3 : (11,1)$, $4 : (8,1)$, $5 : (11,1)$, $6 : (8,1)$, $7 : (9,0)$, $8 : (8,1)$, $9 : (9,1)$, $10 : (8,1)$, $11 : (9,1)$, $12 : (8,1)$, $13 : (9,1)$, $14 : (8,1)$, $15 : (11,0)$, $16 : (8,1)$, $17 : (9,1)$, $18 : (8,1)$, $19 : (9,1)$, $20 : (8,1)$, $21 : (9,1)$, $22 : (8,1)$, $23 : (11,0)$, $24 : (8,1)$, $25 : (9,0)$, $26 : (8,1)$, $27 : (9,0)$, $28 : (8,1)$, $29 : (9,0)$, $30 : (8,1)$, $31 : (11,1)$. Then after ordering according the value of $val[i]$, we choose the row of matrix V as the 5-bit binary expansion of $1, 3, 5, 15$ and 7 with frequency values of 0's as $11, 11, 11, 11, 9$ respectively and $v = (a_1, a_3, a_5, a_{15}, a_7) = (1,1,1,1,0)$. Here the matrix V is a nonsingular matrix. The new function is $g = f(Bx+b)$, where $B = V^{-1}$, $b = V^{-1}v$ and one can check that $|\{x \in \{0,1\}^5 \mid g(x) = 1 \;\&\; wt(x) \leq 2\}| = 16$.

Experiments with this heuristic on different Boolean functions provide very positive results. First of all we have considered the functions which are random affine transformations $g(x)$ of the function [19], $f_s(x) = 1$ for $wt(x) \leq \lfloor \frac{n-1}{2} \rfloor$ and $f_s(x) = 0$ for $wt(x) \geq \lfloor \frac{n+1}{2} \rfloor$, which has no annihilator having degree $\leq \lfloor \frac{n-1}{2} \rfloor$. This experimentation has been done for $n = 5$ to 16. For all the cases running Heuristic 1 on $g(x)$ we could go back to $f_s(x)$. Then we have randomly changed $2^{\zeta n}$ bits on the upper half of $f_s(x)$ ($0.5 \leq \zeta \leq 0.8$ at steps of 0.1) to get $f'_s(x)$ and then put random transformations on $f'_s(x)$ to get $g(x)$. Running Heuristic 1, we could also go back to $f'_s(x)$ easily. For experiments we have taken $maxval = 20$.

The important issue is exactly when this matrix size is asymptotically reduced than the trivial matrix size $wt(f) \times \sum_{i=0}^{d} \binom{n}{i}$ if one writes down the equations by looking at the truth table of the function only. This happens only when μ_f is very close to $\sum_{i=0}^{d} \binom{n}{i}$. Let $\sum_{i=0}^{d} \binom{n}{i} - \mu_f \leq 2^{\zeta n}$, where ζ is a constant such that $0 < \zeta < 1$. In that case the matrix size will be less than or equal to $(wt(f) + 2^{\zeta n} - \sum_{i=0}^{d} \binom{n}{i}) \times 2^{\zeta n}$. When $d = \lfloor \frac{n}{2} \rfloor$ and n odd, $\sum_{i=0}^{d} \binom{n}{i} = 2^{n-1}$. Thus for a balanced function, the size of the matrix becomes as low as $2^{\zeta n} \times 2^{\zeta n}$. We summarize the result as follows.

Theorem 4. *Predetermine a constant ζ, such that $0 < \zeta < 1$. Consider any Boolean function $f(x) \in B_n$ for which there exist a nonsingular binary matrix B and an n-bit vector b such that $\sum_{i=0}^{d} \binom{n}{i} - |\{x | f(Bx+b) = 1, wt(x) \leq d\}| \leq 2^{\zeta n}$. If B and b are known, then the size of the matrix UA^r will be less than or equal to $(wt(f) + 2^{\zeta n} - \sum_{i=0}^{d} \binom{n}{i}) \times 2^{\zeta n}$ which is asymptotically reduced in size than $wt(f) \times \sum_{i=0}^{d} \binom{n}{i}$.*

That B, b can be known is quite likely from the experimental results available running Heuristic 1.

Next we have run our heuristics on randomly chosen balanced functions. The number of inputs up to weight d for a Boolean function is $\sum_{i=0}^{d} \binom{n}{i}$. Thus for a randomly chosen balanced function, it is expected that there will be $\frac{1}{2} \sum_{i=0}^{d} \binom{n}{i}$ many inputs up to weight d for which the outputs are 1. Below we present the improvement (on an average of 100 experiments in each case) we got after running Heuristic 1 with $maxval = 20$ for $n = 12$ to 16.

Table 1. Efficiency of Heuristic 1 on random balanced functions

n	12			13			14			15			16		
d	3	4	5	4	5	6	4	5	6	5	6	7	5	6	7
$\sum_{i=0}^{d} \binom{n}{i}$	299	794	1586	1093	2380	4096	1471	3473	6476	4944	9949	16384	6885	14893	26333
$\lceil \frac{1}{2} \sum_{i=0}^{d} \binom{n}{i} \rceil$	149	397	793	541	1190	2048	735	1736	3238	2472	4974	8192	3442	7446	13166
Heuristic Value	228	535	964	717	1438	2322	957	2051	3648	2917	5525	8811	3995	8194	14114

It should be noted that after running our heuristic on random balanced functions, the improvement is not extremely significant. There are improvements as we find that the the values are significantly more than $\frac{1}{2} \sum_{i=0}^{d} \binom{n}{i}$ (making our algorithm efficient), but the value is not very close to $\sum_{i=0}^{d} \binom{n}{i}$. This is not a problem with the efficiency of the heuristic, but with the inherent property of a random Boolean function that there may not be an affine transformation at all on $f(x)$ such that $|\{x|f(Bx+b)=1, wt(x) \leq d\}|$ is very high. In fact we can show that for highly nonlinear functions $f(x)$, the increment from $|\{x|f(x)=1, wt(x) \leq d\}|$ to $|\{x|f(Bx+b)=1, wt(x) \leq d\}|$ may not be significant for any B, b. The reason for this is as follows.

Proposition 2. *Let $f \in B_n$ be a balanced function (n odd) having nonlinearity $nl(f) = 2^{n-1} - 2^{\frac{n-1}{2}}$. Then for any nonsingular $n \times n$ matrix B and any n-bit vector b, $2^{n-1} - |\{x|f(Bx+b)=1, wt(x) \leq \frac{n-1}{2}\}| \geq \frac{1}{2}\binom{n-1}{\frac{n-1}{2}} - 2^{\frac{n-1}{2}-1}$.*

Proof. Let $f \in B_n$ be a balanced function (n odd) having nonlinearity $nl(f) = 2^{n-1} - 2^{\frac{n-1}{2}}$. Let $g \in B_n$ be the function such that $g(x) = 1$ for $wt(x) \leq \frac{n-1}{2}$. By [19, Theorem 3], $nl(g) = 2^{n-1} - \binom{n-1}{\frac{n-1}{2}}$. Now we like to find out a function $h(x) = f(Bx+b)$ such that $|\{x|h(x)=1, wt(x) \leq \frac{n-1}{2}\}|$ is high. Consider the value $T = |supp(g) \cap supp(h)|$, i.e., $T = |\{x : h(x) = 1 \ \& \ wt(x) \leq \frac{n-1}{2}\}|$. Without loss of generality consider $T \geq 2^{n-2}$. Hence, $d(h,g) = 2(2^{n-1} - T) = 2^n - 2T$. Now, $nl(f) = nl(h) \leq nl(g) + d(h,g) = (2^{n-1} - \binom{n-1}{\frac{n-1}{2}}) + 2^n - 2T$. Thus, $2^{n-1} - 2^{\frac{n-1}{2}} \leq (2^{n-1} - \binom{n-1}{\frac{n-1}{2}}) + 2^n - 2T$, i.e., $2^{n-1} - T \geq \frac{1}{2}\binom{n-1}{\frac{n-1}{2}} - 2^{\frac{n-1}{2}-1}$. □

Thus if one predetermines a ζ, then for a large n we may not satisfy the condition that $\sum_{i=0}^{\frac{n-1}{2}} \binom{n}{i} - |\{x|f(Bx+b)=1, wt(x) \leq d\}| \leq 2^{\zeta n}$. In this direction we present the following general result where the constraint of nonlinearity is removed.

Theorem 5. *Suppose $f \in B_n$ be a randomly chosen balanced function. Then the probability to get an affine transformation such that*

$$|\{x|f(Bx+b)=1, wt(x) \leq \lfloor \frac{n-1}{2} \rfloor\}| > \sum_{i=0}^{\lfloor \frac{n-1}{2} \rfloor} \binom{n}{i} - k \text{ is}$$

1. *less than* $\frac{(n+1)2^n \sum_{i=0}^{k-1} \binom{2^{n-1}}{i}^2}{\binom{2^n}{2^{n-1}}}$ *for n odd.*
2. *less than* $\frac{(n+1)2^n \sum_{i=0}^{k-1} \binom{\sum_{j=0}^{\frac{n}{2}-1}\binom{n}{j}}{i}\binom{2^n - \sum_{j=0}^{\frac{n}{2}-1}\binom{n}{j}}{i+\frac{1}{2}\binom{n}{\frac{n}{2}}}}{\binom{2^n}{2^{n-1}}}$ *for n even.*

Proof. First we prove it for n odd. The number of balanced functions $h \in B_n$ such that $|\{x|h(x)=1, wt(x) \leq \frac{n-1}{2}\}| > 2^{n-1} - k$ is $\sum_{i=0}^{k-1} \binom{2^{n-1}}{i}^2$ (consider the upper and lower half in the truth table of the function). So, there will be at most $\sum_{i=0}^{k-1} \binom{2^{n-1}}{i}^2$ many affinely invariant classes of such functions. Further the total number of balanced function is $\binom{2^n}{2^{n-1}}$. Hence the total number of affinely invariant classes of balanced function is $\geq \frac{\binom{2^n}{2^{n-1}}}{2^n(2^n-1)(2^n-2^1)\ldots(2^n-2^{n-1})} > \frac{\binom{2^n}{2^{n-1}}}{(n+1)2^n}$. Hence the probability of a randomly chosen balanced function will be function type h is bounded by $\frac{(n+1)2^n \sum_{i=0}^{k-1} \binom{2^{n-1}}{i}^2}{\binom{2^n}{2^{n-1}}}$. Similarly, the case for n even can be proved. □

If one takes $k \leq 2^{\frac{3}{4}n}$, then it can be checked easily that the probability decreases fast towards zero as n increases. Thus for a random balanced function f, the probability of getting an affine transformation (which generates the function h from f) such that $|\{x|f(Bx+b)=1, wt(x) \leq \lfloor \frac{n-1}{2} \rfloor\}| > \sum_{i=0}^{\lfloor \frac{n-1}{2} \rfloor} \binom{n}{i} - 2^{\frac{3}{4}n}$ is almost improbable.

Thus when one randomly chosen balanced function is considered, using the strategy of considering the function after affine transformation, one can indeed reduce the matrix size by constant factor, but the reduction may not be significant in asymptotic terms when the annihilators at the degree of $\lfloor \frac{n-1}{2} \rfloor$ are considered for large n.

5 Conclusion

In this paper we study how to reduce the matrix size which is involved in finding the annihilators of a Boolean function. Our results show that considerable reduction in the size of the matrix is achievable. We identify the classes where it provides asymptotic improvement. We also note that for randomly chosen balanced functions, the improvement is rather constant than asymptotic. The reduction in matrix size helps in running the actual annihilator finding steps by Gaussian elimination method. Though our method is less efficient in general than the recently known efficient algorithms [1,20] to find the annihilators, this work helps in theoretically understanding the structure of the matrix involved.

Acknowledgment. We like to acknowledge one of the anonymous reviewers for pointing out some problems in Heuristic 1 in the submitted version of this paper; this is corrected in this final version.

References

1. F. Armknecht, C. Carlet, P. Gaborit, S. Kuenzli, W. Meier and O. Ruatta. Efficient computation of algebraic immunity for algebraic and fast algebraic attacks. In *Eurocrypt 2006*.
2. F. Armknecht. Improving Fast Algebraic Attacks. In *FSE 2004*, number 3017 in Lecture Notes in Computer Science, pages 65–82. Springer Verlag, 2004.
3. G. Ars and J. Faugére. Algebraic Immunities of functions over finite fields. INRIA report, 2005.
4. L. M. Batten. Algebraic Attacks over GF(q). In *Progress in Cryptology - INDOCRYPT 2004*, pages 84–91, number 3348, Lecture Notes in Computer Science, Springer-Verlag.
5. A. Botev and Y. Tarannikov. Lower bounds on algebraic immunity for recursive constructions of nonlinear filters. Preprint 2004.
6. A. Braeken and B. Praneel. Probabilistic algebraic attacks. In *10th IMA international conference on cryptography and coding*, 2005.

7. A. Braeken and B. Praneel. On the Algebraic Immunity of Symmetric Boolean Functions. In *Indocrypt 2005*, number 3797 in LNCS, pages 35–48. Springer Verlag, 2005. Also available at Cryptology ePrint Archive, http://eprint.iacr.org/, No. 2005/245, 26 July, 2005.
8. A. Braeken, J. Lano and B. Praneel. Evaluating the Resistance of Stream Ciphers with Linear Feedback Against Fast Algebraic Attacks. Accepted in *ACISP 2006*.
9. A. Canteaut. Open problems related to algebraic attacks on stream ciphers. In *WCC 2005*, pages 1–10, invited talk.
10. C. Carlet. Improving the algebraic immunity of resilient and nonlinear functions and constructing bent functions. IACR ePrint server, http://eprint.iacr.org, 2004/276.
11. J. H. Cheon and D. H. Lee. Resistance of S-boxes against Algebraic Attacks. In *FSE 2004*, number 3017 in Lecture Notes in Computer Science, pages 83–94. Springer Verlag, 2004.
12. J. Y. Cho and J. Pieprizyk. Algebraic Attacks on SOBER-t32 and SOBER-128. In *FSE 2004*, number 3017 in Lecture Notes in Computer Science, pages 49–64. Springer Verlag, 2004.
13. N. Courtois and J. Pieprzyk. Cryptanalysis of block ciphers with overdefined systems of equations. In *Advances in Cryptology - ASIACRYPT 2002*, number 2501 in Lecture Notes in Computer Science, pages 267–287. Springer Verlag, 2002.
14. N. Courtois and W. Meier. Algebraic attacks on stream ciphers with linear feedback. In *Advances in Cryptology - EUROCRYPT 2003*, number 2656 in Lecture Notes in Computer Science, pages 345–359. Springer Verlag, 2003.
15. N. Courtois. Fast algebraic attacks on stream ciphers with linear feedback. In *Advances in Cryptology - CRYPTO 2003*, number 2729 in Lecture Notes in Computer Science, pages 176–194. Springer Verlag, 2003.
16. N. Courtois, B. Debraize and E. Garrido. On Exact Algebraic [Non-]Immunity of S-boxes Based on Power Functions. Accepted in *ACISP 2006*.
17. D. K. Dalai, K. C. Gupta and S. Maitra. Results on Algebraic Immunity for Cryptographically Significant Boolean Functions. In *INDOCRYPT 2004*, pages 92–106, number 3348, Lecture Notes in Computer Science, Springer-Verlag.
18. D. K. Dalai, K. C. Gupta and S. Maitra. Cryptographically Significant Boolean functions: Construction and Analysis in terms of Algebraic Immunity. In *FSE 2005*, pages 98–111, number 3557, Lecture Notes in Computer Science, Springer-Verlag.
19. D. K. Dalai, S. Maitra and S. Sarkar. Basic Theory in Construction of Boolean Functions with Maximum Possible Annihilator Immunity. Cryptology ePrint Archive, http://eprint.iacr.org/, No. 2005/229, 15 July, 2005.
20. F. Didier and J. Tillich. Computing the Algebraic Immunity Efficiently. In *FSE 2006*.
21. D. H. Lee, J. Kim, J. Hong, J. W. Han and D. Moon. Algebraic Attacks on Summation Generators. In *FSE 2004*, Lecture Notes in Computer Science, pages 34–48. Springer Verlag, 2004.
22. W. Meier, E. Pasalic and C. Carlet. Algebraic attacks and decomposition of Boolean functions. In *Advances in Cryptology - EUROCRYPT 2004*, number 3027 in Lecture Notes in Computer Science, pages 474–491. Springer Verlag, 2004.
23. Y. Nawaz, G. Gong and K. C. Gupta. Upper Bounds on Algebraic Immunity of Power Functions. In *FSE 2006*.

The Algebraic Normal Form, Linear Complexity and k-Error Linear Complexity of Single-Cycle T-Function*

Wenying Zhang and Chuan-Kun Wu

State Key Laboratory of Information Security, Institute of Software, Chinese Academy of Sciences, Beijing, 100080, China
{wenyingzh, ckwu}@is.iscas.ac.cn

Abstract. In this paper, we study single-cycle T-functions which have important applications in new cryptographic algorithms. We present the algebraic normal form (ANF) of all single-cycle T-functions and the enumeration of single-cycle functions, which reveal many mysterious aspects of such functions. We also investigate the linear complexity and the k-error complexity of single-cycle T-functions when $n = 2^t$, the results also reflect the good stability of single-cycle T-functions.

Keywords: Cryptography, Single-cycle T-function, Algebraic normal form, Linear complexity, k-error complexity.

1 Introduction

1.1 Single-Cycle T-Functions

Cryptography is the science of protecting information from unauthorized intruders. In the design of cryptographic transformations such as block ciphers, hash functions and stream ciphers, T-functions are recently found to be useful tools, which help to realize fast encryption under integral and logical instructions. A function from an n-bit input to an n-bit output with the property that the i-th bit of its outputs depend only on the i least significant bits of its inputs is called a *T-function* (short for *triangular function*) [5,6,7]. A typical example of T-function is the polynomial $P(x) = \sum_{i=0}^{d} a_i x_i \pmod{2^n}$ with integral coefficients. A T-function can be treated as a multiple outputs Boolean function, where the i-th coordinate function can be represented in figure 1.

We call a function *invertible* if it is a one-to-one mapping from one set onto itself. If an invertible T-function induces a single cycle as the state graph, then the function is called a *single-cycle T-function*. The states of a single-cycle T-function can be shown in figure 2.

* This work was supported by National Natural Science Foundation of China (90304007) and China Postdoctoral Science Foundation.

G. Gong et al. (Eds.): SETA 2006, LNCS 4086, pp. 391–401, 2006.

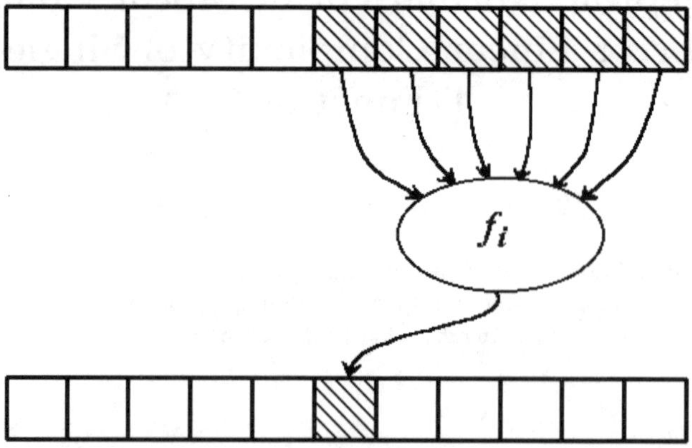

Fig. 1. T-function in Boolean function representation

It should be noted that not all the invertible T-functions are single-cycle ones. Such an example can be found in RC6 encryption algorithm where the invertible map $x \longrightarrow x + 2x^2 \pmod{2^n}$ is not a single-cycle function. This function has very bad properties in terms of the cycle structure. In fact, for any integer c, $2^{\frac{n}{2}}c$ is an invariant when n is even and $2^{\frac{n-1}{2}}c$ is an invariant when n is odd.

The linear feedback shift register (LFSR) based state transform function has the disadvantage that its initial state cannot be $\mathbf{0}$, which may lead to some crucial attacks. One of such example is A5/2 algorithm, which is widely used in global system of mobile (GSM). Compared with normal LFSR, a single-cycle T-function can start from state $\mathbf{0}$, and then goes through all the 2^n consequent states, denoted as $T(\mathbf{0}), T^2(\mathbf{0}), \cdots, T^{2^n-1}(\mathbf{0})$. This is one advantage that an LFSR does not have. If the state transform functions in stream ciphers are replaced by single-cycle T-functions, we do not have to worry about the $\mathbf{0}$ state, and the period of the single-cycle T-function arrives at the maximum-2^n, which is even larger than that of LFSR based ones-$2^n - 1$. Furthermore, single-cycle T-function based state transform function can be nonlinear or even no algebraic. Consequently, Klimov and Shamir proposed the idea to replace linear feedback shift registers (LFSR) with single-cycle T-functions [5].

1.2 Related Work

Due to the advantages of single-cycle T-functions, it has attracted considerable attention in the cryptography research community. Benony et al. have given a method to retrieve internal state of T-function based pseudo-random sequence generators [1]. Building on T-functions, Hong et al. proposed a new stream cipher [4], which is faster than most stream ciphers available today. Unfortunately, only a very small number of single cycle T-functions are known currently, and only characterizations of T-functions in integer residue rings Z_n appeared in the literature [5,6,7].

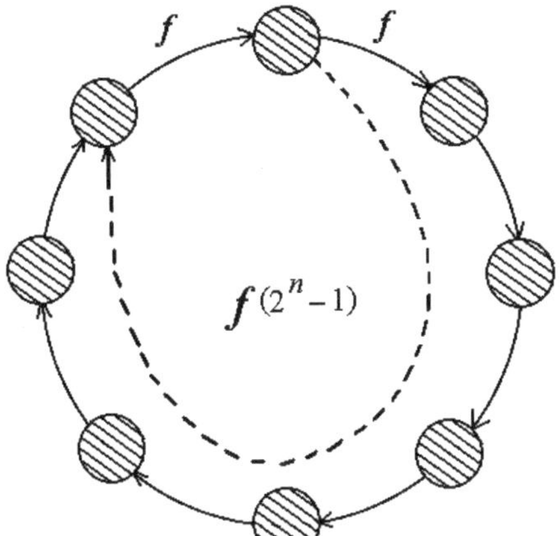

Fig. 2. The states of a single-cycle T-function

1.3 Our Contributions

In this paper, using the techniques of Boolean function, we first give the algebraic normal form (ANF) of all single-cycle T-functions in Boolean function representation. Based on the ANF of single-cycle T-functions, we determine the number of such functions.

Considering that an important application of single-cycle T-functions is in the constructions of synchronous stream ciphers, the concepts of linear complexity and k-error linear complexity are very useful in the study of the security of stream ciphers, we calculate the linear complexity and the k-error complexity of single cycle T-function π with $n = 2^t$. Its linear complexity is about one half of the"length" of the whole sequence, and the minimum value of k such that $LC_k(\pi) < LC(\pi)$ is equal to 2^{n-1}, which shows that single-cycle T-function has good stability.

2 The Algebraic Normal Form (ANF) of Single-Cycle T-Function

We first introduce a few definitions. Let $GF(2)$ be the binary field, n an arbitrary positive integer. A function from $GF^n(2)$ into $GF(2)$ is called a Boolean function of n variables. It can be written as $f(x) = f(x_1, \cdots, x_n)$. For any Boolean function $f(x)$, it can always be written in an algebraic normal form(ANF) as

$$c_0 + c_1x_1 + \cdots + c_nx_n + c_{12}x_1x_2 + c_{13}x_1x_3 + \cdots + c_{12\cdots n}x_1 \cdots x_n,$$

where, each $x_{i_1}, x_{i_2}, \cdots, x_{i_j}$ is called a term of degree j, and $c_{i_1i_2\cdots i_j} \in GF(2)$ is the coefficient of the term in the polynomial expression of $f(x)$. The algebraic degree of $f(x)$, denoted by $deg(f)$, is defined as the highest degree of its terms with a nonzero coefficient. The function f is called balanced if the size of $\{a \in GF^n(2) \mid f(a) = 1\}$ is equal to 2^{n-1}. We will denote $\mathcal{F}_n$ the set of all Boolean functions of n variables.

A mapping from $GF^n(2)$ to $GF^m(2)$ is called a vector Boolean function, or an (n, m)-Boolean function. It is easy to see that a T-function is an (n, n)-Boolean function. We also call it an (n, n) T-function. In this paper, we are mainly interested in single-cycle T-functions, i.e. the outputs of such functions go through all the possible vectors of $GF^n(2)$ from any starting point, including the all-zero vector, see figure 2.

Lemma 1. [11] *Let $\pi = (\varphi_1, \cdots, \varphi_n)$ be a single-cycle T-function. Then the first i coordinates of π forms a new vector Boolean function of i variables $[(\pi)]_{1\to i} = (\varphi_1, \cdots, \varphi_i)$, and it is a single-cycle T-function of period 2^i.*

Lemma 2. [6] *An (n, n) T-function is invertible if and only if its ANF has the form*

$$\begin{aligned}\pi(x) &= (\varphi_1(x), \varphi_2(x), \varphi_3(x), \cdots, \varphi_n(x)) \\ &= (x_1 + a, x_2 + f_2(x_1), x_3 + f_3(x_1, x_2), \cdots, x_n + f_n(x_1, \cdots, x_{n-1})),\end{aligned}$$

where $a = 0$ or 1, $f_i, i = 2, \cdots, n$ are Boolean functions.

Lemma 3. *Let $\psi(x) \in \mathcal{F}_n$. Then $deg(\psi(x)) < n$ if and only if*

$$\sum_{x \in GF^n(2)} \psi(x) = 0 \pmod 2. \tag{1}$$

We are now ready to give the main theorem of this paper on the ANF of single-cycle T-functions.

Theorem 1. *Let $\pi = (\varphi_1, \cdots, \varphi_n)$ be an invertible T-function over $GF^n(2)$. Then π is a single-cycle T-function if and only if its ANF has the following form*

$$\begin{aligned}\pi(x) &= (\varphi_1, \cdots, \varphi_n) \\ &= (x_1 + 1, x_2 + x_1 + \psi_2, x_3 + x_1x_2 + \psi_3, x_4 + x_1x_2x_3 + \psi_4, \cdots, \\ &\quad x_n + x_1x_2 \cdots x_{n-1} + \psi_n(x_1, x_2, \cdots, x_{n-1})),\end{aligned} \tag{2}$$

where $\varphi_i = x_i + x_1 \cdots x_{i-1} + \psi_i(x_1, \cdots, x_{i-1})$, $deg(\psi_i) \le i - 2, i \ge 2$

Proof. (Necessity) It only needs to prove that for any single-cycle T-function π, its coordinate function can be expressed as

$$\varphi_i = x_i + x_1 \cdots x_{i-1} + \psi_i(x_1, \cdots, x_{i-1}),\ deg(\psi_i) \le i - 2.$$

When $n = 1$, the only single-cycle T-function is $[\pi(x)]_1 = x_1 + 1$. Hence the necessity holds for the case of $n = 1$. Assume the conclusion is true for

$n \leq k$, i.e. for any $i \leq k$, φ_i can always be expressed as $x_i + x_1 \cdots x_{i-1} + \psi_i(x_1, \cdots, x_{i-1}), deg(\psi_i \leq i-2)$. We now prove that when $n = k+1$, φ_{k+1} can be expressed as

$$\varphi_{k+1}(x_1, \cdots, x_{k+1}) = x_{k+1} + x_1 \cdots x_k + \psi_{k+1}(x_1, \cdots, x_k),\ deg(\psi_{k+1}) \leq k-1.$$

Since π is single-cycled, by Lemma 1, the first k coordinate functions $(\varphi_1, \cdots, \varphi_k)$ form a new single-cycle T-function of period 2^k. Denote $(\varphi_1, \cdots, \varphi_k)$ by $[\pi]_{1\to k}$. Consider the consequent 2^k states of $[\pi]_{1\to k}$ starting from the initial state of 0:

$$\left(\mathbf{0}, [\pi(\mathbf{0})]_{1\to k}, [\pi^2(\mathbf{0})]_{1\to k}, \cdots, [\pi^{2^k-1}(\mathbf{0})]_{1\to k}\right).$$

Since $[\pi]_{1\to k}$ defines a cycle of length 2^k, we see that the vector $[\pi^j(\mathbf{0})]_{1\to k}(0 \leq j \leq 2^k - 1)$ is a permutation of all the elements in $GF^k(2)$. Let $[\pi^j(\mathbf{0})]_i$ be the i-th bit of $\pi^j(\mathbf{0})$. By Lemma 2, we can write

$$\varphi_{k+1}(x_1, \cdots, x_{k+1}) = x_{k+1} + a x_1 x_2 \cdots x_k + \psi_{k+1}(x_1, x_2, \cdots, x_k),$$

where $a = 0$ or 1, ψ_{k+1} is a Boolean function with $deg(\psi_{k+1}) < k$. Hence

$$\begin{aligned}
[\pi^{2^k}(\mathbf{0})]_{k+1} = \varphi_{k+1}(\pi^{2^k}(\mathbf{0})) &= [\pi^{2^k-1}(\mathbf{0})]_{k+1} + a[\pi^{2^k-1}(\mathbf{0})]_1 \cdots \\
&\quad [\pi^{2^k-1}(\mathbf{0})]_k + \psi_{k+1}([\pi^{2^k-1}(\mathbf{0})]_{1\to k}) \\
[\pi^{2^k-1}(\mathbf{0})]_{k+1} = \varphi_{k+1}(\pi^{2^k-1}(\mathbf{0})) &= [\pi^{2^k-2}(\mathbf{0})]_{k+1} + a[\pi^{2^k-2}(\mathbf{0})]_1 \cdots \\
&\quad [\pi^{2^k-2}(\mathbf{0})]_k + \psi_{k+1}([\pi^{2^k-2}(\mathbf{0})]_{1\to k}) \\
&\quad \cdots
\end{aligned}$$

$$[\pi^2(\mathbf{0})]_{k+1} = \varphi_{k+1}(\pi^2(\mathbf{0})) = [\pi(\mathbf{0})]_{k+1} + a[\pi(\mathbf{0})]_1 \cdots [\pi(\mathbf{0})]_k + \psi_{k+1}([\pi(\mathbf{0})]_{1\to k}).$$

From the above expressions, replace $[\pi^{2^k-1}(\mathbf{0})]_{k+1}$ in the first expression by the second expression, and $[\pi^{2^k-2}(\mathbf{0})]_{k+1}$ in the resulting expression by the third expression, and do the replacement recursively, eventually we get the following expression:

$$\begin{aligned}
&[\pi^{2^k}(\mathbf{0})]_{k+1} \\
&= \varphi_{k+1}(\pi^{2^k-1}(\mathbf{0})) \\
&= [\pi^{2^k-1}(\mathbf{0})]_{k+1} + a[\pi^{2^k-1}(\mathbf{0})]_1 \cdots [\pi^{2^k-1}(\mathbf{0})]_k + \psi_{k+1}([\pi^{2^k-1}(\mathbf{0})]_{1\to k}) \\
&= [\pi^{2^k-2}(\mathbf{0})]_{k+1} + a[\pi^{2^k-2}(\mathbf{0})]_1 \cdots [\pi^{2^k-2}(\mathbf{0})]_k + \psi_{k+1}([\pi^{2^k-2}(\mathbf{0})]_{1\to k}) \\
&\qquad + a[\pi^{2^k-1}(\mathbf{0})]_1 \cdots [\pi^{2^k-1}(\mathbf{0})]_k + \psi_{k+1}([\pi^{2^k-1}(\mathbf{0})]_{1\to k}) \\
&= \cdots \\
&= [\mathbf{0}]_{k+1} + \sum_{j=0}^{2^k-1} \left(a[\pi^j(\mathbf{0})]_1 \cdots [\pi^j(\mathbf{0})]_k + \psi_{k+1}([\pi^j(\mathbf{0})]_{1\to k})\right) \\
&= 0 + a \sum_{s\in GF^k(2)} s_1 \cdots s_k + \sum_{s\in GF^k(2)} \psi_{k+1}(s), \qquad (3)
\end{aligned}$$

where $s = (s_1, \cdots, s_k)$. Note that $s_1 \cdots s_k \neq 0$ holds only when $s_1, \cdots, s_k$ are all equal to 1. Hence we have

$$\sum_{s \in GF^k(2)} s_1 \cdots s_k = 1, \; s = (s_1, \cdots, s_k). \tag{4}$$

Since $deg(\psi(x)) < k$, by Lemma 3 we have

$$\sum_{s \in GF^k(2)} \psi_{k+1}(s) = 0 \pmod 2. \tag{5}$$

Plug (4) and (5) into (3) we obtain that:

$$[\pi^{2^k}(\mathbf{0})]_{k+1} = a.$$

Since $[\pi]_{1\to k}$ is a single-cycle T-function of period 2^k, it is known that its (2^k+1)-th state must be the same as the initial state, i.e.

$$[\pi^{2^k}(\mathbf{0})]_{1\to k} = [\pi^0(\mathbf{0})]_{1\to k} = \underbrace{(0, 0, \cdots, 0)}_{k}.$$

We must have $[\pi^{2^k}(\mathbf{0})]_{k+1} = a \neq 0$ in view of π is single-cycled, otherwise, we would have $[\pi^{2^k}(\mathbf{0})] = \mathbf{0}$, which induces that $[\pi]_{1\to k+1}$ has a period no more than 2^k contradicting with the fact that the period of $\pi_{1\to k+1}$ is 2^{k+1}. Thus $a = 1$. Note that any Boolean function of k variables with algebraic degree less than k satisfies equation (1), it is known that ψ_{k+1} can be chosen at random from those Boolean functions of variables $x_1, \cdots, x_k$ and with algebraic degree less than k. This proves that φ_{k+1} can be written as

$$x_{k+1} + x_1x_2 \cdots x_k + \psi_n(x_1, \cdots, x_k), \; deg(\psi_{k+1}) \leq k - 1.$$

Therefore, for any positive integer n, the necessity holds.

(Sufficiency) Again we use induction on n to prove the sufficiency. When $n = 1$, we must have $\varphi_1(x) = [\pi(x)]_1 = x_1 + 1$, and it is obvious that $f(x)$ is single-cycled in the sense of modulo 2. Assume the conclusion is true when $n = k$, i.e., the first k coordinate functions in equation (2) form a single-cycle T-function of period 2^k. Similar to the proof of necessity, it is easy to prove that

$$[\pi^{2^k}(\mathbf{0})]_{k+1} = \sum_{s \in GF^k(2)} \psi_{k+1}(s) + \sum_{s \in GF^k(2)} s_1 \cdots s_k.$$

Since $deg(\psi_{k+1}) < k$, by Lemma 3 we have $\sum_{s \in GF^k(2)} \psi_{k+1}(s) = 0$. And from $\sum_{s \in GF^k(2)} s_1 \cdots s_k = 1$, we obtain that $[\pi^{2^k}(\mathbf{0})]_{k+1} = 1$. Since $[\pi(x)]_{1\to k}$ is a single-cycle T-function of period 2^k, it is known that the period of $[\pi(x)]_{1\to k+1}$ is at least 2^k. Note that the 2^k-th state is not the same as the initial state $(\mathbf{0})$, and the period of $[\pi(x)]_{1\to k+1}$ must be a multiple of the period of $[\pi(x)]_{1\to k}$ which is 2^k. Thus we can infer that the period of $[\pi(x)]_{1\to k+1}$ must be 2^{k+1}, i.e., $[\pi(x)]_{1\to k+1}$ is a single-cycle T-function. This proves the sufficiency for the case of $n = k+1$, and by the principle of induction, the conclusion is true in general case.

Theorem 1 shows that the i-th output bit of a single-cycle T-function is $x_i + x_1 \cdots x_{i-1} + \cdots$, which is a non-degenerate function of the inputs on positions $1, 2, \cdots, i$, and is independent of the input bits on positions $i+1, \cdots, n$. In this sense we say that an output bit of a single-cycle T-function does not reveal any information about the most significant bits of its inputs, but contains all the information about its least significant bits of its inputs.

In order to demonstrate what a single-cycle T-function looks like, we now give a small example.

Example 1. Let $\pi(x) = (x_1+1, x_2+x_1+1, x_3+x_1x_2, x_4+x_1x_2x_3+1)$, then π is a single-cycle T-function. The state transform of $\pi(x)$ over $GF^4(2)$ is $0 \mapsto 11 \mapsto 6 \mapsto 13 \mapsto 4 \mapsto 15 \mapsto 10 \mapsto 1 \mapsto 8 \mapsto 3 \mapsto 14 \mapsto 5 \mapsto 12 \mapsto 7 \mapsto 2 \mapsto 9 \mapsto 0$.

From the ANF of single-cycle T-functions, we can deduce the enumeration of them.

Theorem 2. *The number of single-cycle (n, n) T-functions is*

$$2^{(2-1)+(2^2-1)+(2^3-1)+\cdots+(2^{n-1}-1)} = 2^{2^n-n-1}.$$

Proof. The conclusion follows from the observation that $\psi_i(x_1, \cdots, x_{i-1}), i = 2, \cdots, n$ can be any of the Boolean functions of $i-1$ variables with algebraic degree being less than $i-1$, and there are 2^{2^i-1} such Boolean functions.

Note that the number of m-sequences with order n is $\varphi(2^n-1)$ ($\varphi(\cdot)$ is Euler function) and the number of M-sequences with order n is $2^{2^{n-1}} - n$. Compared with m-sequences and M-sequences, the number of single-cycle T-functions is much larger. The existence of such functions thwarts against the keystreams by exhaustive search.

3 The Linear Complexity and k-Error Linear Complexity of Single-Cycle T-Function

From an engineering standpoint, because of the compatibility with the binary $\{0, 1\}$ nature of data representation in electronic hardware, the preferred alphabet sizes are $p = 2, 4, 8$ etc, which are all powers of 2. This gives us the motivation to describe the linear complexity of single-cycle T-function, and calculate the k-error linear complexity of single-cycle T-function, when the variables number is a power of 2. In what follows, n is assumed to be a power of 2, that is, there is an positive integer t such that $n = 2^t$.

3.1 The Linear Complexity of Single-Cycle T-Function

Linear complexity is an important cryptographic criterion of stream ciphers. The linear complexity of a sequence (**s**), denoted by $LC(\mathbf{s})$, is defined as the length of the shortest LFSR that generates (**s**). In stream ciphers, periodic sequences that are suitable as keystreams must have large linear complexity to thwart an attack by the Berlekamp-Massey algorithm[9,10].

Given a concrete binary sequence (**s**) of length 2^m, its linear complexity $LC(\mathbf{s})$ can efficiently be determined by the Chan-Games algorithm:

Lemma 4 (Chan-Games algorithm). [3] *Let $\mathbf{s}^m = (s_0, s_1, \cdots, s_{2^m-1})$ be a binary vector of length 2^m, and let the sequence $(\mathbf{s})$ be defined by appending copies of $\mathbf{s}^m$. We decompose $\mathbf{s}^m$ into its left and right halves by*

$$L(\mathbf{s}^m) = (s_0, s_1 \cdots, s_{2^{m-1}-1})$$
$$R(\mathbf{s}^m) = (s_{2^{m-1}}, s_{2^{m-1}+1}, \cdots, s_{2^m-1}).$$

and write $\mathbf{s}^m = (L(\mathbf{s}^m) \vdots R(\mathbf{s}^m))$. Form $d = R(\mathbf{s}^m) + L(\mathbf{s}^m)$, where the addition is performed element wise module 2, so that $(\mathbf{d}) \in B(2^{m-1})$, we have
a)If $\mathbf{d} = 0$, then $LC(\mathbf{s}) = LC(L(\mathbf{s}^m))$;
b)If $\mathbf{d} \neq 0$, then $LC(\mathbf{s}) = 2^{n-1} + LC(\mathbf{d})$.
Where $B(2^{m-1})$ denotes the set of binary sequences of period 2^{m-1}.

Remark 1. It should be mentioned that Lemma 4 can be applied to a periodic sequence with period 2^m directly. Even when the sequence **s** may have a smaller period than 2^m. So we can start the algorithm with any period of s, although it is not the minimum. If 2^m is not the minimum, then the linear complexity of **s** equals to the linear complexity of the 2^{m-1}- periodic sequence $L(\mathbf{s}^m)$. If 2^{m-1} is not the minimum, then the linear complexity of the sequence **s** equals to the linear complexity of the 2^{m-2}- periodic sequence $L(L(\mathbf{s}^m))$, and so on.

Suppose that $\pi(x)$ is a single-cycle T-function of 2^n known consequent states. Write these states as row vectors and form a matrix as follows:

$$A = \begin{pmatrix} a_{11} & a_{12} & a_{13} & \cdots a_{1n} \\ a_{21} & a_{22} & a_{23} & \cdots a_{2n} \\ a_{31} & a_{32} & a_{33} & \cdots a_{3n} \\ \cdots & \cdots & \cdots & \cdots\cdots \\ a_{2^{n-1}1} & a_{2^{n-1}2} & a_{2^{n-1}3} & \cdots a_{2^{n-1}n} \\ \cdots & \cdots & \cdots & \cdots\cdots \\ a_{2^n1} & a_{2^n2} & a_{2^n3} & \cdots a_{2^nn} \end{pmatrix}.$$

Let π^n be a binary vector of length $n \times 2^n$:

$$\pi^n = (a_{11}, \cdots, a_{1n}, a_{21}, \cdots, a_{2n}, \cdots, a_{2^{n-1}1}, \cdots, a_{2^{n-1}n}, \cdots, a_{2^n1}, \cdots, a_{2^nn}),$$

and the sequence (π) be defined by appending copies of π^n, then (π) is an $n \times 2^n$-periodic sequence.

Theorem 3. *If $n = 2^t$, then the linear complexity of (π) is $n \times 2^{n-1} + n$.*

Proof. Since when $n = 2^t$, the period of (π) is a power of 2, hence we can calculate the linear complexity of (π) by Lemma 4 and **Remark** 1. Here

$$L(\pi^n) = (a_{11}, \cdots, a_{1n}, \cdots, a_{2^{n-1}1}, \cdots, a_{2^{n-1}n})$$
$$R(\pi^n) = (a_{2^{n-1}+1\ 1}, \cdots, a_{2^{n-1}+1\ n}, \cdots, a_{2^n 1}, \cdots, a_{2^n n}).$$

$L(\pi^n) + R(\pi^n) = (a_{11} + a_{2^{n-1}+1\ 1}, \cdots, a_{1n} + a_{2^{n-1}+1\ n}, \cdots, a_{2^{n-1}1} + a_{2^n 1}, \cdots, a_{2^{n-1}n} + a_{2^n n})$, hence each indices of $L(\pi^n) + R(\pi^n)$ formed as $a_{ij} + a_{i+2^{n-1}\ j}$. Since π is single-cycled, a_{ij} and $a_{i+2^{n-1}\ j}$ are different only when $j = n$, we have

$$L(\pi^n) + R(\pi^n) = (\underbrace{0, \cdots, 0, 1}_{n\ elements}, \underbrace{0, \cdots, 0, 1}_{n\ elements}, \cdots, \underbrace{0, \cdots, 0, 1}_{n\ elements})$$

and $LC(\pi) = \frac{n \times 2^n}{2} + LC(\underbrace{0, \cdots, 0, 1}_{n\ elements}) = n \times 2^{n-1} + n$ by Lemma 4.

Theorem 3 shows that when n is large enough, the linear complexity of a single-cycle T-function is about one half of its "length".

3.2 The k-Error Linear Complexity of Single-Cycle T-Function

A cryptographically strong sequence should not only have a large linear complexity, but also have a stable linear complexity, by this we mean that by altering a few terms, it should not cause a significant decrease of the linear complexity. According to this requirement, the new notation of k error complexity of periodic sequences was introduced independently by Stamp and Martin [12] and Ding, Xiao, and Shan [2].

Definition 1. [2,12] *The* k-error linear complexity *of the periodic sequence $(s) = (s_0, s_1, \cdots, s_{m-1})$, denoted $LC_k(s)$, is the smallest linear complexity that can be obtained when any k or fewer of the $s_i^{\prime}s$ are altered.*

The k-error linear complexity can be interpreted as a worst case measure of the linear complexity when k or fewer errors occur and hence the terminology k-error linear complexity. We define $minerror(\mathbf{s})$ as the minimum value of k such that

$$LC_k(\mathbf{s}) < LC(\mathbf{s}).$$

In other words, $minerror(\mathbf{s})$ is the smallest Hamming weight of the error vector E such that

$$LC(\mathbf{s} + E) < LC(\mathbf{s}).$$

E is called a critical error vector for $(\mathbf{s})$ [8].

In [8], Kurosawa *et al.* established an efficient algorithm for computing the k-error linear complexity, in the case where $(\mathbf{s})$ is a binary sequence with period 2^m, which is an important tool in the rest of our paper.

Lemma 5. [8] *Let s be a binary sequence with period N. If $N = 2^m$, then*

$$minerror(s) = 2^{w_h(N-LC(s))},$$

where $w_h(C)$ denotes the Hamming weight of the binary representation of an integer C.

Theorem 4. *If $n = 2^t$, then $minerror(\pi) = 2^{n-1}$ and $LC_k(\pi)$ is equal to $n + n \times 2^{n-2}$ for $k = minerror(\pi)$.*

Proof. Here $N = n \times 2^n$. From Theorem 3, we have

$$N - LC(\pi) = n \times 2^n - \frac{n \times 2^n}{2} - n = \frac{n \times 2^n}{2} - n = n \times (2^{n-1} - 1).$$

It is obvious that the hamming weight of $(2^{n-1} - 1)$is $n - 1$. Since $n = 2^t$, the hamming weight of $N - LC(\pi) = n \times (2^{n-1} - 1)$ is $n - 1$, too. Hence, from Lemma 5, $minerror(\pi) = 2^{n-1}$.

Next we determine $LC_k(\pi)$ for $k = minerror(\pi)$. Since $L(\pi^n)+R(\pi^n)$ is 2^{n-1} copies of $0, \cdots, 0, 1$, its length is $\frac{n\times 2^n}{2}$. After changing every n-th bit of $L(\pi^n)+R(\pi^n)$, we obtain a zero vector. So $(\underbrace{0, 0, \cdots, 0}_{n\times 2^{n-1}\ elements}, 0, \cdots, 0, 1, \cdots, 0, \cdots, 0, 1)$ is one of the critical error vectors of (π), and

$$LC_k(\pi) = LC(\pi + E) = LC(L(\pi^n)) = LC(L(L(\pi^n) + R(L(\pi^n)) + \frac{N}{4}$$

We determine $L(L(\pi^n)+R(L(\pi^n)$ first. For π is single cycled, the n-th component of the first 2^{n-1} states of π are the same. Their first $n - 1$ component run over the states of $[(\pi)]_{1\to n-1}$, similar to the reason for the form of $L(\pi^n) + R(\pi^n)$ in the proof Theorem 3, we can get

$$L(L(\pi^n)) + R(L(\pi^n)) = (\underbrace{0, \cdots, 0, 1, 0}_{n\ elements}, \underbrace{0, \cdots, 0, 1, 0}_{n\ elements}, \cdots, \underbrace{0, \cdots, 0, 1, 0}_{n\ elements}).$$

And,

$$LC(\pi+E)=LC(L(\pi^n))=LC(L(L(\pi^n))+R(L(\pi^n))+\frac{N}{4} = n+\frac{N}{4} = n+n\times 2^{n-2}.$$

4 Conclusion

In this paper, we give an algebraic normal form (ANF) of single-cycle T-functions which reveals the algebraic structure of such function. Based on this ANF representation of the single-cycle T-function, we also give the enumeration of such functions. It should be remarkable that such functions are more powerful than m-sequences and M-sequences in stream ciphers. We also determine the linear complexity and the k-error complexity of single-cycle T-function whose variable-number is a power of 2. Either the enumeration or the stability of the single-cycle T-function indicates that the applications of such function cannot be replaced by m or M-sequences.

Acknowledgment

The authors would like to express their deep gratitude to the anonymous referees for their valuable comments on this paper.

References

1. V. Benony, F. Recher, E. Wegrzynowski and C. Fontaine, An Improved Method to Retrieve Internal State of Klimov-Shamir Pseudo-Random Sequence Generators, Third International Conference on Sequence and their Applications-SETA'2004, LNCS Vol.3486, Springer-Verlag,(2004), pp.138-142.
2. C. Ding, G.Xiao, and W.Shan. The Stability Theory of Stream Ciphers, Springer-Verlag press, 1991.
3. R.A. Games and A.H. Chan, A Fast Algorithm for Determining the Complexity of a Binary Sequence with Period 2^n, IEEE Transactions on Information Theory, Vol.29. No.1, pp.144-146, January 1983.
4. J. Hong, D.H. Lee, Y. Yeom, D. Han, A New Class of Single Cycle T-functions, Fast Software Encryption -FSE'2005, LNCS Vol.3557. Springer-Verlag,(2005), pp. 68-82.
5. A. Klimov and A. Shamir, A new Class of Invertible Mappings, Workshop on Cryptographic Hardware and Embeded Systems'2002, LNCS Vol.2523, Springer-Verlag,(2003), pp. 470-483.
6. A. Klimov and A. Shamir, Cryptographic Applications of T-Functions, Ninth workshop on Selected Areas in Cryptography-SAC' 2003, LNCS Vol.3006, Springer-Verlag, (2004), pp. 248-261.
7. A. Klimov and A. Shamir, New Cryptographic Primitives Based on Multi-word T-Functions, Fast Software Encryption-FSE'2004, LNCS Vol.3017, Springer-Verlag,(2004), pp. 1-15.
8. K. KUROSAWA, F. SATO, T. SAKATA and W. KISHMOTO. A Relationship Between Linear Complexity and k-error Linear Complexity, IEEE Transactions on Information Theory, Vol.46. No.2, pp.694-698, July 2000.
9. J.L.Massey, Shift-register syuthosis and BCII decoding. IEEE Transactions on Information Theory, Vol.15, 1969, pp.122-127.
10. W. Meidl, On the Stability of 2^n Periodic Binary Sequences, IEEE Transactions on Information Theory, Vol.51. No.3, pp.1151- 1155, March 2005.
11. H. Molland and T. Helleseth, A linear weakness in the Klimov-Shamir T-function, Proceedings of the 2005 IEEE Int. Symposium on Information Theory, (2005), pp. 1106-1110.
12. M. Stamp and C.F. Martin, An Algorithm for the k-error Linear Complexity of Binary Sequences with Period 2^n, IEEE Transactions on Information Theory, Vol.39. No.4, pp.1398-1401, July 1993.

Partially Perfect Nonlinear Functions and a Construction of Cryptographic Boolean Functions*

Lei Hu[1] and Xiangyong Zeng[1,2]

[1] State Key Laboratory of Information Security,
Graduate School of Chinese Academy of Sciences,
Beijing 100049, P.R. China
hu@is.ac.cn

[2] The Faculty of Mathematics and Computer Science,
Hubei University, Xueyuan Road 11,
Wuhan 430062, P.R. China
xzeng@hubu.edu.cn

Abstract. In this paper the concept of partially perfect nonlinear (PPN) function is introduced as an extension of binary partially Bent function and is used to construct a new class of Boolean functions with good cryptographic properties. The construction is a composition of a PPN function and a Boolean function. The nonlinearity, correlation immunity, propagation criterion, and other cryptographic properties of the constructed functions are analyzed. In particular, new plateaued functions can be obtained by the proposed method and the construction of Khoo and Gong in [1] is improved.

1 Introduction

Boolean functions used in stream cipher systems are required to have good cryptographic properties such as high nonlinearity and high algebraic degree to ensure the systems are resistant against linear cryptanalysis [2], and correlation immunity to offer protection against correlation attack [3]. Another cryptographic property is propagation characteristics, which makes the cipher not prone to differential-like cryptanalysis [4].

Bent functions [5] possess the highest nonlinearity and satisfy the propagation criterion with respect to all nonzero vectors. However, they are neither balanced nor any order correlation immune, and they exist only when the number of input variables is even. These drawbacks prohibit their direct application in practical systems. Partially Bent functions are a generalization of Bent functions, and they can be balanced and have a high level of nonlinearity provided their associated kernels are small [6]. The partially Bent functions include all quadratic functions

* This work is supported in part by the National Science Foundation of China (NSFC) under Grants No.60373041 and No.90104034, and in part by the open foundation of the State Key Laboratory of Information Security.

G. Gong et al. (Eds.): SETA 2006, LNCS 4086, pp. 402–416, 2006.

as their subclass, and they share with the latter some nice cryptographic properties in propagation criterion, balancedness, and correlation immunity. However, partially Bent functions have nonzero linear structures.

In this paper, we introduce the concept of partially perfect nonlinear (PPN) function as an extension of partially Bent functions, and we use it to construct a new class of Boolean functions with good cryptographic properties. The construction is a composition of a PPN function and a Boolean function. By choosing suitable component functions, the composition functions can avoid the drawback of linear structure of partially Bent functions. We study the cryptographical properties of the construction such as spectrum values, nonlinearity, algebraic degree, correlation immunity, and propagation criterion. In particular, the construction is suitable for generating new balanced plateaued functions with high algebraic degree, and is an improvement of the construction of Khoo and Gong [1].

The remainder of this paper is organized as follows. Section 2 gives some necessary definitions and preliminary lemmas. In Section 3 we introduce the concept of PPN function, and in Section 4 we study the cryptographical properties of the Boolean functions composed by PPN functions and other Boolean functions. A preliminary result on the algebraic degrees of composition functions is left to the appendix, since the deduction is long and somewhat technical. The final section is the conclusion.

2 Preliminaries

Let $q = 2^l$, t be a positive integer, and $GF(q^t)$ and $GF(q)$ be the finite fields with q^t and q elements, respectively. Let $\alpha_1, \alpha_2, \cdots, \alpha_t$ be a basis of $GF(q^t)$ over $GF(q)$, then $GF(q^t)$ can be viewed as a $GF(q)$-vector space, and each element x of $GF(q^t)$ is uniquely expressed as a linear combination of $\alpha_1, \alpha_2, \cdots, \alpha_t$ of the form

$$x = x_1\alpha_1 + x_2\alpha_2 + \cdots + x_t\alpha_t,\ x_1, x_2, \cdots, x_t \in GF(q).$$

Identifying x to $(x_1, x_2, \cdots, x_t)$, $GF(q^t)$ is identical to $GF(q)^t$, the t-dimensional $GF(q)$-vector space. Throughout this paper, we assume this identification.

Let n be a multiple of l. The *trace function* $\mathrm{tr}_l^n(\cdot)$ from $GF(2^n)$ to $GF(2^l)$ is defined by

$$\mathrm{tr}_l^n(x) = \sum_{i=0}^{n/l-1} x^{2^{il}}, \forall\ x \in GF(2^n).$$

For details about the properties of the trace function, please refer to [7].

A function f from $GF(q^m)$ to $GF(q)$ has a unique polynomial expression of the form

$$f(x) = \sum_{s=0}^{q^m-1} a_s x^s, \forall x \in GF(q^m), \tag{1}$$

where the coefficients $a_s \in GF(q^m)$. The function f has a unique algebraic expression of the form

$$f(x_1, \cdots, x_m) = \sum_{i_1=0}^{q-1} \cdots \sum_{i_m=0}^{q-1} c_{i_1,\cdots,i_m} x_1^{i_1} \cdots x_m^{i_m}, \forall x_1, \cdots, x_m \in GF(q) \quad (2)$$

with coefficients $c_{i_1,\cdots,i_m} \in GF(q)$. The algebraic degree of f is defined as $\deg(f) = \max\{i_1 + \cdots + i_m : c_{i_1,\cdots,i_m} \neq 0\}$. It is also equal to $\max\{w_q(s) : a_s \neq 0\}$ [8], where $w_q(s)$ is the sum of the coefficients in the q-adic expression of s.

Throughout the paper, Boolean functions always specifically mean functions from $GF(2)^n$ to $GF(2)$ for an integer n (n variable Boolean functions). For any Boolean function g defined on $GF(q)^t$ (it is an tl variable Boolean function), its *Walsh transform* $\widehat{g}$ is defined by

$$\widehat{g}(\lambda) = \sum_{x \in GF(q)^t} (-1)^{g(x) + \mathrm{tr}_1^l(\lambda \cdot x)}, \forall \lambda \in GF(q)^t, \quad (3)$$

where

$$\lambda \cdot x = \lambda_1 x_1 + \cdots + \lambda_t x_t \in GF(q)$$

is the *dot product* of $\lambda = (\lambda_1, \cdots, \lambda_t)$ and $x = (x_1, \cdots, x_t) \in GF(q)^t$. The set of all Walsh transform values of g is called the *spectra* of g. The function g is called *Bent* if $\widehat{g}(\lambda) = \pm q^{t/2}$ for all $\lambda \in GF(q)^t$.

Bent functions achieve optimum nonlinearity, however, being not balanced they are improper for direct cryptographic use. Moreover, they exist only in even dimensions. This led cryptographers to search for new classes of Boolean functions whose elements still have good nonlinearities and can be balanced for both odd and even dimensions. The class of partially Bent functions was first investigated by Carlet [6]. A function $f : GF(2)^t \to GF(2)$ is called *partially bent* if there exist a linear function L on $GF(2)^t$, two $GF(2)$-subspaces E and F of $GF(2)^t$, such that

(1) $GF(2)^t$ is the direct sum of E and F;
(2) $f|_E = p$, the restriction of f to E, is bent; and
(3) for all $x \in E$ and $y \in F$, $f(x + y) = p(x) + L(y)$.

The class of plateaued functions is an extension of the notion of partially Bent function. A function $f : GF(2)^t \to GF(2)$ is called *plateaued* [9] if $\widehat{f}(\lambda) = 0$ or $\pm 2^u$ for all $\lambda \in GF(2)^t$, where u is a positive integer.

Let f and f' be two functions from $GF(q)^t$ to $GF(q)$. The *Hamming distance* between f and f', denoted by $d(f, f')$, is defined by

$$d(f, f') = |\{x \in GF(q) \mid f(x) \neq f'(x)\}|. \quad (4)$$

One way measuring the nonlinearity of f is to use the minimum distance between f and all affine functions from $GF(q)^t$ to $GF(q)$, namely, the nonlinearity of f is defined as

$$N_f = \min_{a \in \mathbf{A}} d(f, a), \quad (5)$$

where **A** is the set of all affine functions from $GF(q)^t$ to $GF(q)$. The measure is related to linear cryptanalysis [2]. When $l = 1$, i.e., $q = 2$, by Eq. (3), one has

$$N_f = 2^{t-1} - 2^{-1} \cdot \max_{\lambda \in GF(2)^t} |\widehat{f}(\lambda)|. \tag{6}$$

Another measure of the nonlinearity of functions is related to differential cryptanalysis [4] and it uses difference operator, $D_\alpha(f(x)) = f(x+\alpha) - f(x)$. Define

$$P_f = \max_{0 \neq \alpha \in GF(q)^t} \max_{\beta \in GF(q)} Pr(D_\alpha f(x) = \beta). \tag{7}$$

f is a *perfect nonlinear function* or has *perfect nonlinearity* if $P_f = q^{-1}$ [10].

Perfect nonlinear functions have the following properties, which are related to our sequel analysis.

Lemma 1 ([10]): Let $f : GF(q)^t \to GF(q)$ be a perfect nonlinear function and $g : GF(q) \to GF(2)$ be an onto linear function. Then the composition of f and g, $g \circ f$, is a perfect nonlinear function from $GF(q)^t$ to $GF(2)$. In particular, for any $v \in GF(q)^*$, the function $\mathrm{tr}_1^l[vf(x)]$ is a perfect nonlinear function.

Lemma 2 ([10]): If $g : GF(q) \to GF(2)$ is a perfect nonlinear function, then $|\widehat{g}(u)| = q^{1/2}$ for all $u \in GF(q)$. In other words, g is a Bent function and l is even.

We list the concepts of correlation immunity, propagation criterion, and linear structure for Boolean functions as follows. See [11,12,13,14,15] for their cryptographic significance.

There exist several equivalent definitions for correlation immune functions [11,12,13]. Let g be a Boolean function defined on $GF(q)$. g is called *k-order correlation immune* (k-CI) if $\widehat{g}(\alpha) = 0$ for all $\alpha \in GF(q)$ with $1 \leq w_2(\alpha) \leq k$, where α is regarded as an l-dimensional $GF(2)$-vector and $w_2(\alpha)$ is the *Hamming weight* of α. Furthermore, g is called *k-resilient* if it is balanced and k-CI.

The concept of propagation criterion is introduced in [14] as a generalization of strict avalanche criterion (SAC) [15]. A function g is said to satisfy the *propagation criterion with respect to* α if $\Delta_g(\alpha) = 0$, where $\alpha \in GF(q)$ and

$$\Delta_g(\alpha) = \sum_{z \in GF(q)} (-1)^{g(z)+g(z+\alpha)}. \tag{8}$$

Furthermore, g is said to satisfy *propagation criterion of degree k* if it satisfies the propagation criterion with respective to every nonzero vector with Hamming weight not exceeding k. By Eq. (8), $\Delta_g(\alpha) = 0$ if and only if $g(z) + g(z+\alpha)$ is a balanced function.

An element α is called a *linear structure* of g if $|\Delta_g(\alpha)| = q$, or equivalently, $g(z) + g(z+\alpha)$ is a constant function. All linear structures of g form a $GF(2)$-linear subspace of $GF(q)$, whose dimension is called the *linearity* of g.

The recent algebraic attacks on stream ciphers based on linear feedback shift registers [16] lead to a new design criterion for cryptographical functions: a Boolean function should not have low degree multiples. The immunity of a

Boolean function against algebraic attacks can be measured by algebraic immunity [17,18], which is extended for functions over any finite fields [19].

The *algebraic immunity* of the function $f : GF(q)^t \to GF(q)$ is defined as

$$AI(f) = \min\{\deg(g) \geq 1 \,|\, g : GF(q)^t \to GF(q),\ f * g = 0 \text{ or } (f^{q-1} - 1) * g = 0\},$$

where $f*g = f(x)g(x)$, and $f*g = 0$ means that $f(x)g(x) = 0$ for all $x \in GF(q)^t$.

Lemma 3 ([19]): Let $f : GF(q)^t \to GF(q)$. Then there is a function $g : GF(q)^t \to GF(q)$, $1 \leq \deg(g) \leq \lceil (q-1)t/2 \rceil$ such that either $f * g = 0$ or $\deg(f * g) \leq \lfloor (q-1)t/2 \rfloor$. In particular, $AI(f) \leq \lfloor (q-1)t/2 \rfloor$.

3 Partially Perfect Nonlinear Functions

Definition 4: A function $f : GF(q)^t \to GF(q)$ is called *partially perfect nonlinear* (PPN) if there exist a linear function L on $GF(q)^t$, two $GF(q)$-subspaces E and F of $GF(q)^t$, such that

(1) $GF(q)^t$ is the direct sum of E and F;
(2) $f|_E = p$, the restriction of f to E, is perfect nonlinear; and
(3) for all $x \in E$ and $y \in F$, $f(x+y) = p(x) + L(y)$.

When $q = 2$, PPN functions are exactly partially Bent functions [6]. When $q > 2$, we can regard PPN functions as an extension of partially Bent Boolean functions.

From Definition 4, one can construct a PPN function from a known perfect nonlinear function by extending the domain of a perfect nonlinear function. There are many constructions for perfect nonlinear functions [10]. One such construction is the following

Proposition 5 ([10]): Let m be even and $\pi = (\pi_1, \pi_2, \cdots, \pi_{m/2})$ be a permutation of $GF(q)^{m/2}$, where π_i is the coordinate function of π. Then

$$\begin{aligned} p(x_1, \cdots, x_m) &= x_1\pi_1(x_{m/2+1}, \cdots, x_m) + x_2\pi_2(x_{m/2+1}, \cdots, x_m) \\ &\quad + \cdots + x_{m/2}\pi_{m/2}(x_{m/2+1}, \cdots, x_m) + p_0(x_{m/2+1}, \cdots, x_m) \end{aligned} \tag{9}$$

is a perfect nonlinear function for any function p_0 from $GF(q)^{m/2}$ to F_q.

For functions defined by Eq. (9), one has $\deg(p) \geq \deg(p_0)$. Taking p_0 as a function of high algebraic degree, perfect nonlinear functions with high algebraic degree is constructed by Proposition 5.

A perfect nonlinear function p from $GF(q)^m$ to $GF(q)$ can be easily extended to a PPN function f defined on $GF(q)^{m+n}$ as follows:

$$f(x+y) = p(x) + L(y), \forall\, x \in GF(q)^m,\ \forall\, y \in GF(q)^n, \tag{10}$$

where L is a linear function defined on $GF(q)^n$.

The following proposition describes the algebraic immunity property of a PPN function, which is suggested by a reviewer of this paper.

Proposition 6: Let $p_1(x)$ and $L_1(x)$ be functions from $GF(q)^t$ to $GF(q)$, and L_1 be affine. Then, the function $f_1(x) = p_1(x) + L_1(x)$ satisfies

$$AI(f_1) \leq AI(p_1) + q - 1.$$

In particular, for a PPN function defined as Eq. (10), one has

$$AI(p) - (q-1) \leq AI(f) \leq \lfloor (q-1)m/2 \rfloor + (q-1).$$

Proof: Assume that a function $p'(x) : GF(q)^t \to GF(q)$ satisfies $p_1 * p' = 0$ or $(p_1^{q-1} - 1) * p' = 0$. Then, for any affine function L_1, if $p_1 * p' = 0$ then one has $(p_1 + L_1) * [(L_1^{q-1} - 1) * p'] = 0$, and if $(p_1^{q-1} - 1) * p' = 0$ then one has

$$\begin{aligned} & [(p_1 + L_1)^{q-1} - 1] * [(L_1^{q-1} - 1) * p'] \\ &= [(p_1 + L_1)^{q-1} - 1] * [(p_1 + L_1) * (L_1^{q-1} - 1) * p_1^{q-2} * p'] . \\ &= 0 \end{aligned}$$

Therefore, $AI(f_1) \leq AI(p_1) + q - 1$.

For the PPN function defined by Eq. (10), one has $f(x+y) = p(x) + L(y)$ and then $p(x) = f(x+y) - L(y)$. Thus, it is true that

$$AI(p) - (q-1) \leq AI(f) \leq AI(p) + (q-1).$$

By Lemma 3, the proof is finished. □

When f is a PPN function, by Proposition 6,

$$AI(f) \leq \lfloor (q-1)m/2 \rfloor + (q-1).$$

Thus, as pointed out by the reviewer (also see Lemma 3), the algebraic immunity of a PPN function will not be optimal if n takes a large value, for example, $n \geq 3$. Theorems 8 and 12 below also show that, in order to obtain a composition function without nonzero linear structure and with high nonlinearity, the condition $n = 1$ must be imposed on the PPN function f defined by Eq. (10). By Proposition 6, the PPN function f has good algebraic immunity property if $n = 1$ and the perfect nonlinear function p has high algebraic immunity.

For $q = 2$, a PPN function is also a partially Bent function, whose algebraic immunity satisfies $AI(p) \leq AI(f) \leq AI(p) + 1$ [18], which gives a tighter lower bound than that in Proposition 6.

We establish the following relation between PPN functions and partially Bent functions.

Proposition 7: Let $f(x) : GF(q)^t \to GF(q)$ be a PPN function, then $g \circ f$ is partially Bent for any nonzero linear function $g : GF(q) \to GF(2)$.

Proof: By Definition 4, there are two subspaces E and F such that $GF(q)^t = E \oplus F$ and $f(x+y) = p(x) + L(y)$ for all $x \in E$ and $y \in F$, where p is perfect nonlinear on E and L is linear on F. For any nonzero linear function $g : GF(q) \to GF(2)$, one has

$$(g \circ f)(x + y) = (g \circ p)(x) + (g \circ L)(y).$$

By Lemmas 1 and 2, $g \circ p$ is Bent. Since $g \circ p$ is the restriction of $g \circ f$ to E and $g \circ L$ is linear, by the definition of partially Bent function, $g \circ f$ is a partially Bent function. □

Proposition 7 is applicable to derive partially Bent functions from PPN functions. Inspired by Proposition 7, an interesting problem is to study cryptographic properties of the composition $g \circ f$ for a non-linear function $g : GF(q) \to GF(2)$. We leave this to the next section.

4 A Class of Boolean Functions and Their Cryptographic Properties

This section studies a class of Boolean functions constructed by composing a PPN function and another Boolean function.

Let f be a PPN function from $GF(q)^t$ to $GF(q)$ and g be a function from $GF(q)$ to $GF(2)$. We define a tl variable Boolean function h as

$$h(x) = g(f(x)). \tag{11}$$

By Definition 4, we assume without loss of generality that $t = m + n$,

$$f(x, y) = p(x) + L(y)$$

for all $x \in GF(q)^m$, $y \in GF(q)^n$, a perfect nonlinear function $p(x)$ and a linear function $L(y)$. We always assume below L is a nonzero function.

Theorem 8: For any $u \in GF(q)^m$ and $v \in GF(q)^n$, $\widehat{h}(u, v)$ takes a value in the set

$$\{0,\ q^{m+n-1}\widehat{g}(0),\ \pm q^{m/2+n-1}\widehat{g}(\lambda) : \lambda \in GF(q)^*\}.$$

Proof: Let $L(y) = a_1 y_1 + a_2 y_2 + \cdots + a_n y_n$, and $a_{i_0} \neq 0$ for some $1 \leq i_0 \leq n$. Making an invertible linear transformation

$$y_i \mapsto y'_i \text{ for } i \neq i_0,\ y_{i_0} \mapsto a_{i_0}^{-1}(y'_{i_0} + \sum_{i \neq i_0} a_i y'_i)$$

on y, one has $L(y) = y'_{i_0}$. Since an invertible linear transformation does not change the spectra of $h(x, y)$, without loss of generality, we assume $L(y) = y_1$.

For $v, y \in GF(q)^n$, one has

$$\mathrm{tr}_1^l(v \cdot y) = \mathrm{tr}_1^l(\sum_{i=1}^{n} v_i y_i) = \mathrm{tr}_1^l[v_1(z - p(x))] + \mathrm{tr}_1^l(\sum_{i=2}^{n} v_i y_i), \tag{12}$$

where $z = p(x) + L(y) = p(x) + y_1$. For fixed x and $(y_2, \cdots, y_n)$, when y_1 runs through all elements of $GF(q)$, z runs through all elements of $GF(q)$ for exactly once. Thus, from Eq. (12), one has

$$\begin{aligned}
&\widehat{h}(u,v)\\
&= \sum_{x\in GF(q)^m}\sum_{y\in GF(q)^n}(-1)^{g(p(x)+L(y))+\mathrm{tr}_1^l(u\cdot x+v\cdot y)}\\
&= \sum_{x\in GF(q)^m}\sum_{z\in GF(q)}(-1)^{g(z)+\mathrm{tr}_1^l(u\cdot x)+\mathrm{tr}_1^l[v_1(z-p(x))]}\sum_{y_2,\cdots,y_n}(-1)^{\mathrm{tr}_1^l(\sum_{i=2}^{n}v_iy_i)} \qquad (13)\\
&= \widehat{p_{v_1}}(u)\,\widehat{g}(v_1)\sum_{y_2}(-1)^{\mathrm{tr}_1^l(v_2y_2)}\cdots\sum_{y_n}(-1)^{\mathrm{tr}_1^l(v_ny_n)},
\end{aligned}$$

where $p_{v_1}(x) = \mathrm{tr}_1^l[v_1p(x)]$ is a Boolean function from $GF(q)^m$ to $GF(2)$.

The calculation of $\widehat{h}(u,v)$ is now divided into four cases.

Case 1: $u = v = 0$. Since $p_{v_1}(x) = 0$ and $\sum_{y_i}(-1)^{\mathrm{tr}_1^l(v_iy_i)} = q$ for $2 \le i \le n$, by Eq. (13), one has

$$\widehat{h}(u,v) = q^{m+n-1}\widehat{g}(0).$$

Case 2: $u \neq 0$ and $v = 0$. Since

$$\sum_{x\in GF(q)^m}(-1)^{\mathrm{tr}_1^l[v_1p(x)+u\cdot x]} = \sum_{x\in GF(q)^m}(-1)^{\mathrm{tr}_1^l(u\cdot x)} = 0,$$

by Eq. (13), one has $\widehat{h}(u,v) = 0$.

Case 3: There is an integer i_1 with $2 \le i_1 \le n$ such that $v_{i_1} \neq 0$. Then $\sum_{y_{i_1}}(-1)^{tr_1^l(v_{i_1}y_{i_1})} = 0$, and $\widehat{h}(u,v) = 0$.

Case 4: $v_1 \neq 0$ and $v_i = 0$ for $2 \le i \le n$. By Lemmas 1 and 2, one has $\widehat{p_{v_1}}(u) = \pm q^{m/2}$, which implies

$$\widehat{h}(u,v) = \pm q^{m/2+n-1}\widehat{g}(v_1).$$

□

When g is balanced, i.e., $\widehat{g}(0) = 0$, we have the following

Corollary 9: Assume g is balanced. Then $h(x,y)$ is balanced and its spectra takes values in $\{0, \pm q^{m/2+n-1}\widehat{g}(\lambda) : \lambda \in GF(q)^*\}$. In particular, if $n = 1$ and $|\widehat{g}(\lambda)| = 0$ or $2^{(l+c)/2}$ for all $\lambda \in GF(q)$, where c is 1 if l is odd and 2 otherwise, then $|\widehat{h}(u,v)| = 0$ or $2^{(lt+c)/2}$.

According to Corollary 9, the composition construction defined by Eq. (11) provides a method to construct new balanced plateaued functions [1] from known ones provided a PPN function from $GF(q)^t$ to $GF(q)$ exists. This construction is a generalization of that in [1], since the PPN functions used in [1] are only quadratic functions. Our method can generate functions with higher algebraic degree if the PPN functions with high algebraic degree are used, as shown below in Theorem 10.

Theorem 10: There is a PPN function f such that the function h defined by Eq. (11) has algebraic degree $m/2 \cdot \deg(g)$.

Proof: With the same notions as in Proposition 5, let

$$p_0(x_{m/2+1}, \cdots, x_m) = x_{m/2+1} \cdots x_m, \;\; \pi_i(x_{m/2+1}, \cdots, x_m) = x_{m/2+i}$$

for $1 \leq i \leq m/2$. Thus, $p(x_1, \cdots, x_m)$ is perfect nonlinear. By Definition 4, for any linear function L over $GF(q)^n$, $f(x, y) = p(x) + L(y)$ is a PPN function over $GF(q)^{m+n}$.

Let $\{\alpha_1, \alpha_2, \cdots, \alpha_{m+n}\}$ be a basis of $GF(q)^{m+n}$ over $GF(q)$ and $\{\beta_1, \beta_2, \cdots, \beta_{m+n}\}$ be its dual basis. Let $z = \sum_{i=1}^{m} x_i\alpha_i + \sum_{j=1}^{n} y_i\alpha_{m+i} \in GF(q)^{m+n}$, $x_i, y_j \in GF(q)$, then $x_i = tr_l^{l(m+n)}(\beta_i z)$. Thus, the monomial $p_0(x_{m/2+1}, \cdots, x_m) = x_{m/2+1} \cdots x_m$ has a polynomial representation as follows

$$x_{m/2+1} \cdots x_m = \prod_{j=1}^{m/2} tr_l^{l(m+n)}(\beta_{j+m/2} z) = \sum_{(d_1, \cdots, d_{m/2}) \in V} \Big(\prod_{j=1}^{m/2} \beta_{j+m/2}^{q^{d_i}} \Big) z^{\sum_{i=1}^{m/2} q^{d_i}}$$

where $V = \{0, 1, \cdots, m+n-1\}^{m/2}$. Let $V = V_1 \cup V_2$, where

$$V_1 = \{\underline{d} = (d_1, \cdots, d_{m/2}) \in V \,|\, d_i \neq d_j \text{ for } 1 \leq i \neq j \leq m/2\},$$

and $V_2 = V \setminus V_1$. Then, for any $\underline{d} \in V_1$,

$$\sum_{i=0}^{m/2} q^{d_i} < q^{m+n} \text{ and } w_q(\sum_{i=0}^{m/2} q^{d_i}) = w_2(\sum_{i=0}^{m/2} q^{d_i}) = m/2.$$

Moreover, for different $\underline{d} \in V_1$, the exponents $\sum_{i=0}^{m/2} q^{d_i}$ are different.

For $\underline{d} \in V_2$, let $\Phi(\underline{d}) = \sum_{i=0}^{m/2} q^{d_i} \bmod (q^{m+n} - 1)$ if $\sum_{i=0}^{m/2} q^{d_i} \geq q^{m+n}$ and $\Phi(\underline{d}) = \sum_{i=0}^{m/2} q^{d_i}$ otherwise. Then, one has $w_q(\Phi(\underline{d})) \leq w_q(\sum_{i=0}^{m/2} q^{d_i}) \leq m/2$.

Thus, the polynomial representation of $f(x, y) = p(x) + L(y)$ satisfies the condition in Eq. (22) in Appendix A, and by Proposition 17, $\deg(h) = m/2 \cdot \deg(g)$. □

Some concrete examples for h obtained by choosing different f and g show that, even if f does not satisfies the condition in Eq. (22), the constructed function h can also achieve higher algebraic degree. However in this moment, we do not know how to determine the exact value of $\deg(h)$ for this case.

From the proof of Theorem 8, one can also easily determine the correlation immunity and resiliency of $h(x, y)$ as follows.

Proposition 11: Assume g is k-CI or k-resilient and $L(y) = a_1 y_1 + \cdots + a_n y_n$. If $a_i = 1$ for some $i \in \{1, 2, \cdots, n\}$, then h is k-CI or k-resilient, respectively.

The following theorem characterizes the propagation characteristics of h.

Theorem 12: Assume g is a nonzero function with linearity d. Then

(1) The linearity of h is $(n-1)l + d$; and
(2) Assume g is balanced and $L(y) = a_1 y_1 + \cdots + a_n y_n$ with binary coefficients $a_i = 0$ or 1. If g satisfies the propagation criterion of degree k, then so does h.

Proof: For $\alpha \in GF(q)^m$ and $\beta \in GF(q)^n$, one has

$$\begin{aligned} \Delta_h(\alpha, \beta) &= \sum_{x \in GF(q)^m,\, y \in GF(q)^n} (-1)^{h(x+\alpha, y+\beta)+h(x,y)} \\ &= \sum_{x \in GF(q)^m,\, y \in GF(q)^n} (-1)^{g(p(x+\alpha)+L(y+\beta))+g(p(x)+L(y))}. \end{aligned} \tag{14}$$

Since $p(x)$ is a perfect nonlinear function over $GF(q)^m$, then for any $0 \neq \alpha \in GF(q)^m$, the function $p(z+\alpha) + p(z)$ is balanced over $GF(q)^m$. Thus, there is an even partition $\{U_1, U_2, \cdots, U_q\}$ of $GF(q)^m$ (i.e., $|U_i| = q^{m-1}$) such that $p(x+\alpha) + p(x) = c_i$ for any $x \in U_i$ and $\{c_1, c_2, \cdots, c_q\} = GF(q)$.

Therefore, for $\alpha \neq 0$ and a nonzero $L(y)$, Eq. (14) can be written as

$$\begin{aligned} \Delta_h(\alpha, \beta) &= \sum_{i=1}^{q} \sum_{x \in U_i} \sum_{y \in GF(q)^n} (-1)^{g[p(x)+L(y)+c_i+L(\beta)]+g(p(x)+L(y))} \\ &= \sum_{i=1}^{q} q^{m-1} q^{n-1} \Delta_g(c_i + L(\beta)) \\ &= q^{m+n-2} \sum_{z \in GF(q)} \Delta_g(z) \\ &= q^{m+n-2} \sum_{z \in GF(q)} \sum_{c \in GF(q)} (-1)^{g(z+c)+g(c)} \\ &= q^{m+n-2} \sum_{c \in GF(q)} (-1)^{g(c)} \sum_{z \in GF(q)} (-1)^{g(z+c)} \\ &= q^{m+n-2} I(g)^2 \end{aligned} \tag{15}$$

where $I(g) = \sum_{c \in GF(q)} (-1)^{g(c)}$.

When $\alpha = 0$, since $p(x) + L(y)$ is balanced, one has

$$\begin{aligned} \Delta_h(\alpha, \beta) &= \sum_{x \in GF(q)^m,\, y \in GF(q)^n} (-1)^{g(p(x)+L(y+\beta))+g(p(x)+L(y))} \\ &= q^{m+n-1} \sum_{z \in GF(q)} (-1)^{g[z+L(\beta)]+g(z)} \\ &= q^{m+n-1} \Delta_g[L(\beta)]. \end{aligned} \tag{16}$$

Thus, by Eq. (15) and Eq. (16), for a non-constant function g, (α, β) is a linear structure of h if and only if $\alpha = 0$ and $L(\beta)$ is a linear structure of $g(z)$. For each fixed $L(\beta)$, there are $2^{(n-1)l}$ elements y in $GF(q)^n$ such that $L(y) = L(\beta)$. Thus, the linearity of h is $(n-1)l+d$. If g is balanced, then $I(g) = 0$. Therefore, Theorem 12 (2) holds since $w_2(L(\beta)) \leq \sum_{i=1}^{n}(w_2(\beta_i))$ where $\beta = (\beta_1, \beta_2, \cdots, \beta_n)$ and $w_2(\beta)$ is the Hamming weight of β. □

By Theorem 12 (1), h has no nonzero linear structure if $d = 0$ and $n = 1$.

Let $n = 1$ and $L(x) = x$, by choosing g with good trade-offs between nonlinearity, resiliency, propagation characteristics, and linearity, h also has these desired properties. Two concrete examples are constructed by the proposed method as follows.

Example 13: The function

$$\begin{aligned}
&f(x_1, x_2, x_3, x_4, y_1, y_2, y_3, y_4, z_1, z_2, z_3, z_4)\\
&= (x_1y_1 + x_2y_4 + x_3y_3 + x_4y_2 + z_1, x_1y_2 + x_2y_1 + x_2y_4 + x_3y_3\\
&\quad +x_4y_2 + x_4y_3 + z_2, x_1y_3 + x_2y_2 + x_3y_1 + x_3y_4 + x_4y_3 + x_4y_4\\
&\quad +z_3, x_1y_4 + x_2y_3 + x_3y_2 + x_4y_1 + x_4y_4 + y_1y_2y_3y_4 + z_4)
\end{aligned}$$

is a PPN function from $GF(2^{12})$ to $GF(2^4)$. The quadratic Boolean function

$$g(z_1, z_2, z_3, z_4) = z_1 + z_2 + z_3z_4 + z_2z_3 + z_1z_4 + z_1z_2$$

is a $(4, 2, 1, 1, 4, 2)$-function, where (n, d, k, r, N_g, L_g) denotes an n-variable function with algebraic degree d, resiliency of order k, propagation criterion of degree r, nonlinearity N_g and linearity L_g. Then, $h = g \circ f$ is a $(12, 5, 1, 1, 1984, 2)$-Boolean function, whose algebraic expression is listed in Appendix B. For a $(4, 3, 0, 1, 4, 0)$-Boolean function

$$g'(z_1, z_2, z_3, z_4) = z_1z_3 + z_2z_3 + z_1z_4 + z_2z_4 + z_3z_4 + z_1z_2z_3,$$

the composition function $h' = g' \circ f$ is a $(12, 6, 0, 1, 1984, 0)$-Boolean function, whose algebraic expression is too long and omitted here.

5 Conclusion

In this paper we introduced the concept of partially perfect nonlinear (PPN) function to construct a new class of Boolean (in particular, plateaued) functions with good cryptographic properties. The new functions are constructed by composing a PPN function and a Boolean function. The nonlinearity, correlation immunity, propagation criterion, and other cryptographic properties of the resulting functions are good if suitable PPN functions and Boolean functions are employed.

Acknowledgment. The authors thank anonymous reviewers for their useful comments.

References

1. Khoo, K., Gong, G.: New Constructions for Resilient and Highly Nonlinear Boolean Functions. In: Matsui, M. (ed.): Information Security and Privacy: 8th Australasian Conference, ACISP 2003, Lecture Notes in Computer Science, Vol. 2727. Springer-Verlag, Berlin Heidelberg New York (2003) 498–509
2. Matsui, M.: Linear Cryptanalysis Method for DES Cipher. In: Helleseth, T. (ed.): Eurocrypt 1993, Lecture Notes in Computer Science, Vol. 765, Springer-Verlag Berlin Heidelberg New York (1994) 386–397
3. Rueppel, R.: Analysis and Design of Stream Ciphers. Springer-Verlag, Berlin Heidelberg New York (1986)
4. Biham, E., Shamir, A.: Differential Cryptanalysis of DES-like Cryptosystems. J. Cryptology. **4** (1991) 3–72
5. Rothaus, O.: On Bent Functions. J. Combinatorial Theory. **20** (1976) 300–305
6. Carlet, C.: Partially-Bent Functions. In: Brickell, E. (ed): Crypto 1992. Lecture Notes in Computer Science, Vol. 740, Springer-Verlag, Berlin Heidelberg New York (1993) 280–291
7. Lidl, R., Niederreiter, H.: Finite Fields. Encyclopedia of Mathematics and Its Applications. Vol. 20, Reading MA Addison-Wesley (1983)
8. McWilliams, F., Solane, N.: The Theory of Error-Correcting Codes. North-Holland Amsterdam (1977)
9. Zheng, Y., Zhang, X.: Plateaued Functions. In: Varadharajan, V., Mu, Y. (eds.): Information and Communication Security: Second International Conference, ICICS'99, Lecture Notes in Computer Science, Vol. 1726. Springer-Verlag, Berlin Heidelberg New York (1999) 284–300
10. Carlet, C., Ding, C.: Highly Nonlinear Mappings. J. Complexity. **20** (2004) 205–244
11. Siegenthaler, T.: Correlation Immunity of Nonlinear Combining Functions for Cryptographic Applications. IEEE Trans. Inform. Theory. **30** (1984) 776–780
12. Xiao, G., Massey, J.: A Spectral Characterization of Correlation Immune Combining Functions. IEEE Trans. Inform. Theory. **34** (1988) 569–571
13. Camion, P., Carlet, C., Charpin, P., Sendrier, N.: On Correlation-Immune Functions. In: Feigenbaum, J. (ed.): Crypto 1991. Lecture Notes in Computer Science, Vol. 576, Springer-Verlag, Berlin Heidelberg New York (1992) 86–100
14. Preneel, B., Leekwijck, W., Linden, L., Govaerts, R., Vandewalle, J.: Propagation Characteristics of Boolean Functions. In: Damgard, I. (ed.): Eurocrypt 1990. Lecture Notes in Computer Science, Vol. 473. Springer-Verlag, Berlin Heidelberg New York (1990) 161–173
15. Webster, A., Tavares, S.: On the Design of S-box. In: Williams, H. (ed.): Crypto 1985. Lecture Notes in Computer Science, Vol. 218. Springer-Verlag, Berlin Heidelberg New York (1986) 523–534
16. Courtois, N., Meier, W.: Algebriac Attack on Stream Ciphers with Linear Feeback. In: Biham, E. (ed.): Eurocrypt 2003. Lecture Notes in Computer Science, Vol. 2656. Springer-Verlag, Berlin Heidelberg New York (2003) 345–359
17. Meier, W., Pasalic, E., Carlet, C.: Algebriac Attack and Decomposition of Boolean Functions. In: Cachin, C., Camenisch, J. (eds.): Eurocrypt 2004. Lecture Notes in Computer Science, Vol. 3027. Springer-Verlag, Berlin Heidelberg New York (2004) 474–491
18. Dalai, D., Gupta, K., Maitra, S.: Results on Algebraic Immunity for Cryptographivally Signficant Boolean Functions. In: Canteaut, A., Viswanathan, K. (eds.): Indocrypt 2004. Lecture Notes in Computer Science, Vol. 3348. Springer-Verlag, Berlin Heidelberg New York (2004) 92–106

19. Batten, L.: Algebraic Attack over $GF(q)$. In: Canteaut, A., Viswanathan, K. (eds.): Indocrypt 2004. Lecture Notes in Computer Science, Vol. 3348. Springer-Verlag, Berlin Heidelberg New York (2004) 84–91

Appendix A: Algebraic Degree of Composition Functions

In this appendix, the algebraic degree of composition functions over a finite field is considered.

Let s be a nonnegative integer less than 2^{ml}, and write it as

$$s = \sum_{i=0}^{ml-1} a_i 2^i = \sum_{i=0}^{m-1} s_i q^i, \tag{17}$$

where $a_i \in \{0,1\}$ and $s_i \in \{0,1,\cdots,q-1\}$. Define $w_2(s) := \sum_{i=0}^{ml-1} a_i$ and $w_q(s) := \sum_{i=0}^{m-1} s_i$.

Lemma 14:

$$w_2(s) \leq w_q(s). \tag{18}$$

The equality holds if and only if

$$s = \sum_{i=0}^{m-1} s_i q^i, s_i \in \{0,1\}. \tag{19}$$

Proof: By Eq. (18), one has $w_2(s) = \sum_{i=0}^{m-1} w_2(s_i)$. Since $w_2(s_i) \leq s_i$ and that $w_2(s_i) = s_i$ if and only if $s_i = 0$ or 1, it is clear that

$$w_2(s) = \sum_{i=0}^{m-1} w_2(s_i) \leq \sum_{i=0}^{m-1} s_i = w_q(s). \tag{20}$$

The equality in Eq. (21) holds if and only if $s_i \in \{0,1\}$ for all $0 \leq i \leq m-1$. □

Let $f(x)$ be defined as in Eq. (1). Denote the set of all nonzero coefficients a_s by

$$C = \{0 \leq s \leq q^m - 1 \,|\, a_s \neq 0\}. \tag{21}$$

If there is an integer $s' \in C$ such that

$$\deg(f) = w_q(s') = w_2(s'), \tag{22}$$

then by Lemma 14, one has $s' = \sum_{i=0}^{m-1} s'_i q^i \in C$ where $s'_i \in \{0,1\}$. Furthermore, for any $s \in C$, $w_2(s) \leq w_2(s') \leq m$, and $w_2(s) = w_2(s')$ implies

$$w_2(s) = w_q(s) = w_q(s') = w_2(s'),$$

i.e., $s = \sum_{i=0}^{m-1} s_i q^i$ with $s_i \in \{0, 1\}$.

Partition C into two subsets:

$$C_1 = \{s \in C \mid w_2(s) = w_2(s')\}, \quad C_2 = \{s \in C \mid w_2(s) < w_2(s')\}. \tag{23}$$

For any integer of the form $r = 2^{r_1} + \cdots + 2^{r_{w_2(r)}} \leq q - 1$, where $0 \leq r_1 < \cdots < r_{w_2(r)} < l$, the expansion of $(f(x))^r$ can be written as

$$[f(x)]^r = \prod_{j=1}^{w_2(r)} (\sum_{s=0}^{q^m-1} a_s x^s)^{2^{r_j}} = \prod_{j=1}^{w_2(r)} (\sum_{s=0}^{q^m-1} a_s^{2^{r_j}} x^{s \cdot 2^{r_j}}) = \sum_{\underline{e} \in C^{w_2(r)}} (\prod_{j=1}^{w_2(r)} a_{e_j}^{2^{r_j}}) x^{\Phi_{r,\underline{e}}} \tag{24}$$

where $\underline{e} = (e_1, e_2, \cdots, e_{w_2(r)})$ and $\Phi_{r,\underline{e}} = \sum_{j=1}^{w_2(r)} 2^{r_j} e_j$.

The value and Hamming weight of the integer $\Phi_{r,\underline{e}}$ are measured by the following two lemmas.

Lemma 15: Assume that f satisfies the condition in Eq. (22). Then for $\underline{e} \in C_1^{w_2(r)}$, one has

$$\Phi_{r,\underline{e}} < q^m, \text{ and } w_2(\Phi_{r,\underline{e}}) = w_2(r) \cdot w_2(s'). \tag{25}$$

If $\underline{e}' \in C_1^{w_2(r)}$ and $\underline{e}' \neq \underline{e}$, then $\Phi_{r,\underline{e}'} \neq \Phi_{r,\underline{e}}$.

Proof: Since $e_j \leq 1 + q + q^2 + \cdots + q^{m-1} = (q^m - 1)/(q - 1)$, one has

$$\Phi_{r,\underline{e}} = \sum_{j=1}^{w_2(r)} 2^{r_j} e_j \leq \sum_{j=1}^{w_2(r)} 2^{r_j} (q^m - 1)/(q - 1) \leq q^m - 1. \tag{26}$$

For any $1 \leq j \leq w_2(r)$, e_j can be written as

$$e_j = q^{j_1} + q^{j_2} + \cdots + q^{j_{w_2(s')}}.$$

By the fact that $2^i q^u = 2^{i'} q^{u'}$ holds for $0 \leq i, i' < l$ is equivalent to that $i = i'$ and $u = u'$, one has

$$w_2(\Phi_{r,\underline{e}}) = w_2(r) \cdot w_2(s').$$

If $\underline{e}' \neq \underline{e}$, it is easy to verify $\Phi_{r,\underline{e}'} \neq \Phi_{r,\underline{e}}$. This finishes the proof. □

Lemma 16: For $\underline{e}' \in C^{w_2(r)} \backslash C_1^{w_2(r)}$, set $\Phi'_{r,\underline{e}'} = \Phi_{r,\underline{e}'}$ if $\Phi_{r,\underline{e}'} < q^m$ and $\Phi'_{r,\underline{e}'} = \Phi_{r,\underline{e}'} \bmod (q^m - 1)$ if $\Phi_{r,\underline{e}'} \geq q^m$. Then

$$w_2(\Phi'_{r,\underline{e}'}) < w_2(r) \cdot w_2(s'). \tag{27}$$

Proof: For $\underline{e}' \in C^{w_2(r)} \backslash C_1^{w_2(r)}$, one has

$$w_2(\Phi_{r,\underline{e}'}) = w_2(\sum_{j=1}^{w_2(r)} 2^{r_j} e'_j) \leq \sum_{j=1}^{w_2(r)} w_2(e'_j) < \sum_{j=1}^{w_2(r)} w_2(s') = w_2(r) \cdot w_2(s'). \tag{28}$$

The proof follows the fact that $w_2(\Phi'_{r,\underline{e}'}) \leq w_2(\Phi_{r,\underline{e}'})$. □

Applying Lemmas 15 and 16, the algebraic degree of a composition function is determined as follows.

Proposition 17: For functions $f : GF(q)^m \to GF(q)$ and $g(x) : GF(q) \to GF(2)$, if f satisfies the condition in Eq. (22), then $\deg(g \circ f) = \deg(f) \cdot \deg(g)$.

Proof: Let f be expressed as Eq. (1) and $g(x) = \sum_{r=0}^{q-1} b_r x^r$. Then, by Eq. (25), one has

$$g(f(x)) = \sum_{r=0}^{q-1} b_r [f(x)]^r = \sum_{r=0}^{q-1} \sum_{\underline{e} \in C^{w_2(r)}} b_r \Big(\prod_{j=1}^{w_2(r)} a_{e_j}^{2^{r_j}} \Big) x^{\Phi_{r,\underline{e}}}. \tag{29}$$

By Lemmas 15 and 16, for any $1 \leq r \leq q-1$, one has

$$\deg([f(x)]^r) = w_2(r) \cdot w_2(s') \leq \deg(g) \cdot \deg(f),$$

where the equality holds if and only if $w_2(r) = \deg(g)$.

For integers r_1 and r_2 with $w_2(r_1) = w_2(r_2) = \deg(g)$, and for $\underline{e}, \underline{e}' \in C_1^{\deg(g)}$, the equality $\Phi_{r_1,\underline{e}} = \Phi_{r_2,\underline{e}'}$ holds if and only if

$$r_1 = r_2 \quad \text{and} \quad \underline{e} = \underline{e}'.$$

This together with Lemmas 15 and 16 finish the proof of Proposition 17. □

Appendix B: Algebraic Expression of a Composition Function in Example 13

$$\begin{aligned}
&h(x_1, x_2, x_3, x_4, x_5, x_6, x_7, x_8, x_9, x_{10}, x_{11}, x_{12}) \\
= \; & x_1x_2x_5x_8 + x_1x_2x_5 + x_1x_2x_6 + x_1x_2x_7x_8 + x_1x_2x_7 + x_1x_2x_8 \\
&+ x_1x_3x_5x_7 + x_1x_3x_6x_8 + x_1x_3x_7 + x_1x_3x_8 + x_1x_4x_5x_6 + x_1x_4x_5x_8 \\
&+ x_1x_4x_5 + x_1x_4x_6 + x_1x_4x_7 + x_1x_4x_8 + x_1x_5x_6 + x_1x_5x_8 + x_1x_5x_{10} \\
&+ x_1x_5x_{12} + x_1x_5 + x_1x_6x_7 + x_1x_6x_9 + x_1x_6x_{11} + x_1x_6 + x_1x_7x_8 \\
&+ x_1x_7x_{10} + x_1x_7x_{12} + x_1x_8x_9 + x_1x_8x_{11} + x_2x_3x_5 + x_2x_3x_6x_7 + x_2x_3x_6 \\
&+ x_2x_3x_7x_8 + x_2x_3x_7 + x_2x_4x_5x_7 + x_2x_4x_6x_8 + x_2x_4x_6 + x_2x_4x_7x_8 \\
&+ x_2x_4x_7 + x_2x_5x_6 + x_2x_5x_8 + x_2x_5x_9 + x_2x_5x_{11} + x_2x_5 + x_2x_6x_7 \\
&+ x_2x_6x_8 + x_2x_6x_{10} + x_2x_6x_{12} + x_2x_7x_8 + x_2x_7x_9 + x_2x_7x_{11} + x_2x_8x_9 \\
&+ x_2x_8x_{10} + x_2x_8x_{11} + x_2x_8x_{12} + x_2x_8 + x_3x_4x_5x_6 + x_3x_4x_5 + x_3x_4x_6x_7 \\
&+ x_3x_4x_6x_8 + x_3x_4x_6 + x_3x_5x_6x_7x_8 + x_3x_5x_6 + x_3x_5x_7 + x_3x_5x_8 \\
&+ x_3x_5x_{10} + x_3x_5x_{12} + x_3x_6x_7 + x_3x_6x_8 + x_3x_6x_9 + x_3x_6x_{11} + x_3x_7x_9 \\
&+ x_3x_7x_{10} + x_3x_7x_{11} + x_3x_7x_{12} + x_3x_7 + x_3x_8x_9 + x_3x_8x_{10} + x_3x_8x_{11} \\
&+ x_3x_8x_{12} + x_4x_5x_6x_7x_8 + x_4x_5x_6 + x_4x_5x_7 + x_4x_5x_8 + x_4x_5x_9 \\
&+ x_4x_5x_{11} + x_4x_6x_9 + x_4x_6x_{10} + x_4x_6x_{11} + x_4x_6x_{12} + x_4x_6 + x_4x_7x_9 \\
&+ x_4x_7x_{10} + x_4x_7x_{11} + x_4x_7x_{12} + x_4x_8x_9 + x_4x_8x_{10} + x_4x_8x_{11} + x_4x_8x_{12} \\
&+ x_4x_8 + x_5x_6x_7x_8x_9 + x_5x_6x_7x_8x_{11} + x_9x_{10} + x_9x_{12} + x_9 + x_{10}x_{11} \\
&+ x_{10} + x_{11}x_{12}.
\end{aligned}$$

Construction of 1-Resilient Boolean Functions with Very Good Nonlinearity

Soumen Maity[1], Chrisil Arackaparambil[1], and Kezhasono Meyase[2]

[1] Department of Mathematics, Indian Institute of Technology Guwahati
Guwahati 781 039, Assam, India
soumen@iitg.ernet.in, joseph@iitg.ernet.in

[2] Tata Consultancy Services Limited, Abhilash Software Development Centre
Plot No. 96, EPIP Industrial Area, Whitefield
Bangalore - 560066, Karnataka, India
kezhasono.meyase@tcs.com

Abstract. In this paper we present a strategy to construct 1-resilient Boolean functions with very good nonlinearity and autocorrelation. Our strategy to construct a 1-resilient function is based on modifying a bent function, by toggling some of its output bits. Two natural questions that arise in this context are "at least how many bits and which bits in the output of a bent function need to be changed to construct a 1-resilient Boolean function". We present an algorithm which determines a minimum number of bits of a bent function that need to be changed to construct a 1-resilient Boolean function. We also present a technique to compute points whose output in the bent function need to be modified to get a 1-resilient function. In particular, the technique is applied upto 14-variable functions and we show that the construction provides 1-resilient functions reaching currently best known nonlinearity and achieving very low autocorrelation absolute indicator values which were not known earlier.

Keywords: Autocorrelation, Bent Function, Boolean Function, Nonlinearity, Resiliency.

1 Introduction

Boolean functions are extensively used in stream cipher systems. Important necessary properties of Boolean functions used in these systems are balancedness, high order resiliency, high algebraic degree, and high nonlinearity. Constructions of Boolean functions possessing a good combination of these properties have been proposed in [8,10]. In [9,3], it had been shown how bent functions can be modified to construct highly nonlinear balanced Boolean functions. A recent construction method [5,6] presents modification of some output points of a bent function to construct highly nonlinear 1-resilient functions. In [6], a lower bound on the minimum number of bits of a bent function that need to be modified is given. However the bound is not tight for functions with more than 10 variables. In this paper, we give a better lower bound on the minimum number of bits of a bent function that need to be changed. The bound is proved to be

G. Gong et al. (Eds.): SETA 2006, LNCS 4086, pp. 417–431, 2006.

tight for functions up to 14 variables. Further [6] does not provide any technique to select the points whose output in the bent function need to be modified and the points are selected by computer simulation. Our main contribution here is a construction to select those points whose output in the bent function need to be modified to get a 1-resilient function. For the first time, we give a combinatorial construction which can be used to obtain a 1-resilient function for any n. In particular, we concentrate on construction of 1-resilient Boolean functions up to 14 variables with best known nonlinearity and autocorrelation. *Throughout the paper we consider the number of input variables* (n) *is even.* Here, we identify Maiorana-McFarland type bent functions which can be modified to get 1-resilient functions with currently best known parameters. We get 1-resilient functions with better nonlinearity and autocorrelation absolute indicator values that were not known earlier for $n = 12, 14$ variables.

1.1 Preliminaries

A Boolean function on n variables may be viewed as a mapping from $\{0,1\}^n$ into $\{0,1\}$. The *Hamming distance* between two binary strings S_1, S_2 is denoted by $d(S_1, S_2)$, i.e., $d(S_1, S_2) = \#(S_1 \neq S_2)$. Also the *Hamming weight* or simply the weight of a binary string S is the number of ones in S. This is denoted by $wt(S)$. An n-variable function f is said to be *balanced* if its output column in the truth table contains equal number of 0s and 1s (i.e., $wt(f) = 2^{n-1}$).

Denote addition operator over $GF(2)$ by $\oplus$. An n-variable Boolean function $f(x_1, \ldots, x_n)$ can be considered to be a multivariate polynomial over $GF(2)$. This polynomial can be expressed as a sum of product representation of all distinct k-th order products $(0 \leq k \leq n)$ of the variables. More precisely, $f(x_1, \ldots, x_n)$ can be written as $a_0 \oplus \bigoplus_{1 \leq i \leq n} a_i x_i \oplus \bigoplus_{1 \leq i < j \leq n} a_{ij} x_i x_j \oplus \ldots \oplus a_{12\ldots n} x_1 x_2 \ldots x_n$, where the coefficients $a_0, a_{ij}, \ldots, a_{12\ldots n} \in \{0,1\}$. This representation of f is called the *algebraic normal form* (ANF) of f. The number of variables in the highest order product term with nonzero coefficient is called the *algebraic degree*, or simply the degree of f and denoted by $deg(f)$.

Functions of degree at most one are called *affine* functions. An affine function with constant term equal to zero is called a *linear* function. The set of all n-variable affine functions is denoted by $A(n)$. The nonlinearity of an n-variable function f is $nl(f) = \min_{g \in A(n)} d(f, g)$, i.e., the distance from the set of all n-variable affine functions.

Let $x = (x_1, \ldots, x_n)$ and $\omega = (\omega_1, \ldots, \omega_n)$ both belong to $\{0,1\}^n$ and $x \cdot \omega = x_1\omega_1 \oplus \ldots \oplus x_n\omega_n$. Let $f(x)$ be a Boolean function on n variables. Then the *Walsh transform* of $f(x)$ is a real valued function over $\{0,1\}^n$ which is defined as $W_f(\omega) = \sum_{x \in \{0,1\}^n} (-1)^{f(x) \oplus x \cdot \omega}$. In terms of Walsh spectrum, the nonlinearity of f is given by $nl(f) = 2^{n-1} - \frac{1}{2} \max_{\omega \in \{0,1\}^n} |W_f(\omega)|$. For n even, the maximum nonlinearity of a Boolean function can be $2^{n-1} - 2^{\frac{n}{2}-1}$ and the functions possessing this nonlinearity are called bent functions [7]. Further, for a bent function f on n variables, $W_f(\omega) = \pm 2^{\frac{n}{2}}$ for all ω.

In [4], an important characterization of correlation immune and resilient functions has been presented, which we use as the definition here. An n-variable function f is m-resilient (respectively m-th order correlation immune) iff its Walsh transform satisfies $W_f(\omega) = 0$, for $0 \le wt(\omega) \le m$ (respectively $W_f(\omega) = 0$, for $1 \le wt(\omega) \le m$).

We will now define *restricted Walsh transform* which will be frequently used in this text. The *restricted Walsh transform* of $f(x)$ on a subset S of $\{0,1\}^n$ is a real valued function over $\{0,1\}^n$ which is defined as $W_f(\omega)|_S = \sum_{x \in S}(-1)^{f(x) \oplus x \cdot \omega}$. Now we present the following technical result.

Proposition 1. *[6] Let $S \subset \{0,1\}^n$ and $b(x), f(x)$ be two n-variable Boolean functions such that $f(x) = 1 \oplus b(x)$ when $x \in S$ and $f(x) = b(x)$ otherwise. Then $W_f(\omega) = W_b(\omega) - 2W_b(\omega)|_S$.*

Let $\alpha \in \{0,1\}^n$ and f be an n-variable Boolean function. Define the autocorrelation value of f with respect to the vector α as $\Delta_f(\alpha) = \sum_{x \in \{0,1\}^n}(-1)^{f(x) \oplus f(x \oplus \alpha)}$ and the absolute indicator $\Delta_f = \max_{\alpha \in \{0,1\}^n, \alpha \neq \bar{0}} |\Delta_f(\alpha)|$. Note that, for a bent function f on n variables, $\Delta_f(\alpha) = 0$ for all nonzero α, i.e., $\Delta_f = 0$.

Analysis of autocorrelation properties of correlation immune and resilient Boolean functions has gained substantial interest recently. In [1], it has been identified that some well known constructions of resilient Boolean functions are not good in terms of autocorrelation properties. Since the present construction is modification of bent functions which possess the best possible autocorrelation properties, we get very good autocorrelation properties of the 1-resilient functions.

2 Main Results

In this section, we present an algorithm which determines a minimum number of bits of a bent function that need to be changed to construct a 1-resilient Boolean function. We also provide a construction that computes the points whose output in the bent function need to be modified to get a 1-resilient function.

Let $\ell(n)$ be the minimum distance between n-variable bent and 1-resilient functions, i.e., $\ell(n) = \min\{d(b, f) : b$ is a bent function, f is a 1-resilient function$\}$. Then it is easy to note that $\ell(n) \ge 2^{\frac{n}{2}-1}$. For a bent function b on n variables the Walsh spectrum values are $+2^{\frac{n}{2}}$ or $-2^{\frac{n}{2}}$. *In this paper, we consider the bent functions b with $W_b(\omega) = +2^{\frac{n}{2}}$ for $0 \le wt(\omega) \le 1$.* Let S be a subset of $\{0,1\}^n$ and $f(x)$ be an n-variable Boolean function obtained by modifying the $b(x)$ values for $x \in S$ and keeping the other bits unchanged. That is,

$$\begin{aligned} f(x) &= 1 \oplus b(x), \quad \text{if } x \in S \\ &= b(x), \qquad\quad \text{otherwise.} \end{aligned}$$

Then from Proposition 1, $W_f(\omega) = W_b(\omega) - 2W_b(\omega)|_S \ \forall \ \omega$, and in particular, $W_f(\omega) = 2^{\frac{n}{2}} - 2W_b(\omega)|_S$ for $0 \le wt(\omega) \le 1$. It is known that f is 1-resilient iff $W_f(\omega) = 0$ for $0 \le wt(\omega) \le 1$, i.e., iff $W_b(\omega)|_S = 2^{\frac{n}{2}-1}$ for $0 \le wt(\omega) \le 1$. Thus the problem is to find a subset S of $\{0,1\}^n$ of minimum cardinality and a suitable bent function $b(x)$ that satisfy the following conditions:

$$W_b(\omega)|_S = 2^{\frac{n}{2}-1} \qquad \text{for} \quad 0 \le wt(\omega) \le 1 \tag{1}$$

$$W_b(\omega) = +2^{\frac{n}{2}} \qquad \text{for} \quad 0 \le wt(\omega) \le 1 \tag{2}$$

2.1 Determining Minimum Number of Bits of an n-Variable Bent Function That Need to Be Modified to Construct a 1-Resilient Function

For the convenience of the reader, we would like to write subset S as matrix $\mathbf{S}$ whose rows are the elements of S. Formally, given $S = \{x^{i_1}, x^{i_2}, \ldots, x^{i_k}\} \subseteq \{0,1\}^n$, consider the matrices

$$\mathbf{S}^{k\times n} = (x^{i_1}, x^{i_2}, \ldots, x^{i_k})^T,\ b(\mathbf{S})^{k\times 1} = (b(x^{i_1}), b(x^{i_2}), \ldots, b(x^{i_k}))^T, \text{ and}$$

$$(\mathbf{S} \oplus b(\mathbf{S}))^{k\times n} = (x^{i_1} \oplus b(x^{i_1}), x^{i_2} \oplus b(x^{i_2}), \ldots, x^{i_k} \oplus b(x^{i_k}))^T .$$

By A^T we mean transpose of a matrix A. Also by abuse of notation, $x^{i_j} \oplus b(x^{i_j})$ means the GF(2) addition (XOR) of the bit $b(x^{i_j})$ with each of the bits of x^{i_j}.

Consider Condition 1 with $wt(\omega) = 0$. If k_0 is the number of 0s in $b(S)$ and k_1 is the number of 1s in $b(S)$, then we have that, $k_0 - k_1 = 2^{\frac{n}{2}-1}$. Also, $k_0 + k_1 = k$. Solving these two equations, $k_0 = \frac{k}{2} + 2^{\frac{n}{2}-2}$ and $k_1 = \frac{k}{2} - 2^{\frac{n}{2}-2}$. Now consider Condition 1 with $wt(\omega) = 1$. Let ω be the unit vector having a 1 in position j and 0 in all other places. Then the jth column $\left(x_j^{i_1} \oplus b(x^{i_1}), x_j^{i_2} \oplus b(x^{i_2}), \ldots x_j^{i_k} \oplus b(x^{i_k})\right)^T$ of $\mathbf{S} \oplus b(\mathbf{S})$ has $\frac{k}{2} + 2^{\frac{n}{2}-2}$ 0s and $\frac{k}{2} - 2^{\frac{n}{2}-2}$ 1s. Thus by Condition 1, we have that there are exactly $\frac{k}{2} + 2^{\frac{n}{2}-2}$ many 0's and $\frac{k}{2} - 2^{\frac{n}{2}-2}$ many 1's in $b(\mathbf{S})$ and in each column of $\mathbf{S} \oplus b(\mathbf{S})$.

Without loss of generality we assume that the first $\frac{k}{2} + 2^{\frac{n}{2}-2}$ entries of $b(\mathbf{S})$ are 0s and the last $\frac{k}{2} - 2^{\frac{n}{2}-2}$ entries are 1s. Denote the sub-matrix consisting of the first $\frac{k}{2} + 2^{\frac{n}{2}-2}$ rows of $\mathbf{S}$ as block $\mathbf{S}_0$ (the corresponding elements are in S_0) and the sub-matrix consisting of the last $\frac{k}{2} - 2^{\frac{n}{2}-2}$ rows of $\mathbf{S}$ as block $\mathbf{S}_1$ (the corresponding elements are in S_1). Thus $S = S_0 \cup S_1$ and $\mathbf{S} = (\mathbf{S}_0, \mathbf{S}_1)^T$. Since S is a set, all the rows of $\mathbf{S}$ are distinct and furthermore, as the first $\frac{k}{2} + 2^{\frac{n}{2}-2}$ entries of $b(\mathbf{S})$ are 0s and the last $\frac{k}{2} - 2^{\frac{n}{2}-2}$ entries are 1s, the rows of $\mathbf{S}_0 \oplus b(\mathbf{S}_0)$ are distinct among themselves as are the rows of $\mathbf{S}_1 \oplus b(\mathbf{S}_1)$. Further, the Boolean complement of any row of $\mathbf{S}_0 \oplus b(\mathbf{S}_0)$ is not a row in $\mathbf{S}_1 \oplus b(\mathbf{S}_1)$.

Our problem is now to construct a matrix $\mathbf{S} \oplus b(\mathbf{S}) = (\mathbf{S}_0 \oplus b(\mathbf{S}_0), \mathbf{S}_1 \oplus b(\mathbf{S}_1))^T$ satisfying the conditions (Condition 1 in matrix notation):

(a) The number of rows in $\mathbf{S}_0 \oplus b(\mathbf{S}_0)$ is $\frac{k}{2} + 2^{\frac{n}{2}-2}$, and the number of rows in $\mathbf{S}_1 \oplus b(\mathbf{S}_1)$ is $\frac{k}{2} - 2^{\frac{n}{2}-2}$.
(b) Weight of each column of $\mathbf{S} \oplus b(\mathbf{S})$ is $\frac{k}{2} - 2^{\frac{n}{2}-2}$.
(c) Rows of $\mathbf{S}_0 \oplus b(\mathbf{S}_0)$ are distinct among themselves and so are the rows of $\mathbf{S}_1 \oplus b(\mathbf{S}_1)$. Further, the Boolean complement of any row of $\mathbf{S}_0 \oplus b(\mathbf{S}_0)$ is not in $\mathbf{S}_1 \oplus b(\mathbf{S}_1)$.

Note that, for the above condition to be satisfied, weight of each column of $\mathbf{S} \oplus b(\mathbf{S})$ must be at least one for, if it is zero all rows of $\mathbf{S} \oplus b(\mathbf{S})$ will be zero row vectors and hence identical. So, $\frac{k}{2} - 2^{\frac{n}{2}-2} \geq 1$ which gives $k \geq 2^{\frac{n}{2}-1} + 2$.

Suppose that one such matrix $\mathbf{S}\oplus b(\mathbf{S})$ is constructed. Note that the minimum number of 1s required for the distinct rows of $\mathbf{S}_0 \oplus b(\mathbf{S}_0)$ is at least $\sum_{i=1}^{r_0} i\binom{n}{i} + (r_0+1)\left(\frac{k}{2} + 2^{\frac{n}{2}-2} - \sum_{i=0}^{r_0}\binom{n}{i}\right)$, where r_0 is such that

$$\sum_{i=0}^{r_0}\binom{n}{i} \le \frac{k}{2} + 2^{\frac{n}{2}-2} < \sum_{i=0}^{r_0+1}\binom{n}{i}$$

is satisfied (using all the rows upto weight r_0 and some of the rows with weight r_0+1). Similarly the minimum number of 1s required for the distinct rows of $\mathbf{S}_1 \oplus b(\mathbf{S}_1)$ is at least $\sum_{i=1}^{r_1} i\binom{n}{i} + (r_1+1)\left(\frac{k}{2} - 2^{\frac{n}{2}-2} - \sum_{i=0}^{r_1}\binom{n}{i}\right)$, where r_1 is such that

$$\sum_{i=0}^{r_1}\binom{n}{i} \le \frac{k}{2} - 2^{\frac{n}{2}-2} < \sum_{i=0}^{r_1+1}\binom{n}{i}$$

is satisfied. So the minimum number of 1s required to form $\mathbf{S} \oplus b(\mathbf{S})$ is at least

$$\sum_{i=1}^{r_0} i\binom{n}{i} + (r_0+1)\left(\frac{k}{2} + 2^{\frac{n}{2}-2} - \sum_{i=0}^{r_0}\binom{n}{i}\right) + \sum_{i=1}^{r_1} i\binom{n}{i} + (r_1+1)\left(\frac{k}{2} - 2^{\frac{n}{2}-2} - \sum_{i=0}^{r_1}\binom{n}{i}\right),$$

where r_0 and r_1 are as above.

On the other hand, Condition 1b says there would be exactly $n \times (\frac{k}{2} - 2^{\frac{n}{2}-2})$ many 1s in $\mathbf{S} \oplus b(\mathbf{S})$ as each column contains exactly $\frac{k}{2} - 2^{\frac{n}{2}-2}$ many 1s and there are n columns. If using rows of lower weight we obtain columns of weight less than $\frac{k}{2} - 2^{\frac{n}{2}-2}$ then we may increase the weight of our rows. However, if the weight of some column is greater than $\frac{k}{2} - 2^{\frac{n}{2}-2}$ then we cannot do with k rows and must increase k. This is the basis of the next algorithm which computes a lower bound on $\ell(n)$. The above arguments tell us that k must satisfy the following condition:

$$\sum_{i=1}^{r_0} i\binom{n}{i} + (r_0+1)\left(\frac{k}{2} + 2^{\frac{n}{2}-2} - \sum_{i=0}^{r_0}\binom{n}{i}\right) + \sum_{i=1}^{r_1} i\binom{n}{i} + (r_1+1)\left(\frac{k}{2} - 2^{\frac{n}{2}-2} - \sum_{i=0}^{r_1}\binom{n}{i}\right) \le n \times \left(\frac{k}{2} - 2^{\frac{n}{2}-2}\right).$$

Here is an algorithm to compute the minimum k satisfying this condition.

Algorithm 1
Input: number of variables n.
Output: number of points required k, r_0 and r_1.

1. Set $k = 2^{\frac{n}{2}-1} + 2$ and $w = 1$ where w is the weight of columns of $\mathbf{S} \oplus b(\mathbf{S})$. $(w = \frac{k}{2} - 2^{\frac{n}{2}-2})$

2. Compute r_0 and r_1 such that

$$\sum_{i=0}^{r_0} \binom{n}{i} \leq \frac{k}{2} + 2^{\frac{n}{2}-2} < \sum_{i=0}^{r_0+1} \binom{n}{i} \text{ and } \sum_{i=0}^{r_1} \binom{n}{i} \leq \frac{k}{2} - 2^{\frac{n}{2}-2} < \sum_{i=0}^{r_1+1} \binom{n}{i}$$

are satisfied.
3. Set minimum number of 1s in $\mathbf{S} \oplus b(\mathbf{S})$,

$$z = \sum_{i=0}^{r_0} i \binom{n}{i} + (r_0 + 1) \left(\frac{k}{2} + 2^{\frac{n}{2}-2} - \sum_{i=0}^{r_0} \binom{n}{i} \right) + \sum_{i=0}^{r_1} i \binom{n}{i} + (r_1 + 1) \left(\frac{k}{2} - 2^{\frac{n}{2}-2} - \sum_{i=0}^{r_1} \binom{n}{i} \right).$$

4. If $z \leq n \cdot w$, stop. k is the required number of points.
5. $k = k + 2$, $w = w + 1$. ($w = \frac{k}{2} - 2^{\frac{n}{2}-2}$, so that when k increases by 2, w increases by 1.)
6. Go to step 2.

The following table illustrates the number of points k, as computed by the above algorithm for different values of n.

n	8	10	12	14	16	18	20	22	24	26
k	10	22	44	86	168	342	684	1350	2662	5430

Theorem 1. *The above algorithm gives a lower bound on the number of bits of an n-variable bent function that need to be modified to construct a 1-resilient function, that is $k \leq \ell(n)$.*

Proof. Let b be a bent function and f a 1-resilient function such the distance between b and f is $\ell(n)$, that is the number of points where b and f give different values is $\ell(n)$. Let S be the set of points where b and f give different values ($|S| = \ell(n)$). That is, $S = \{x \in \{0,1\}^n : b(x) \neq f(x)\}$. Then by modifying the bits of b corresponding to elements of S we obtain f. Hence $\ell(n)$ must satisfy the necessary Conditions 1a, 1b & 1c for k. The above algorithm computes the minimum k that satisfies Conditions 1a, 1b & 1c, and hence $k \leq \ell(n)$. □

The algorithm in the next section computes k points satisfying Conditions 1a, 1b & 1c. Then, to get a 1-resilient function, we will need a bent function b which has the desired output values at the points given by the next algorithm, namely, b must be such that

$$b(x) = \begin{cases} 0, & \text{for } x \in S_0 \\ 1, & \text{for } x \in S_1 \end{cases}$$

Since the class of bent functions is very large, it may be conjectured that we can find a bent function satisfying the above condition. Then, the above algorithm gives us the minimum distance since it is already a lower bound. For $n = 10, 12, 14$ we identify Maiorana-McFarland type bent functions which can be modified to get 1-resilient functions, using the points given by the next algorithm. This shows that the bound given by the above algorithm is tight and is the minimum distance for these values of n.

2.2 Finding Points Whose Output Bits in the Bent Function Need to Be Modified to Get 1-Resilient Function

Basic Idea and Approach

To find an n variable 1-resilient function from a bent function, we modify output for certain points of the bent function. Essentially, we look only at the $\mathbf{S} \oplus b(\mathbf{S})$ matrix where S is the set of points to be modified and b is the bent function. Our aim is to find a set of points S satisfying Condition 1. Here we give a construction of $\mathbf{S}$ with the number of rows k given by Algorithm 1, $\frac{k}{2} + 2^{\frac{n}{2}-2}$ rows in $\mathbf{S}_0 \oplus b(\mathbf{S}_0)$ and $\frac{k}{2} - 2^{\frac{n}{2}-2}$ in $\mathbf{S}_1 \oplus b(\mathbf{S}_1)$. Our technique is as suggested by Algorithm 1.

Since we want $\mathbf{S}$ satisfying Conditions 1a, 1b & 1c with minimum number of rows, we use rows of minimum weight. If we use rows of higher weight, column weight $\frac{k}{2} - 2^{\frac{n}{2}-2}$ also increases so that we need more number of points k. First we construct the matrix $\mathbf{S}_0 \oplus b(\mathbf{S}_0)$. Matrix $\mathbf{S}_1 \oplus b(\mathbf{S}_1)$ is also constructed in a similar manner. As in Algorithm 1, to construct $\mathbf{S}_0 \oplus b(\mathbf{S}_0)$ we use all points of weight $\leq r_0$. These rows will have a uniform column weight $\frac{\sum_{i=0}^{r_0} i\binom{n}{i}}{n}$. Further we need $m_0 = \frac{k}{2} + 2^{\frac{n}{2}-2} - \sum_{i=0}^{r_0} \binom{n}{i}$ rows of weight $r_0 + 1$.

We want to select the remaining rows of weight $r_0 + 1$ such that the weight of all n columns is more or less uniform to keep the total weight of each column in $\mathbf{S} \oplus b(\mathbf{S})$ as $\frac{k}{2} - 2^{\frac{n}{2}-2}$. Let $w_0 = \left\lfloor \frac{m_0 \times (r_0+1)}{n} \right\rfloor$ and t_0 be the remainder so that $m_0 \times (r_0 + 1) = n \times w_0 + t_0$. By a careful selection of m_0 rows of weight $r_0 + 1$, we can get t_0 columns of weight $w_0 + 1$ and $n - t_0$ columns of weight w_0, that is the column weights do not differ by more than 1. We now need a few definitions.

The *circular shift operator* ROT rotates the Boolean vector x by d positions. That is, if $y = \mathrm{ROT}(x, d)$ then y is the vector obtained by a circular shift of the bits in x by d positions. For example, $\mathrm{ROT}(x, 2) = (0,0,0,1,0,1,0,0)$, where $x = (0,0,0,0,0,1,0,1)$.

A set C of Boolean row vectors is called a *circular block* if for any $x \in C$, $C = \{\ \mathrm{ROT}(x, d) : d \text{ is an integer}\}$. That is, the vectors in C are identical up to circular shifting and C is closed under circular shifting.

Example 1. When $x = (0,0,0,0,0,0,1,1)$ in the above definition, we get the circular block

$$\begin{aligned} C = \ &\{(0,0,0,0,0,0,1,1),(0,0,0,0,0,1,1,0),(0,0,0,0,1,1,0,0),\\ &(0,0,0,1,1,0,0,0),(0,0,1,1,0,0,0,0),(0,1,1,0,0,0,0,0),\\ &(1,1,0,0,0,0,0,0),(1,0,0,0,0,0,0,1)\}. \end{aligned}$$

An important characteristic of circular blocks is that all columns are of equal weight. Note that $|C| \leq n$. Also, a circular block may not have n vectors.

Example 2. When $x = (0,0,0,1,0,0,0,1)$, we get the circular block $C = \{(0,0,0,1,0,0,0,1),(0,0,1,0,0,0,1,0),\ (0,1,0,0,0,1,0,0),\ (1,0,0,0,1,0,0,0)\}$ with $|C| = 4$.

A *generator* of a circular block C is the vector $g \in C$ which appears first in lexicographic order. In other words, it is the smallest number when the vectors in C are interpreted as binary numbers. For the circular block in Example 1 the generator is $(0,0,0,0,0,0,1,1)$ while for that in Example 2 the generator is $(0,0,0,1,0,0,0,1)$. Note that the Least Significant Bit (LSB) of a generator is always 1. We will obtain circular block C by $\leq n$ circular shifts of it's generator g. That is $C = \{\ \mathrm{ROT}(g,d) : d \leq n\}$

The next algorithm constructs a matrix T with m rows of weight r.

We can represent a point x by a set containing, the positions of the r 1s in the point. For example, $x = (0,0,1,0,0,0,1,0)$ is represented by the set $\{2,6\}$, which we denote by the ordered list $\hat{x} = [2,6]$.

Since the LSB of a generator is always 1, the number of generators is $\leq \binom{n-1}{r-1}$ and their ordered list representations will be a selection of $r-1$ positions from the set $\{2,3,\ldots,n\}$ in addition to the LSB. However, all such selections will not give a generator. Still, we can easily check if such a selection gives a generator or not.

First, note that the ROT operation for the ordered list representation is just addition modulo n to each of the list elements.

To check if a selection $[p_1,p_2,\ldots,p_{r-1}]$ gives a generator, note that the corresponding vector with the LSB is $\hat{x} = [1,p_1,p_2,\ldots,p_{r-1}]$. Now, if this vector is not a generator then it must have a generator g in it's circular block. Then $g < x$ in the lexicographic ordering. Also, g can be obtained from x by circular shift. Since LSB in g is 1, when rotating x to get g, a 1 in x initially at position $p_i, 1 \leq i \leq (r-1)$ will come at the LSB. This corresponds to a rotation by $n - p_i + 1$ bit shifts.

So to check if a vector x given by the selection is a generator or not we just have to rotate each of the $(r-1)$ 1s at positions $p_i, 1 \leq i \leq (r-1)$ by $n - p_i + 1$ and check if the resulting vector $y_i < x$ in lexicographic ordering. If no $y_i < x$ then x is a generator. Thus checking if x is a generator requires $O(r)$ operations.

Now to get a circular block, we get a selection $[p_1,p_2,\ldots,p_{r-1}]$ from $\{2,3,\ldots,n\}$ and check if the vector x corresponding to $\hat{x} = [1,p_1,p_2,\ldots,p_{r-1}]$ is a generator. Then we construct a circular block using the generator.

The issue is, when constructing m rows of weight r, after using some number of circular blocks, we may find that the number of rows required is less than the number of rows in the next circular block. If we use only some rows of the next circular block to complete m rows we may find that the column weights differ by more than 1. To overcome this, we reserve a generator g_r and do not use it at first. Only when we find that the remaining number of rows to be constructed is $\leq n$ we use g_r. g_r is chosen as follows:

1. **If** r **divides** n**:** $\hat{g_r} = [1,2,\ldots,r]$. We generate points from this generator as shown in the next example:

   Example 3. For $n = 8$ with $r = 2$, $\hat{g_r} = [1,2] = (0,0,0,0,0,0,1,1)$ and the points are generated in the following sequence:
   $(0,0,0,0,0,0,1,1)$, $(0,0,0,0,1,1,0,0)$, $(0,0,1,1,0,0,0,0)$, $(1,1,0,0,0,0,0,0)$, $(0,0,0,0,0,1,1,0)$, $(0,0,0,1,1,0,0,0)$, $(0,1,1,0,0,0,0,0)$, $(1,0,0,0,0,0,0,1)$.

The following pseudocode generates these n rows, $\{x[1], x[2], \ldots, x[n]\}$

```
for i in {0,1, ..., r-1}:
        for j in {0, 1, ..., n/r-1}:
                x[i+j] = ROT(g, j*r + i)
```

2. **If r does not divide n:** g_r is chosen by distributing the $n-r$ 0s equally among the 1s. Points are generated by successive circular shifts as shown below:

   Example 4. For $n = 8$ with $r = 3$, $\hat{g_r} = [1, 3, 6]$ and the points are generated in the following sequence:
   $(0,0,1,0,0,1,0,1)$, $(0,1,0,0,1,0,1,0)$, $(1,0,0,1,0,1,0,0)$, $(0,0,1,0,1,0,0,1)$, $(0,1,0,1,0,0,1,0)$, $(1,0,1,0,0,1,0,0)$, $(0,1,0,0,1,0,0,1)$, $(1,0,0,1,0,0,1,0)$.

Note that in each case, after every row, column weights do not differ by more than 1.

Algorithm 2
Input: number of variables n, number of rows m, row weight r.
Output: $m \times n$ matrix T having t columns of weight $w+1$ and $n-t$ columns of weight w, with $w = \lfloor \frac{m \times r}{n} \rfloor$ and t the remainder so that $m \times r = n \times w + t$.

1. Initialize T as the empty matrix. $T = ()$.
2. Compute the reserved generator g_r accordingly as r divides n or not.
3. Initialize $m' = 0$, the number of rows of T constructed so far.
4. If $m - m' \leq n$, go to Step (8).
5. Compute a new generator g, $g \neq g_r$.
6. Construct the circular block C by repeatedly circular shifting g (Let the number of vectors in C be d).
7. $T = (T, C)^T$, $m' = m' + d$ and go to Step (4).
8. Use the reserved generator g_r to construct the partial block D with $m - m'$ rows.
9. $T = (T, D)^T$.

Theorem 2. *Algorithm 2 runs correctly in $O(r \cdot \binom{n-1}{r-1})$ time.*

Proof. Since we ensure that at the end of the algorithm, column weights do not differ by more than 1, and we use rows of minimum possible weights, we get the column weights as desired and the algorithm runs correctly. Further, the number of generators is $\leq \binom{n-1}{r-1}$, obtained by a selection of $r-1$ positions from the set $\{2, 3, \ldots, n\}$ in addition to the LSB. Checking if a selection gives a generator or not requires $O(r)$ operations. So Algorithm 2 requires $O(r \cdot \binom{n-1}{r-1})$ operations. □

Now, we construct $\mathbf{S}_0 \oplus b(\mathbf{S}_0)$ by first including all points of weight upto r_0 and then using Algorithm 2 to find the remaining $m_0 = \frac{k}{2} + 2^{\frac{n}{2}-2} - \sum_{i=0}^{r_0} \binom{n}{i}$ points of weight r_0+1. Similarly for $\mathbf{S}_1 \oplus b(\mathbf{S}_1)$, include all points of weight upto r_1 and then use Algorithm 2 to find the remaining $m_1 = \frac{k}{2} - 2^{\frac{n}{2}-2} - \sum_{i=0}^{r_1} \binom{n}{i}$ points of weight $r_1 + 1$.

After constructing $\mathbf{S}_0 \oplus b(\mathbf{S}_0)$ and $\mathbf{S}_1 \oplus b(\mathbf{S}_1)$ in this manner, column weights in the two matrices do not differ by more than 1. But the $\mathbf{S} \oplus b(\mathbf{S})$ matrix thus obtained may have column weights differing by more than 1. To avoid this we permute the columns of $\mathbf{S}_1 \oplus b(\mathbf{S}_1)$ so that the columns of higher weight are identified with the columns of lower weight of $\mathbf{S}_0 \oplus b(\mathbf{S}_0)$. Then in the resulting $\mathbf{S} \oplus b(\mathbf{S})$ matrix, column weights do not differ by more than 1.

Now to satisfy Conditions 1a, 1b & 1c we need only that the columns weights are equal. To do this we need to add exactly one 1 in certain columns, z' in number (note that $z' < n$). This is not too difficult since we have a large number of rows ($> 2^{\frac{n}{2}-1}$).

Construction 1
Input: number of variables n, number of points k, r_0 and r_1 from Algorithm 1.
Output: $k \times n$ matrix $\mathbf{S}$ satisfying Conditions 1a, 1b & 1c.

1. Add all rows of weight r_0 and r_1 to the matrices $\mathbf{S}_0 \oplus b(\mathbf{S}_0)$ and $\mathbf{S}_1 \oplus b(\mathbf{S}_1)$ respectively.
2. Compute $m_0 = \frac{k}{2} + 2^{\frac{n}{2}-2} - \sum_{i=0}^{r_0} \binom{n}{i}$ and $m_1 = \frac{k}{2} - 2^{\frac{n}{2}-2} - \sum_{i=0}^{r_1} \binom{n}{i}$.
3. Use Algorithm 2 with inputs n, m_0, $r_0 + 1$ to get matrix T_0.
   $\mathbf{S}_0 \oplus b(\mathbf{S}_0) = (\mathbf{S}_0 \oplus b(\mathbf{S}_0), T_0)^T$.
4. Use Algorithm 2 with inputs n, m_1, $r_1 + 1$ to get matrix T_1.
   $\mathbf{S}_1 \oplus b(\mathbf{S}_1) = (\mathbf{S}_1 \oplus b(\mathbf{S}_1), T_1)^T$.
5. Permute columns of $\mathbf{S}_1 \oplus b(\mathbf{S}_1)$ suitably so that columns of higher weight of the $\mathbf{S}_1 \oplus b(\mathbf{S}_1)$ matrix are identified with that of lower weight columns of $\mathbf{S}_0 \oplus b(\mathbf{S}_0)$.
6. Accommodate the remaining z' ones in the two matrices in a suitable manner.
7. $\mathbf{S}_0 = \mathbf{S}_0 \oplus b(\mathbf{S}_0)$ and $\mathbf{S}_1 = 1 \oplus (\mathbf{S}_1 \oplus b(\mathbf{S}_1))$.
8. $\mathbf{S} = (\mathbf{S}_0, \mathbf{S}_1)^T$.

Theorem 3. *Construction 1 finds inputs whose output in the bent function need to be modified to get* 1*-resilient function.*

Proof. To show that Conditions 1a, 1b & 1c hold for the matrix $\mathbf{S}$ constructed as above, note that k is obtained from Algorithm 1, at the end of which $z \leq w \cdot n$. Algorithm 2 ensures that column weights of $\mathbf{S} \oplus b(\mathbf{S})$ do not differ by more than 1, using rows of minimum possible weights. So in Construction 1 after adding the remaining z' ones, each column has weight exactly $\frac{k}{2} - 2^{\frac{n}{2}-2}$.

Algorithm 2 constructs the matrices using distinct rows. We now only need to show that the Boolean complement of any row of $\mathbf{S}_0 \oplus b(\mathbf{S}_0)$ is not in $\mathbf{S}_1 \oplus b(\mathbf{S}_1)$. Weight of any row in $\mathbf{S}_0 \oplus b(\mathbf{S}_0)$ is $\leq r_0 + 1$ so it's complement must have weight $\geq n - r_0 - 1$. So if the rows in $\mathbf{S}_1 \oplus b(\mathbf{S}_1)$ are of weight $< n - r_0 - 1$ then we are through. Here we assume that $r_1 < n - r_0$ or equivalently $r_0 + r_1 < n$. That this is a reasonable assumption can be observed from the table giving values for r_0 and r_1 up to $n = 26$. We can see that r_0 and r_1 grow very slowly as compared to n. □

Since $r_1 \leq r_0$, the next theorem holds.

Theorem 4. *Construction 1 requires* $O(r_0 \cdot \binom{n-1}{r_0-1})$ *time.*

3 Construction of the 1-Resilient Function

Now that we have the set S we need to construct the bent function b satisfying Conditions 1 & 2.

The original Maiorana-McFarland class of bent function is as follows [2]. Consider n-variable Boolean functions on (X, Y), where $X, Y \in \{0,1\}^{\frac{n}{2}}$ of the form $b(X, Y) = X \cdot \pi(Y) + g(Y)$ where π is a permutation on $\{0,1\}^{\frac{n}{2}}$ and g is any Boolean function on $\frac{n}{2}$ variables. Then b is a bent function. For a fixed value of Y, $X \cdot \pi(Y)$ can be seen as a linear function on X and $g(Y)$ is constant either 0 or 1 over all X. So that the function b can be seen as a concatenation of $2^{\frac{n}{2}}$ distinct (upto complementation) affine function on $\frac{n}{2}$ variables.

We require a bent function $b(x)$ on n variables satisfying the condition that $b(x) = 0$ for $x \in S_0$ and $b(x) = 1$ for $x \in S_1$. We have to decide what permutations π on $\{0,1\}^{\frac{n}{2}}$ and what kind of functions g on $\{0,1\}^{\frac{n}{2}}$ we can take such that the conditions on b are satisfied. Let us fix the notation and ordering of input variables as $x = (x_1, x_2, \ldots, x_n)$, $X = (X_1, X_2, \ldots, X_{\frac{n}{2}}) = (x_1, x_2, \ldots, x_{\frac{n}{2}})$, and $Y = (Y_1, Y_2, \ldots, Y_{\frac{n}{2}}) = (x_{\frac{n}{2}+1}, x_{\frac{n}{2}+2}, \ldots, x_n)$.

Now, we look at Condition 2. It is easy to see that for $0 \leq wt(\omega) \leq 1$ a bent function will have the restricted Walsh spectrum value $W_b(\omega)|_{(X,Y), X \in \{0,1\}^{\frac{n}{2}}} = 0$ for all values of Y except for one Y where it is $\pm 2^{\frac{n}{2}}$. We want $W_b(\omega) = +2^{\frac{n}{2}}$ at that Y. This will happen only when $X \cdot \pi(Y) \oplus g(Y) \oplus x \cdot \omega = 0$ or $X \cdot \pi(Y) \oplus g(Y) = x \cdot \omega$ at that Y. We ensure this by conditions as below:

1. For $wt(\omega) = 0$ we want for one Y, $X \cdot \pi(Y) \oplus g(Y) = x \cdot \omega$. That is $X \cdot \pi(Y) \oplus g(Y) = 0$. Not that here, X is variable and takes all possible values. Equating the constant parts, we get $g(Y) = 0$. Equating the variable parts, we get $X \cdot \pi(Y) = 0$ so that $\pi(Y) = (0, 0, \ldots, 0)$.
   So we require for a particular Y, $\pi(Y) = (0, 0, \ldots, 0)$ and $g(Y) = 0$.
2. For ω having a 1 in the latter half, we want for one Y, $X \cdot \pi(Y) \oplus g(Y) = x \cdot \omega$. But $x \cdot \omega = x_i$, with $\frac{n}{2} < i \leq n$, which is constant. So $X \cdot \pi(Y)$ must be constant giving $\pi(Y) = (0, 0, \ldots, 0)$. This must hold for each such value of ω so that $g(Y) = x_{\frac{n}{2}+1} = x_{\frac{n}{2}+2} = \ldots = x_n$
   So we require either $Y = (0, 0, \ldots 0)$ with $\pi(Y) = (0, 0, \ldots, 0)$ and $g(Y) = 0$ OR $Y = (1, 1, \ldots 1)$ with $\pi(Y) = (0, 0, \ldots, 0)$ and $g(Y) = 1$.
3. The last case is for ω having a 1 in the former half, we want for one Y, $X \cdot \pi(Y) \oplus g(Y) = x \cdot \omega$. But $x \cdot \omega = x_i$, with $1 \leq i \leq \frac{n}{2}$. Equating constant parts, $g(Y) = 0$, so that $X \cdot \pi(Y) = x_i$. We get $wt(\pi(Y)) = 1$ with the 1 in the i'th position.
   So our condition is: for $\pi(Y) \in \{(1, 0, \ldots, 0), (0, 1, \ldots, 0), (0, 0, \ldots, 1))\}$, $g(Y) = 0$.

We can combine the first two parts above to give the following two conditions:

$$\pi(0, 0, \ldots, 0) = (0, 0, \ldots, 0) \text{ and } g(0, 0, \ldots, 0) = 0 \ . \tag{3}$$

$$\text{For } \ \pi(Y) \in \{(1, 0, \ldots, 0), (0, 1, \ldots, 0), (0, 0, \ldots, 1))\}, g(Y) = 0 \ . \tag{4}$$

Construction of 1-resilient functions for $n = 10, 12, 14$ are placed in the Appendices.

4 Conclusions

In this paper we present a strategy to construct highly nonlinear 1-resilient functions by modifying some output bits of a bent function. We present a good lower bound on the minimum number of bits of a bent function needed to be modified. We have shown that the bound is tight for functions upto 14-input variables. One interesting problem is to study whether Algorithm 1 provides the minimum distance between n-variable bent and 1-resilient functions for all even n. We present an algorithm to generate the points whose output in the bent function require to be modified. For $n = 10, 12, 14$ we identify Maiorana-McFarland type bent functions which can be modified to get 1-resilient functions, using the points given by Construction 1. This shows that the bound given by Algorithm 1 is tight and is the minimum distance for these values of n. Further our construction is superior to [6] in terms of the nonlinearity (we get better nonlinearity for 14 variables) and autocorrelation absolute indicator (we get 1-resilient functions with absolute indicator value that was not known earlier for 12 variable). Since the class of bent functions is very large, it may be conjectured that it is always possible to identify bent functions which can be modified to get 1-resilient functions, using the points given by Construction 1.

References

1. P. Charpin and E. Pasalic. On propagation characteristics of resilient functions. In *SAC 2002*, number 2595 in Lecture Notes in Computer Science, pages 175–195. Springer-Verlag, 2003.
2. J. F. Dillon. Elementary Hadamard Difference sets. PhD Thesis, University of Maryland, 1974.
3. H. Dobbertin. Construction of bent functions and balanced Boolean functions with high nonlinearity. In *Fast Software Encryption - FSE 1994*, number 1008 in Lecture Notes in Computer Science, pages 61–74. Springer-Verlag, 1994.
4. X. Guo-Zhen and J. Massey. A spectral characterization of correlation immune combining functions. *IEEE Transactions on Information Theory*, 34(3): 569–571, May 1988.
5. S. Maity and T. Johansson. Construction of Cryptographically important Boolean functions. In *INDOCRYPT 2002*, number 2551 in Lecture Notes in Computer Science, pages 234–245, Springer Verlag, 2002.
6. S. Maity and S. Maitra. Minimum distance between bent and 1-resilient functions. In *Fast Software Encryption - FSE 2004*, number 3017 in Lecture Notes in Computer Science, pages 143-160, Springer Verlag, 2004.
7. O. S. Rothaus. On bent functions. *Journal of Combinatorial Theory, Series A*, 20: 300–305, 1976.
8. P. Sarkar and S. Maitra. Nonlinearity bounds and constructions of resilient Boolean functions. In *Advances in Cryptology - CRYPTO 2000*, number 1880 in Lecture Notes in Computer Science, pages 515–532. Springer Verlag, 2000.

9. J. Seberry, X. M. Zhang, and Y. Zheng. Nonlinearly balanced Boolean functions and their propagation characteristics. In *Advances in Cryptology - CRYPTO'93*, number 773 in Lecture Notes in Computer Science, pages 49–60. Springer-Verlag, 1994.
10. Y. V. Tarannikov. New constructions of resilient Boolean functions with maximal nonlinearity. In *Fast Software Encryption - FSE 2001*, number 2355 in Lecture Notes in Computer Science, pages 66–77. Springer Verlag, 2001.

Appendices

A The 10-Variable 1-Resilient Functions

Algorithm 1 gives us $k = 22$. We compute $r_0 = 1$ and $r_1 = 0$ so that $m_0 = 8$ and $m_1 = 2$. Using all points of weight ≤ 1 (as $r_0 = 1$) and the reserved generator $[12] = (0000000011)$ (as $m_0 - m' = 8 < 10$ and 2 divides n in Algorithm 2) we get

$$\mathbf{S_0} = \mathbf{S_0} \oplus b(\mathbf{S_0}) = \begin{pmatrix} 0&0&0&0&0&0&0&0&0&0 \\ 0&0&0&0&0&0&0&0&0&1 \\ 0&0&0&0&0&0&0&0&1&0 \\ 0&0&0&0&0&0&0&1&0&0 \\ 0&0&0&0&0&0&1&0&0&0 \\ 0&0&0&0&0&1&0&0&0&0 \\ 0&0&0&0&1&0&0&0&0&0 \\ 0&0&0&1&0&0&0&0&0&0 \\ 0&0&1&0&0&0&0&0&0&0 \\ 0&1&0&0&0&0&0&0&0&0 \\ 1&0&0&0&0&0&0&0&0&0 \\ 0&0&0&0&0&0&0&0&1&1 \\ 0&0&0&0&0&0&1&1&0&0 \\ 0&0&0&0&1&1&0&0&0&0 \\ 0&0&1&1&0&0&0&0&0&0 \\ 1&1&0&0&0&0&0&0&0&0 \\ 0&0&0&0&0&0&0&1&1&0 \\ 0&0&0&0&0&1&1&0&0&0 \\ 0&0&0&1&1&0&0&0&0&0 \end{pmatrix}.$$

Using the point of weight zero and the reserved generator $[1] = (0000000001)$ (as $m_1 - m' = 2 < 10 = n$ and 1 divides n) we get

$$\mathbf{S_1} \oplus b(\mathbf{S_1}) = \begin{pmatrix} 0\,0\,0\,0\,0\,0\,0\,0\,0\,0 \\ 0\,0\,0\,0\,0\,0\,0\,0\,0\,1 \\ 0\,0\,1\,0\,0\,0\,0\,0\,0\,0 \end{pmatrix}.$$

We find that $z' = 2$. We add these in the rows of $\mathbf{S_1} \oplus b(\mathbf{S_1})$ to get

$$\mathbf{S_1} \oplus b(\mathbf{S_1}) = \begin{pmatrix} 0\,0\,0\,0\,0\,0\,0\,0\,0\,0 \\ 1\,0\,0\,0\,0\,0\,0\,0\,0\,1 \\ 0\,1\,1\,0\,0\,0\,0\,0\,0\,0 \end{pmatrix}$$

and

$$\mathbf{S_1} = 1 \oplus \mathbf{S_1} \oplus b(\mathbf{S_1}) = \begin{pmatrix} 1\,1\,1\,1\,1\,1\,1\,1\,1\,1 \\ 0\,1\,1\,1\,1\,1\,1\,1\,1\,0 \\ 1\,0\,0\,1\,1\,1\,1\,1\,1\,1 \end{pmatrix}.$$

Taking $g(Y) = 0$ and $\pi(Y) = Y$. we get the value of $b(x) = b(X, Y) = X \cdot \pi(Y) + g(Y)$ to be zero when $x \in S_0$ and one when $x \in S_1$. Also π and g satisfy Conditions 3 & 4. A 1-resilient function $f(x)$ is obtained as before. The nonlinearity of f is 488, algebraic degree is 8 and $\triangle_f = 48$.

B The 12-Variable 1-Resilient Functions

Algorithm 1 gives us $k = 44$. We compute $r_0 = 1$ and $r_1 = 0$ so that $m_0 = 25$ and $m_1 = 5$. Using all points of weight ≤ 1, generators $[1, 3]$, $[1, 4]$ and the reserved generator $[1, 2]$ (2 divides n) for points of weight 2 we get $\mathbf{S_0} = \mathbf{S_0} \oplus b(\mathbf{S_0})$ with 38 rows.

Using the point of weight zero and the reserved generator $[1] = (000000000001)$ (1 divides n) we get after permutation

$$\mathbf{S_1} \oplus b(\mathbf{S_1}) = \begin{pmatrix} 0\,0\,0\,0\,0\,0\,0\,0\,0\,0\,0\,0 \\ 0\,0\,0\,0\,0\,0\,0\,0\,0\,1\,0\,0 \\ 0\,0\,0\,0\,0\,0\,0\,0\,1\,0\,0\,0 \\ 0\,0\,0\,0\,0\,0\,0\,1\,0\,0\,0\,0 \\ 0\,0\,0\,0\,0\,0\,1\,0\,0\,0\,0\,0 \\ 0\,0\,0\,0\,0\,1\,0\,0\,0\,0\,0\,0 \end{pmatrix}.$$

We find that $z' = 5$. We add these in the rows of $\mathbf{S_1} \oplus b(\mathbf{S_1})$ to get

$$\mathbf{S_1} \oplus b(\mathbf{S_1}) = \begin{pmatrix} 0\,0\,1\,1\,1\,0\,0\,0\,0\,0\,0\,0 \\ 0\,0\,0\,0\,0\,0\,0\,0\,0\,1\,0\,0 \\ 0\,0\,0\,0\,0\,0\,0\,0\,1\,0\,0\,0 \\ 0\,0\,0\,0\,0\,0\,0\,1\,0\,0\,0\,0 \\ 0\,0\,0\,0\,0\,0\,1\,0\,0\,0\,0\,0 \\ 1\,1\,0\,0\,0\,1\,0\,0\,0\,0\,0\,0 \end{pmatrix}$$

and

$$\mathbf{S_1} = 1 \oplus \mathbf{S_1} \oplus b(\mathbf{S_1}) = \begin{pmatrix} 1\,1\,0\,0\,0\,1\,1\,1\,1\,1\,1\,1 \\ 1\,1\,1\,1\,1\,1\,1\,1\,1\,0\,1\,1 \\ 1\,1\,1\,1\,1\,1\,1\,1\,0\,1\,1\,1 \\ 1\,1\,1\,1\,1\,1\,1\,0\,1\,1\,1\,1 \\ 1\,1\,1\,1\,1\,1\,0\,1\,1\,1\,1\,1 \\ 0\,0\,1\,1\,1\,0\,1\,1\,1\,1\,1\,1 \end{pmatrix}.$$

Taking $g(Y) = 0$ and $\pi(Y) = Y$. we get the value of $b(x) = b(X, Y) = X \cdot \pi(Y) + g(Y)$ to be zero when $x \in S_0$ and one when $x \in S_1$. Also π and g satisfy Conditions 3 & 4. A 1-resilient function $f(x)$ is obtained as before. The nonlinearity of f is 2000 and algebraic degree is 10. The function f we constructed here has $\Delta_f = 104$ and this is the best known value which is achieved for the first time here.

C The 14-Variable 1-Resilient Functions

Algorithm 1 gives us $k = 86$. We compute $r_0 = 1$ and $r_1 = 0$ so that $m_0 = 60$ and $m_1 = 10$. Using all points of weight ≤ 1, generators $[1, 3]$, $[1, 4]$, $[1, 5]$, $[1, 6]$ and the reserved generator $[1, 2]$ (2 divides n) for points of weight 2 we get $\mathbf{S_0} = \mathbf{S_0} \oplus b(\mathbf{S_0})$ with 75 rows.

Using the point of weight zero and the reserved generator $[1]$ (1 divides n) we get after permutation

$$\mathbf{S_1} \oplus b(\mathbf{S_1}) = \begin{pmatrix} 0&0&0&0&0&0&0&0&0&0&0&0&0&0 \\ 0&0&0&0&0&1&0&0&0&0&0&0&0&0 \\ 0&0&0&0&1&0&0&0&0&0&0&0&0&0 \\ 0&0&0&1&0&0&0&0&0&0&0&0&0&0 \\ 0&0&1&0&0&0&0&0&0&0&0&0&0&0 \\ 0&1&0&0&0&0&0&0&0&0&0&0&0&0 \\ 1&0&0&0&0&0&0&0&0&0&0&0&0&0 \\ 0&0&0&0&0&0&0&0&0&0&0&0&0&1 \\ 0&0&0&0&0&0&0&0&0&0&0&0&1&0 \\ 0&0&0&0&0&0&0&0&0&0&0&1&0&0 \\ 0&0&0&0&0&0&0&0&0&0&1&0&0&0 \end{pmatrix}.$$

We find that $z' = 3$. We add these in the rows of $\mathbf{S_1} \oplus b(\mathbf{S_1})$ to get

$$\mathbf{S_1} = 1 \oplus \mathbf{S_1} \oplus b(\mathbf{S_1}) = \begin{pmatrix} 1&1&1&1&1&1&1&0&0&0&1&1&1&1 \\ 1&1&1&1&1&0&1&1&1&1&1&1&1&1 \\ 1&1&1&1&0&1&1&1&1&1&1&1&1&1 \\ 1&1&1&0&1&1&1&1&1&1&1&1&1&1 \\ 1&1&0&1&1&1&1&1&1&1&1&1&1&1 \\ 1&0&1&1&1&1&1&1&1&1&1&1&1&1 \\ 0&1&1&1&1&1&1&1&1&1&1&1&1&1 \\ 1&1&1&1&1&1&1&1&1&1&1&1&1&0 \\ 1&1&1&1&1&1&1&1&1&1&1&1&0&1 \\ 1&1&1&1&1&1&1&1&1&1&1&0&1&1 \\ 1&1&1&1&1&1&1&1&1&1&0&1&1&1 \end{pmatrix}.$$

We define g and π below which satisfies the conditions for construction of the required bent function.

1. $g(Y) = \begin{cases} 1, & \text{if } Y \in A \\ 0, & \text{otherwise.} \end{cases}$

   where

$$\begin{aligned} A = \ & \{(0,0,0,1,1,1,1),(1,1,1,1,1,1,0),(1,1,1,1,1,0,1), \\ & (1,1,1,1,0,1,1),(1,1,1,0,1,1,1),(1,1,0,1,1,1,1), \\ & (1,0,1,1,1,1,1),(0,1,1,1,1,1,1),(1,1,1,1,1,1,1)\} \end{aligned}$$

2. $\pi(Y) = Y$.

If we take π and g as above then we get the value of $b(x) = b(X,Y) = X \cdot \pi(Y) + g(Y)$ to be zero when $x \in S_0$ and one when $x \in S_1$. Also π and g satisfy Conditions 3 & 4. A 1-resilient function $f(x)$ is obtained as before. The function f we constructed here has nonlinearity 8098 and this is the best known nonlinearity value which is achieved for the first time here.

Author Index

Lecture Notes in Computer Science

For information about Vols. 1–4085

please contact your bookseller or Springer

Vol. 4228: D.E. Lightfoot, C.A. Szyperski (Eds.), Modular Programming Languages. X, 415 pages. 2006.

Vol. 4208: M. Gerndt, D. Kranzlmüller (Eds.), High Performance Computing and Communications. XXII, 938 pages. 2006.

Vol. 4206: P. Dourish, A. Friday (Eds.), UbiComp 2006: Ubiquitous Computing. XIX, 526 pages. 2006.

Vol. 4193: T.P. Runarsson, H.-G. Beyer, E. Burke, J.J. Merelo-Guervós, L. D. Whitley, X. Yao (Eds.), Parallel Problem Solving from Nature - PPSN IX. XIX, 1061 pages. 2006.

Vol. 4192: B. Mohr, J.L. Träff, J. Worringen, J. Dongarra (Eds.), Recent Advances in Parallel Virtual Machine and Message Passing Interface. XVI, 414 pages. 2006.

Vol. 4188: P. Sojka, I. Kopeček, K. Pala (Eds.), Text, Speech and Dialogue. XIV, 721 pages. 2006. (Sublibrary LNAI).

Vol. 4187: J.J. Alferes, J. Bailey, W. May, U. Schwertel (Eds.), Principles and Practice of Semantic Web Reasoning. XI, 277 pages. 2006.

Vol. 4186: C. Jesshope, C. Egan (Eds.), Advances in Computer Systems Architecture. XIV, 605 pages. 2006.

Vol. 4185: R. Mizoguchi, Z. Shi, F. Giunchiglia (Eds.), The Semantic Web – ASWC 2006. XX, 778 pages. 2006.

Vol. 4184: M. Bravetti, M. Núñez, G. Zavattaro (Eds.), Web Services and Formal Methods. X, 289 pages. 2006.

Vol. 4183: J. Euzenat, J. Domingue (Eds.), Artificial Intelligence: Methodology, Systems, and Applications. XIII, 291 pages. 2006. (Sublibrary LNAI).

Vol. 4180: M. Kohlhase, OMDoc – An Open Markup Format for Mathematical Documents [version 1.2]. XIX, 428 pages. 2006. (Sublibrary LNAI).

Vol. 4178: A. Corradini, H. Ehrig, U. Montanari, L. Ribeiro, G. Rozenberg (Eds.), Graph Transformations. XII, 473 pages. 2006.

Vol. 4176: S.K. Katsikas, J. Lopez, M. Backes, S. Gritzalis, B. Preneel (Eds.), Information Security. XIV, 548 pages. 2006.

Vol. 4175: P. Bücher, B.M.E. Moret (Eds.), Algorithms in Bioinformatics. XII, 402 pages. 2006. (Sublibrary LNBI).

Vol. 4174: K. Franke, K.-R. Müller, B. Nickolay, R. Schäfer (Eds.), Pattern Recognition. XX, 773 pages. 2006.

Vol. 4169: H.L. Bodlaender, M.A. Langston (Eds.), Parameterized and Exact Computation. XI, 279 pages. 2006.

Vol. 4168: Y. Azar, T. Erlebach (Eds.), Algorithms – ESA 2006. XVIII, 843 pages. 2006.

Vol. 4165: W. Jonker, M. Petković (Eds.), Secure, Data Management. X, 185 pages. 2006.

Vol. 4163: H. Bersini, J. Carneiro (Eds.), Artificial Immune Systems. XII, 460 pages. 2006.

Vol. 4162: R. Královič, P. Urzyczyn (Eds.), Mathematical Foundations of Computer Science 2006. XV, 814 pages. 2006.

Vol. 4160: M. Fisher, W.v.d. Hoek, B. Konev, A. Lisitsa (Eds.), Logics in Artificial Intelligence. XII, 516 pages. 2006. (Sublibrary LNAI).

Vol. 4159: J. Ma, H. Jin, L.T. Yang, J.J.-P. Tsai (Eds.), Ubiquitous Intelligence and Computing. XXII, 1190 pages. 2006.

Vol. 4158: L.T. Yang, H. Jin, J. Ma, T. Ungerer (Eds.), Autonomic and Trusted Computing. XIV, 613 pages. 2006.

Vol. 4156: S. Amer-Yahia, Z. Bellahsène, E. Hunt, R. Unland, J.X. Yu (Eds.), Database and XML Technologies. IX, 123 pages. 2006.

Vol. 4155: O. Stock, M. Schaerf (Eds.), Reasoning, Action and Interaction in AI Theories and Systems. XVIII, 343 pages. 2006. (Sublibrary LNAI).

Vol. 4153: N. Zheng, X. Jiang, X. Lan (Eds.), Advances in Machine Vision, Image Processing, and Pattern Analysis. XIII, 506 pages. 2006.

Vol. 4152: Y. Manolopoulos, J. Pokorný, T. Sellis (Eds.), Advances in Databases and Information Systems. XV, 448 pages. 2006.

Vol. 4151: A. Iglesias, N. Takayama (Eds.), Mathematical Software - ICMS 2006. XVII, 452 pages. 2006.

Vol. 4150: M. Dorigo, L.M. Gambardella, M. Birattari, A. Martinoli, R. Poli, T. Stützle (Eds.), Ant Colony Optimization and Swarm Intelligence. XVI, 526 pages. 2006.

Vol. 4149: M. Klusch, M. Rovatsos, T.R. Payne (Eds.), Cooperative Information Agents X. XII, 477 pages. 2006. (Sublibrary LNAI).

Vol. 4148: J. Vounckx, N. Azemard, P. Maurine (Eds.), Integrated Circuit and System Design. XVI, 677 pages. 2006.

Vol. 4146: J.C. Rajapakse, L. Wong, R. Acharya (Eds.), Pattern Recognition in Bioinformatics. XIV, 186 pages. 2006. (Sublibrary LNBI).

Vol. 4144: T. Ball, R.B. Jones (Eds.), Computer Aided Verification. XV, 564 pages. 2006.

Vol. 4139: T. Salakoski, F. Ginter, S. Pyysalo, T. Pahikkala, Advances in Natural Language Processing. XVI, 771 pages. 2006. (Sublibrary LNAI).

Vol. 4138: X. Cheng, W. Li, T. Znati (Eds.), Wireless Algorithms, Systems, and Applications. XVI, 709 pages. 2006.

Vol. 4137: C. Baier, H. Hermanns (Eds.), CONCUR 2006 – Concurrency Theory. XIII, 525 pages. 2006.

Vol. 4136: R.A. Schmidt (Ed.), Relations and Kleene Algebra in Computer Science. XI, 433 pages. 2006.

Vol. 4135: C.S. Calude, M.J. Dinneen, G. Păun, G. Rozenberg, S. Stepney (Eds.), Unconventional Computation. X, 267 pages. 2006.

Vol. 4134: K. Yi (Ed.), Static Analysis. XIII, 443 pages. 2006.

Vol. 4133: J. Gratch, M. Young, R. Aylett, D. Ballin, P. Olivier (Eds.), Intelligent Virtual Agents. XIV, 472 pages. 2006. (Sublibrary LNAI).

Vol. 4132: S. Kollias, A. Stafylopatis, W. Duch, E. Oja (Eds.), Artificial Neural Networks – ICANN 2006, Part II. XXXIV, 1028 pages. 2006.

Vol. 4131: S. Kollias, A. Stafylopatis, W. Duch, E. Oja (Eds.), Artificial Neural Networks – ICANN 2006, Part I. XXXIV, 1008 pages. 2006.

Vol. 4130: U. Furbach, N. Shankar (Eds.), Automated Reasoning. XV, 680 pages. 2006. (Sublibrary LNAI).

Vol. 4129: D. McGookin, S. Brewster (Eds.), Haptic and Audio Interaction Design. XII, 167 pages. 2006.

Vol. 4128: W.E. Nagel, W.V. Walter, W. Lehner (Eds.), Euro-Par 2006 Parallel Processing. XXXIII, 1221 pages. 2006.

Vol. 4127: E. Damiani, P. Liu (Eds.), Data and Applications Security XX. X, 319 pages. 2006.

Vol. 4126: P. Barahona, F. Bry, E. Franconi, N. Henze, U. Sattler, Reasoning Web. X, 269 pages. 2006.

Vol. 4124: H. de Meer, J.P. G. Sterbenz (Eds.), Self-Organizing Systems. XIV, 261 pages. 2006.

Vol. 4121: A. Biere, C.P. Gomes (Eds.), Theory and Applications of Satisfiability Testing - SAT 2006. XII, 438 pages. 2006.

Vol. 4119: C. Dony, J.L. Knudsen, A. Romanovsky, A. Tripathi (Eds.), Advanced Topics in Exception Handling Components. X, 302 pages. 2006.

Vol. 4117: C. Dwork (Ed.), Advances in Cryptology - CRYPTO 2006. XIII, 621 pages. 2006.

Vol. 4116: R. De Prisco, M. Yung (Eds.), Security and Cryptography for Networks. XI, 366 pages. 2006.

Vol. 4115: D.-S. Huang, K. Li, G.W. Irwin (Eds.), Computational Intelligence and Bioinformatics, Part III. XXI, 803 pages. 2006. (Sublibrary LNBI).

Vol. 4114: D.-S. Huang, K. Li, G.W. Irwin (Eds.), Computational Intelligence, Part II. XXVII, 1337 pages. 2006. (Sublibrary LNAI).

Vol. 4113: D.-S. Huang, K. Li, G.W. Irwin (Eds.), Intelligent Computing, Part I. XXVII, 1331 pages. 2006.

Vol. 4112: D.Z. Chen, D. T. Lee (Eds.), Computing and Combinatorics. XIV, 528 pages. 2006.

Vol. 4111: F.S. de Boer, M.M. Bonsangue, S. Graf, W.-P. de Roever (Eds.), Formal Methods for Components and Objects. VIII, 447 pages. 2006.

Vol. 4110: J. Díaz, K. Jansen, J.D.P. Rolim, U. Zwick (Eds.), Approximation, Randomization, and Combinatorial Optimization. XII, 522 pages. 2006.

Vol. 4109: D.-Y. Yeung, J.T. Kwok, A. Fred, F. Roli, D. de Ridder (Eds.), Structural, Syntactic, and Statistical Pattern Recognition. XXI, 939 pages. 2006.

Vol. 4108: J.M. Borwein, W.M. Farmer (Eds.), Mathematical Knowledge Management. VIII, 295 pages. 2006. (Sublibrary LNAI).

Vol. 4106: T.R. Roth-Berghofer, M.H. Göker, H. A. Güvenir (Eds.), Advances in Case-Based Reasoning. XIV, 566 pages. 2006. (Sublibrary LNAI).

Vol. 4105: B. Gunsel, A.K. Jain, A. M. Tekalp, B. Sankur (Eds.), Multimedia, Content Representation, Classification and Security. XIX, 804 pages. 2006.

Vol. 4104: T. Kunz, S.S. Ravi (Eds.), Ad-Hoc, Mobile, and Wireless Networks. XII, 474 pages. 2006.

Vol. 4103: J. Eder, S. Dustdar (Eds.), Business Process Management Workshops. XI, 508 pages. 2006.

Vol. 4102: S. Dustdar, J.L. Fiadeiro, A. Sheth (Eds.), Business Process Management. XV, 486 pages. 2006.

Vol. 4099: Q. Yang, G. Webb (Eds.), PRICAI 2006: Trends in Artificial Intelligence. XXVIII, 1263 pages. 2006. (Sublibrary LNAI).

Vol. 4098: F. Pfenning (Ed.), Term Rewriting and Applications. XIII, 415 pages. 2006.

Vol. 4097: X. Zhou, O. Sokolsky, L. Yan, E.-S. Jung, Z. Shao, Y. Mu, D.C. Lee, D. Kim, Y.-S. Jeong, C.-Z. Xu (Eds.), Emerging Directions in Embedded and Ubiquitous Computing. XXVII, 1034 pages. 2006.

Vol. 4096: E. Sha, S.-K. Han, C.-Z. Xu, M.H. Kim, L.T. Yang, B. Xiao (Eds.), Embedded and Ubiquitous Computing. XXIV, 1170 pages. 2006.

Vol. 4095: S. Nolfi, G. Baldassarre, R. Calabretta, J.C. T. Hallam, D. Marocco, J.-A. Meyer, O. Miglino, D. Parisi (Eds.), From Animals to Animats 9. XV, 869 pages. 2006. (Sublibrary LNAI).

Vol. 4094: O. H. Ibarra, H.-C. Yen (Eds.), Implementation and Application of Automata. XIII, 291 pages. 2006.

Vol. 4093: X. Li, O.R. Zaïane, Z. Li (Eds.), Advanced Data Mining and Applications. XXI, 1110 pages. 2006. (Sublibrary LNAI).

Vol. 4092: J. Lang, F. Lin, J. Wang (Eds.), Knowledge Science, Engineering and Management. XV, 664 pages. 2006. (Sublibrary LNAI).

Vol. 4091: G.-Z. Yang, T. Jiang, D. Shen, L. Gu, J. Yang (Eds.), Medical Imaging and Augmented Reality. XIII, 399 pages. 2006.

Vol. 4090: S. Spaccapietra, K. Aberer, P. Cudré-Mauroux (Eds.), Journal on Data Semantics VI. XI, 211 pages. 2006.

Vol. 4089: W. Löwe, M. Südholt (Eds.), Software Composition. X, 339 pages. 2006.

Vol. 4088: Z.-Z. Shi, R. Sadananda (Eds.), Agent Computing and Multi-Agent Systems. XVII, 827 pages. 2006. (Sublibrary LNAI).

Vol. 4087: F. Schwenker, S. Marinai (Eds.), Artificial Neural Networks in Pattern Recognition. IX, 299 pages. 2006. (Sublibrary LNAI).

Vol. 4086: G. Gong, T. Helleseth, H.-Y. Song, K. Yang (Eds.), Sequences and Their Applications – SETA 2006. XII, 433 pages. 2006.